FORMULAS FOR VOLUME

Figure		*Volume*
Rectangular solid	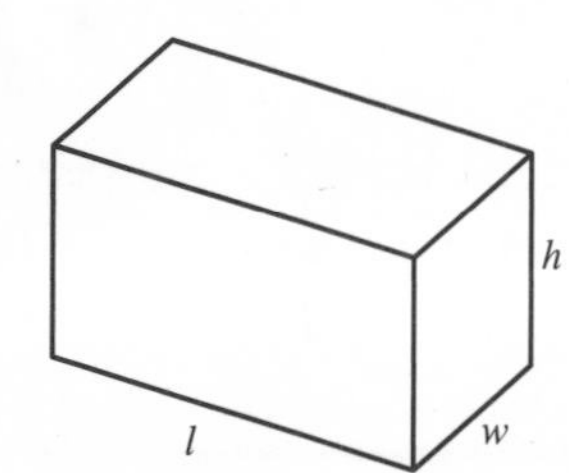	$V = lwh$
Cube	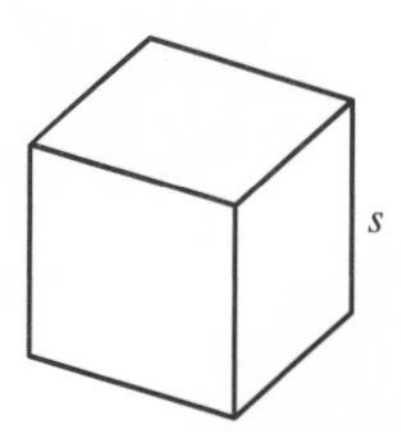	$V = s^3$
Right circular cylinder	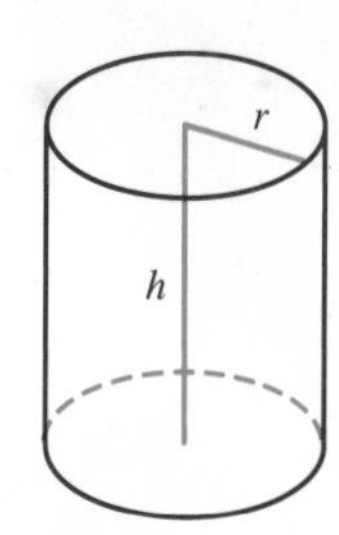	$V = \pi r^2 h$
Right circular cone	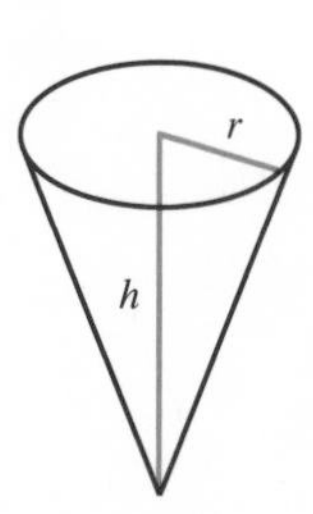	$V = \frac{1}{3}\pi r^2 h$
Sphere	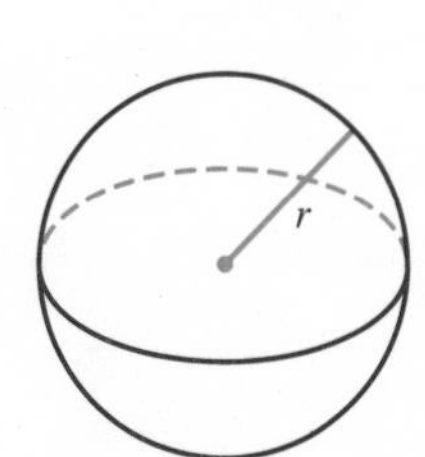	$V = \frac{4}{3}\pi r^3$

Beginning Algebra

with applications

Beginning Algebra
with applications

LINDA L. EXLEY
VINCENT K SMITH
DeKalb College
Clarkston, Georgia

PRENTICE HALL, Englewood Cliffs, New Jersey 07632

Library of Congress Cataloging-in-Publication Data

Exley, Linda L.
Beginning algebra: with applications / Linda L. Exley and Vincent K Smith.
p. cm.
ISBN 0-13-072448-3
1. Algebra. I. Smith, Vincent K II. Title.
QA152.2.E95 1990
512.9—dc20 89-37983
CIP

Editorial/production supervision: Cristina Ferrari
Interior design: Anne T. Bonanno
Cover design: Anne T. Bonanno
Cover art: Network Graphics
Manufacturing buyer: Paula Massenaro
Photo editor: Lori Morris-Nantz
Photo research: Rhoda Sidney

Photo credits: 1, Cary Wolinsky/Stock, Boston · 51, Bob Daemmrich/The Image Works · 113, NASA · 171, Georg Gerster/Photo Researchers · 229, Balloon Excelsior, Inc. · 267, Margot Granitsas/The Image Works · 321, Thomas Wanstall/The Image Works · 375, Dan Burns/Monkmeyer Press · 413, Omar Noor/Super Stock · 445, George Goodwin/Monkmeyer Press

A Division of Simon & Schuster
Englewood Cliffs, New Jersey 07632

Printed in the United States of America
10 9 8 7 6 5 4 3

ISBN 0-13-072448-3

Prentice-Hall International (UK) Limited, *London*
Prentice-Hall of Australia Pty. Limited, *Sydney*
Prentice-Hall Canada Inc., *Toronto*
Prentice-Hall Hispanoamericana, S.A., *Mexico*
Prentice-Hall of India Private Limited, *New Delhi*
Prentice-Hall of Japan, Inc., *Tokyo*
Simon & Schuster Asia Pts. Ltd., *Singapore*
Editora Prentice-Hall do Brazil, Ltda., *Rio de Janeiro*

Dedicated to Our Children
Amanda and Chuck Exley
Martin and Rowland Smith

Contents

OUR
COMMITMENT
TO EXCELLENCE

Preface

This book is a beginning algebra textbook with a review of arithmetic and geometry. It can be used as a developmental text to prepare students for intermediate algebra or liberal arts mathematics.

GETTING STARTED

Chapter 0 reviews topics from arithmetic and geometry. The sections in this chapter may be covered in detail, treated as a review, or omitted entirely. If the chapter is omitted, the individual topics can be covered as required with the material of the later chapters. A short review of geometry is included in Chapter 0 and geometric problems are integrated throughout the book.

PEDAGOGY

This book is written in a language that students understand. Wordy statements and mathematical jargon are kept to a minimum while methods and techniques are illustrated with worked examples whenever possible. Although simple language is used and rigor is not stressed, careful mathematics is.

The heart of the book is found in over 5500 carefully selected exercises that develop and illustrate the ideas of algebra in a modern setting. Extensive problem sets occur at the end of each section. Each set starts with a section titled "Warm-ups." Warm-ups are a collection of problems graded in difficulty from easy to medium, which are keyed to worked examples in the text. Warm-ups are followed by a set of exercises called "Practice Exercises." This longer set of problems contains the drill and practice exercises so necessary for students to develop their manipulative algebra skills. The Practice Exercises are also graded in difficulty but are mixed and not keyed to worked examples. The problem sets also have a few "Challenge Problems" that allow the student to probe into the natural extensions of the ideas presented in the text. The problem sets contain sufficient numbers of problems to provide the instructor considerable flexibility. The pedagogy is never dictated by a limited selection of problems.

The new material and all the teaching points of each section are presented with worked and annotated examples. Problem-solving steps are identified and explained. The students see what is to be done and how to do it. Most examples and all of the teaching points are reinforced with Warm-up problems keyed to the examples. The topics are arranged in such a manner as to allow the student to progress through the book developing skills as they are needed. The student is not put into a position of using a concept that has not already been developed.

The topics are developed so that instructors can tailor a course to fit their needs.

OTHER FEATURES OF THE TEXT

A Problem Solving Approach A problem-solving approach is used throughout the book. Worked examples with explanatory text provide the key to an extensive collection of applied problems.

Connections Each chapter opens with an introductory paragraph putting the upcoming material in context, and showing its connections with other chapters, other courses, and historical development of mathematical ideas.

Key Problems The Warm-ups and Practice Exercises contain key problems, marked in color. These problems illustrate the necessary teaching points of the section. The instructor can choose to work some or all of them in class as blackboard examples. They can be assigned as hand-in homework to be graded or they can be used for any purpose where the instructor wishes to be sure of a limited but representative cross-section of the material.

Let's Not Forget The review problems in every chapter, except Chapter 0, end with a novel feature called "Let's Not Forget." This segment contains a few carefully selected problems that repeat earlier themes, particularly certain sticky ideas and problems that historically give students trouble. The Let's Not Forget problems expand with each chapter.

Checkups Each chapter summary ends with a section called "Checkups." These are worked examples from the text stated as problems. These can be used as a self-test for students to check their understanding of the main points in the chapter. Each problem is keyed to an example number so that students can check their work by referring back to the text.

Be Careful! Common Student Errors are prominently flagged in the margin with the admonition, "Be Careful!" The adjoining text explains the caution.

In Your Own Words The problem sets in each section end with a few questions marked, "In Your Own Words." These may be used to encourage good writing in mathematics and to test comprehension of certain ideas.

Calculator Boxes Calculators are introduced where appropriate and instructional material on the proper use of calculators is included. The calculator boxes are not mere window dressing but provide detailed instructions on meaningful problems for new calculator users. However, the calculator material, including exercises, is segregated and clearly marked for those who wish to omit it. In the calculator boxes and elsewhere in the text, the distinction between exact, approximated, and estimated answers is emphasized.

Chapter Summary Each chapter contains a summary of the material and glossary of new terms.

Chapter Review Each chapter ends with a set of review problems and a chapter test to reinforce the material.

SUPPLEMENTS

In order to make both the teaching and the learning process easier, the following supplements are available to accompany *Beginning Algebra with Applications:*

Annotated Instructor's Edition The Annotated Instuctor's Edition is a real time-saving device for the experienced teacher and a near necessity for the new or part-time instructor. Copious margin notes provide teaching tips and alternate approaches to the material. Historical notes provide the instructor with the background material to liven up the classroom and humanize the material. Common student errors are identified and suggestions given. Places where classroom discussion is appropriate are identified. Points to Emphasize, Discovery Problems, and short True-False quizzes called "Temptations" are also scattered throughout the margins.

Supplements Guide and Demo Instructor's Disk, (by Joan Dykes)—a unique supplement which facilitates your maximum use of the complete package provided with the book, lists alternative syllabi, possible homework assignments, as well as a section-by-section listing of appropriate supplements for use by both the student and the professor. An *"Instructor's Disk"* contains much of this material on disk, so that you can easily edit and adapt it to your own course.

Transparencies and Transparency Masters provide the instructors with an additional teaching tool around which lectures can be built—worked out examples and diagrams.

Instructor's Manual with Tests (by Mary Jean and Shannon Brod) provides nine tests which were designed specifically for this text and which can be photocopied directly from the manual, thus eliminating typing or cutting and pasting of test questions. Five tests per chapter are short answer with workspace, four are multiple choice tests which have been carefully crafted to include realistic altenate answers.

Computerized Test Generator produces multiple versions of a test with minimum effort. Tests can be randomly generated by the computer to your specifications or you can choose items from a printed *Test Item File,* which lists all the items consecutively. You can add your own test items as well, by using the editing function.

Instructor's Solutions Manual (by Elizabeth Sirjani) contains fully worked-out solutions to all the even-numbered problems in the book. (The odd-numbered solutions are provided in the Student Solutions Manual.) These can also be photocopied and given to students who need extra help, or made available in the library or math lab; however, they will not be sold to students.

FOR THE STUDENT:

Interactive Algebra Tutor Software provides the student with instant access to tutorial help, additional practice problems, immediate feedback in the form of diagnostic comments, and sample quizzes. Site License available with qualified adoption.

Algebra Problem Solver Software (by H & N Software): Through the use of artificial intelligence, gives students another chance to practice the major topics in algebra, and provides fully worked-out solutions to problems supplied by either the student or the computer.

Student Solutions Manual (by Virginia Parks): Provides fully worked-out solutions to the odd-numbered problems in the book, and to all Chapter Test questions.

"How to Study Math" (by Helen Burrier): This unique booklet is provided free in quantity to adopting instructors to hand out to each student. Contact your local Prentice Hall rep for more information.

Study Guide Keyed directly to the book and software, the Study Guide provides an additional reference for the student who needs more help.

Videotapes Keyed to each section of the book, the videotapes are a reinforcement of the text and lecture or can be used in a math lab as a lecture substitute. Available with a qualified adoption.

Audiotapes (by Peg Greene): Giving additional help in the form of worked-out examples, audiocassettes are available with a qualified adoption.

The supplements package, in combination with an excellent and well-developed text, will provide students and instructors with a state-of-the-art approach to learning Beginning Algebra and proceed to Intermediate Algebra.

We would like to thank the following reviewers:

Loyde Beam *Garland County Community College*
James D. Blackburn *Tulsa Junior College (text)*
Louis F. Bush *San Diego City College*
Alan Chutsky *Queensborough Community College*
Terrance W. Cremeans *Oakland Community College*
Patricia Deamer *Skyline College*
Michael Divinia *San Jose City College*
Jack Drucker *Borough of Manhattan College*
H. Joan Dykes *Edison Community College*
Elayn Gay *University of New Orleans*
Margaret Greene *Florida Community College*
Adelbert Hackert *Sacramento City College*
Thomas Ray Hamel *Austin Peay State University*
Virginia Lee *Brookdale Community College*
Dr. Peter Lindstrom *North Lake College*
Thomas A. McCready *California State University at Chico*
Michael Mears *Manatee Community College*
Thomas Mowry *Diablo Valley College*
Catherine Pace *Louisiana Technical University*
Matthew Pickard *University of Puget Sound*
Faye Thames *Lamar University*

Finally, we would like to thank the staff and faculty of DeKalb College for their continued support; Peggy Estes who insisted we do this work; Cristina Ferrari, our diligent production editor; designer Anne Bonanno; development editor Christine Peckaitis; and lastly, our senior editor and friend, Priscilla McGeehon.

Linda L. Exley
Vincent K Smith

Advice to the Successful Math Student

Throughout the years, we have seen a parade of algebra students pass through the classroom. Many succeed, but some do not. We would be poor students ourselves if we did not notice the things that make the difference. Three essentials are evident: a will to succeed, homework discipline, and test-taking skills.

Math Success

The first day or two of class, some students freely admit, "I hate math," or "I can't do math," or perhaps, "Math was alway my worst subject." We can't make students like math and we realize that many subjects are easier. However, the mental state that such statements indicate leads almost certainly to failure. Regardless of your past experience, you must decide to succeed. You must make up your mind *to do* whatever it takes to succeed. No one can do that for you. You must decide on your own.

Math Homework

Once you have decided to succeed, you must realize that homework is the key to your success. This book has large problem sets to allow plenty of drill problems. Do all of the assigned problems. Don't stop because you already know how to do that kind of problem or because you have figured it out. Working drill problems that you know how to do develops the speed and accuracy you need at test time, and most importantly, gives you the self confidence you need to combat test anxiety.

HOW TO DO HOMEWORK IN MATH

1. Do all homework every day. Plan on 1–3 hours studying for each hour in class.
2. Read the instructions to the problem. Make sure that you understand what it is that you are doing with each problem.
3. Start by working examples in the book. Work them until you can do them without looking at the book.
4. Do homework neatly just as if it were to be turned in. Don't be sloppy and write down things which make no sense. Keep it in a notebook in an organized way.
5. Don't just memorize. Try to understand why you do each step. Don't expect to understand everything the first time you see it. Sometimes you must mimic steps to solve a problem. Understanding may come gradually.
6. Do the problems enough times so that you don't rely on notes, the book, or a friend to give you hints. You

must practice until you can work the problems with no help.

7. Develop speed as well as accuracy. This comes with much practice.
8. Don't be afraid of words. Often a word problem is very easy. Follow the procedure outlined in the book for solving word problems and don't just try to figure out the answer.
9. Review old material often.
10. Test yourself by working the keyed problems with no help. Do this after you have done all the problems assigned.
11. Read the material in the book before it is covered in class and again after it is covered.
12. Do the review problems in each chapter.

Math Tests

Test anxiety, indicated by statements like, "I just go blank on tests" or "I understand in class, but can't do the test problems," is usually caused by lack of self confidence. The cure is practice, practice, practice.

HOW TO TAKE A MATH TEST

1. Be rested.
2. Be prepared.
3. Watch the time. Don't rush, but pace yourself so that you can try every problem. If a problem seems hard, skip it. Do problems that you can solve quickly first.
4. Read the instructions for each problem carefully. Make sure that you know *what* you are to do with the problem.

See Problem Set 0.4, Exercise 114.

Natural Numbers, Fractions, and Decimals

CONNECTIONS

Arithmetic books dating back to the Middle Ages have attempted to teach students how to calculate with counting numbers and fractions. For thousands of years, people have sought ways to calculate more rapidly and accurately. Various devices have been invented to carry out computations automatically. The oldest such device is the abacus which often consisted of a series of beads strung on rows of wire each representing a power of 10.

Today calculators and computers are used to make calculations in fields ranging from engineering to coaching basketball. Computers make thousands of arithmetic calculations for meteorologists in preparing a weather forecast for the National Weather Service.

So, why should we study arithmetic? Understanding how to add, subtract, multiply, and divide with natural numbers, fractions, and decimals is the foundation upon which algebra is developed. Algebra can be described as arithmetic with letters used to represent numbers. To understand the connection between arithmetic and algebra, we must first understand the concepts of arithmetic. This chapter contains a review of arithmetic.

0.1 ORDER OF OPERATIONS

Arithmetic is a study of numbers. The symbols that we use to name numbers are called **numerals.** For example, 2 and II are names for the number two. The symbol 2 is an Arabic numeral, whereas II is a Roman numeral. We use Arabic numerals almost exclusively in mathematics.

There are many kinds of numbers. In this chapter we will use the counting numbers, or natural numbers. They are 1, 2, 3, ... The three dots (called an ellipsis) indicate that the numbers continue.

In arithmetic, we perform four operations (addition, subtraction, multiplication, and division) with numbers. We use symbols to indicate these operations. The following chart indicates the symbols that we use for each of the operations with the numbers 7 and 5.

OPERATION	SYMBOL(S)	READ
Addition	$7 + 5$	7 plus 5
Subtraction	$7 - 5$	7 minus 5
Multiplication	$7 \cdot 5$; $7(5)$; $(7)5$; $(7)(5)$	7 times 5
Division	$7 \div 5$; $\frac{7}{5}$; $7/5$; $5\overline{)7}$	7 divided by 5

The **sum** of two numbers is the result of adding them. The **difference** in two numbers is the result of subtracting one from the other. The **product** of two numbers is the result of multiplying them. The **quotient** is the result of dividing one number (dividend) by another (divisor).

EXAMPLE 1. Express each phrase in symbols.

(a) The sum of 8 and 3.

(b) The difference when 6 is subtracted from 9.

(c) The difference when 9 is subtracted from 6.

(d) The product of 2 and 7.

(e) The quotient when 12 is divided by 19.

(f) The quotient when 19 is divided by 12.

Solutions:

(a) $8 + 3$ (b) $9 - 6$ (c) $6 - 9$

(d) $2 \cdot 7$ (e) $12 \div 19$ (f) $19 \div 12$ □

Mathematicians have developed a shorthand notation for writing products with the same number. Instead of writing $2 \cdot 2 \cdot 2$, we write 2^3. When we write 3^5, we mean $3 \cdot 3 \cdot 3 \cdot 3 \cdot 3$.

In an expression such as 3^5, we call 3 the **base,** and 5 the **exponent.** The exponent tells us how many times to use the base in the product. We read 3^5 as 3 to the fifth **power.** We use the second and third powers so often that we read 7^2 as 7 *squared* and 4^3 as 4 *cubed.*

EXAMPLE 2. Find the value of each expression.

(a) 3^2 (b) 2^3

Solutions:

(a) $3^2 = 3 \cdot 3$
$= 9$

(b) $2^3 = 2 \cdot 2 \cdot 2$
$= 8$ □

Many times more than one operation is involved in simplifying an expression. For example, $2 + 3 \cdot 4$ has multiplication and addition. We must decide which operation to perform first. **Grouping symbols** such as parentheses ( ), brackets [ ], or braces { } are used to make the order of operations clear. We perform operations inside grouping symbols first.

If we write $(2 + 3) \cdot 4$, then we perform the addition first and then the multiplication. So,

$$\begin{aligned}(2 + 3) \cdot 4 &= 5 \cdot 4 \\ &= 20\end{aligned}$$

However, if we write $2 + (3 \cdot 4)$, then we perform the multiplication first followed by the addition. So,

$$\begin{aligned}2 + (3 \cdot 4) &= 2 + 12 \\ &= 14\end{aligned}$$

If there are no grouping symbols, then we do multiplications and divisions from left to right before we do additions and subtractions. So,

$$\begin{aligned}2 + 3 \cdot 4 &= 2 + 12 \\ &= 14\end{aligned}$$

Fraction bars are also grouping symbols. For example,

$$\frac{2+10}{2+4} = \frac{12}{6}$$
$$= 2$$

This expression is the same as $(2 + 10) \div (2 + 4)$. The fraction allows us to omit the parentheses.

The following rules determine which operation to perform first.

Order of Operations

If grouping symbols are present, perform operations inside them, starting with the innermost grouping symbol, in the following order.

1. Perform any exponentiations.
2. Perform all multiplications and divisions in order from left to right.
3. Perform all additions and subtractions in order from left to right.

If grouping symbols are *not* present, perform operations in the above order.

A commonly used memory aid is the sentence, "Please excuse my dear Aunt Sally." The first letters of each word give the order of operations.

Parentheses
Exponents
Multiplication and Division from left to right
Addition and Subtraction from left to right

Be Careful!

Note that multiplication does not have to be done before division, nor addition before subtraction. Multiplications and divisions are done in order as they occur working from left to right. Additions and subtractions are done in order as they occur working from left to right.

EXAMPLE 3. Perform the operations indicated.

(a) $6 \cdot 3 - \frac{4}{2}$ (b) $10 - 3 + 2^2$

(c) $14/2 \cdot 3$ (d) $12 - 2(6 - 3)$

(e) $(3^2 - 4)^2$ (f) $\frac{2 \cdot 3 + 8}{10 - 3}$

Solutions:

(a) We perform the multiplication and division first.

$$6 \cdot 3 - \frac{4}{2} = 18 - 2$$
$$= 16$$

(b) We work with the exponent first.

$$10 - 3 + 2^2 = 10 - 3 + 4$$
$$= 11$$

(c) We perform multiplications and divisions from left to right in the order they occur. So, we divide first and then multiply.

$$14/2 \cdot 3 = 7 \cdot 3$$
$$= 21$$

(d) We work inside the parentheses first.

$$12 - 2(6 - 3) = 12 - 2 \cdot 3$$
$$= 12 - 6$$
$$= 6$$

(e) $(3^2 - 4)^2 = (9 - 4)^2$
$= 5^2$
$= 25$

(f) We simplify the numerator and denominator first.

$$\frac{2 \cdot 3 + 8}{10 - 3} = \frac{6 + 8}{10 - 3}$$
$$= \frac{14}{7}$$
$$= 2$$

□

PROBLEM SET 0.1

Warm-ups

In Problems 1 through 15, express each phrase in symbols. See Example 1.

1. The sum of 3 and 4.
2. 7 more than 5.
3. 3 times 9.
4. 6 subtracted from 11.
5. The quotient when 21 is divided by 8.
6. Twice 14.
7. One-half of 17.
8. 8 squared.
9. 5 more than 1.
10. The quotient when 5 is divided by 9.
11. The product of 2 and 5.
12. 4 less than 6.
13. 2 cubed.
14. 11 squared plus 6.
15. One less than twice 13.

Find the value of each. See Example 2.

16. 2^3
17. 3^2
18. 4^2
19. 5^3
20. 10^2

In Problems 21 through 36, perform the operations indicated. See Example 3.

21. $6 + 3 \cdot 2$
22. $12 - 2^2$
23. $\frac{16 - 2}{7}$
24. $(3 + 4)(6 - 4)$
25. $(3 + 4)6 - 4$
26. $3 + 4(6 - 4)$
27. $12 - 2(8 - 4)$
28. $\frac{8^2 - 6 \cdot 7}{2 \cdot 3 + 5}$
29. $9 + 42/7 - 2^2$

30. $2 \cdot 3^2 - 2^2$

31. $\dfrac{2\left(\frac{12}{4}\right) - 2}{2^2}$

32. $(6 - 3)^2$

33. $45 - (2 \cdot 3)^2$

34. $18/6 \cdot 2$

35. $28 \div 4 - 2 \cdot 3$

36. $2^2 \cdot (2 \cdot 3)^2$

Practice Exercises

In Problems 37 through 50, find each value.

37. The product of 3 and 4.

38. 5 less than 17.

39. 3 added to 53.

40. 1 subtracted from 6.

41. The quotient when 77 is divided by 11.

42. One-third of 12.

43. 12 squared.

44. 11 less than 15.

45. The quotient when 6 is divided by 2.

46. The square of the sum of 5 and 3.

47. Twice the sum of 8 and 7.

48. 4 cubed.

49. 13 squared minus 6.

50. One more than twice 7.

In Problems 51 through 76, perform the operations indicated.

51. $6 \cdot 3 - 2$

52. $2^2 - 3$

53. $\dfrac{16 - 9}{7}$

54. $(6 - 4)(6 + 4)$

55. $(7 - 4)6 - 4$

56. $12 - 4(6 - 4)$

57. $12 + 2(8 - 7)$

58. $\dfrac{8 \cdot 7 - 6^2}{2 \cdot 5 - 6}$

59. $22 - (20 - 14)$

60. $6^2 - \dfrac{16 - 4}{4}$

61. $23 - 63/7 - 3^2$

62. $2^2 \cdot 3^2 - 2^2$

63. $\dfrac{2\left(\frac{48}{16}\right) - 2}{2^2}$

64. $(6 + 3)^2$

65. $25 - 2 \cdot 3^2$

66. $3 \cdot 8/6$

67. $36 \div 9 \cdot 2 - 3$

68. $2 \cdot (4 \cdot 3)^2$

69. $2(3^3 - 3^2)$

70. $4^2 + 5 \div 5 - 2 \cdot 3$

71. $2 + (3 \cdot 2^3 - 2 \cdot 3^2)$

72. $2 + 3[2 + 3(7 - 6)]$

73. $19 - 3[2 + 2(11 - 3^2)]$

74. $[13 - 2(3 \cdot 5 - 12)] - 2^2$

75. $3(4^2 - 2^3)$

76. $3^2 - 3 \div 3 + 3 \cdot 3$

In Problems 77 through 96, answer each question.

77. Mae has 9 books and Sue has 5 books. How many books do they have together?

78. Jo has 4 times 3 skirts and Geisla has 2 times 8 skirts. How many skirts do they have together?

79. Tony has 24 quarters and Ellen has 10 fewer quarters than Tony. How many quarters does Ellen have?

80. Jorge has 9 sodas and Manuel has 3 fewer sodas than Jorge. How many sodas does Manuel have?

81. John has 6 records and Mary has 10 records. How many records do they have together?

82. Susan has 3 times 9 dollars and Judy has 2 times 8 dollars. How many dollars do they have?

83. Wei has 26 nickels and Choi has 10 more nickels than Wei. How many nickels does Choi have?

84. Juan has 12 marbles and Maria has 7 fewer marbles than Juan. How many marbles does Maria have?

IN YOUR OWN WORDS . . .

85. State the order of operations.

Write in words what each expression means.

86. $8 + 7$

87. $11 - 14$

88. $5 \cdot 23$

89. $6 \div 7$

90. $9/22$

91. $\dfrac{7}{6}$

92. Explain two ways to interpret the statement "The sum of 7 and 2 squared."

0.2 FACTORING

In much of our work in algebra, it is necessary to write a number as a product of numbers. This process is called **factoring.** The numbers in the product are called **factors.** Most often we deal with natural numbers.

A **prime** number is a natural number larger than 1 whose only natural number factors are itself and 1. For example, 5 is prime because $5 = 5 \cdot 1$. This is the only way to write 5 as a product of natural numbers. Likewise, 2, 3, 7, and 11 are prime. We don't call 1 a prime number. The smallest prime number is 2.

Sometimes we need to factor a natural number that is not prime. Often there are several ways to do this. For example 12 can be factored as follows:

$$12 = 2 \cdot 2 \cdot 3 \qquad 12 = 2 \cdot 6 \qquad 12 = 3 \cdot 4 \qquad 12 = 12 \cdot 1$$

Frequently, the most useful of these is $2 \cdot 2 \cdot 3$ because it is a product of prime numbers. There is only one way to factor a natural number using only prime factors. When we write a number as a product of prime numbers, we say that we have **factored it completely,** or that we have the **prime factorization.**

In factoring large numbers, it is helpful to know when a number is divisible by a prime.

Some Tests for Divisibility

1. A natural number is divisible by 2 if it is even — that is, if its last digit is 0, 2, 4, 6, or 8.
2. A natural number is divisible by 3 if the sum of the digits of the number is divisible by 3.
3. A natural number is divisible by 5 if its last digit is 0 or 5.

To factor a number into prime factors, we start with the smallest prime that will divide into the number and continue until we have all prime factors.

EXAMPLE 1. Write each number as a product of prime factors.

(a) 45 (b) 108 (c) 392

Solutions:

(a) We notice that 45 is not even, so it is not divisible by 2. However, the sum of its digits is 9, so it is divisible by 3.

$$\begin{aligned} 45 &= 3 \cdot 15 \\ &= 3 \cdot 3 \cdot 5 \\ &= 3^2 \cdot 5 \end{aligned}$$

(b) Since 108 is even, it is divisible by 2.

$$108 = 2 \cdot 54$$

Now 54 is also divisible by 2. So,

$$108 = 2 \cdot 2 \cdot 27$$

(continued)

$$= 2 \cdot 2 \cdot 3 \cdot 3 \cdot 3$$
$$= 2^2 \cdot 3^3$$

(c) Since 392 is even, we divide by 2.

$$\begin{aligned} 392 &= 2 \cdot 196 \\ &= 2 \cdot 2 \cdot 98 \\ &= 2 \cdot 2 \cdot 2 \cdot 49 \\ &= 2^3 \cdot 7^2 \end{aligned}$$ □

Short division is long division without all of the arithmetic. It is a convenient way to factor numbers. Let's factor 588 using short division. We begin with the smallest prime that will divide into 588. It is 2:

$$\begin{array}{r} 294 \\ 2\overline{)588} \end{array}$$

We do the same thing with the number 294. The smallest prime that will divide into 294 is 2:

$$\begin{array}{r} 147 \\ 2\overline{)294} \\ 2\overline{)\ 588} \end{array}$$

Now 147 is not divisible by 2. However, since the sum of its digits is 12, it is divisible by 3:

$$\begin{array}{r} 49 \\ 3\overline{)147} \\ 2\overline{)\ 294} \\ 2\overline{)\quad 588} \end{array}$$

The smallest prime that will divide into 49 is 7:

$$\begin{array}{r} 7 \\ 7\overline{)49} \\ 3\overline{)147} \\ 2\overline{)\ 294} \\ 2\overline{)\quad 588} \end{array}$$

Since 7 is prime, we stop the process. Thus, $588 = 2^2 \cdot 3 \cdot 7^2$.

PROBLEM SET 0.2

Warm-ups

In Problems 1 through 10, write each number as a product of prime factors. See Example 1.

1. 24 **2.** 36 **3.** 25
4. 16 **5.** 8 **6.** 20
7. 35 **8.** 18 **9.** 45
10. 75

Practice Exercises

In Problems 11 through 40, write each number as a product of prime factors.

11. 12
12. 48
13. 50
14. 32
15. 81
16. 40
17. 70
18. 63
19. 90
20. 125
21. 200
22. 243
23. 56
24. 196
25. 150
26. 100
27. 96
28. 72
29. 108
30. 180
31. 441
32. 144
33. 3969
34. 576
35. 972
36. 625
37. 169
38. 900
39. 960
40. 1008

Answer each question.

41. Martin bought 6 pictures and Rowland bought three times as many as Martin. How many pictures did Rowland buy?

42. Vincent and Kim each ate 2 candy bars every month for one year. How many candy bars did they eat in the year?

43. Ken has 3 rare coins in his collection and Branden has double that many in his collection. How many rare coins does Branden have?

44. Susie has 41 stamps in her collection. If she triples the number of stamps that she has, how many stamps will she have?

45. Tim bought 7 pencils and Irene bought twice as many as Tim. How many pencils did Irene buy?

46. Ky has 6 sandwiches and Nyugen has seven times as many sandwiches as Ky. How many sandwiches does Nyugen have?

47. Pierre cuts a board into three pieces each of length 5 m. How long was the board originally?

48. A basketball team had 43 points at halftime. At the end of the game they had doubled the halftime score. How many points did the team have at the end of the game?

IN YOUR OWN WORDS...

49. What does it mean to factor a natural number?

50. Describe the steps in finding the prime factors of 294.

0.3 FRACTIONS

Fractions are very common in algebra. This section will review operations with fractions in arithmetic.

Throughout this section, we will use the letters p, q, r, and s to represent natural numbers. We do this in order to state the basic rules and principles of operations with fractions. If we divide 5 by 9, we can write the quotient as the fraction $\frac{5}{9}$. In general we say if we divide a *number* p by a *number* q, we can write the quotient as the fraction, $\frac{p}{q}$. We call p the **numerator** and q the **denominator.**

When working problems with fractions, we sometimes see fractions in which the numerator is larger than or equal to the denominator. Such fractions are called **improper fractions.** In *arithmetic* it is common to change improper fractions into mixed numbers. For example,

$$\frac{22}{9} = 2\frac{4}{9}$$

Notice that the mixed number $2\frac{4}{9}$ means $2 + \frac{4}{9}$.

In algebra we prefer to do arithmetic with a fraction rather than a mixed number (because we indicate *multiplication* when we write numbers next to each other, not addition.) So in algebra, when improper fractions arise, we leave them. When we find mixed numbers, we write them as fractions.

If a problem is stated using mixed numbers, we convert the mixed numbers to fractions and solve the problem. We answer the question using mixed numbers, since we usually answer questions in the form in which they are stated.

The fractions $\frac{1}{2}$, $\frac{2}{4}$, and $\frac{3}{6}$ are different ways of writing the same number. This is the idea behind the Fundamental Principle of Fractions.

Fundamental Principle of Fractions

$$\frac{p}{q} = \frac{p \cdot r}{q \cdot r}$$

If we say $\frac{3}{6} = \frac{1}{2}$, then we are using the Fundamental Principle to reduce the fraction. It tells us that we can remove **factors** that are common to both the numerator and the denominator. If there are no factors common to both the numerator and the denominator, then we say the fraction is **reduced to lowest terms.**

EXAMPLE 1. Reduce $\dfrac{48}{60}$ to lowest terms.

Solution:

First we factor the numerator and denominator.

$$\frac{48}{60} = \frac{2 \cdot 2 \cdot 2 \cdot 2 \cdot 3}{2 \cdot 2 \cdot 3 \cdot 5}$$

The common factors in numerator and denominator are 3 and $2 \cdot 2$.

$$\begin{aligned} \frac{48}{60} &= \frac{2 \cdot 2}{5} \\ &= \frac{4}{5} \end{aligned}$$

□

If we see that 12 is a factor of both 48 and 60, then we can do Example 1 without factoring completely.

$$\frac{48}{60} = \frac{12 \cdot 4}{12 \cdot 5} = \frac{4}{5}$$

If we say $\frac{1}{2} = \frac{3}{6}$, then we are using the Fundamental Principle to build up a fraction. Often we must do this when we add or subtract fractions. The Fundamental Principle tells us that we can multiply both numerator and denominator by the same number and not change the value of a fraction.

EXAMPLE 2. Write $\frac{2}{15}$ with a denominator of 45.

Solution:

Since $45 = 15 \cdot 3$, we must multiply both numerator and denominator by 3.

$$\frac{2}{15} = \frac{2 \cdot 3}{15 \cdot 3}$$

$$= \frac{6}{45}$$

□

To multiply two fractions, we multiply numerators and multiply denominators.

Multiplication of Fractions

$$\frac{p}{q} \cdot \frac{r}{s} = \frac{p \cdot r}{q \cdot s}$$

EXAMPLE 3. Find the products indicated and reduce answers to lowest terms.

(a) $\left(\frac{2}{3}\right)^3$ (b) $\frac{12}{35} \cdot 1\frac{1}{14}$

Solutions:

(a) $\left(\frac{2}{3}\right)^3 = \frac{2}{3} \cdot \frac{2}{3} \cdot \frac{2}{3}$

$$= \frac{2 \cdot 2 \cdot 2}{3 \cdot 3 \cdot 3}$$

$$= \frac{8}{27}$$

(b) $\frac{12}{35} \cdot 1\frac{1}{14} = \frac{12}{35} \cdot \frac{15}{14}$ Express as an improper fraction.

$$= \frac{12 \cdot 15}{35 \cdot 14}$$

Instead of multiplying the numbers in the numerator and in the denominator, we factor so that we can look for common factors to reduce.

$$\frac{12}{35} \cdot 1\frac{1}{14} = \frac{2^2 \cdot 3 \cdot 3 \cdot 5}{7 \cdot 5 \cdot 7 \cdot 2}$$

$$= \frac{2 \cdot 3 \cdot 3}{7 \cdot 7}$$ Common factors are 2 and 5.

$$= \frac{18}{49}$$

□

Notice that when we build up a fraction using the Fundamental Principle, we are actually multiplying the fraction by 1. Suppose we wish to write the fraction $\frac{5}{8}$ with a denominator of 56. Since $56 = 8 \cdot 7$, we need to multiply the denominator by 7.

$$\begin{aligned} \frac{5}{8} &= \frac{5}{8} \cdot 1 \\ &= \frac{5}{8} \cdot \frac{7}{7} && 1 = \frac{7}{7} \\ &= \frac{35}{56} && \text{Multiplication of fractions} \end{aligned}$$

To divide two fractions, multiply the first fraction by the reciprocal of the divisor.

Division of Fractions

$$\frac{p}{q} \div \frac{r}{s} = \frac{p}{q} \cdot \frac{s}{r} = \frac{p \cdot s}{q \cdot r}$$

EXAMPLE 4. Perform the division indicated $\frac{27}{16} \div \frac{9}{8}$, and reduce the answer to lowest terms.

Solution:

$$\begin{aligned} \frac{27}{16} \div \frac{9}{8} &= \frac{27}{16} \cdot \frac{8}{9} \\ &= \frac{27 \cdot 8}{16 \cdot 9} \\ &= \frac{3}{2} \end{aligned}$$

□

We add fractions that have the same denominator by adding the numerators and using the common denominator as the new denominator. Subtraction of fractions is done in a similar manner.

Addition and Subtraction of Fractions

$$\frac{p}{q} + \frac{r}{q} = \frac{p + r}{q}$$

$$\frac{p}{q} - \frac{r}{q} = \frac{p - r}{q}$$

EXAMPLE 5. Perform the operations indicated.

(a) $\frac{7}{13}+\frac{5}{13}$ (b) $\frac{16}{23}-\frac{11}{23}$

Solutions:

(a) $\frac{7}{13}+\frac{5}{13}=\frac{7+5}{13}$

$=\frac{12}{13}$

(b) $\frac{16}{23}-\frac{11}{23}=\frac{16-11}{23}$

$=\frac{5}{23}$ □

When the denominators are not the same, we must find an appropriate denominator and write each fraction with this denominator. The denominator should be the smallest number divisible by each denominator. We call this denominator the **least common denominator.**

Procedure for Finding the Least Common Denominator (LCD)

1. Factor each denominator completely, using exponents.
2. List all the different prime factors from all the denominators.
3. The LCD is the product of the factors in step 2 each raised to the largest power of that factor in any single denominator.

EXAMPLE 6. Find the least common denominator for each pair of denominators.

(a) 12 and 15 (b) 8 and 24 (c) 72 and 108

Solutions:

(a) First we factor each number.

$$12 = 2^2 \cdot 3$$

$$15 = 3 \cdot 5$$

The different prime factors are 2, 3, and 5.

2 occurs twice in 12 and no times in 15.
So, the LCD must contain 2 as a factor twice.

3 occurs one time in each number.
The LCD must contain one factor of 3.

5 occurs no times in 12 and 1 time in 15.
The LCD must contain one factor of 5.

The LCD is $2^2 \cdot 3 \cdot 5$, or 60.

(continued)

(b) $24 = 2^3 \cdot 3$

$8 = 2^3$

The different prime factors are 2 and 3.

2 occurs three times in both numbers.
The LCD must contain three factors of 2.

3 occurs one time in 24 and no times in 8.
The LCD must contain one factor of 3.

The LCD is $2^3 \cdot 3$, or 24.

(c) $72 = 2^3 \cdot 3^2$

$108 = 2^2 \cdot 3^3$

The different prime factors are 2 and 3.

2 occurs three times in 72 and two times in 108.
The LCD must contain 2^3.

3 occurs two times in 72 and three times in 108.
The LCD must contain 3^3.

The LCD is $2^3 \cdot 3^3$, or 216. □

EXAMPLE 7. Perform the operations indicated and reduce the answers to lowest terms.

(a) $\frac{5}{24} + 1\frac{5}{8}$ (b) $\frac{7}{12} - \frac{8}{15}$.

Solutions:

(a) We must first find the LCD. From Example 6(b), it is 24.

$$\frac{5}{24} + 1\frac{5}{8} = \frac{5}{24} + \frac{13}{8} \quad \text{Express as an improper fraction.}$$

$$= \frac{5}{24} + \frac{13 \cdot 3}{8 \cdot 3}$$

$$= \frac{5}{24} + \frac{39}{24}$$

$$= \frac{44}{24}$$

$$= \frac{11}{6} \quad \text{Reduce.}$$

$$= 1\frac{5}{6} \quad \text{Express as a mixed number.}$$

(b) From Example 6(a), the LCD is 60.

$$\frac{7}{12} - \frac{8}{15} = \frac{7 \cdot 5}{12 \cdot 5} - \frac{8 \cdot 4}{15 \cdot 4}$$

$$= \frac{35}{60} - \frac{32}{60}$$

$$= \frac{3}{60}$$

$$= \frac{1}{20}$$ □

The order of operations that we learned in Section 0.1 applies to fractions.

EXAMPLE 8. Perform the operations indicated and reduce to lowest terms.

(a) $\frac{3}{4} + \frac{2}{3} \cdot \frac{1}{2}$ (b) $\left(\frac{2}{3} - \frac{1}{5}\right) \div \frac{7}{3}$

Solutions:

(a) We multiply before we add.

$$\frac{3}{4} + \frac{2}{3} \cdot \frac{1}{2} = \frac{3}{4} + \frac{1}{3}$$

$$= \frac{9}{12} + \frac{4}{12}$$

$$= \frac{13}{12}$$

(b) Subtract inside the parentheses first.

$$\left(\frac{2}{3} - \frac{1}{5}\right) \div \frac{7}{3} = \left(\frac{10}{15} - \frac{3}{15}\right) \div \frac{7}{3}$$

$$= \frac{7}{15} \div \frac{7}{3}$$

$$= \frac{7}{15} \cdot \frac{3}{7}$$

$$= \frac{7 \cdot 3}{15 \cdot 7}$$

$$= \frac{1}{5}$$ □

EXAMPLE 9. Jean has $5\frac{1}{2}$ bushels of peaches, which she wants to divide evenly among her three children. How many bushels will each child receive?

Solution:

We must divide $5\frac{1}{2}$ by 3.

$$5\frac{1}{2} \div 3 = \frac{11}{2} \div 3 \qquad \text{Express as a fraction.}$$

$$= \frac{11}{2} \cdot \frac{1}{3} \qquad \text{Multiply by the reciprocal.}$$

(continued)

$$= \frac{11}{6}$$

$$= 1\frac{5}{6} \qquad \text{Express as a mixed number.}$$

Each child will receive $1\frac{5}{6}$ bushels of peaches. □

PROBLEM SET 0.3

Warm-ups

In Problems 1 and 2, use the Fundamental Principle to reduce each fraction to lowest terms. See Example 1.

1. $\frac{12}{18}$ **2.** $\frac{36}{72}$

In Problems 3 through 6, use the Fundamental Principle to write each fraction with the given denominator. See Example 2.

3. $\frac{1}{2}$; denominator of 10 **4.** $\frac{3}{4}$; denominator of 12

5. $\frac{4}{3}$; denominator of 18 **6.** $\frac{11}{36}$; denominator of 144

In Problems 7 through 41, perform the operations indicated. Reduce answers to lowest terms.
In Problems 7 through 12, see Example 3.

7. $\frac{3}{4} \cdot \frac{1}{3}$ **8.** $\left(\frac{1}{3}\right)^3$ **9.** $\frac{15}{33} \cdot \frac{44}{35}$

10. $\left(1\frac{5}{6}\right) \cdot \left(1\frac{7}{11}\right)$ **11.** $\left(1\frac{3}{4}\right)^2$ **12.** $\frac{2}{3} \cdot 6$

In Problems 13 through 16, see Example 4.

13. $\frac{3}{4} \div \frac{1}{3}$ **14.** $\frac{4}{7} \div 2$ **15.** $3\left(\frac{7}{9}\right) \div \left(3\frac{2}{3}\right)$ **16.** $\frac{25}{24} \div \frac{45}{42}$

In Problems 17 through 22, find the LCD for each pair of numbers. See Example 6.

17. 3 and 7 **18.** 8 and 16 **19.** 6 and 8
20. 20 and 24 **21.** 36 and 27 **22.** 135 and 54

In Problems 23 through 34, see Examples 5 and 7.

23. $\frac{1}{3} + \frac{4}{15}$ **24.** $\frac{1}{4} + \frac{1}{3}$ **25.** $\frac{3}{4} - \frac{1}{3}$

26. $\frac{3}{4} - \frac{1}{8}$ **27.** $\frac{2}{3} - \frac{3}{8}$ **28.** $\frac{5}{6} - \frac{5}{8}$

29. $\frac{5}{12} + \frac{7}{18}$ **30.** $1\frac{3}{5} + 2\frac{1}{5}$ **31.** $5\frac{2}{7} - 3\frac{3}{7}$

32. $2\frac{11}{12} - 1\frac{13}{18}$ **33.** $\frac{3}{7} + 1$ **34.** $2 - \frac{4}{5}$

In Problems 35 through 41, see Example 8.

35. $\frac{1}{18} + \frac{7}{12} - \frac{5}{9}$

36. $\frac{3}{8} \cdot \frac{2}{3} \div 4$

37. $\frac{3}{2} \cdot \frac{1}{3} \cdot \frac{4}{5}$

38. $\frac{3}{2} \cdot \frac{1}{3} \div \frac{4}{5}$

39. $\frac{3}{2} \div \frac{1}{3} + \frac{4}{5}$

40. $\left(\frac{3}{2} + \frac{1}{3}\right) \cdot \frac{4}{5}$

41. $5\left(\frac{3}{2}\right)^2 \div \frac{5}{12}$

In Problems 42 through 45, answer each question with a statement. See Example 9.

42. Subtract $3\frac{1}{3}$ from the product of $8\frac{1}{3}$ and $\frac{3}{5}$.

43. Add $\left(\frac{2}{3}\right)^2$ to $6\frac{1}{3}$.

44. Joan cuts a wire that is $2\frac{1}{3}$ yards (yd) long in half. How long will each half be?

45. Marge is in charge of making pancakes at her club's breakfast. She must have 15 pounds (lb) of flour. If Virginia donated $5\frac{1}{2}$ lb, Jane gave $8\frac{2}{3}$ lb, and Madelyn gave $5\frac{1}{6}$ lb of flour, does Marge have enough flour?

Practice Exercises

In Problems 46 through 84, perform the operations indicated. Reduce answers to lowest terms.

46. $\frac{3}{5} + \frac{1}{3}$

47. $\frac{3}{5} - \frac{1}{3}$

48. $\frac{3}{5} \cdot \frac{1}{3}$

49. $\frac{3}{5} \div \frac{1}{3}$

50. $\left(\frac{1}{3}\right)^2$

51. $\left(\frac{1}{2}\right)^3$

52. $\left(\frac{2}{5}\right)^3$

53. $\left(\frac{1}{10}\right)^2$

54. $\frac{1}{3} + \frac{5}{6}$

55. $\frac{3}{7} - \frac{1}{14}$

56. $\frac{2}{3} - \frac{3}{7}$

57. $\frac{5}{6} - \frac{5}{9}$

58. $\frac{7}{12} + \frac{5}{18}$

59. $\frac{24}{50} \cdot \frac{40}{27}$

60. $\frac{36}{35} \div \frac{42}{5}$

61. $2\frac{5}{6} + 1\frac{1}{6}$

62. $4\frac{1}{8} - 3\frac{3}{8}$

63. $\left(1\frac{3}{7}\right) \cdot \left(2\frac{4}{5}\right)$

64. $\left(3\frac{3}{8}\right) \div \left(2\frac{1}{4}\right)$

65. $5\frac{3}{4} - 3\frac{1}{2}$

66. $\left(1\frac{4}{5}\right)^2$

67. $\frac{3}{4} + 1$

68. $3 - \frac{3}{2}$

69. $\frac{4}{5} \cdot 10$

70. $\frac{6}{7} \div 3$

71. $\frac{2}{27} + \frac{35}{54} - \frac{5}{9}$

72. $\frac{5}{8} \cdot \frac{2}{5} \div 6$

73. $\frac{1}{4} \cdot \frac{4}{3} \cdot \frac{3}{4}$

74. $\frac{1}{4} \cdot \frac{4}{3} \div \frac{3}{4}$

75. $\frac{1}{4} \div \frac{4}{3} + \frac{3}{4}$

76. $\frac{1}{4} + \frac{4}{3} \cdot \frac{3}{4}$

77. $\frac{1}{4} \div \left(\frac{4}{3} + \frac{3}{4}\right)$

78. $\left(\frac{1}{4} + \frac{4}{3}\right) \cdot \frac{3}{4}$

79. $\left(\frac{3}{4} - \frac{1}{5}\right) \div \frac{11}{5}$

80. $5\left(\frac{2}{3}\right)^2 \div \frac{10}{3}$

81. $\left(\frac{1}{2}\right)^2 + \left(\frac{1}{3}\right)^2$

82. $\left(\frac{2}{3} - \frac{1}{6}\right)^2$

83. $\left(\frac{1}{3} + \frac{1}{4}\right)\left(\frac{1}{3} - \frac{1}{4}\right)$

84. $\left(\frac{3}{2} - 1\right)^2$

In Problems 85 through 88, answer each question with a statement.

85. Spencer bought a 25-lb bag of fertilizer. If Linda gave $12\frac{3}{4}$ lb of it to her friend, how much fertilizer does Spencer have left?

86. What is $7\frac{2}{3}$ squared?

87. Last month Bonnie's phone bill showed that she talked $2\frac{1}{2}$ minutes (min) to her mother, $6\frac{1}{2}$ min to her sister, 30 seconds (s) to her boss, and 12 min to her son. For how many minutes did she talk last month?

88. Shirley wishes to cut a cord that is $2\frac{1}{3}$ yd long into three equal parts. How long will each part be?

IN YOUR OWN WORDS...

Write out in words what each expression means.

89. $\frac{1}{5} \cdot 6 \cdot 7$

90. $\frac{1}{5} \cdot 6 + 7$

91. $\frac{1}{5}(6 + 7)$

92. $8 + 9 \div 10$

93. $(8 + 9) \div 10$

94. $\frac{8 + 9}{10}$

0.4 DECIMALS AND PERCENTS

Fractions are quotients. The fraction $\frac{2}{5}$ is another way of writing $2 \div 5$. If we perform long division, we obtain a decimal.

$$\begin{array}{r} 0.4 \\ 5\overline{)2.0} \\ \underline{2\ 0} \end{array}$$

Every fraction can be represented as a decimal by dividing the denominator into the numerator.

EXAMPLE 1. Express each fraction as a decimal.

(a) $\frac{1}{2}$ (b) $\frac{1}{3}$ (c) $\frac{1}{6}$

Solutions:

(a)
$$\begin{array}{r} 0.5 \\ 2\overline{)1.0} \\ \underline{1\ 0} \end{array}$$

$$\frac{1}{2} = 0.5$$

Notice that we write 0 before the decimal point. This is a good practice. We prefer to write a digit on each side of the decimal point.

(b)
$$\begin{array}{r} 0.3333 \\ 3\overline{)1.0000} \\ \underline{9} \\ 10 \\ \underline{9} \\ 10 \\ \underline{9} \\ 10 \\ \underline{9} \\ 1 \end{array}$$

To indicate that the 3 continues without end, we write $\frac{1}{3} = 0.\overline{3}$, or $\frac{1}{3} = 0.333\ldots.$

(c)
$$\begin{array}{r} 0.1666 \\ 6\overline{)1.0000} \\ \underline{6} \\ 40 \\ \underline{36} \\ 40 \\ \underline{36} \\ 40 \\ \underline{36} \\ 4 \end{array}$$

$\frac{1}{6} = 0.1\overline{6}$, or $\frac{1}{6} = 0.1666\ldots$ □

In Example 1, notice that each problem involved the quotient of two natural numbers. In each case, we wrote each fraction as a decimal, which either terminated or repeated. Any time that we divide a natural number by a natural number, the decimal will either repeat or terminate.

We perform operations with decimals as we do with other numbers. The next example illustrates the operations of addition, subtraction, multiplication, and division with decimals.

EXAMPLE 2. Perform the operations indicated.

(a) $127.32 + 24.1 + 7.003$

(b) $128.73 - 38.658$

(c) $(22.3)(12.25)$

(d) $32.4 \div 8.1$

Solutions:

(a) We line up the decimal points and add.

$$\begin{array}{r} 127.320 \\ 24.100 \\ +\quad 7.003 \\ \hline 158.423 \end{array}$$

(b) Again, we line up the decimal points.

$$\begin{array}{r} 128.730 \\ -\quad 38.658 \\ \hline 90.072 \end{array}$$

(c)

$$\begin{array}{rl} 12.25 & \text{(2 decimal places)} \\ \times\ 22.3 & \text{(1 decimal place)} \\ \hline 3675 & \\ 2450 & \\ 2450 & \\ \hline 273.175 & \text{(3 decimal places)} \end{array}$$

Notice that the answer has three decimal places because there are two decimal places in 12.25 and one in 22.3.

$$(22.3)(12.25) = 273.175$$

(continued)

(d) First we move the decimal place to the right in both the dividend and the divisor until the divisor is a natural number.

$$\begin{array}{r} 4 \\ 8.1\overline{)32.4} \\ \underline{32\ 4} \end{array}$$

$$32.4 \div 8.1 = 4 \quad \square$$

EXAMPLE 3. Subtract 8.73 from the sum of 5.2 and 6.9.

Solution:

Since we are to subtract a number from the *sum* of 5.2 and 6.9, we add them first; then we subtract.

$$\begin{aligned}(5.2 + 6.9) - 8.73 &= 12.1 - 8.73 \\ &= 3.37\end{aligned} \quad \square$$

Fractions and decimals are often written as *percents.* Writing 15% is another way of writing $\frac{15}{100}$, or 0.15. Although percents are common in writing and talking, they are usually written as decimals or fractions in computation.

Percent

$$1\% = 0.01 = \frac{1}{100}$$

First, let's look at how to write a decimal as a percent. To write 0.25 as a percent, we notice that

$$\begin{aligned}0.25 &= 25(0.01) \\ &= 25\%\end{aligned}$$

Notice that we wrote 0.25 with a factor of 0.01 and that we replaced 0.01 with the % symbol. Another observation is that when we wrote 0.25 as 25%, we *moved the decimal point* ***two*** *places to the right.*

EXAMPLE 4. Write each decimal as a percent.

(a) 0.35 (b) 0.5 (c) 0.125
(d) $0.\overline{3}$ (e) 2.0

Solutions:

(a) We move the decimal point two places to the right and write the % symbol.

$$\begin{aligned}0.35 &= 0.35\% \\ &= 35\%\end{aligned}$$

(b) $0.5 = 0.50\%$

$= 50\%$

(c) $0.125 = 0.125\%$

$= 12.5\%$

(d) First, we write three digits of the repeating decimal so that we can move the decimal two places to the right.

$$0.\overline{3} = 0.33\overline{3}$$

Now we move the decimal point and write the %.

$$0.33\overline{3} = 0.33\overline{3}\,\%$$

$$= 33.\overline{3}\%$$

(e) $2.0 = 2.00\%$

$= 200\%$ □

To write a percent as a decimal we reverse the procedure. That is, we replace the % symbol with the factor 0.01 and multiply.

$$41\% = 41(0.01)$$

$$= 0.41$$

Notice that we *moved the decimal point* ***two*** *places to the left.*

EXAMPLE 5. Write the following percents as decimals.

(a) 17% (b) 88%

(c) 7% (d) 41.6%

(e) $99\frac{1}{2}\%$ (f) $33\frac{1}{3}\%$

Solutions:

(a) We move the decimal point two places to the left and remove the % symbol.

$$17\% = 17.$$

$$= 0.17$$

(b) $88\% = 88.$

$= 0.88$

(c) $7\% = 07.$

$= 0.07$

Notice it was necessary to place a leading zero in front of the 7 in order to move the decimal point *two* places.

(continued)

(d) $41.6\% = 4_\wedge 1.6$

$= 0.416$

(e) If the percent includes a fraction, we convert it to a decimal.

$$99\frac{1}{2}\% = 99.5\%$$

$$= _\wedge 99.5$$

$$= 0.995$$

(f) $33\frac{1}{3}\% = 33.\overline{3}\%$

$= _\wedge 33.\overline{3}$

$= 0.33\overline{3}$

□

Percent-Decimal Conversion

Decimal to Percent:

1. Move the decimal point *two* places to the right.
2. Write the % symbol.

Percent to Decimal:

1. Replace any fraction with an equivalent decimal.
2. Move the decimal point *two* places to the left.
3. Remove the % symbol.

To write a fraction as a percent, we write the fraction as a decimal and then follow the procedure given.

EXAMPLE 6. Write each fraction as a percent.

(a) $\frac{1}{4}$ (b) $\frac{1}{8}$ (c) $\frac{1}{3}$

Solutions:

(a) $\frac{1}{4} = 0.25$

$= 25\%$

(b) $\frac{1}{8} = 0.125$

$= 12.5\%$

(c) $\frac{1}{3} = 0.33\overline{3}$

$= 33.\overline{3}\%$

□

Another useful way to write percents is as fractions. Since

$$0.01 = \frac{1}{100}$$

we can replace the % symbol with the fraction $\frac{1}{100}$ and multiply.

EXAMPLE 7. Write the following percents as fractions.

(a) 17% (b) 88%
(c) 7% (d) 41.6%
(e) $99\frac{1}{2}\%$ (f) $33\frac{1}{3}\%$

Solutions:

(a) $17\% = 17\left(\frac{1}{100}\right) = \frac{17}{100}$

(b) $88\% = 88\left(\frac{1}{100}\right) = \frac{88}{100}$

However, the fraction $\frac{88}{100}$ can be reduced to $\frac{22}{25}$.

$$88\% = \frac{22}{25}$$

(c) $7\% = 7\left(\frac{1}{100}\right) = \frac{7}{100}$

(d) $41.6\% = 41.6\left(\frac{1}{100}\right) = \frac{41.6}{100}$

We should not leave a decimal point in a fraction. So, we multiply numerator and denominator by 10.

$$\frac{41.6}{100} = \frac{416}{1000}$$

Since 8 is a common factor, this fraction can be reduced.

$$\frac{416}{1000} = \frac{52}{125}$$

(e) We can work with $99\frac{1}{2}\%$ using fractions or we can work with 99.5% using decimals.

Using fractions,

$$99\frac{1}{2}\% = \left(99\frac{1}{2}\right)\left(\frac{1}{100}\right) = \frac{199}{2} \cdot \frac{1}{100} = \frac{199}{200}$$

Using decimals,

$$99\frac{1}{2}\% = (99.5)\left(\frac{1}{100}\right) = \frac{99.5}{100} = \frac{995}{1000} = \frac{199}{200}$$

(continued)

(f) Again we may use decimals or fractions. Using fractions,

$$33\frac{1}{3}\% = \left(33\frac{1}{3}\right)\left(\frac{1}{100}\right) = \frac{100}{3} \cdot \frac{1}{100} = \frac{1}{3}$$ □

To say a solution is 32% alcohol is to say $\frac{32}{100}$ of the solution is alcohol. The amount of alcohol in 500 liters (L) of a 32% alcohol solution is given by

$$\frac{32}{100} \cdot 500 = 32 \cdot 5 = 160$$

There are 160 L of alcohol in the solution.

To find a specified percent of a number, we change the percent to a fraction or a decimal and then multiply.

EXAMPLE 8. Find each of the following.

(a) 23% of 3300 (b) $16\frac{1}{2}\%$ of 98

(c) $66\frac{2}{3}\%$ of 291 (d) $9\frac{1}{4}\%$ of \$24,000

Solutions:

(a) Since $23\% = \frac{23}{100}$, we have

$$\frac{23}{100} \cdot 3300 = 23 \cdot 33 = 759$$

23% of 3300 is 759.

(b) Because $16\frac{1}{2}\% = 0.165$, we have

$$(0.165)(98) = 16.17$$

$16\frac{1}{2}\%$ of 98 is 16.17.

(c) Since $66\frac{2}{3}\% = \frac{200}{3} \cdot \frac{1}{100} = \frac{2}{3}$, we have

$$\frac{2}{3} \cdot 291 = \frac{582}{3} = 194$$

$66\frac{2}{3}\%$ of 291 is 194.

(d) Since $9\frac{1}{4}\% = 9.25\% = 0.0925$, we have

$$(0.0925)(24{,}000) = 2220$$

$9\frac{1}{4}\%$ of \$24,000 is \$2220. □

PROBLEM SET 0.4

Warm-ups

In Problems 1 through 5, write each fraction as a decimal. See Example 1.

1. $\frac{3}{5}$ **2.** $\frac{3}{8}$ 3. $\frac{7}{6}$ **4.** $3\frac{4}{5}$ 5. $\frac{8}{3}$

In Problems 6 through 9, perform the operations indicated. See Example 2.

6. $12.9 + 23.67$
7. $234.9 - 123.45$
8. $(23.45)(2.03)$
9. $100.4 \div 50.2$

In Problems 10 through 12, see Example 3.

10. Subtract 13.8752 from the sum of 24.6 and 8.2.
11. Subtract the sum of 22.09 and 5.345 from 50.
12. One-half of 48.6 added to one third of 66.9.

In Problems 13 through 21, write the given numbers as percents. See Examples 4 and 6.

13. $\frac{91}{100}$
14. 0.77
15. 0.09
16. 0.8
17. $\frac{7}{50}$
18. 0.08
19. $0.\overline{6}$
20. 0.888
21. $\frac{3}{8}$

In Problems 22 through 30, write the given percents as fractions and *as decimals. See Examples 5 and 7.*

22. 87%
23. 11%
24. 50%
25. 8%
26. 98%
27. $22\frac{1}{2}\%$
28. 0.3%
29. $3\frac{1}{3}\%$
30. 125%

In Problems 31 through 39, find the quantity described. See Example 8.

31. 10% of 400
32. 28% of 10,000
33. 80% of 245
34. 5% of 720
35. 99% of 5280
36. 1% of 550
37. $8\frac{1}{2}\%$ of \$50,000
38. $33\frac{1}{3}\%$ of 963
39. 0.7% of 21,200

Practice Exercises

In Problems 40 through 44, write each fraction as a decimal.

40. $\frac{7}{8}$
41. $\frac{9}{4}$
42. $\frac{1}{7}$
43. $\frac{11}{6}$
44. $5\frac{2}{5}$

In Problems 45 through 56, write the given percents as fractions and *as decimals.*

45. 78%
46. 33%
47. 10%
48. 55%
49. 2%
50. 25%
51. 11%
52. 89%
53. $11\frac{1}{2}\%$
54. 0.9%
55. $6\frac{2}{3}\%$
56. 250%

In Problems 57 through 65, write the given numbers as percents.

57. $\frac{97}{100}$
58. 0.99
59. 0.03

60. $\frac{1}{25}$

61. $\frac{11}{50}$

62. 0.008

63. $0.\overline{2}$

64. 0.555

65. $\frac{5}{8}$

In Problems 66 through 74, find the quantity described.

66. 15% of 400

67. 34% of 10,000

68. 80% of 555

69. 3% of 700

70. 77% of 5280

71. 1% of 470

72. $\frac{1}{2}$% of $3,000

73. $66\frac{2}{3}$% of 963

74. 0.4% of 21,200

Perform the operations indicated.

75. 28.91 − 20.07

76. 34.9 + 123.45

77. (23.15)(2.03)

78. 84.3572 ÷ 68.03

79. (12.5 − 6.2) + 3.8

80. 4.8(2.1) − 3.2

81. 12.4 ÷ 0.062 − 6.3

82. (34.25 − 24.5) − (23.01 − 16.7)

83. 0.8/0.04 + 0.2/0.4

84. 6.3 − (5.7 − 1.2)

85. 2.1(2.5 + 6.7)

86. (8.8 − 2.5) ÷ 0.7

87. 12 − (2 − 1.5) − 8.5

88. 4(1.2) ÷ 1.6 − 2.1

89. $(2.1)^2$

90. $(0.2)^3$

91. $(0.1)^2$

92. $4(1.2)^2 - (0.1)^3$

93. $(2.01)(10.2)^2$

94. $(1.01)^2(2.01)^2$

95. Subtract 3.2345 from the sum of 4.6 and 8.022.

96. Subtract the sum of 12.09 and 5.009 from 25.

97. Find one-half of 48.06 added to one third of 66.009.

Answer each question.

98. Greg has 50 nickels and six $5 bills. How much money does he have?

99. Jamie has 8 dimes, 3 quarters, and five $10 bills. How much money does he have?

100. Jennifer has 24 pennies, 33 nickels, 15 dimes, and 7 quarters. How much money does she have?

101. Kristen wanted to buy $5.35 worth of chicken, $.89 worth of onions, $1.89 worth of potatoes, and $.99 worth of broccoli at the grocery store. She had $7.50 in her purse. Did she have enough money?

102. Sondra paid phone bills of $34.24, $45.89, $29.00, $57.90. Her budget allows for $175. Did she stay within her budget?

103. Suzanne bought a diamond ring for $5000. She discovered that its value has increased by 18%. What is the ring worth now?

104. A variety store advertised a 10% discount on drills. If the original price of a drill was $59.95, what is the sale price?

105. Joseph receives an allowance of $50 per week. If his father increases his allowance by 5%, what will his new allowance be?

106. Ray has 22 nickels and twelve $5 bills. How much money does he have?

107. Frank has 2 dimes, 4 quarters, and seven $20 dollar bills. How much money does he have?

108. Joan has 4 pennies, 3 nickels, 12 dimes and 5 quarters. How much money does she have?

109. Deli ham sells for $2.29 per $\frac{1}{4}$ lb. What is the price of deli ham per pound?

110. Jo Ann bought $4.19 worth of chicken, $.79 worth of onions, $1.49 worth of potatoes, and $1.19 worth of broccoli at the grocery store. She had $7.50 in her purse. Did she have enough money?

111. Carolyn is buying hamburger every week. For four weeks she has bought 2.12 lb, 4.65 lb, 3.15 lb, and 5.05 lb. Has she bought more than 15 lb?

112. Sara bought 2.01 lb of grapes, 6.9 lb of apples, 3 lb of pears, 4.33 lb of bananas, and 3.05 lb of nectarines. She cannot carry more than 20 lb because of a bad back. Will she be able to carry all her purchases?

113. A dress was marked down 20%. If the original price was $59, what was the sale price?

114. Jack decided to sell his house for 7% more than he paid for it. If he paid $62,000 for it, what will his selling price be?

115. John studied his math for 2 hours (hr) every day. He thinks he can make an A if he increases his daily study time by 25%. How much longer each day will he study?

IN YOUR OWN WORDS...

116. Describe how to convert 24% into a fraction.

117. How do we add decimals?

118. How do we multiply decimals?

119. How do we divide decimals?

0.5 SQUARE ROOTS

If we square 3, we get 9. That is,

$$3^2 = 9$$

The reverse of squaring is taking a square root. We write $\sqrt{9}$ to indicate the square root of 9. The square root of 9 is 3 because $3^2 = 9$.

EXAMPLE 1. Find the square roots indicated.

(a) $\sqrt{4}$ (b) $\sqrt{16}$

(c) $\sqrt{25}$ (d) $\sqrt{18^2}$

Solutions:

(a) $\sqrt{4} = 2$ because $2^2 = 4$

(b) $\sqrt{16} = 4$ because $4^2 = 16$

(c) $\sqrt{25} = 5$ because $5^2 = 25$

(d) $\sqrt{18^2} = 18$ because $18^2 = 18^2$ □

A number that is the square of a natural number is called a **perfect square.** For example, 9 and 16 are perfect squares because $3^2 = 9$ and $4^2 = 16$.

In Example 1, we found square roots of perfect squares. If we want to find square roots of natural numbers that are not perfect squares, we use a calculator or a square root table to *approximate* them. For example, $\sqrt{2}$ is approximately 1.414. We write $\sqrt{2} \approx 1.414$ to indicate that 1.414 is an approximation.

The $\sqrt{}$ symbol is another grouping symbol. That is, we perform operations *inside* the $\sqrt{}$ *before* we evaluate it. The next example illustrates this.

EXAMPLE 2. Evaluate each expression.

(a) $\sqrt{144 + 25}$ (b) $\sqrt{144} + \sqrt{25}$

Solutions:

(a) Since $\sqrt{}$ is a grouping symbol, we *must* perform the operations inside first.

$$\sqrt{144 + 25} = \sqrt{169}$$

$$= 13 \qquad \text{Because } 13^2 = 169$$

(b) $\sqrt{144} + \sqrt{25} = 12 + 5$ $\quad 12^2 = 144$ and $5^2 = 25$

$= 17$ □

CALCULATOR BOX

Square Roots with a Calculator

To find $\sqrt{5}$ using a calculator, press the following keys.

[5] [√]

Read **2.236067977** on the display.

$\sqrt{5} \approx 2.236$, rounding to the nearest thousandth. It is important to estimate the answer so that we can determine that our answer is reasonable. $\sqrt{5}$ should be a little larger than 2, since $\sqrt{4} = 2$. So, our answer is reasonable.

Calculator Exercises

Approximate the square roots indicated to three decimal places. Be sure to use the ≈ symbol to indicate an approximation. Estimate the answer first.

1. $\sqrt{3}$ **2.** $\sqrt{10}$ **3.** $\sqrt{47}$ **4.** $\sqrt{82}$

Answers:

1. Estimate is between 1 and 2; $\sqrt{3} \approx 1.732$

2. Estimate is about 3; $\sqrt{10} \approx 3.162$

3. Estimate is a little less than 7; $\sqrt{47} \approx 6.856$

4. Estimate is a little more than 9; $\sqrt{82} \approx 9.055$

PROBLEM SET 0.5

Warm-ups

In Problems 1 through 8, find the square roots indicated. See Example 1.

1. $\sqrt{25}$ **2.** $\sqrt{36}$ **3.** $\sqrt{100}$ **4.** $\sqrt{400}$

5. $\sqrt{49}$ 6. $\sqrt{121}$ **7.** $\sqrt{13^2}$ 8. $\sqrt{15^2}$

Evaluate each expression. See Example 2.

9. $\sqrt{9 + 16}$ **10.** $\sqrt{9} + \sqrt{16}$ **11.** $\sqrt{8^2} + \sqrt{6^2}$

12. $\sqrt{8^2 + 6^2}$ **13.** $\sqrt{289 - 64}$ **14.** $\sqrt{289} - \sqrt{64}$

15. $\sqrt{26^2} - \sqrt{10^2}$ **16.** $\sqrt{26^2 - 10^2}$

Practice Exercises

Evaluate each of the following expressions. If necessary, approximate the square roots indicated to three decimal places. Be sure to use the ≈ symbol to indicate an approximation.

17. $\sqrt{64}$ **18.** $\sqrt{81}$ 19. $\sqrt{900}$

20. $\sqrt{144}$

21. $\sqrt{10,000}$

22. $\sqrt{99}$

23. $\sqrt{3600}$

24. $\sqrt{1600}$

25. $\sqrt{17}$

26. $\sqrt{50^2}$

27. $\sqrt{48}$

28. $\sqrt{16 - 9}$

29. $\sqrt{16} - \sqrt{9}$

30. $\sqrt{13 + 3}$

Answer each question.

31. The hypotenuse of a right triangle is $\sqrt{5^2 + 12^2}$ units long. What is the length of the hypotenuse of this triangle?

32. One leg of a right triangle is $\sqrt{11^2 - 7^2}$ units long. What is its length to the nearest tenth?

33. What is the sum of the square root of 64 and the square of 14?

34. Approximate to the nearest hundredth the square root of 50 subtracted from the square root of 60.

Challenge Problems

35. Find $\sqrt{200}$ to two decimal places. Compare it to the value of $\sqrt{2}$.

36. Find $\sqrt{2000}$ to two decimal places. Compare it to the value of $\sqrt{20}$.

IN YOUR OWN WORDS...

37. What do we mean by $\sqrt{21}$?

38. What do we mean by 21^2?

0.6 ANGLES, LINES, PLANE FIGURES (OPTIONAL)

This section is a review of some basic ideas and terminology from geometry. Such fundamental ideas as point, line, angle, and length are left undefined. No attempt is made to use precise notation. Angles with the same measure are said to be equal. The symbol "∠" indicates the measure of an angle. Lowercase letters are used to designate the length of line segments.

Angles

An *acute angle* is an angle of less than 90°.

A *right angle* is an angle of exactly 90°. The symbol "⌉" is used to indicate a right angle.

An *obtuse angle* is an angle of greater than 90°, but less than 180°.

A *straight angle* is an angle of exactly 180°. Its sides form a line.

Complementary angles are two angles whose sum is 90°. Each is called the *complement* of the other.

Supplementary angles are two angles whose sum is 180°. Each is called the *supplement* of the other.

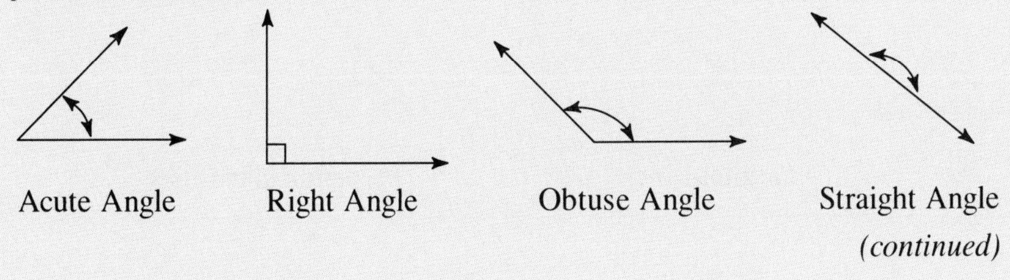

(continued)

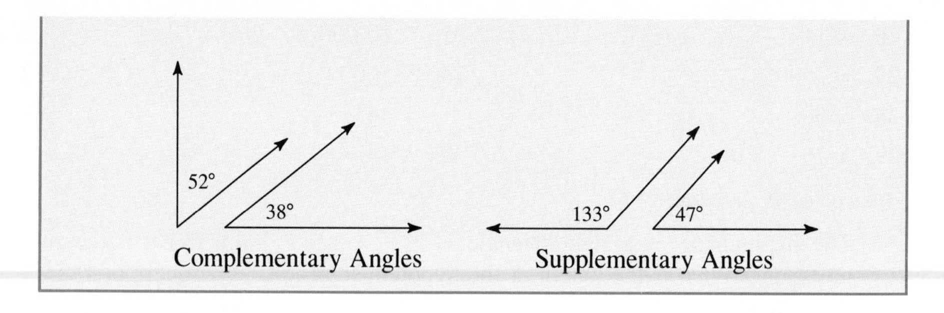

Complementary Angles Supplementary Angles

Two intersecting lines form the angles numbered 1, 2, 3, and 4.

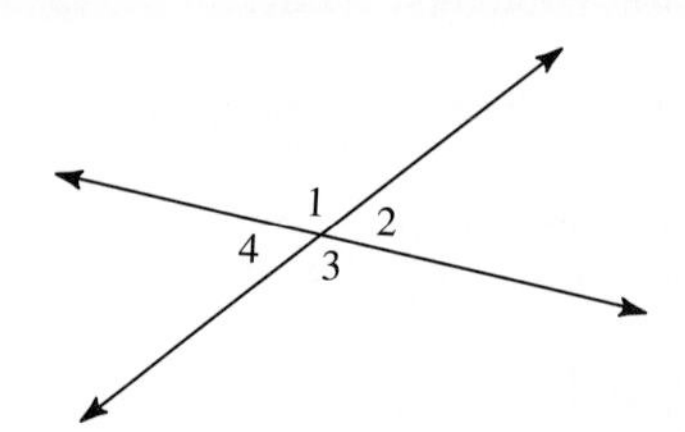

Angles 1 and 3 are *vertical* angles as are angles 2 and 4. Vertical angles are equal. Angles which have a common vertex and a common angle are called *adjacent* angles. Angles 1 and 2 are adjacent angles as are angles 2 and 3, and angles 3 and 4, and angles 1 and 4.

EXAMPLE 1. Are the angles below complementary or supplementary?

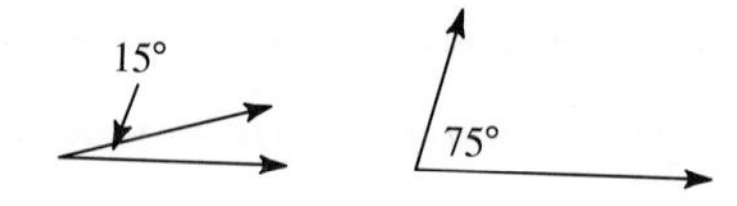

Solution:

The sum of the measures of the two angles is 90°. The angles are complementary. □

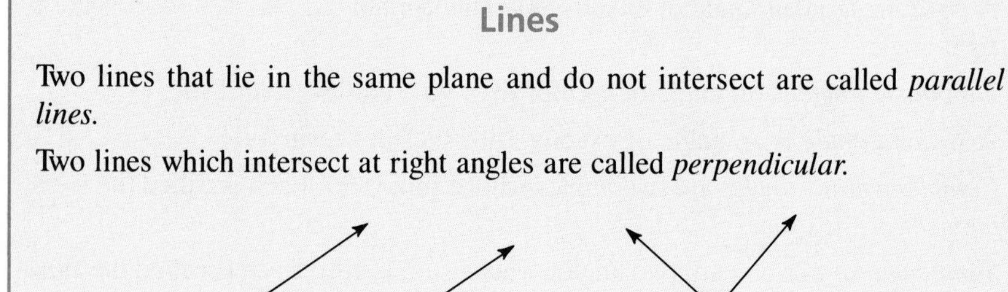

Lines

Two lines that lie in the same plane and do not intersect are called *parallel lines.*

Two lines which intersect at right angles are called *perpendicular.*

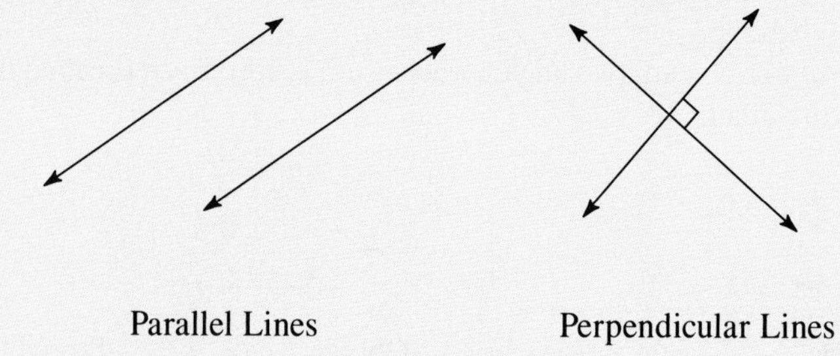

Parallel Lines Perpendicular Lines

A *transversal* is a line that intersects two or more lines at different points. A transversal of two lines forms some angles that are given special names.

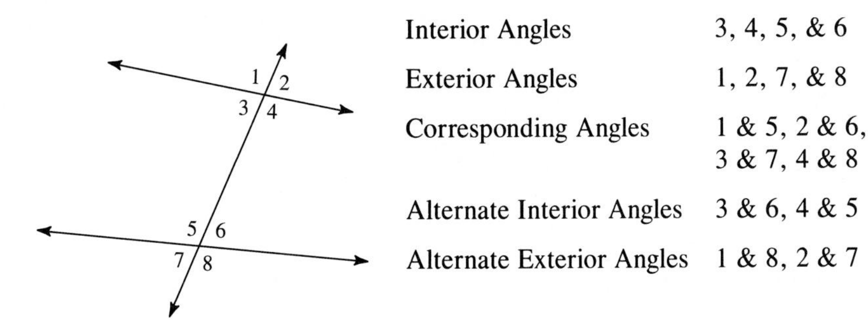

Interior Angles	3, 4, 5, & 6
Exterior Angles	1, 2, 7, & 8
Corresponding Angles	1 & 5, 2 & 6, 3 & 7, 4 & 8
Alternate Interior Angles	3 & 6, 4 & 5
Alternate Exterior Angles	1 & 8, 2 & 7

Parallel Lines Cut by a Transversal

If two lines are cut by a transversal, the lines are parallel if any of the following statements are true.

1. Corresponding angles are equal.
2. Alternate interior angles are equal.
3. Alternate exterior angles are equal.

If two lines are parallel, then all of the statements are true.

EXAMPLE 2. Determine the measure of each angle in the figure. Lines l_1 and l_2 are parallel. $\angle 1 = 120°$.

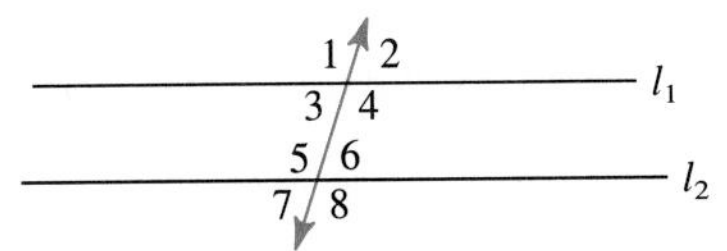

Solution:

Angles 1 and 2 are supplementary angles. So, $\angle 2$ is 60°.
Angles 1, 4, 5, and 8 are equal. So, each is 120°.
Angles 2, 3, 6, and 7 are equal. Thus, each is 60°. □

A *polygon* is a closed plane figure bounded by line segments. Triangles, quadrilaterals, pentagons, hexagons, and octagons are polygons with 3, 4, 5, 6, and 8 sides respectively. If all the sides are of equal length, a polygon is called *equilateral*. If all the angles are equal, it is said to be *equiangular*. A *regular* polygon is one that is both equilateral and equiangular. Several polygons are shown below. The dotted lines are called *diagonals*.

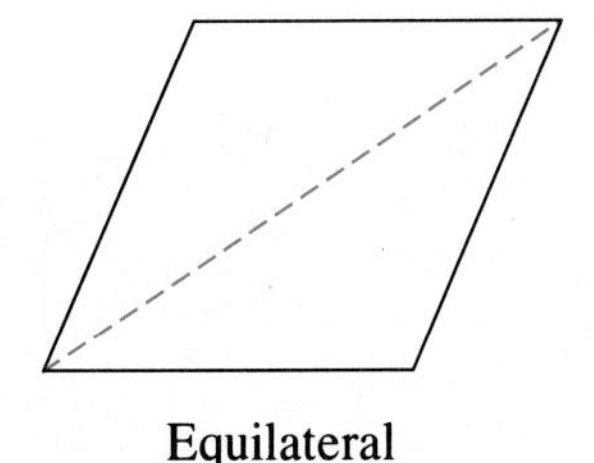
Equilateral

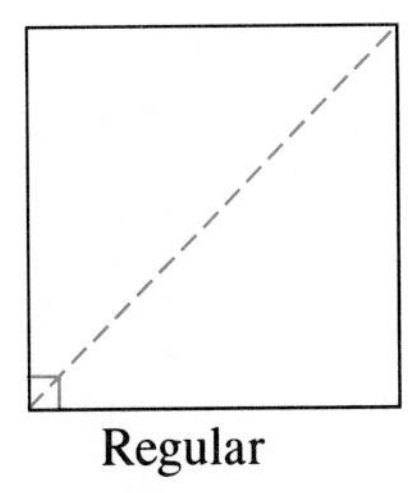
Equiangular

Regular

A *triangle* is a polygon with three sides.

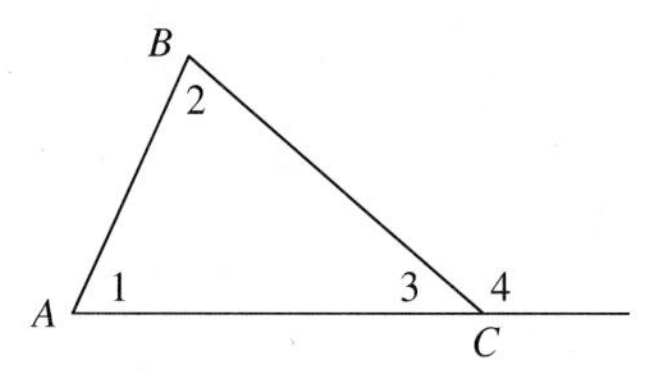

The three points, A, B, and C are the *vertices* (singular, *vertex*) of the triangle. The three angles, 1, 2, and 3 are the *interior angles* of the triangle, or simply the angles of the triangle. The three sides of a triangle and the three angles are sometimes called the *parts* of a triangle. Angle 4 is called an *exterior angle* of the triangle.

Two important relationships exist for these angles. The sum of the measures of the interior angles is 180°. The measure of an exterior angle is equal to the sum of the measures of the opposite interior angles.

$$\angle 1 + \angle 2 + \angle 3 = 180°$$

$$\angle 4 = \angle 1 + \angle 2$$

EXAMPLE 3. Determine $\angle A$ in each figure.

(a) (b)

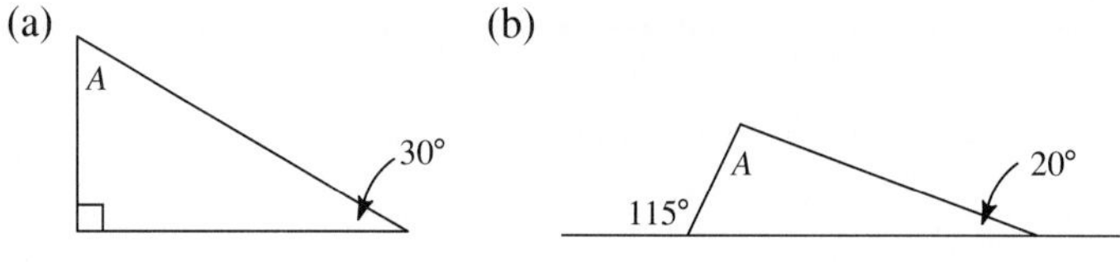

Solutions:

(a) Since this is a right triangle, $\angle A$ must be 60°.

(b) $\angle A + 20° = 115°$

$\angle A = 95°$

□

Triangles

An *isosceles triangle* has two sides of equal length. The angles opposite the equal sides are equal.

An *equilateral triangle* has all three sides equal. It is also equiangular.

An *acute triangle* has three acute angles.

An *obtuse triangle* has an obtuse angle.

A *right triangle* contains a right angle. The longest side of a right triangle (the side opposite the right angle) is called the *hypotenuse*. The other two sides are called the *legs* of a right triangle.

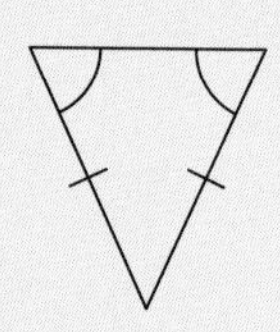

Isosceles Triangle

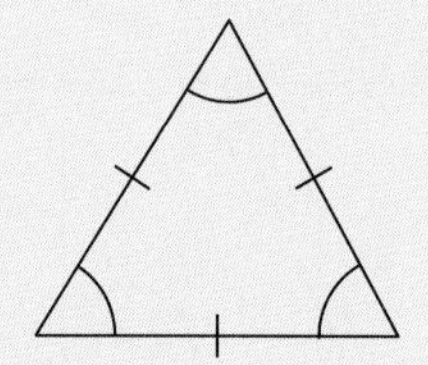

Equilateral Triangle

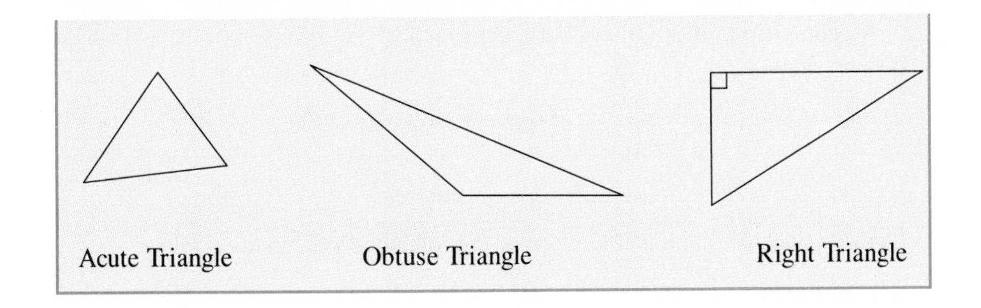

The following result is one of the most useful relationships in all of mathematics.

Pythagorean Theorem

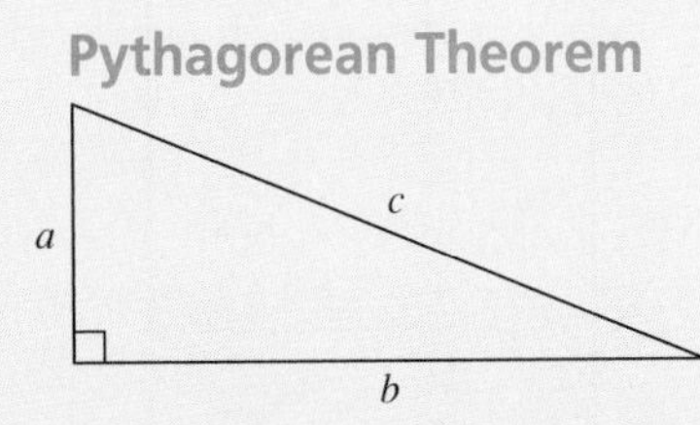

In the right triangle above, the side of length c units is the hypotenuse and the sides of lengths a and b units are the legs. The square of the length of the hypotenuse equals the sum of the squares of the lengths of the legs. That is,

$$c^2 = a^2 + b^2$$

EXAMPLE 4. Verify the Pythagorean Theorem for a right triangle with sides of 3, 4, and 5 inches.

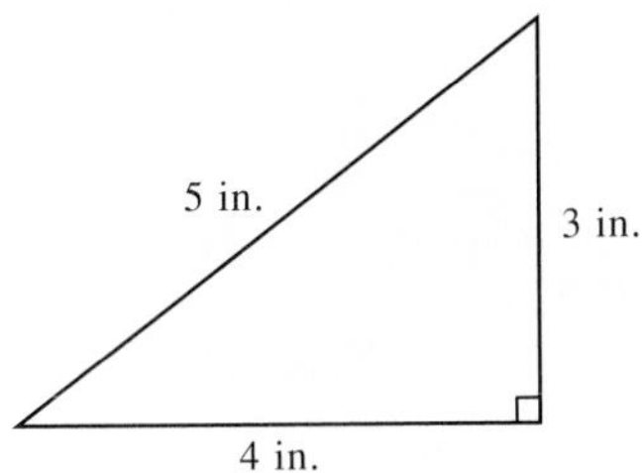

Solution:

The hypotenuse is the longest side. So, it must be 5.

$$5^2 = 25$$

The legs are 3 and 4.

$$\begin{aligned} 3^2 + 4^2 &= 9 + 16 \\ &= 25 \end{aligned}$$

Thus, the square of the length of the hypotenuse equals the sum of the squares of the lengths of the two legs. □

A geometric property that is of great practical use involves triangles that are the same shape.

Similar Triangles

Two triangles are similar if any one of the following statements is true.

1. Two angles of one triangle equal two angles of the other.

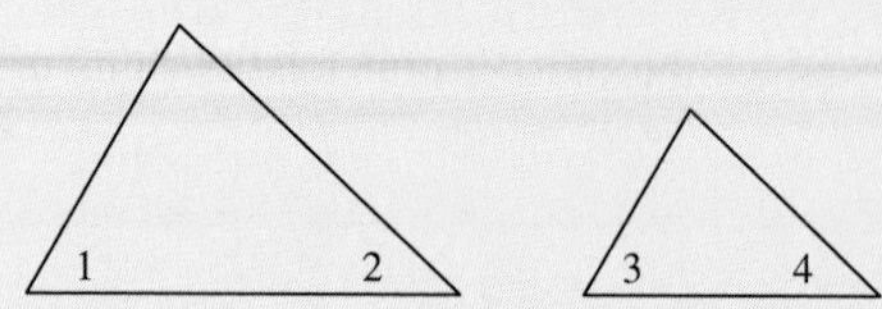

$\angle 1 = \angle 3$
$\angle 2 = \angle 4$

2. Corresponding sides are proportional.

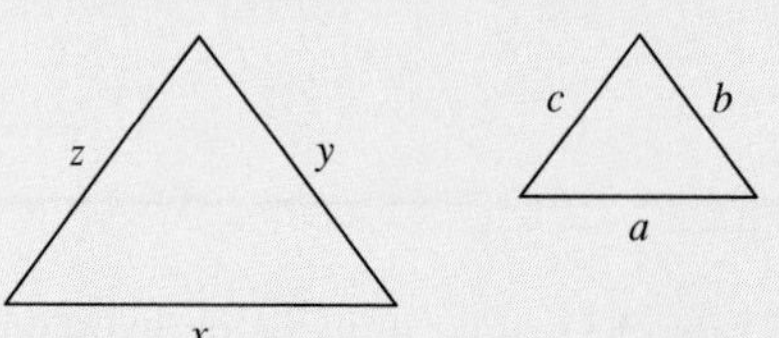

$$\frac{x}{a} = \frac{y}{b} = \frac{z}{c}$$

3. Two corresponding pairs of sides are proportional and the angle between them is equal.

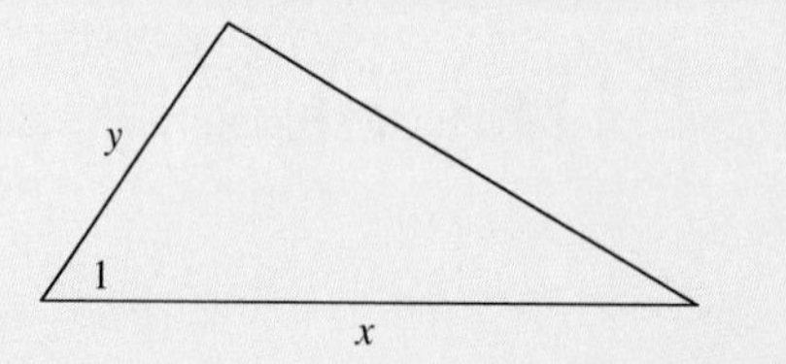

$\angle 1 = \angle 2$ and
$$\frac{x}{a} = \frac{y}{b}$$

If two triangles are similar, then all of these statements are true.

EXAMPLE 5. Are the triangles in the figure similar?

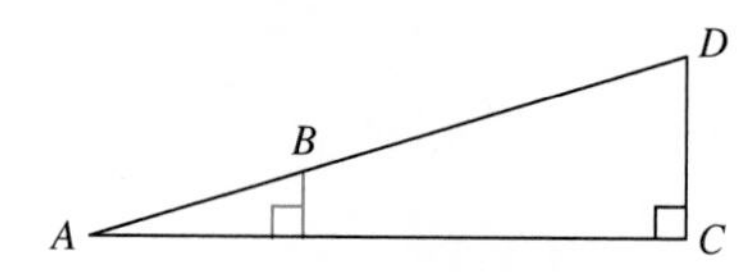

Solution:

As we examine the figure, we notice a large triangle and a small triangle. Both are right triangles and they share the angle at the vertex A. This makes the angles at B and D equal. We see that they have two equal angles. Thus, by 1 above, they are similar triangles. □

EXAMPLE 6. The triangles are similar. Set up proportions for corresponding sides.

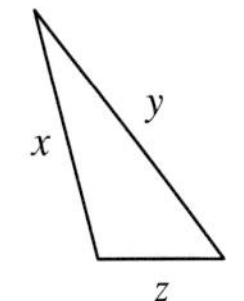

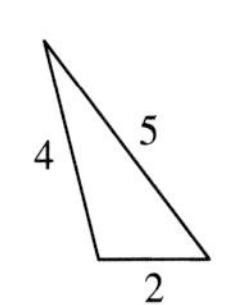

Solution:

$$\frac{x}{4} = \frac{y}{5} = \frac{z}{2}$$ □

Another important situation arises when two triangles have the same shape and the same size.

Congruent Triangles

Two triangles are congruent if any one of the following statements is true.

1. Two angles of one triangle equal two angles of the other triangle and the lengths of the sides between each pair of angles are equal. (Angle-Side-Angle)

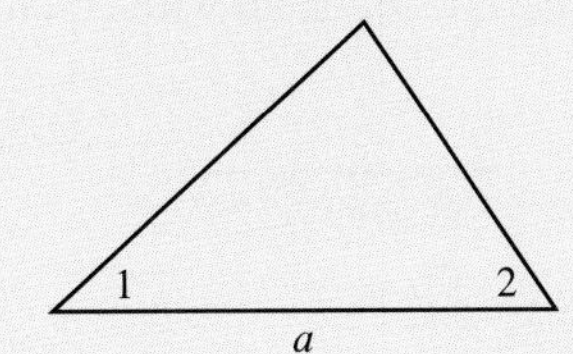

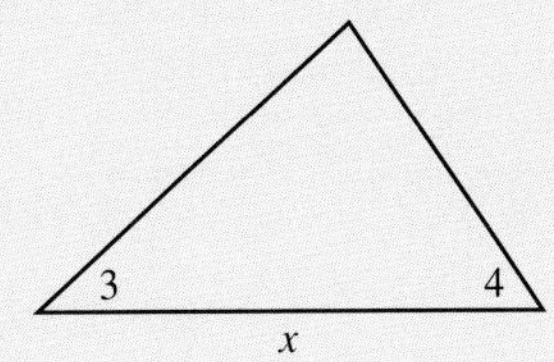

$\angle 1 = \angle 3$
$\angle 2 = \angle 4$
$a = x$

2. Corresponding sides are equal. (Side-Side-Side)

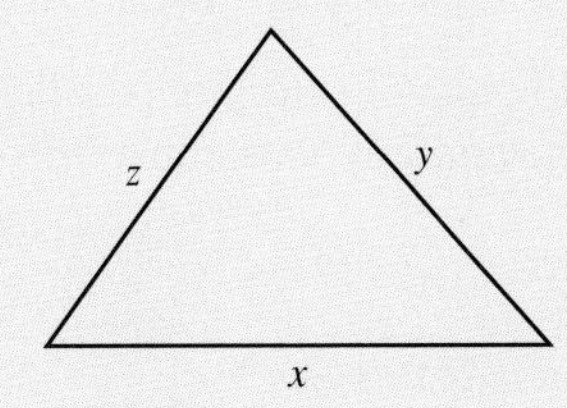

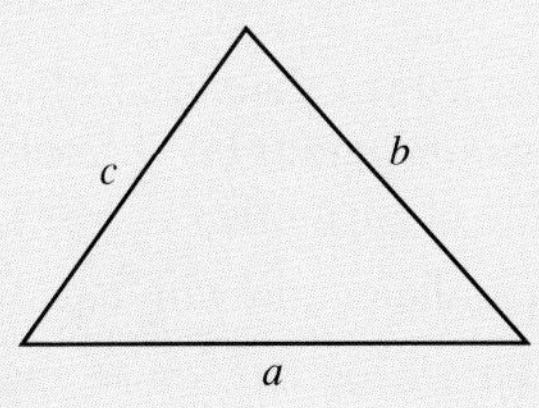

$a = x$
$b = y$
$c = z$

3. Two corresponding pairs of sides are equal and the angle between them is equal. (Side-Angle-Side)

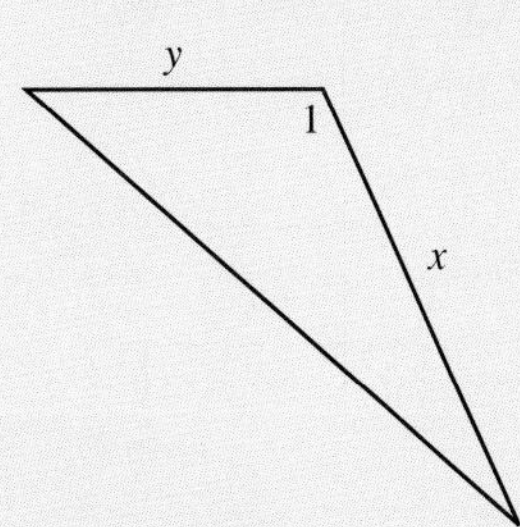

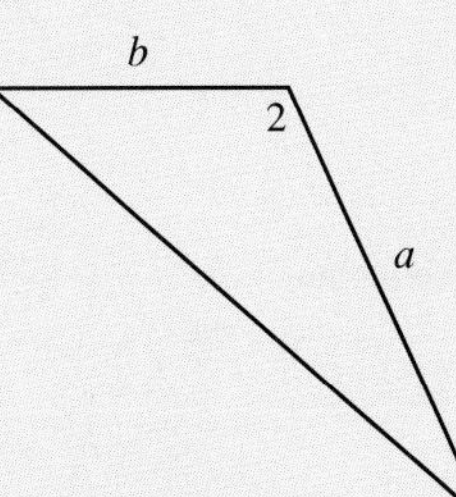

$a = x$
$b = y$
$\angle 1 = \angle 2$

If two triangles are congruent then all of these statements are true.

EXAMPLE 7. Are the triangles in the figure congruent?

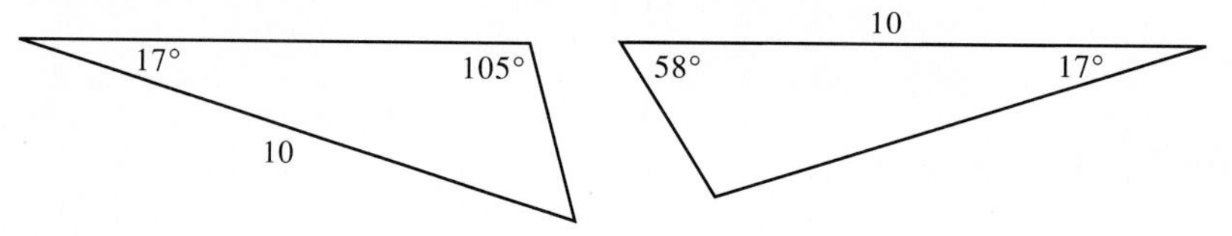

Solution:

We notice that the third angle in the left triangle can be calculated.

$$180° - 105° - 17° = 58°$$

So, the left triangle has a side of length 10 between angles of 17° and 58°. As the triangle on the right also has a side of length 10 between angles of measure 17° and 58°, we conclude that the triangles are congruent. (Angle-Side-Angle) □

A *quadrilateral* is a four-sided plane figure.

Quadrilaterals

A *trapezoid* is a quadrilateral with exactly two sides parallel. The parallel sides are called *bases* and the other sides are called *legs*. A trapezoid with equal legs is called an *isosceles trapezoid*.

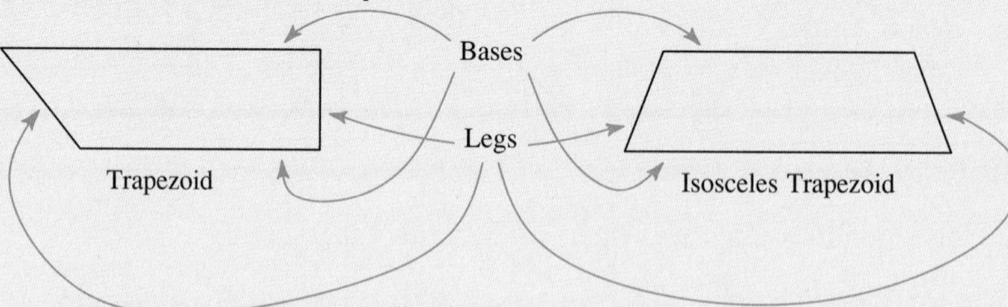

A *parallelogram* is a quadrilateral in which both pairs of sides are parallel. The diagonals of a parallelogram bisect each other, and any two consecutive angles are supplements of each other.

A *rhombus* is a parallelogram with equal sides.

A *rectangle* is a parallelogram whose angles are right angles.

A *square* is a rectangle with equal sides.

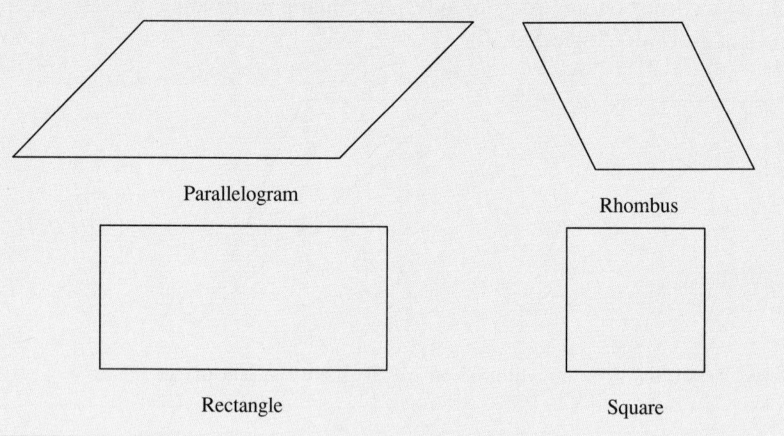

Circles

A *circle* is the set of all points in a plane the same distance from a given point called the *center.* A *radius* of a circle is a line segment connecting the center with a point on the circle. A *chord* is a line segment connecting two points on the circle. A *diameter* is a chord that contains the center. A *tangent* to a circle is a line in the plane of the circle that intersects the circle in exactly one point.

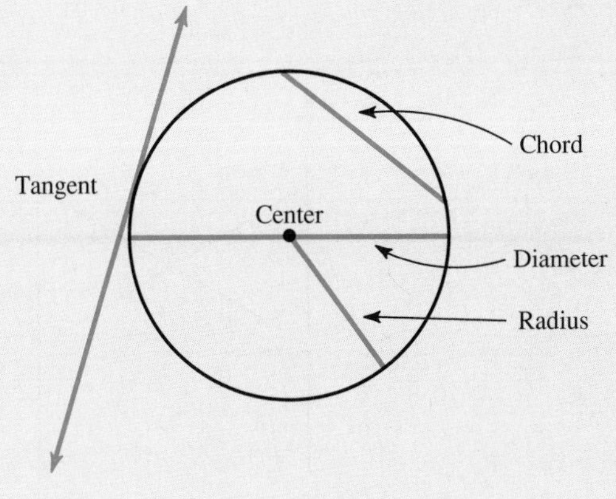

PROBLEM SET 0.6

Warm-ups

1. Find the measure of each angle in the figure. Lines l_1 and l_2 are parallel. Lines l_3 and l_4 are perpendicular. $\angle b = 135°$. See Examples 1 and 2.

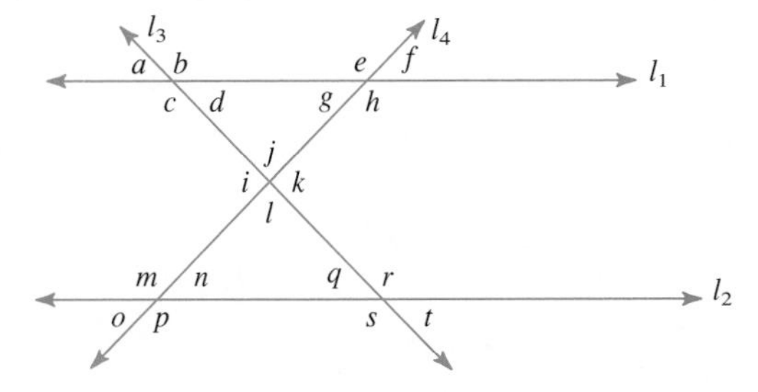

In Problems 2 through 5, find $\angle A$. *See Example 3.*

2.

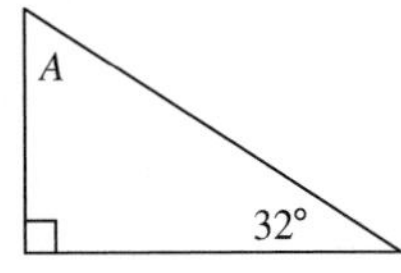

3.

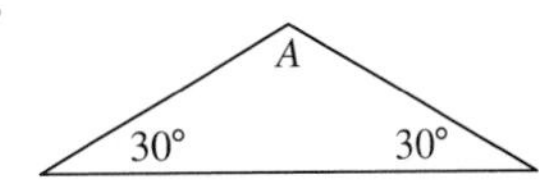

4.

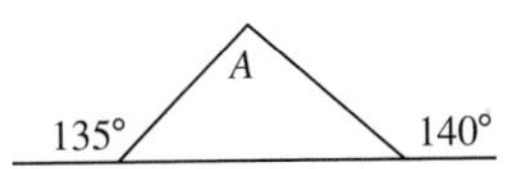

5.

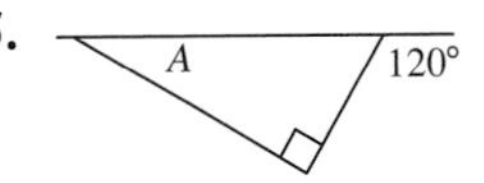

6. Verify the Pythagorean Theorem for a right triangle with sides of lengths 6, 8, and 10 units. See Example 4.

7. Why is each pair of triangles similar? See Example 5.

(a)

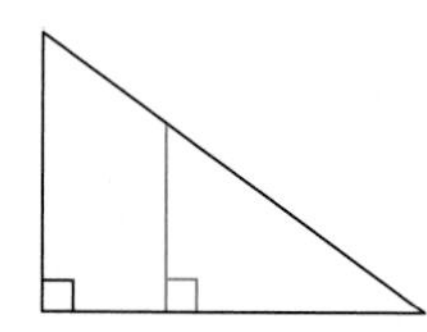

(b) $l_2 \parallel l_2$

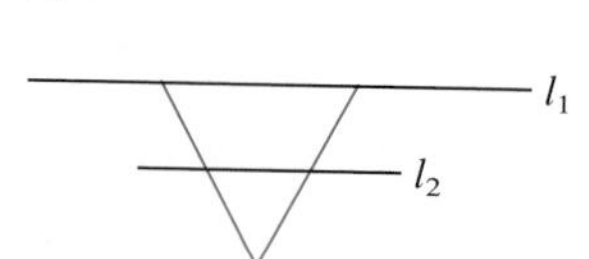

8. Why is each pair of triangles congruent? See Example 7.

(a)

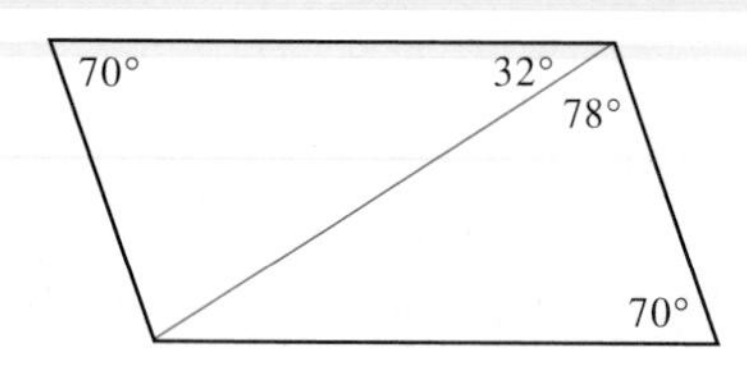

(b) $l_1 \parallel l_2$

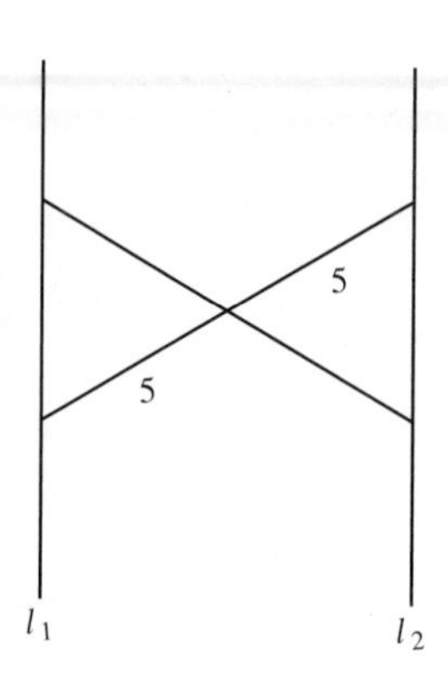

Practice Exercises

In Problems 9 through 20, determine if each statement is true or false. Use the figure below. Lines l_1 and l_2 are parallel. $\angle d = 60°$ and $\angle g = 75°$

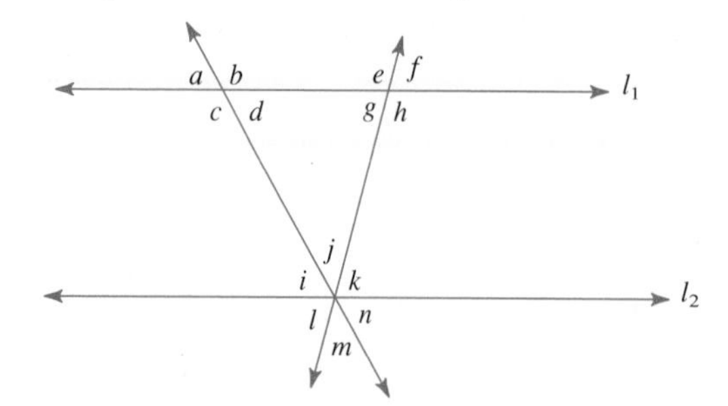

9. $\angle i = 75°$
10. $\angle j + \angle n = 180°$
11. $\angle g + \angle j = 105°$
12. $\angle g = 45°$
13. $\angle m + \angle n = 135°$
14. Angles h and k are supplementary.
15. $\angle l = 45°$
16. Angle b is an acute angle.
17. Angle f is obtuse.
18. Angles k and n are complementary.
19. $\angle b = 105°$
20. $\angle m + \angle n = 105°$

In Problems 21 through 26, find $\angle A$.

21.

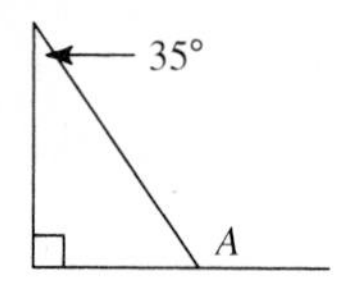

22.

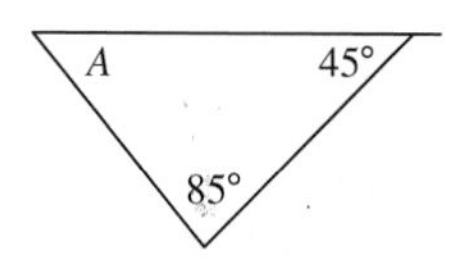

23.

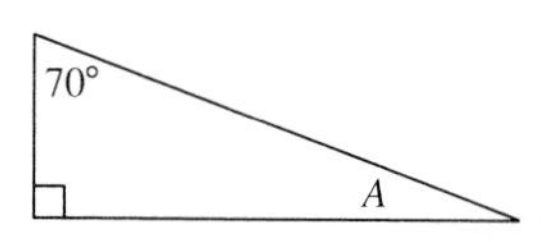

24.

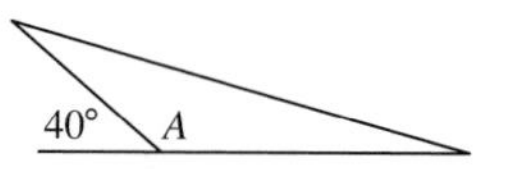

25.

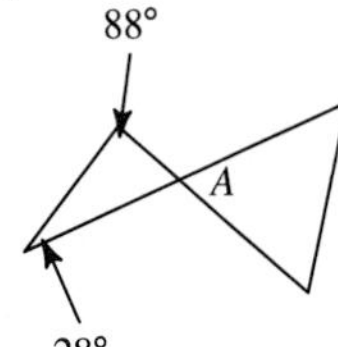

26.

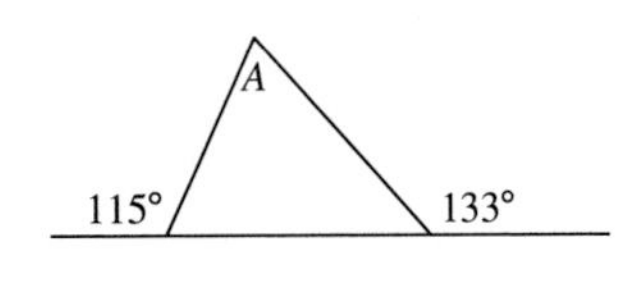

27. Verify the Pythagorean Theorem for a right triangle with sides of lengths 5, 12, and 13 units.

In Problems 28 through 31, determine if each pair of triangles is congruent or similar but not congruent. Why?

28.

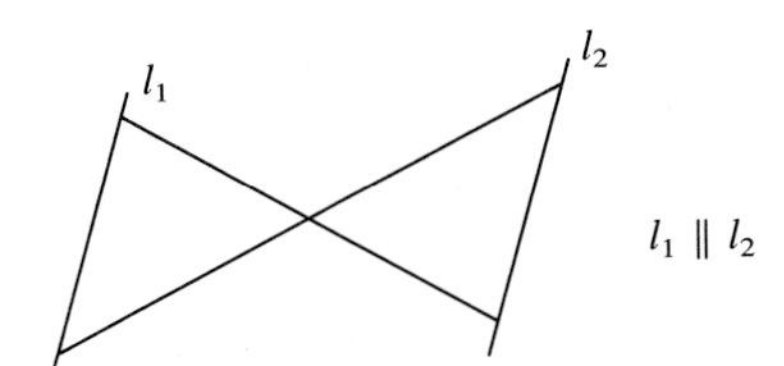

29.

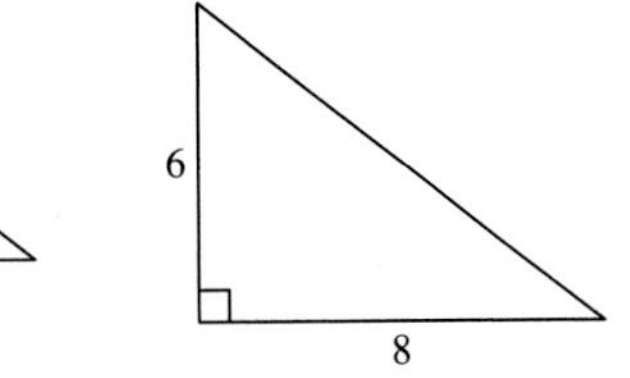

30.

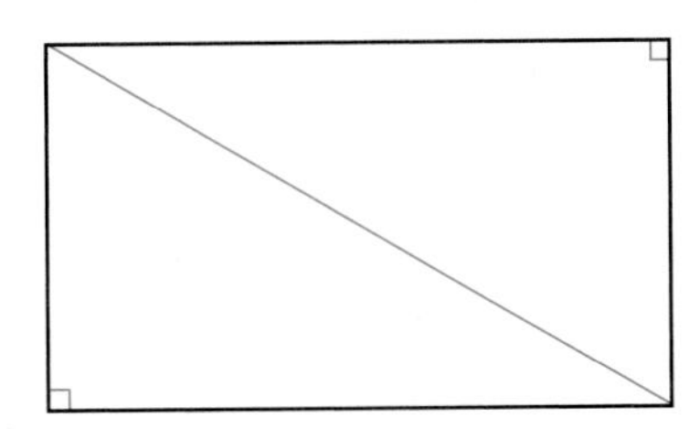

31.

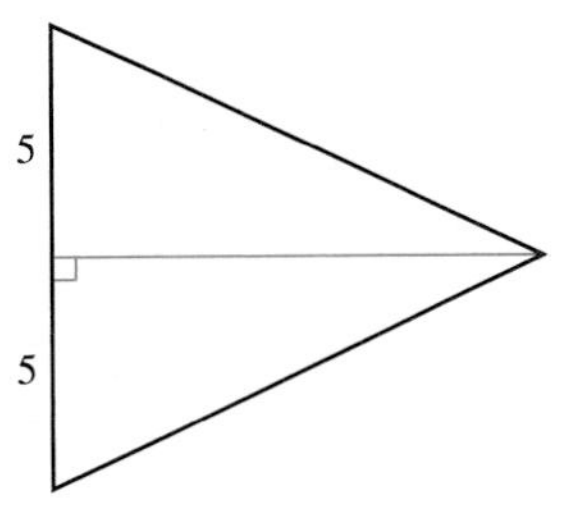

In Problems 32 throuth 53, determine whether each statement is true or false.

32. A rectangle is equiangular.

33. An isosceles triangle is an acute triangle.

34. A square is a rhombus.

35. A square is a regular polygon.

36. A triangle is a polygon.

37. The acute angles in a right triangle are complementary.

38. A rhombus is an equilateral polygon.

Challenge Problems

39. Figure out a formula for the number of diagonals that can be drawn from one vertex of a regular polygon.

IN YOUR OWN WORDS . . .

40. What are congruent triangles?

41. What does the Pythagorean Theorem say?

0.7 APPLICATIONS IN GEOMETRY

This section contains formulas for perimeter, area, and volume of geometric figures. The **perimeter** of a figure is the distance around the figure. The perimeter of a circle is called **circumference.**

Perimeter and Area Formulas

Figure	*Sketch*	*Perimeter*	*Area*
Triangle	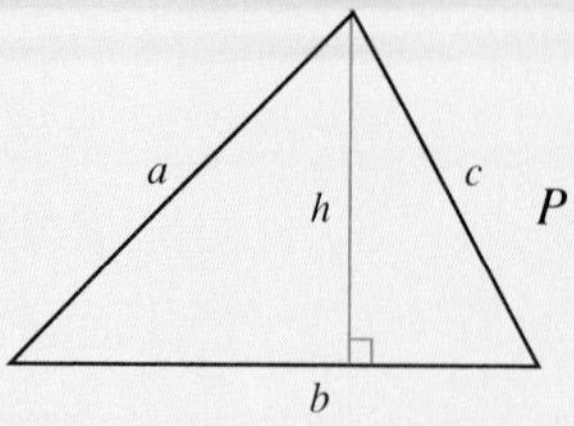 sides a, c, base b, height h	$P = a + b + c$	$A = \frac{1}{2}bh$
Rectangle	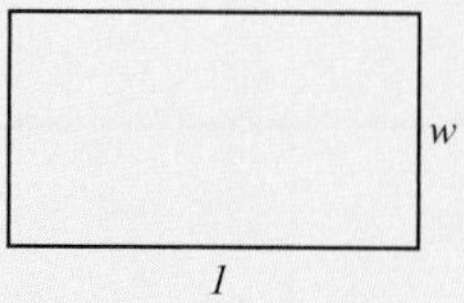 length l, width w	$P = 2l + 2w$	$A = lw$
Parallelogram	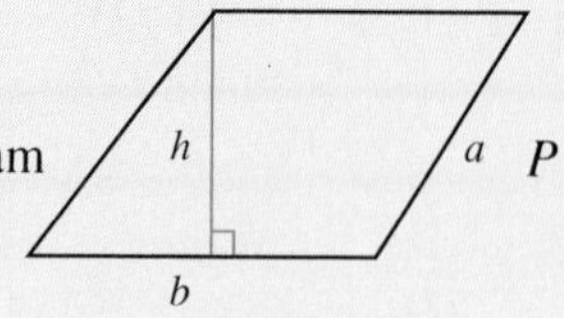 side a, base b, height h	$P = 2a + 2b$	$A = bh$
Square	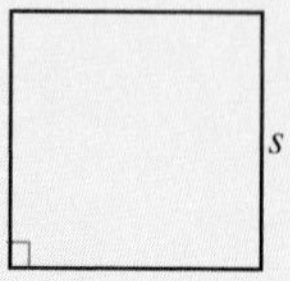 side s	$P = 4s$	$A = s^2$
Trapezoid	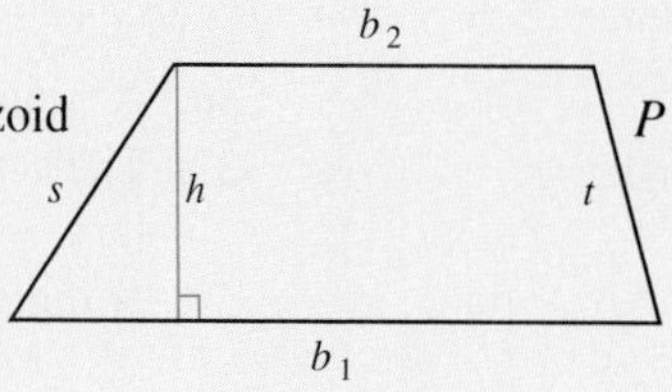 bases b_1, b_2, sides s, t, height h	$P = b_1 + s + t + b_2$	$A = \frac{1}{2}h(b_1 + b_2)$
Circle	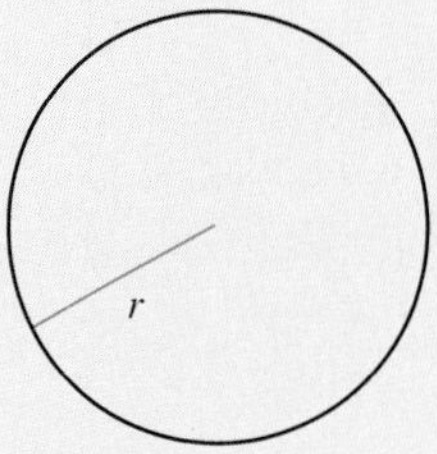 radius r	$C = 2\pi r$	$A = \pi r^2$

EXAMPLE 1. Find the area and perimeter of a rectangle with dimensions $6\frac{1}{4}$ feet (ft) by $2\frac{1}{3}$ ft.

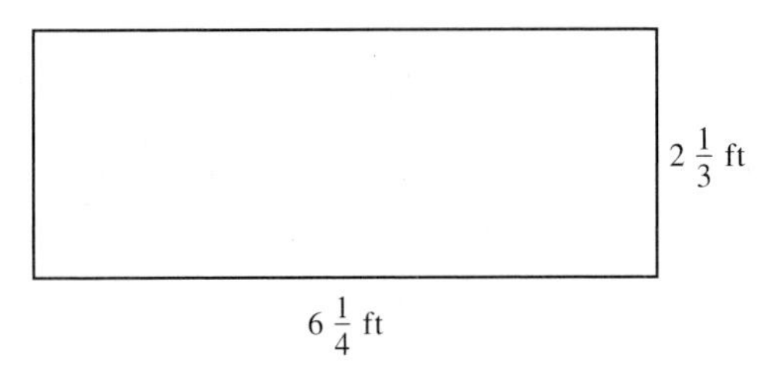

Solution:

The area of a rectangle is the product of its length and width:

$$A = 6\frac{1}{4} \cdot 2\frac{1}{3} \qquad \text{Express mixed numbers as fractions.}$$

$$= \frac{25}{4} \cdot \frac{7}{3} \qquad \text{Multiply fractions.}$$

$$= \frac{25 \cdot 7}{4 \cdot 3}$$

$$= \frac{175}{12}$$

Since the problem is stated using mixed numbers, we change this fraction to a mixed number to answer the first part of the question.

The area is $14\frac{7}{12}$ square feet (ft^2).

The perimeter is the distance around the rectangle. So,

$$P = 6\frac{1}{4} + 6\frac{1}{4} + 2\frac{1}{3} + 2\frac{1}{3}$$

$$= 12\frac{1}{2} + 4\frac{2}{3}$$

$$= \frac{25}{2} + \frac{14}{3} \qquad \text{Express mixed numbers as fractions.}$$

$$= \frac{75}{6} + \frac{28}{6} \qquad \text{Find the LCD.}$$

$$= \frac{103}{6} \qquad \text{Add the fractions.}$$

$$= 17\frac{1}{6} \qquad \text{Change to a mixed number.}$$

The perimeter is $17\frac{1}{6}$ ft. □

EXAMPLE 2. Find the perimeter of each figure.

(a)

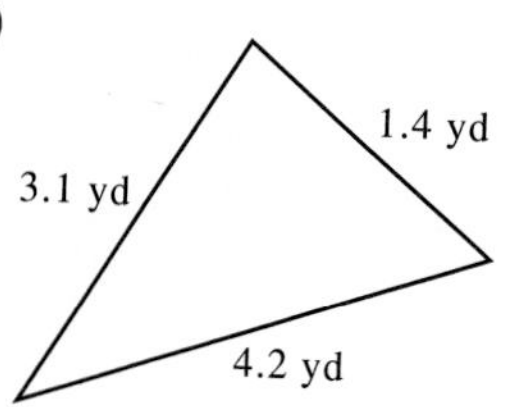

(b)

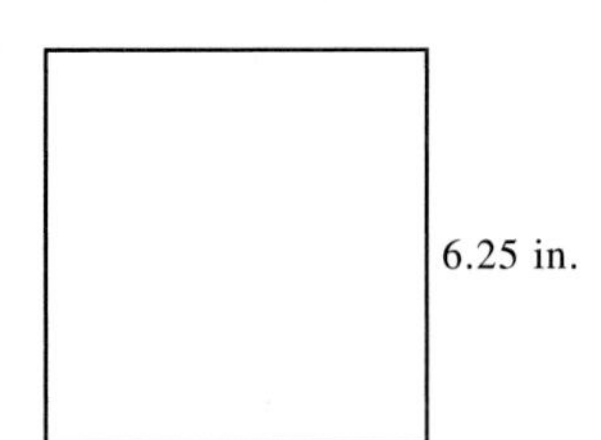

(c)

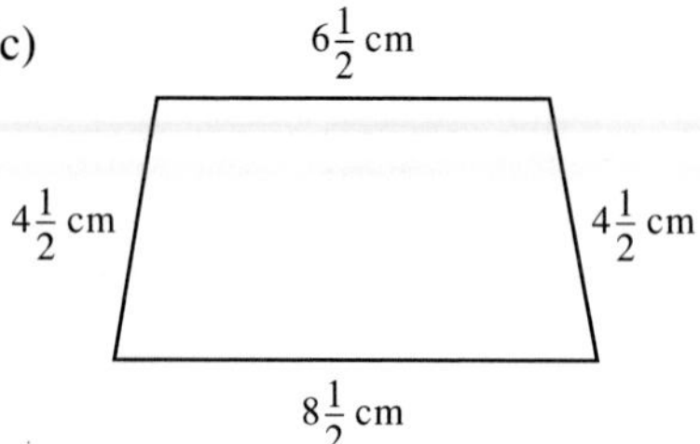

(d)

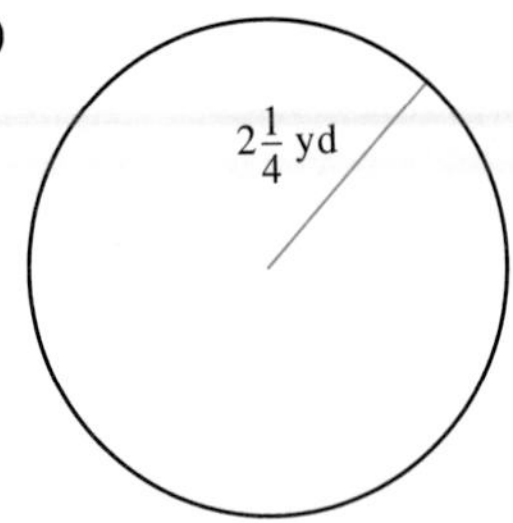

Solutions:

(a) The perimeter is the sum of the three sides.

$$P = 3.1 + 4.2 + 1.4$$
$$= 8.7$$

The perimeter is 8.7 yd.

(b) The perimeter of a square is 4 times the length of a side.

$$P = 4(6.25)$$
$$= 25$$

The perimeter is 25 inches (in.).

(c) The perimeter is the sum of the lengths of the four sides.

$$P = 4\frac{1}{2} + 6\frac{1}{2} + 4\frac{1}{2} + 8\frac{1}{2}$$
$$= 24$$

The perimeter is 24 centimeters (cm).

(d) The circumference of a circle is 2π times the radius.

$$C = 2 \cdot \pi \cdot 2\frac{1}{4}$$
$$= 2\pi \cdot \frac{9}{4} \qquad \text{Change to fraction.}$$
$$= \frac{18}{4}\pi$$
$$= \frac{9}{2}\pi$$

The circumference is $4\frac{1}{2}\pi$ yd. □

EXAMPLE 3. Find the area of each figure.

(a) A parallelogram with base 7 in. and height 21 in.

(b) A $3\frac{1}{2}$-ft by $8\frac{1}{4}$-ft rectangle.

(c) A triangle with height 5 cm and base $10\frac{1}{3}$ cm.

(d) A circle of diameter 8 ft 4 in.

Solutions:

(a) The formula for the area of a parallelogram is $A = bh$:

$$A = bh$$
$$= 7 \cdot 21$$
$$= 147$$

The area is 147 in^2.

(b) The area of a rectangle is given by $A = lw$:

$$A = lw$$
$$A = 3\frac{1}{2} \cdot 8\frac{1}{4}$$
$$= \frac{7}{2} \cdot \frac{33}{4} \qquad \text{Express as fractions.}$$
$$= \frac{231}{8}$$
$$= 28\frac{7}{8} \qquad \text{Express as a mixed number.}$$

The area is $28\frac{7}{8}$ ft^2.

(c) The area of a triangle is one-half the product of the height and the base.

$$A = \frac{1}{2}bh$$
$$= \frac{1}{2} \cdot 10\frac{1}{3} \cdot 5$$
$$= \frac{1}{2} \cdot \frac{31}{3} \cdot 5 \qquad \text{Express as a fraction.}$$
$$= \frac{155}{6}$$
$$= 25\frac{5}{6} \qquad \text{Express as a mixed number.}$$

The area is $25\frac{5}{6}$ cm^2.

(continued)

(d) To find the area of a circle of diameter 8 ft 4 in., we first must write the diameter in one unit of measure. Since 8 feet is $8 \cdot 12 = 96$ in., the diameter of the circle is $96 + 4 = 100$ in. Thus the radius is 50 in. The area is given by

$$\begin{aligned} A &= \pi r^2 \\ &= \pi(50)^2 \\ &= 2500\pi \end{aligned}$$

The area is 2500π in^2. □

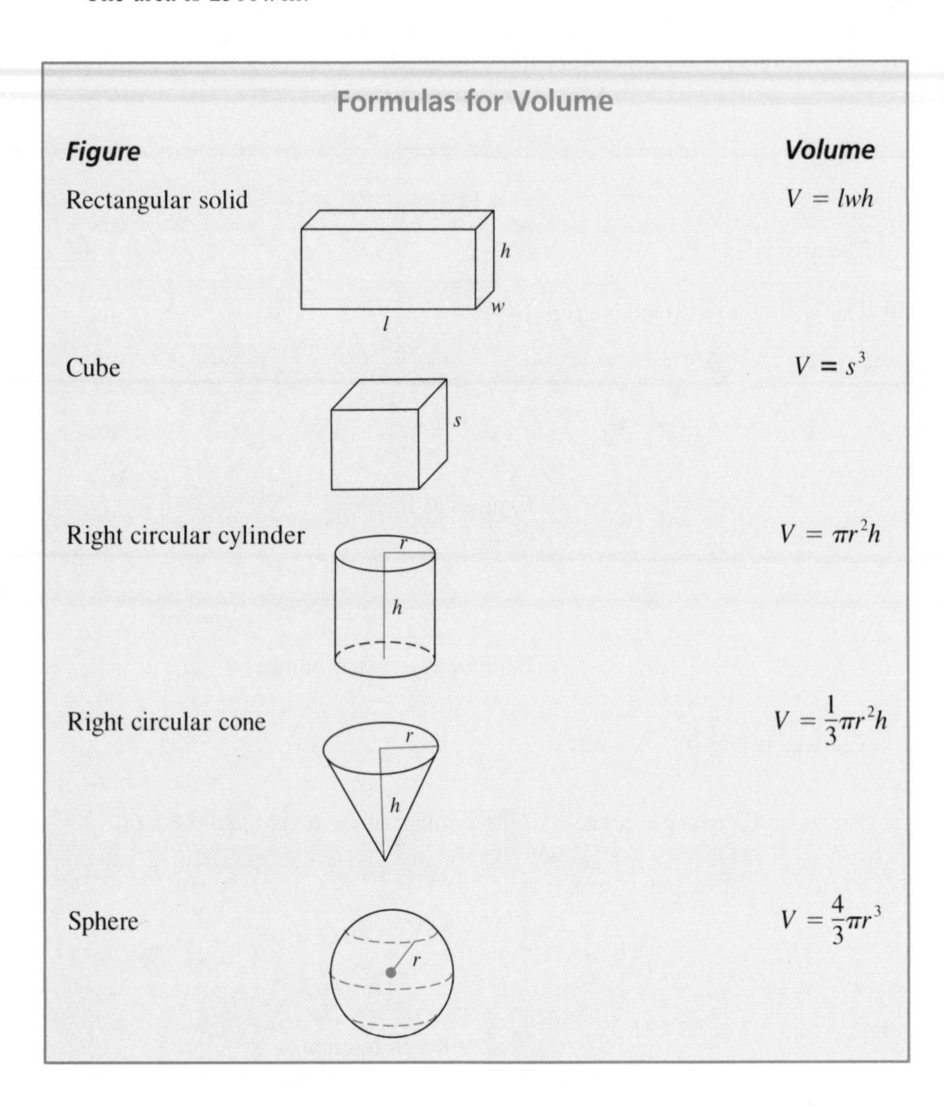

Formulas for Volume

Figure	*Volume*
Rectangular solid	$V = lwh$
Cube	$V = s^3$
Right circular cylinder	$V = \pi r^2 h$
Right circular cone	$V = \frac{1}{3}\pi r^2 h$
Sphere	$V = \frac{4}{3}\pi r^3$

EXAMPLE 4. Find the volume of each of the following figures.

(a) A rectangular solid with length 5 in., width 2 in., and height 3 in.

(b) A right circular cylinder with radius 5 ft and height 22 ft.

(c) A sphere of radius 30 miles.

Solutions:

(a) The formula for the volume of a rectangular solid is $V = lwh$.

$$V = lwh$$
$$= 5 \cdot 2 \cdot 3$$
$$= 30$$

The volume is 30 in^3.

(b) The volume of a right circular cylinder is given by $V = \pi r^2 h$.

$$V = \pi r^2 h$$
$$= \pi 5^2 \cdot 22$$
$$= 550\pi$$

The volume is $550\pi\, ft^3$.

(c) The volume of a sphere is $\frac{4}{3}\pi$ times the cube of the radius.

$$V = \frac{4}{3}\pi r^3$$
$$= \frac{4}{3}\pi (30)^3$$
$$= \frac{4}{3}\pi \cdot 27000$$
$$= 36000\pi$$

The volume is $36{,}000\pi$ cubic miles. □

PROBLEM SET 0.7

Warm-ups

In Problems 1 through 5, answer each question with a statement. See Examples 1, 2, 3, and 4.

1. The length of a rectangle is $4\frac{1}{2}$ ft and its width is $1\frac{1}{4}$ ft. Find its area and perimeter.
2. Find the perimeter and area of the trapezoid.
3. Find the perimeter and area of the triangle.

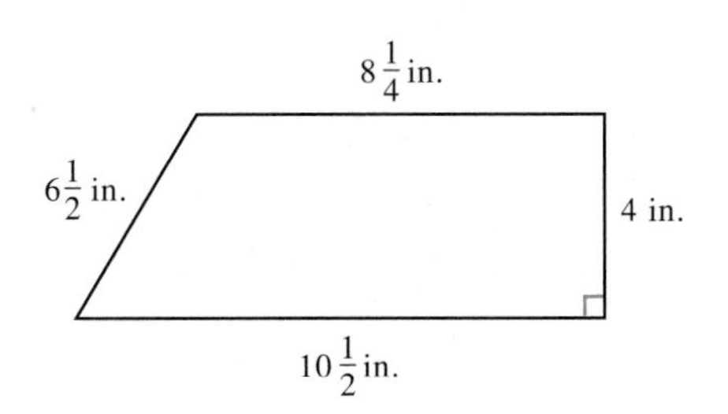

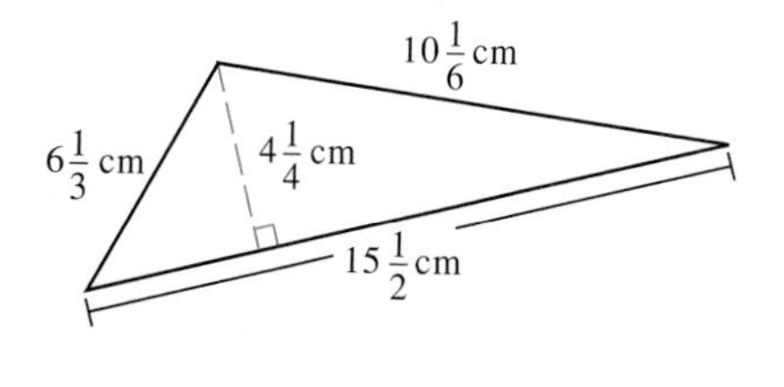

4. Find the circumference and area of the circle.

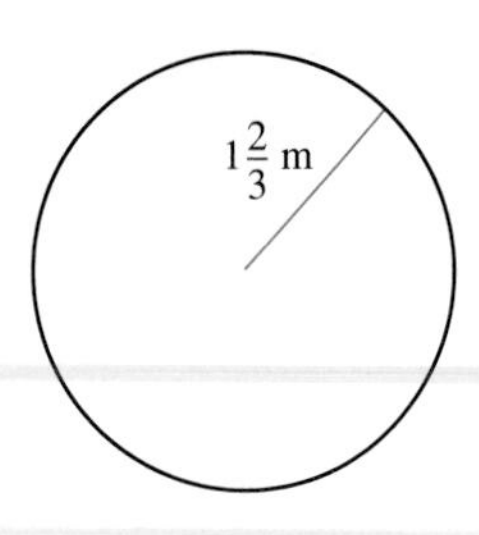

5. Find the perimeter and area of the following figure.

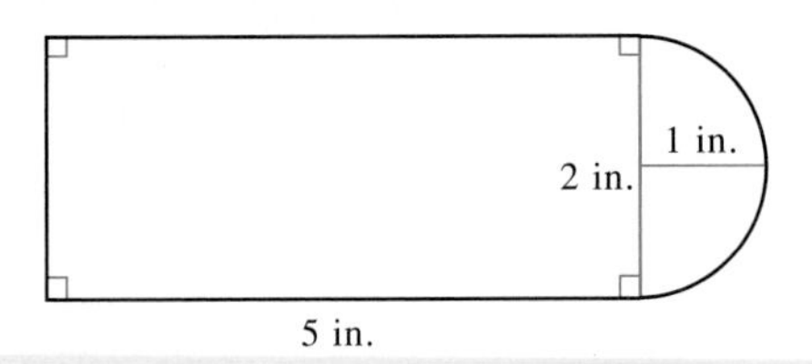

6. Find the volume of a cube with side 4 yd.
7. Find the volume of a right circular cone of radius 2 m and height 6 m.

Practice Exercises

8. The length of a rectangle is $16\frac{1}{2}$ ft and its width is $12\frac{1}{3}$ ft. Find its area and perimeter.
9. Find the perimeter and area of the following parallelogram.

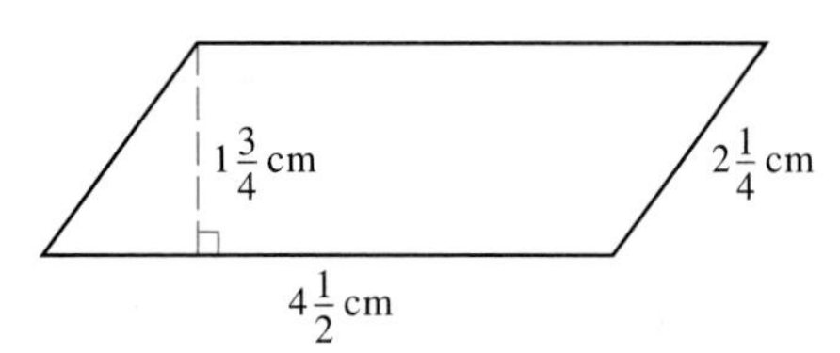

10. Find the perimeter and area of the triangle.

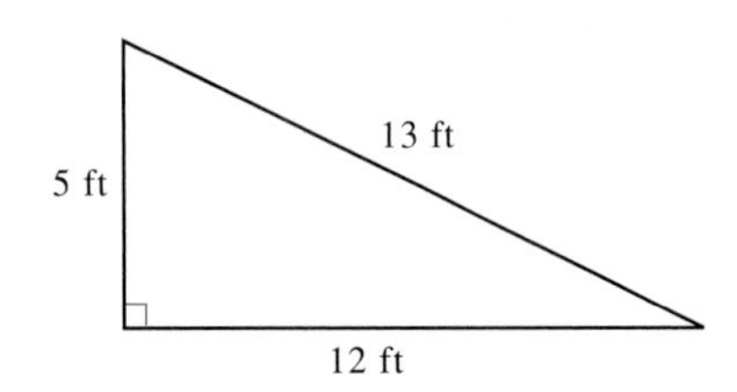

11. Allen is painting the ceiling of his den, which measures $15\frac{1}{2}$ ft by 20 ft. How many square feet does he need to paint?
12. Find the circumference of a circular flower pot of diameter $7\frac{2}{3}$ in.
13. Alex has 250 meters (m) of chain link fence. He plans to fence his rectangular garden plot with dimensions $30\frac{1}{2}$ by $85\frac{3}{4}$ m. Does he have enough fencing?
14. Alice is making a circular table cloth. The radius of the table cloth is to be 45 in. How many yards of trim should she buy to sew around the edge?
15. Carolyn is putting wallpaper on three walls of a room, each of which measures $15\frac{1}{2}$ ft by 8 ft. How many square feet of wallpaper must she buy?
16. Amanda is making cookies that are semicircles. If the diameter of a cookie is 3 in., find the distance around each cookie.
17. Jerome made a triangular poster, which has a base of 28.2 in. and a height of 23 in. What is the area of the poster?
18. Find the perimeter of a monopoly board if a side is $19\frac{1}{2}$ in.
19. What is the perimeter of a triangular patch with all sides of length $3\frac{1}{3}$ cm?
20. Find the area of the square cover of a floppy disk with sides of $5\frac{1}{4}$ in.
21. Find the area of a nickel. (Its diameter is 0.8 in.)
22. The perimeter of a square is 5 m. Find its area.
23. Find the volume of a spherical globe of diameter 12 inches.
24. How many cubic feet of silage will a silo of radius 12 ft and height 50 ft contain? (Consider a silo to be a right circular cylinder.)

IN YOUR OWN WORDS...

25. Explain the formula for finding the area of a trapezoid.

CHAPTER SUMMARY

GLOSSARY

Numeral: A symbol used to name a number.

Sum: The result of adding two or more numbers.

Difference: The result of subtracting one number from another number.

Product: The result of multiplying two or more numbers.

Quotient: The result of dividing one number by another number.

Exponent: The number of repeated multiplications of the same number (the base).

Base: The number being multiplied the number of times given by the exponent.

Grouping symbols: Symbols such as ( ), [ ], { }, $\sqrt{\ }$, and fraction bars used to make the order of operations clear.

Factors: Numbers that are multiplied together.

Prime number: A natural number larger than 1 whose only factors are itself and 1.

Reduced to lowest terms: A fraction with no factors common to the numerator and denominator.

Square root: A positive number that when squared is a given number.

ORDER OF OPERATIONS

If grouping symbols are present, perform operations inside them, starting with the innermost grouping symbol, in the following order.

1. Perform any exponentiations.
2. Perform all multiplications and divisions in order from left to right.
3. Perform all additions and subtractions in order from left to right.

If grouping symbols are not present, perform operations in the above order.

OPERATIONS WITH FRACTIONS

Operations with fractions follow these rules (p, q, r, and s are natural numbers.)

Fundamental Principle of Fractions

$$\frac{p}{q} = \frac{p \cdot r}{q \cdot r}$$

Multiplication of Fractions

$$\frac{p}{q} \cdot \frac{r}{s} = \frac{p \cdot r}{q \cdot s}$$

Division of Fractions

$$\frac{p}{q} \div \frac{r}{s} = \frac{p}{q} \cdot \frac{s}{r} = \frac{p \cdot s}{q \cdot r}$$

Addition and Subtraction of Fractions

$$\frac{p}{q} + \frac{r}{q} = \frac{p + r}{q} \qquad \frac{p}{q} - \frac{r}{q} = \frac{p - r}{q}$$

PROCEDURE FOR FINDING THE LEAST COMMON DENOMINATOR (LCD)

1. Factor each denominator completely, using exponents.
2. List all the different prime factors from all the denominators.
3. The LCD is the product of the factors in step 2 each raised to the largest power of that factor in any single denominator.

PERCENT-DECIMAL CONVERSION

Decimal to Percent:

1. Move the decimal point *two* places to the right.
2. Write the % symbol.

Percent to Decimal:

1. Replace any fraction with an equivalent decimal.
2. Move the decimal point *two* places to the left.
3. Remove the % symbol.

PERIMETER AND AREA FORMULAS

FIGURE	SKETCH	PERIMETER	AREA
Triangle		$P = a + b + c$	$A = \frac{1}{2}bh$
Rectangle		$P = 2l + 2w$	$A = lw$
Parallelogram		$P = 2a + 2b$	$A = bh$
Square		$P = 4s$	$A = s^2$
Trapezoid		$P = b_1 + s + t + b_2$	$A = \frac{1}{2}h(b_1 + b_2)$
Circle		$C = 2\pi r$	$A = \pi r^2$

CHECKUPS

In Problems 1 and 2, express each phrase in symbols.

1. The product of 2 and 7. — Section 0.1; Example 1d
2. The quotient when 12 is divided by 19. — Section 0.1; Example 1e

In Problems 3 and 4, perform the operations indicated.

3. $10 - 3 + 2^2$ — Section 0.1; Example 3b

4. $2 - 2(6 - 3)$ — Section 0.1; Example 3d

5. Factor 392 into prime factors. — Section 0.2; Example 1c

In Problems 6 and 7, perform the operations indicated.

6. $\frac{27}{16} \div \frac{9}{8}$ — Section 0.3; Example 4

7. $\frac{3}{4} + \frac{2}{3} \cdot \frac{1}{2}$ — Section 0.3; Example 8a

8. Write 41.6% as a decimal. — Section 0.4; Example 5d

9. Find 23% of 3300. — Section 0.4; Example 8a

In Problems 10 and 11, evaluate the expressions.

10. $\sqrt{25}$ — Section 0.5; Example 1c

11. $\sqrt{144 + 25}$ — Section 0.5; Example 2a

12. Find the area and perimeter of a rectangle with dimensions of $6\frac{1}{4}$ ft by $2\frac{1}{3}$ ft. — Section 0.7; Example 1

REVIEW PROBLEMS

In Problems 1 through 10, perform the operations indicated.

1. $\frac{6}{7} - \frac{2}{3}$

2. $\frac{1}{2} \cdot \frac{4}{9} + \frac{5}{6}$

3. $18/6 \cdot 5$

4. $2^2 - (10 - 8)$

5. $\frac{6(3) - 3^2}{6 - 4}$

6. $12 - 2 \cdot 3 + 6 \div 3$

7. $\left(\frac{1}{2}\right)^2 - \frac{4}{9} \cdot \frac{9}{16}$

8. $6.3(2.1) + 8.4/0.04$

9. $7 - \frac{2}{3}$

10. $\left(1 + \frac{1}{2}\right)^2$

In Problems 11 through 15, express each phrase in symbols.

11. Twice 5 increased by 3.

12. One less than 9.

13. The sum of one-half 7 and twice 4.

14. The difference when twice 2 is subtracted from twice 8.

15. The sum of 4 and 6 subtracted from 15.

In Problems 16 through 20, answer each question.

16. Jack has 7 baseball cards. Roberto has 2 fewer baseball cards than Jack. How many cards does Roberto have?

17. Sue bought 4.32 lb of chicken, 3.72 lb of hamburger, 8.18 lb of steak, and 1 lb of bacon. How many pounds of meat did she buy?

18. Wendell budgeted 17% of his money for entertainment. If he has $212, how much did he budget for entertainment?

19. Find 50% of 300.

20. Bill, Joe, and Sam need 50 ft of rope. Bill has $12\frac{1}{2}$ ft, Joe has $20\frac{1}{4}$ ft, and Sam has $15\frac{7}{8}$ ft. Do they have enough rope?

In Problems 21 through 25, express each percent as a fraction and a decimal.

21. 15%

22. 2%

23. 11.5%

24. 12 1/2%

25. 83 1/3%

In Problems 26 through 30, express each number as a percent.

26. $\frac{13}{100}$

27. 0.18

28. 0.08

29. $\frac{5}{6}$

30. 2.5

In Problems 31 through 35, factor each number completely.

31. 16

32. 84

33. 100

34. 1944

35. 400

In Problems 36 through 43, evaluate the given expressions. If necessary, use a table or a calculator. Write approximations to three decimal places.

36. $\sqrt{4}$

37. $\sqrt{4900}$

38. $\sqrt{86}$

39. $\sqrt{6400}$

40. $\sqrt{400}$

41. $\sqrt{40}$

42. $\sqrt{144 + 81}$

43. $\sqrt{144} + \sqrt{81}$

44. Find the area and perimeter of a rectangle with dimensions $12\frac{1}{2}$ cm by $8\frac{1}{3}$ cm.

45. Find the area and circumference of a circle with radius $6\frac{3}{4}$ m.

CHAPTER O TEST

In Problems 1 through 5, choose the correct answer.

1. When reduced, $\frac{15}{60} = (?)$.

A. $\frac{1}{4}$ B. $\frac{1}{5}$

C. $\frac{1}{2}$ D. $\frac{1}{3}$

2. When factored completely, $96 = (?)$.

A. $16 \cdot 6$ B. $24 \cdot 4$

C. $32 \cdot 3$ D. $2^5 \cdot 3$

3. $12\frac{1}{2}\% = (?)$

A. 0.125 B. 1.25

C. 12.5 D. 125

4. $1.2/0.06 - (1.3)(0.05) = (?)$.

A. 1.935 B. 19.935

C. 199.935 D. 374

5. $\sqrt{25 - 9} = (?)$

A. 2 B. 4

C. 8 D. 16

In Problems 6 through 10, perform the operations indicated.

6. $2 + 3^2$

7. $23.01 - 6.1(2.0)$

8. $\left(\frac{2}{3}\right)^3 + \frac{1}{3} - \frac{2}{9}$

9. $\sqrt{25 + 144}$

10. $\frac{3 \cdot 2^2 - (6 - 3)}{8 - 5}$

11. Find $22\frac{1}{2}\%$ of 60.

12. Find the area of a rectangle with dimensions $2\frac{3}{4}$ in. by $1\frac{1}{5}$ in.

13. Write $\frac{3}{8}$ as a percent.

14. Simplify $\frac{2}{3} \div \frac{8}{15} + 2$

15. Katy bought $4\frac{1}{2}$ yd of cotton, $2\frac{7}{8}$ yd of polyester, and $6\frac{1}{4}$ yd of denim. How many yards of fabric did she buy?

CHAPTER 1

See Problem Set 1.2, Exercise 43.

Real Numbers

CONNECTIONS

For over 6000 years, people have been using numbers and the properties of numbers. The need to record information and transmit knowledge from one place to another forced the development of a written system of numeration. Today, most of the world uses the Hindu-Arabic numeral system, named for the Hindus, who probably invented it, and for the Arabs, who introduced it to Western Europe.

At first, small counting numbers like 1, 2, and 3 were sufficient for barter. As society developed, the growth of commerce dictated a need for larger numbers, negative numbers, and fractions. Numbers like $\frac{5}{8}$ were used by merchants to indicate portions of goods, and numbers like -60 were used to indicate debts.

In today's world, stock market gains and losses are reported using signed numbers like $+5\frac{1}{8}$, and -4, and weather forecasts routinely use negative integers to report temperatures. Scientists and engineers use numbers like $\sqrt{2}$ and π in their daily work.

The system of real numbers discussed in this chapter is a natural extension of the arithmetic of natural numbers discussed in Chapter 0 and provides the basic number system of algebra. We begin by learning the arithmetic of real numbers and their properties. The idea of using a letter for a number is introduced and we form expressions and equations. These ideas are essential concepts of algebra which are examined in more detail in later chapters.

1.1 SETS OF NUMBERS

A set of is a collection of objects. The objects are called **elements,** or **members.** We usually use capital letters to name sets and list the elements in braces. For example, $A = \{a, b, c, d, e\}$ is a set containing the first five letters of the alphabet. $B = \{1, 2, 3, \ldots, 10\}$ is a set containing the counting numbers 1 through 10. (The three dots, called an ellipsis, mean "continue on in the pattern that has been established.") $N = \{1, 2, 3, \ldots\}$ is a set containing all the counting numbers.

If the number of elements in a set is a counting number, the set is called a **finite** set. If there are no elements at all in a set, it is the **empty set.** Otherwise it is called an **infinite** set. Of the sets listed above, A and B are finite, and N is infinite.

Be Careful!

The empty set is sometimes called the **null set.** It is written $\{\ \}$ or $\varnothing$. It would be incorrect to write the empty set as $\{\varnothing\}$, as the set $\{\varnothing\}$ contains one element.

The symbol $\in$ is used to mean **is an element of,** and the symbol $\notin$ is used to mean **is not an element of.** If $D = \{a, b, c, d, e\}$, then we see that $a \in D$, but $2 \notin D$.

Sometimes we don't list the elements of the set. Instead, we use a notation called **set-builder notation.** $\{x \mid x \text{ is a vowel}\}$ describes the "set of all x such that x is a vowel." The bar is read, "such that." It is the same set as {a, e, i, o, u}.

Set A is a **subset** of a set B if every element in A is also an element in B. We use the symbol $\subseteq$ to mean **is a subset of.** If $A = \{1, 5, 7\}$ and $B = \{1, 3, 5, 7, 9\}$, then $A \subseteq B$.

EXAMPLE 1. List all the subsets of $\{a, b, c\}$.

Solution:

The subsets are:

$$\begin{array}{llll} \{a, b, c\} & \{a, b\} & \{a\} & \varnothing \\ & \{a, c\} & \{b\} & \\ & \{b, c\} & \{c\} & \end{array}$$

□

Notice that the set is a subset of itself and that the empty set is a subset of every set.

Subsets of Every Set

If A is any set, $A \subseteq A$ and $\varnothing \subseteq A$.

Many times we combine sets using the operations of **union** ($\cup$) and **intersection** ($\cap$).

Union of Two Sets

If A and B are sets,

$$A \cup B = \{x \mid x \in A \text{ or } x \in B\}$$

That is, the union of two sets is the set of all elements that belong to either of the two sets (or both of them).

Intersection of Two Sets

If A and B are sets,

$$A \cap B = \{x \mid x \in A \text{ and } x \in B\}$$

The intersection of two sets is the set of all elements that belong to both of the two sets.

EXAMPLE 2. If $A = \{1, 2, 7\}$ and $B = \{1, 2, 3, 4, 5\}$, find $A \cup B$ and $A \cap B$.

Solutions:

$$A \cup B = \{1, 2, 3, 4, 5, 7\}$$

$$A \cap B = \{1, 2\}$$

□

John Venn, an English mathematician, popularized the use of diagrams to help visualize sets. Called **Venn diagrams,** these diagrams use circles to indicate the contents of sets. For example, if circle A indicates the contents of set A and circle B shows the contents of set B, we can use shading to indicate the contents of various sets.

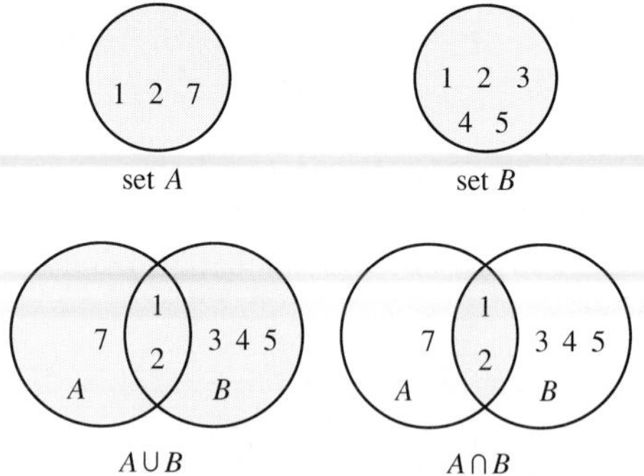

(The numbers in the sets are from Example 2.)

In algebra we are concerned with sets of numbers. We worked with counting numbers, or natural numbers, in Chapter 0, and in the beginning of this section we named the set of counting numbers N.

> **Natural Numbers, or Counting Numbers**
>
> $N = \{1, 2, 3, \ldots\}$

If we put the number 0 into this set, we have a set called the set of whole numbers. We name the set W and write it as follows.

> **Whole Numbers**
>
> $W = \{0, 1, 2, 3, \ldots\}$

It is much easier to visualize sets of numbers if we draw a picture of them. This is done by using a **number line.** We draw a line and pick a starting point called the **origin** and label it with the number 0. We move to the right of 0 and represent natural numbers in increasing order. We space the numbers at equal distances with a convenient unit of measure.

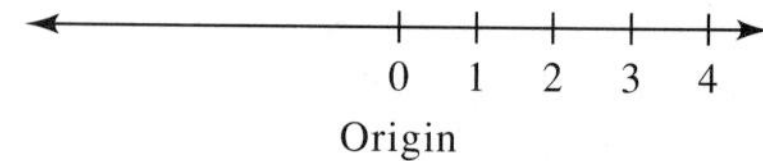

The number associated with a point on a number line is called the point's **coordinate.**

To **graph** a set of numbers means to locate them on a number line.

There are points to the left of 0 as well as to the right. If we move to the left of 0 with the same unit of measure, we name the points associated with negative numbers. The number one unit to the left of 0 is called -1 (read "negative one"). The number two units to the left of 0 is called -2, and so on.

$-2 \quad -1 \quad 0 \quad 1 \quad 2$

We call this set of numbers the integers and name it Z.

Integers

$$Z = \{\ldots, -3, -2, -1, 0, 1, 2, 3, \ldots\}$$

The integers 0, 2, 4, 6 and 8 are called **even.** The integers 1, 3, 5, 7 and 9 are called **odd.** An **even integer** is an integer that ends in an even digit, whereas, an **odd integer** is one that ends in an odd digit.

Fractions are placed on the number line using the same unit of measure.

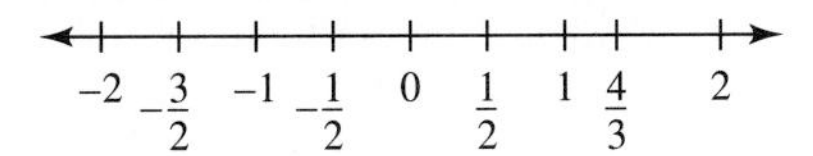

This set of numbers is called the **rational numbers.** We name it Q and write it in set notation as follows.

Rational Numbers

$$Q = \{x \mid x \text{ can be written as } \frac{p}{q}, \text{ where } p \text{ and } q \text{ are integers and } q \neq 0\}$$

Since a rational number is the quotient of two integers, we often refer to rational numbers as fractions. We saw in Chapter 0 that every quotient of natural numbers can be written as a decimal that repeats or terminates. So, every rational number can be written as a fraction or as a terminating or repeating decimal.

There are other points on the number line that we have not named. When written in decimal form, they do not repeat or terminate. They are called the **irrational numbers.** Some examples of irrational numbers are $\sqrt{2}$, $\sqrt{3}$, and π. In fact, $\sqrt{x}$, where x is a positive integer but not a perfect square, is an irrational number.

Irrational Numbers

$$I = \{x \mid x \text{ is on the number line but is not a rational number}\}$$

We have now named all the numbers on the number line. We call all these numbers the **real numbers.** We name the set of real numbers R, and with set notation write them as follows.

Real Numbers

$$R = \{x \mid x \text{ is a rational or an irrational number}\}$$

The following diagram shows how these sets are related to each other.

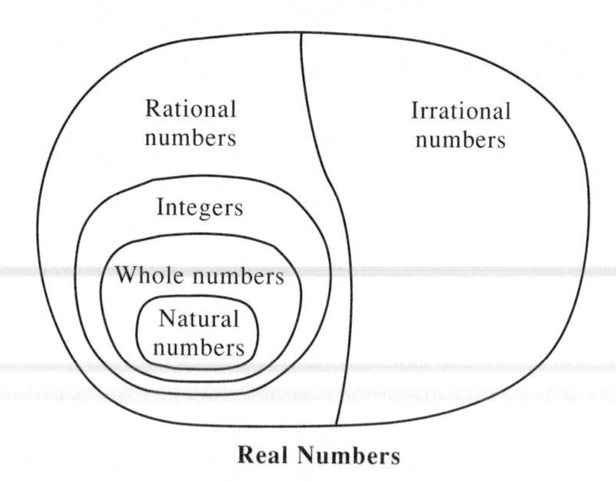

Real Numbers

Notice that since $5 = \frac{5}{1}$, the integer 5 is a rational number. Likewise all integers are rational numbers.

EXAMPLE 3. Classify each number as rational or irrational.

(a) $\dfrac{3}{7}$ (b) $0.\overline{78}$ (c) $\sqrt{7}$ (d) 1.7

Solutions:

(a) $\dfrac{3}{7}$ is rational.

(b) $0.\overline{78}$ is a repeating decimal; thus it is rational.

(c) $\sqrt{7}$ is irrational, since 7 is not a perfect square.

(d) 1.7 is rational because $1.7 = \dfrac{17}{10}$. □

The number line helps us to understand many concepts of numbers. If a number p lies to the left of another number q on a number line, we say that p is **less than** q.

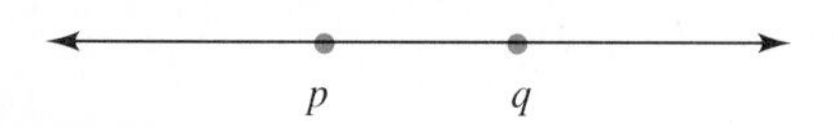

Less Than

If p lies to the left of q on a number line,

$$p < q$$

We read this "p is less than q."

Likewise if a number r lies to the right of a number s on a number line, we say that r is **greater than** s, which we write as $r > s$.

Greater Than

If r lies to the right of s on a number line,

$$r > s$$

We read this "r is greater than s."

Sometimes we combine equals with less than or greater than, using the symbols $\le$ for **less than or equal to** and $\ge$ for **greater than or equal to.**

Less Than or Equal To–Greater Than or Equal To

$p \le q$ means that p is less than or equal to q.
$r \ge s$ means that r is greater than or equal to s.

EXAMPLE 4. Indicate whether each statement is true or false.

(a) $-\frac{1}{2} < 0$ (b) $\frac{3}{5} < \frac{2}{3}$

(c) $-1.4 > -1.2$ (d) $5 \le 5$

(e) $-\frac{1}{2} \le \frac{1}{2}$

Solutions:

(a) $-\frac{1}{2} < 0$

Because $-\frac{1}{2}$ lies to the left of 0, the statement is true.

(b) $\frac{3}{5} < \frac{2}{3}$

To see where these numbers fall on a number line, we write them with the same denominator.

$$\frac{3}{5} = \frac{9}{15} \quad \text{and} \quad \frac{2}{3} = \frac{10}{15}$$

So, $\frac{9}{15}$ lies to the left of $\frac{10}{15}$. Thus $\frac{3}{5} < \frac{2}{3}$ is a true statement.

(c) $-1.4 > -1.2$

Since -1.4 lies to the left of -1.2 on a number line, the statement is false.

(d) $5 \le 5$

Since $5 = 5$, the statement is true.

(e) $-\frac{1}{2} \le \frac{1}{2}$

Because $-\frac{1}{2}$ lies to the left of $\frac{1}{2}$ on a number line, the statement is true. □

Suppose that the numbers p and q are graphed on a number line:

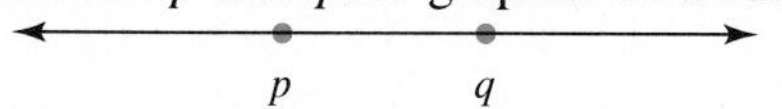

We can see that $p < q$ and also that $q > p$. When we write $p < q$ and read it as p is less than q, we are reading from left to right. If we read $p < q$ from right to left, it says q is greater than p. We usually read from left to right.

The numbers to the right of 0 are called the **positive** numbers and the numbers to the left of 0 are called the **negative** numbers. Zero is neither positive nor negative.

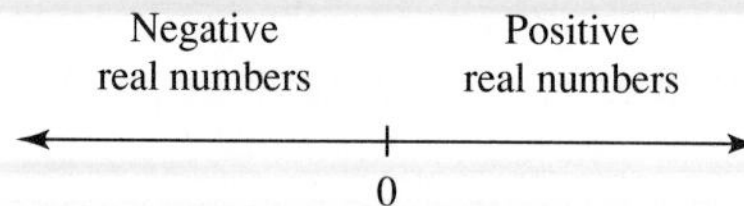

In set notation we write these sets as follows.

> **Positive Real Numbers**
>
> $\{x \mid x > 0\}$
>
> **Negative Real Numbers**
>
> $\{x \mid x < 0\}$

EXAMPLE 5. Locate each number on a number line.

(a) x, where x is a positive real number.
(b) y, where y is a negative real number.
(c) x and y, where $x > 0$ and $y < 0$.

Solutions:

(a)

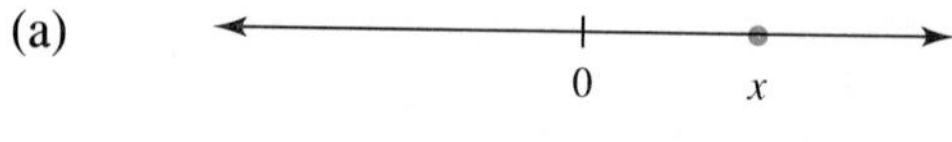

(b)

y 0

(c)

y 0 x

□

The points on the number line whose coordinates are 3 and -3 are on opposite sides of 0 but are the same distance from 0.

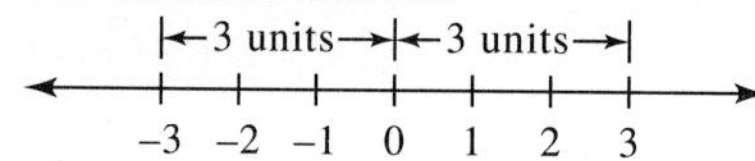

We call 3 and -3 *opposites.*

> **Additive Inverse, or Opposite**
>
> If p is a real number, then $-p$ is the real number that is on the opposite side of 0 and the same distance from 0 as p. We call $-p$ the **additive inverse,** or **opposite,** of p.

EXAMPLE 6. Find the additive inverse of each number.

(a) $\frac{2}{3}$ (b) -0.2 (c) 0 (d) x

Solutions:

(a) The additive inverse of $\frac{2}{3}$ is $-\frac{2}{3}$.

(b) The additive inverse of -0.2 is 0.2.

(c) The additive inverse of 0 is 0.

(d) The additive inverse of x is $-x$. □

If p is a positive number, then p is to the right of 0. The opposite of p, which we call $-p$, is to the left of 0.

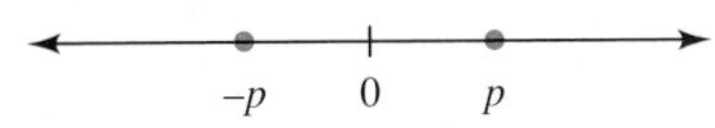

If q is a negative number, then q lies to the left of 0. The opposite of q, which we call $-q$, is to the right of 0. Notice that $-q$ is a positive number because q is negative.

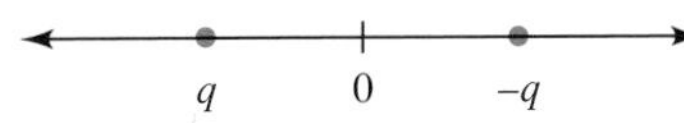

The opposite of -3 is 3. We can write this by saying

$$-(-3) = 3$$

What is the opposite of the opposite of a number p? That question sounds confusing. However, if we write it in symbols, it is less confusing. The opposite of p is $-p$. So, the opposite of the opposite of p is $-(-p)$. What number lies on the opposite side of 0 from $-p$? It is p.

Double Negative Property

If p is a real number, $-(-p) = p$.

EXAMPLE 7. Simplify each of the following.

(a) $-(-5)$ (b) $-\left(-\frac{3}{4}\right)$ (c) $-(-z)$

Solutions:

(a) $-(-5) = 5$

(b) $-\left(-\frac{3}{4}\right) = \frac{3}{4}$

(c) $-(-z) = z$ □

We have seen three uses of the symbol $-$. We have used it to indicate the *operation of subtraction,* as part of the *name of a number,* and to indicate the *opposite,* or *additive inverse,* of a number. Sometimes we read it as "minus," or sometimes as "subtract," sometimes as "negative," and sometimes as "the opposite of."

$$\boxed{1}\ 6-2 \qquad \boxed{2}\ -5 \qquad \boxed{3}\ -p$$

In $\boxed{1}$ the $-$ symbol indicates subtraction and we read it "6 minus 2," or "6 subtract 2." In $\boxed{2}$ it indicates the name of a number, negative 5. In $\boxed{3}$ it indicates the opposite of p and is read "the opposite of p," or "negative p."

A number and its opposite have the common property that each is the same distance from 0. Both 3 and -3 are 3 units from 0. We name this distance from 0 the **absolute value** of a number and indicate it by vertical bars, $|\ |$. $|3|$ is the distance between 0 and 3. So, $|3| = 3$. Likewise, $|-3|$ is the distance between 0 and -3. So, $|-3| = 3$. Thus, we see that $|3| = |-3|$.

> **Absolute Value**
>
> If p is a real number, the absolute value of p, $|p|$, is its distance from 0.

EXAMPLE 8. Simplify each of the following.

(a) $|-5|$ (b) $2|17|$ (c) $-|-7|$ (d) $|-\pi|$

Solutions:

(a) On the number line we see that -5 is 5 units from 0.

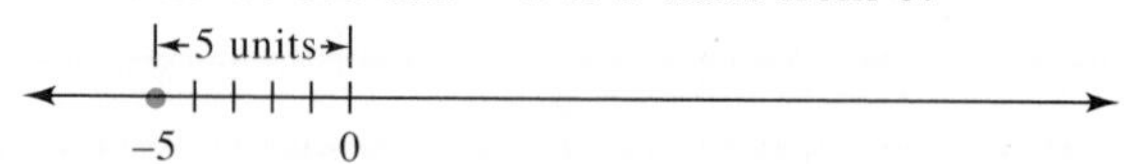

Thus $|-5| = 5$.

(b) Again we look at the number line.

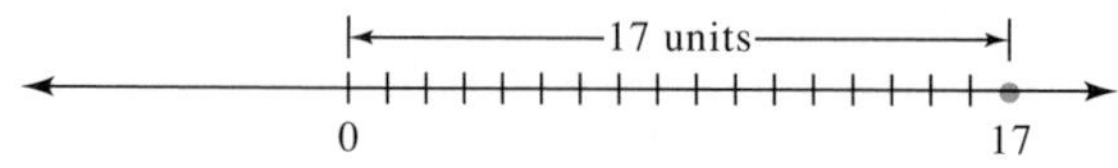

So $|17| = 17$; then

$$\begin{aligned} 2|17| &= 2 \cdot 17 \\ &= 34 \end{aligned}$$

(c)

7 units

–7 0

Thus $|-7| = 7$, and so $-|-7| = -7$.

(d)

π units

$-\pi$ 0

And so $|-\pi| = \pi$. □

PROBLEM SET 1.1

Warm-ups

In Problems 1 through 6, $A = \{2, 4, 6, 8\}$, $B = \{1, 4\}$ *and* $C = \{-2, 0, 2\}$.
In Problems 1 and 2, see Example 1.

1. List all the subsets of B.
2. List all the subsets of C.

In Problems 3 through 6, see Example 2.

3. List the elements in $A \cup B$.
4. List the elements in $A \cup C$.
5. List the elements in $A \cap B$.
6. List the elements in $A \cap C$.

In Problems 7 through 10, classify each number as rational or irrational. See Example 3.

7. 5.41
8. $\frac{3}{8}$
9. $\sqrt{11}$
10. $2.\overline{12}$

In Problems 11 through 16, indicate whether each statement is true or false. See Example 4.

11. $-\frac{1}{3} < -\frac{1}{4}$
12. $1.1 \geq 1.1$
13. $-5 \leq -10$
14. $5.7 > 0$
15. $-1.12 \leq -1.25$
16. $-0.111 > 2.333$

In Problems 17 through 20, graph each number on a number line. Assume that x is a positive *real number and that y is a* negative *real number. See Example 5.*

17. x
18. y
19. $-x$
20. $-y$

In Problems 21 through 24, find the additive inverse (opposite) of each number. See Example 6.

21. -3.5
22. $\sqrt{5}$
23. $\frac{11}{13}$
24. z

Simplify each of the following. See Examples 7 and 8.

25. $-(-6)$
26. $-(-35)$
27. $-(-\pi)$
28. $-(-q)$
29. $|-12|$
30. $|29|$
31. $2|-8|$
32. $3|6|$
33. $-|4|$
34. $-|-5|$
35. $-|-16|$
36. $-|-3|$

Practice Exercises

In Problems 37 through 39, $A = \left\{\frac{1}{2}, 2, \frac{9}{4}, 5\right\}$ and $B = \left\{\frac{9}{4}, 5\right\}$.

37. List the elements in $A \cup B$.
38. List the elements in $A \cap B$.
39. List all the subsets of B.

In Problems 40 through 49, indicate whether each statement is true or false.

40. The set of natural numbers is a subset of the set of whole numbers.

41. Every negative integer is a rational number.

42. $\frac{1}{2}$ is a real number.

43. 0 is a rational number.

44. The positive integers are a subset of the whole numbers.

45. A whole number is a nonnegative integer.

46. $\sqrt{16}$ is an irrational number.

47. The empty set is written as $\{\varnothing\}$.

48. Every rational number is an integer.

49. $\sqrt{11}$ is real.

50. Fill in the chart to indicate the set(s) to which each number belongs. N = natural numbers, W = whole numbers, Z = integers, Q = rational numbers, I = irrational numbers, R = real numbers.

Number	N	W	Z	Q	I	R
$\frac{4}{3}$						
-1						
$\sqrt{7}$						
π						
$\sqrt{25}$						
10						
$1.\overline{32}$						
-2.3						
$\frac{20}{10}$						

In Problems 51 through 54, graph each number on a number line. Assume that c is a positive *real number and that d is a* negative *real number.*

51. c

52. d

53. $-c$

54. $-d$

In Problems 55 through 60, indicate whether each statement is true or false.

55. $-\frac{2}{3} < -\frac{5}{8}$

56. $7 \leq 7$

57. $-5 \leq -1$

58. $-5.7 > 0$

59. $-3.415 < 2.333$

60. $-3.14 \leq -3.18$

In Problems 61 through 72, simplify each.

61. $|-2|$

62. $|29|$

63. $2|-4|$

64. $|9|$

65. $-(-5)$

66. $-(-98)$

67. $-(-3.14)$

68. $-(-k)$

69. $-|8|$

70. $-|-11|$

71. $-|-37|$

72. $-|-6|$

73. Write three consecutive integers if the smallest is 48.

74. Write three consecutive *even* integers if the smallest is 88.

75. Write three consecutive odd integers if the smallest is 37.

76. Write three consecutive integers if the largest is 20.

Challenge Problems

77. Express 0.5 as a quotient of two integers.

78. Express 1.2 as a quotient of two integers.

In Problems 79 through 82, list the elements of A.

79. $A = \{x \mid x \text{ is a natural number less than } 10\}$

80. $A = \{x \mid x \text{ is an integer between } -5 \text{ and } 2\}$

81. $A = \{x \mid x \text{ is a natural number divisible by } 5\}$

82. $A = \{x \mid x \text{ is an integer divisible by } 5\}$

In Problems 83 through 86, write A using set-builder notation.

83. $A = \{2, 4, 6, \ldots\}$

84. $A = \{2, 4, 6, 8, 10, 12\}$

85. $A = \{-3, -2, -1, 0, 1, 2, 3\}$

86. $A = \{3, 4, 5, 6, 7, 8\}$

IN YOUR OWN WORDS...

87. Describe the set of integers.

88. What does it mean to say that the number x is less than 4?

89. What do we mean by the absolute value of a number?

90. What do we mean by the opposite of a number?

1.2 ADDITION

In this section we will study addition of real numbers. In arithmetic we add natural numbers. In algebra we must learn how to add real numbers. We know that $2 + 3 = 5$, but what about $-2 + 3$?

To help us understand how the rules for addition work, imagine that we are statisticians for a football game. We use a positive number to represent yards gained and a negative number to represent yards lost. If two plays gained 2 yd and 3 yd, then the total yardage gained would be 5 yd, which is the sum of 2 and 3. Now suppose that the first play lost 2 yd and the second play gained 3 yd. The total yardage gained would be $-2 + 3$, which is 1 yd. So,

$$-2 + 3 = 1$$

Likewise, a gain of 2 yd followed by a loss of 3 yd would give a total loss of 1 yd. So, $2 + -3$ must equal -1. That is,

$$2 + -3 = -1$$

This looks awkward, so we use parentheses around the -3 to separate the two symbols $+$ and $-$.

$$2 + (-3) = -1$$

A loss of 2 yd and another loss of 3 yd would give a total loss of 5 yd. So,

$$-2 + (-3) = -5$$

These ideas can be pictured using a number line. We think of moving to the right as a gain and moving to the left as a loss. So, to illustrate that $-2 + 3 = 1$, we start at 0 and move to the left two units. This puts us at -2.

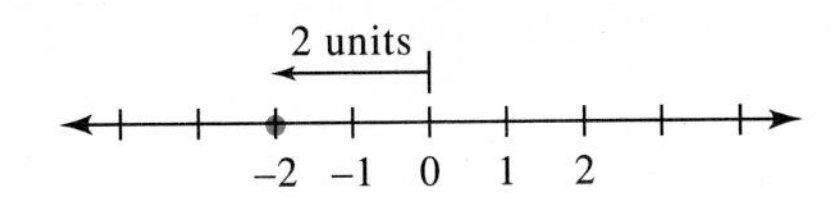

From there we move 3 units to the right.

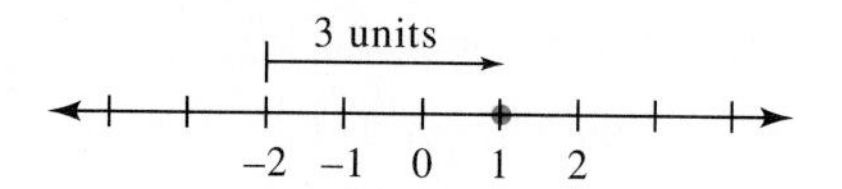

We arrive at 1.

We have two rules for the addition of real numbers. One rule is for addition of numbers that are both positive or both negative and the other is for numbers that are not both positive or not both negative. We must use the idea of absolute value to write the rules.

Addition of Real Numbers

1. To add two numbers with *like signs* (both positive or both negative), add their absolute values and keep the common sign.
2. To add two numbers with *unlike signs* (one positive and one negative), subtract the smaller absolute value from the larger absolute value and keep the sign of the number with the greater absolute value.

Using the rules is easier than reading the rules. Practicing with many problems is the secret to success.

The next example uses the first rule. Notice how the rule is used. Each problem can be verified using a number line.

EXAMPLE 1. Find the following sums.

(a) $-5 + (-9)$ (b) $-7 + (-12)$

Solutions:

(a) We use the first rule, since both numbers are negative. $|-5| = 5$ and $|-9| = 9$. We add 5 and 9 and keep the negative sign.

$$-5 + (-9) = -14$$

(b) Again, we use the first rule. $|-7| = 7$ and $|-12| = 12$. We add 7 and 12 and keep the negative sign.

$$-7 + (-12) = -19$$

□

The next example uses the second rule.

EXAMPLE 2. Find the sums.

(a) $-5 + 9$ (b) $5 + (-9)$

Solutions:

(a) We use the second rule, since one number is positive and the other is negative.

$$|-5| = 5 \quad \text{and} \quad |9| = 9$$

So, we subtract 5 from 9 and keep the positive sign, since $|9|$ is greater than $|-5|$.

$$-5 + 9 = 4$$

(b) Again we use the second rule.

$$|5| = 5 \quad \text{and} \quad |-9| = 9$$

We subtract 5 from 9 and keep the negative sign, since $|-9|$ is greater than $|5|$.

$$5 + (-9) = -4$$ □

EXAMPLE 3. Find each sum.

(a) $4 + (-8)$ (b) $-4 + (-7)$ (c) $-7 + 5$

Solutions:

(a) $4 + (-8) = -4$

(b) $-4 + (-7) = -11$

(c) $-7 + 5 = -2$ □

When we add 0 to any real number, we get the same number.

Addition Property of Zero

If p is a real number,

$$p + 0 = 0 + p = p$$

The order of operations that we learned in Chapter 0 applies to real numbers as well as natural numbers.

Order of Operations

If grouping symbols are present, perform operations inside them, starting with the innermost grouping symbol, in the following order.

1. Perform any exponentiations.
2. Perform all multiplications and divisions in order from left to right.
3. Perform all additions and subtractions in order from left to right.

If grouping symbols are *not* present, perform operations in the above order.

EXAMPLE 4. Perform the operations indicated.

(a) $5 + (-8 + 6)$ (b) $[5 + (-13)] + [-6 + (-8)]$

(c) $\{-1.2 + [2.7 + (-6.1)]\}$

Solutions:

(a) We work inside the parentheses first.

$$5 + (-8 + 6) = 5 + (-2)$$
$$= 3$$

(continued)

(b) We work inside the brackets first.

$$[5 + (-13)] + [-6 + (-8)] = [-8] + [-14]$$
$$= -22$$

(c) We begin with the brackets.

$$\{-1.2 + [2.7 + (-6.1)]\} = \{-1.2 + (-3.4)\}$$
$$= -4.6$$

□

Fraction bars, absolute value bars, and square root symbols are also grouping symbols. That is, we perform operations *over* and *under* the fraction bar, *inside* the absolute value symbols, and *inside* the square root symbol *before* we evaluate further.

EXAMPLE 5. Evaluate the following:

(a) $|-4 + 9|$ (b) $|-4| + |9|$ (c) $\sqrt{25 + 144}$

(d) $\sqrt{25} + \sqrt{144}$ (e) $\dfrac{9 + 5}{9 - 2}$

Solutions:

(a) Since absolute value bars are a grouping symbol, we must perform the operations inside first.

$$|-4 + 9| = |5|$$
$$= 5$$

(b) $|-4| + |9| = 4 + 9$
$= 13$

(c) The $\sqrt{\ }$ is another grouping symbol.

$$\sqrt{25 + 144} = \sqrt{169}$$
$$= 13$$

(d) $\sqrt{25} + \sqrt{144} = 5 + 12$
$= 17$

(e) The fraction bar is also a grouping symbol. We evaluate both the numerator and the denominator *before* we divide.

$$\frac{9 + 5}{9 - 2} = \frac{14}{7}$$
$$= 2$$

□

EXAMPLE 6. Answer each of the following questions.

(a) What is the sum of 12 and -5, increased by 9?

(b) The low temperature this morning was $-3°$. If the temperature rose 21° by noon, what was the noontime temperature?

Solutions:

(a) To increase a number by 9 we add 9 to it.

$$[12 + (-5)] + 9 = 7 + 9$$
$$= 16$$

(b) Since the temperature *rose* 21° from $-3°$, we must add 21 to -3.

$$-3 + 21 = 18$$

The noontime temperature was 18°. □

PROBLEM SET 1.2

Warm-ups

Perform the operations indicated.
In Problems 1 through 21, see Examples 1, 2, and 3.

1. $5 + 3$

2. $5 + (-8)$

3. $-3 + 7$

4. $6 + 0$

5. $-5 + (-8)$

6. $10 + 1$

7. $0 + (-1)$

8. $-1 + (-1)$

9. $8 + (-4)$

10. $-11 + (-3)$

11. $7 + (-7)$

12. $\frac{1}{4} + \left(-\frac{1}{3}\right)$

13. $-\frac{1}{2} + \left(-\frac{1}{2}\right)$

14. $2 + \left(-\frac{1}{8}\right)$

15. $-\frac{1}{8} + \frac{1}{3}$

16. $-\frac{5}{6} + \frac{5}{6}$

17. $-\frac{2}{3} + (-1)$

18. $-\frac{3}{5} + \left(-\frac{1}{10}\right)$

19. $-1.2 + 3.4$

20. $-9.8 + (-2.7)$

21. $5.3 + 8.7$

In Problems 22 through 35, see Example 4.

22. $-7 + (-5 + 9)$

23. $(-9 + 12) + (-5 + 3)$

24. $\{-9 + [-8 + (-3)]\} - 3$

25. $[13 + (-8)] + [-2 + (-7)]$

26. $\frac{1}{4} + \left(-\frac{1}{2} + \frac{1}{8}\right)$

27. $[1.2 + (-3.6)] + (-2.5)$

28. $7 + \{-12 + [-5 + (-3)]\}$

29. $-3 + \{[-2 + 4] + \{5 + (-8)\}\}$

30. $-3.6 + [-2.4 + (-3.1)]$

31. $\left(-\frac{2}{3} + \frac{1}{6}\right) + \left(-\frac{3}{8}\right)$

32. $-3 + \{[-5 + (-5)] + [7 + (-13)]\}$

33. $(15 + 4) + \{[-3 + (-8)] + [6 + (-4)]\}$

34. $-\frac{5}{6} + \left[-\frac{3}{4} + \left(-\frac{2}{3}\right)\right]$

35. $-9.9 + [-3.3 + (-6.6)]$

In Problems 36 and 37, evaluate the given expressions. See Example 5.

36. (a) $|-2 + 7|$ (b) $|-2| + |7|$

37. (a) $\sqrt{225 + 64}$ (b) $\sqrt{225} + \sqrt{64}$

In Problems 38 through 44, answer each question. See Example 6.

38. Find the sum of -6 and 3.

39. What is the sum of -9 and 6 increased by 5?

40. Find the sum of 11 and -6 increased by 5.

41. Last night the low temperature was $-8°$. Tonight's low is predicted to be 12° higher than last night's low. What is the predicted low temperature for tonight?

42. The high temperature for today was $-3°$. If the temperature dropped 27° to reach the low temperature, find the low temperature for the day.

43. A football team had the following yardage on its first four downs, a gain of 1 yd, a loss of 5 yd, a loss of 6 yd, and a gain of 15 yd. Did they make a first down? (A first down means that they gained more than 10 yd.)

44. Henry owes Gerald \$35. If he pays him \$24 today and \$9 next week, how much will Henry still owe Gerald?

Practice Exercises

In Problems 45 through 100, perform the operations indicated.

45. $7 + 2$

46. $4 + (-3)$

47. $-2 + 6$

48. $-6 + (-7)$

49. $5 + 0$

50. $3 + (-5)$

51. $-12 + 7$

52. $32 + (-9)$

53. $-9 + 9$

54. $-13 + 5$

55. $21 + (-5)$

56. $-8 + 8$

57. $6 + (-2)$

58. $-3 + (-3)$

59. $8 + (-5)$

60. $5.5 + (-5.8)$

61. $-6.3 + 6.3$

62. $-5.0 + 1.0$

63. $25 + 25$

64. $-30 + 20$

65. $14 + (-17)$

66. $-13 + (-5)$

67. $8 + (-8)$

68. $\frac{1}{2} + \left(-\frac{1}{3}\right)$

69. $-\frac{1}{5} + \left(-\frac{1}{5}\right)$

70. $3 + \left(-\frac{1}{4}\right)$

71. $-\frac{2}{5} + \frac{1}{3}$

72. $-\frac{7}{8} + \frac{7}{8}$

73. $-\frac{3}{4} + (-1)$

74. $-\frac{3}{7} + \left(-\frac{1}{14}\right)$

75. $-1.5 + 2.1$

76. $-7.7 + (-2.3)$

77. $6.8 + 3.7$

78. $6.5 + (-6.8)$

79. $-4.3 + 4.3$

80. $-7.0 + 3.0$

81. $-5 + (-7 + 8)$

82. $(-5 + 11) + (-7 + 2)$

83. $-8 + [-9 + (-2)]$

84. $[17 + (-9)] + [-3 + (-6)]$

85. $\frac{1}{3} + \left(-\frac{1}{2} + \frac{5}{6}\right)$

86. $[7.2 + (-4.6)] + (-8.5)$

87. $9 + \{-10 + [-3 + (-3)]\}$

88. $-1 + \{[-3 + 7] + [1 + (-1)]\}$

89. $[-4 + (-8)] + (-6)$

90. $[2 + (-22)] + (-12)$

91. $-5.6 + [-1.4 + (-3.1)]$

92. $\left(-\frac{2}{3} + \frac{1}{2}\right) + \left(-\frac{5}{8}\right)$

93. $5 + [3 + (-7)]$

94. $[-1 + (-1)] + (-6 + 7)$

95. $-1 + \{[-4 + (-4)] + [7 + (-12)]\}$

96. $[-5 + (-3)] + [3 + (-8)]$

97. $[-8 + (-3)] + \{7 + [1 + (-9)]\}$

98. $-\frac{7}{15} + \left[-\frac{3}{5} + \left(-\frac{2}{3}\right)\right]$

99. $[1 + (-2)] + (-3 + 2)$

100. $-0.9 + [-2.4 + (-4.6)]$

In Problems 101 and 102, evaluate each of the given expressions.

101. (a) $|-6 + 17|$ (b) $|-6| + |17|$

102. (a) $\sqrt{15^2 + 8^2}$ (b) $\sqrt{15^2} + \sqrt{8^2}$

In Problems 103 through 110, answer each question.

103. Find the sum of -11 and -11.

104. Find the sum of 16 and -8 decreased by 9.

105. What is the sum of -25 and -10 increased by 50.

106. The high temperature yesterday was $-10°$. If today's high is 5° more than yesterday's high, what is today's high?

107. If the temperature is $-15°$ at 5 A.M. and rises at the rate of 3° per hour, what will the temperature be at noon?

108. Bill Barber got a raise of $200 a month. If his federal income tax increased by $59 and his state income tax rose by $12, how much was his net raise?

109. Chuck makes a deposit of $450 and withdrawals of $235 and $350. If he started with a balance of $195, what is his new balance?

110. A football team gains 3 yd, loses 12 yd, gains 9 yd, and gains 5 yd. What is the team's total yardage?

IN YOUR OWN WORDS...

111. How do we add real numbers?

Write in words the meaning of each of the following expressions.

112. $7 + (-3)$

113. $-11 + 8$

114. $-4 + 0$

115. $-2 + (-2)$

1.3 SUBTRACTION

We remember from arithmetic that $5 - 3$ is 2. In the last section we noticed that $5 + (-3)$ is also 2. That is, subtracting 3 is the same as adding the opposite of 3. We use this relationship for our definition of subtraction.

Subtraction of Real Numbers

$$p - q = p + (-q)$$

1. Rewrite the problem as an addition problem by adding the opposite of the number being subtracted.
2. Follow the rules for addition.

EXAMPLE 1. Perform the subtractions indicated.

(a) $7 - 2$ (b) $-7 - 2$ (c) $-7 - (-2)$

Solutions:

(a) $7 - 2 = 7 + (-2)$ Add the opposite of 2.
$= 5$

(b) $-7 - 2 = -7 + (-2)$ Add the opposite of 2.
$= -9$

(c) $-7 - (-2) = -7 + 2$ Add the opposite of -2.
$= -5$

□

EXAMPLE 2. Perform the operations indicated.

(a) $[(9 - 4) - (14 - 18)]$ (b) $\frac{1}{2} - \left(\frac{1}{3} - 2\right)$

(c) $14 - |3 - 8|$ (d) $\sqrt{225 + 8^2} - 8^2$

Solutions:

(a) We begin with the innermost grouping symbol.

$$
\begin{aligned}
[(9-4)-(14-18)] &= [5-(-4)] \\
&= 5+4 \\
&= 9
\end{aligned}
$$

(b)
$$
\begin{aligned}
\frac{1}{2}-\left(\frac{1}{3}-2\right) &= \frac{1}{2}-\left(\frac{1}{3}-\frac{6}{3}\right) \\
&= \frac{1}{2}-\left(-\frac{5}{3}\right) \\
&= \frac{1}{2}+\frac{5}{3} \\
&= \frac{3}{6}+\frac{10}{6} \\
&= \frac{13}{6}
\end{aligned}
$$

(c) Absolute value bars are a grouping symbol.

$$
\begin{aligned}
14-|3-8| &= 14-|-5| \\
&= 14-5 \\
&= 9
\end{aligned}
$$

(d) The $\sqrt{\ }$ symbol is another grouping symbol.

$$
\begin{aligned}
\sqrt{225+8^2}-8^2 &= \sqrt{225+64}-64 \\
&= \sqrt{289}-64 \\
&= 17-64 \\
&= -47
\end{aligned}
$$

□

EXAMPLE 3. Answer each question.

(a) Subtract 8 from 5.

(b) Subtract -11 from 2.

(c) What is -8 decreased by the sum of 3 and -5?

(d) If today's high temperature is 7° and the temperature falls 24° for the nighttime low, what is the nighttime low?

Solutions:

(a) To subtract 8 from 5, we add the opposite of 8 to 5.

$$
\begin{aligned}
5-8 &= 5+(-8) \\
&= -3
\end{aligned}
$$

(b) To subtract -11 from 2, we add the opposite of -11 to 2.

$$
\begin{aligned}
2-(-11) &= 2+11 \\
&= 13
\end{aligned}
$$

(c) To *decrease* -8 by a quantity, we subtract the quantity from -8. So our problem is to evaluate,

$$-8 - [3 + (-5)]$$

The order of operations tells us to work inside the brackets first.

$$\begin{aligned} -8 - [3 + (-5)] &= -8 - (-2) \\ &= -8 + 2 \\ &= -6 \end{aligned}$$

(d) If today's high temperature is 7° and the temperature falls 24°, then we must subtract 24 from 7 to obtain the nighttime low.

$$\begin{aligned} 7 - 24 &= 7 + (-24) \\ &= -17 \end{aligned}$$

The nighttime low is $-17°$. □

PROBLEM SET 1.3

Warm-ups

Perform the operations indicated.
In Problems 1 through 24, See Example 1.

1. $5 - 3$
2. $5 - (-8)$
3. $-3 - 7$
4. $-13 - 5$
5. $21 - (-5)$
6. $-8 - 8$
7. $-5 - (-8)$
8. $6 - 0$
9. $0 - 1$
10. $5 - (-1)$
11. $-1 - (-1)$
12. $8 - (-4)$
13. $-15 - 10$
14. $-20 - 10$
15. $18 - (-3)$
16. $-\frac{1}{2} - \left(-\frac{1}{2}\right)$
17. $2 - \left(-\frac{1}{8}\right)$
18. $-\frac{1}{8} - \frac{1}{3}$
19. $-\frac{5}{6} - \frac{5}{6}$
20. $-\frac{2}{3} - (-1)$
21. $-\frac{3}{5} - \left(-\frac{1}{10}\right)$
22. $-1.2 - 3.4$
23. $-9.8 - (-2.7)$
24. $5.3 - 8.7$

In Problems 25 through 38, see Example 2.

25. $-7 - (-5 - 9)$
26. $(-9 - 12) - (-5 - 3)$
27. $-9 - [-8 - (-3)]$
28. $[13 - (-8)] - [-2 - (-7)]$
29. $\frac{1}{4} - \left(-\frac{1}{2} - \frac{1}{8}\right)$
30. $[1.2 - (-3.6)] - (-2.5)$
31. $2 - [(-8 - 7) - (-3 + 1)]$
32. $[-11 + (-3)] - 5$
33. $\frac{2}{5} + \left(\frac{1}{3} - \frac{3}{5}\right)$
34. $(-2.7 + 5.1) - 4.8$
35. $-5 - \sqrt{-3 - (-7)}$
36. $[-6 - 7] - |-6 + 13|$
37. $12 - |7 - 11| - 7$
38. $23 - \sqrt{11 - (3 - 8)}$

In Problems 39 through 44, answer the question in each. See Example 3.

39. Subtract -6 from 6.
40. Subtract 8 from the sum of -11 and -4.
41. What is -12 decreased by 5?
42. What is the sum of 4 and -9, decreased by 1?

43. A grocer has his freezer set at $-5°$. If he decreases the temperature by 10°, what will the temperature be?

44. The high temperature for today was 15°. If the temperature dropped 27° to reach the low temperature, find the low temperature for the day.

Practice Exercises

Perform the operations indicated.

45. $8 - 5$
46. $-7 - 1$
47. $-8 - (-3)$
48. $-6 - 2$
49. $3 - (-1)$
50. $5 - (-5)$
51. $-5 - (-5)$
52. $4 - (-6)$
53. $-3 - 1$
54. $-9 - (-7)$
55. $-3 - (-3)$
56. $-5 - (-8)$
57. $6 - (-6)$
58. $-6 - (-6)$
59. $6 - (-1)$
60. $-10 - 10$
61. $-12 - (-7)$
62. $13 - 7$
63. $15 - (-12)$
64. $-17 - 17$
65. $-17 - (-17)$
66. $18 - 32$
67. $18 - (-32)$
68. $-18 - 32$
69. $-18 - (-32)$
70. $-32 - 18$
71. $-32 - (-18)$
72. $-\frac{2}{5} - \frac{2}{3}$
73. $\frac{7}{8} - \frac{3}{4}$
74. $-\frac{1}{6} - \left(-\frac{1}{3}\right)$
75. $-\frac{4}{5} - \frac{4}{5}$
76. $3 - \frac{3}{2}$
77. $\frac{1}{2} - 1$
78. $1.2 - 2.4$
79. $-2.5 - 3.4$
80. $-3.3 - (-6.6)$
81. $7 - (6 - 12)$
82. $(-7 - 9) - 3$
83. $[(7 - 14) - (-5 - 8)]$
84. $[11 - (-6 - 6)]$
85. $[-9 - (-6)] - [13 - (-7)]$
86. $-\frac{2}{3} - \left(-\frac{5}{6} - \frac{1}{2}\right)$
87. $(-4 + |-4 - 15|) + \sqrt{16 - (-9)}$
88. $\{[8 + (-12)] - (4 - 7)\} - 8$
89. $12 - [(-3 - 20) + (-33 + 10)]$
90. $\{-11 + [-9 - (-20)]\} - \{9 + [-3 - (-6)]\}$

In Problems 91 through 95, answer each question.

91. Find the difference when -6 is subtracted from 3.
92. Subtract -5 from -2.
93. Subtract -7 from the sum of -6 and -5.
94. Last night the low temperature was 8°. Tonight's low is predicted to be 12° less than last night's low. What is the predicted low temperature for tonight?
95. The temperature was $-3°$ at 7:00 P.M. If the temperature dropped 15° by midnight, what was the temperature at midnight?

IN YOUR OWN WORDS...

96. How is subtraction of real numbers done?

Write the meaning of the following expressions.

97. $3 - 7$
98. $7 - (-3)$
99. $-7 - (-3)$
100. $16 - (4 + 10)$

1.4 MULTIPLICATION AND DIVISION

Multiplication and division with real numbers follow two rules. The rules are the same for both operations, since division is just a form of multiplication. Writing $\frac{6}{3}$ is the same as writing $6 \cdot \frac{1}{3}$. That is, a division problem can be written as a multiplication problem, as long as the divisor is not 0.

Division

$$\frac{p}{q} = p \cdot \frac{1}{q}; \qquad q \neq 0$$

If we look at all the possibilities for multiplying two real numbers, we see that both numbers can be positive, both numbers can be negative, or one number can be positive and the other number negative.

We already know how to multiply two positive numbers. For example,

$$2 \cdot 3 = 6$$

That is, a positive number times a positive number is a positive number.

To understand what $2(-3)$ is, remember that multiplication of -3 by 2 is the same as adding $-3 + (-3)$, which is -6. This gives us the idea that a positive number times a negative number is a negative number.

The product of two negative numbers is a positive number. To gain some understanding why this must be so, consider the following.

$$(-2) \cdot 3 = -6$$
$$(-2) \cdot 2 = -4$$
$$(-2) \cdot 1 = -2$$
$$(-2) \cdot 0 = 0$$
$$(-2) \cdot -1 = 2$$
$$(-2) \cdot -2 = 4$$

Notice that the answer increases by 2 each time. So, the product of two negative numbers must be positive.

Multiplication and Division of Real Numbers

1. To multiply or divide two numbers with *like signs* (both positive or both negative), multiply or divide their absolute values. The sign will be positive.
2. To multiply or divide two numbers with *unlike signs* (one positive and one negative), multiply or divide their absolute values. The sign will be negative.

Remember, we **cannot** divide by 0.

EXAMPLE 1. Perform the multiplications indicated.

(a) $(-3)(5)$ (b) $(-4)(-2)$

(c) $\frac{1}{3}\left(-\frac{3}{2}\right)$ (d) $(-2)^2$

Solutions:

(a) $(-3)(5) = -15$ (b) $(-4)(-2) = 8$

(c) $\frac{1}{3}\left(-\frac{3}{2}\right) = -\frac{1}{2}$ (d) $(-2)^2 = (-2)(-2)$
$= 4$

□

EXAMPLE 2. Perform the divisions indicated.

(a) $-6 \div 3$ (b) $\frac{15}{-5}$ (c) $10/(-2)$

Solutions:

(a) $-6 \div 3 = -2$ (b) $\frac{15}{-5} = -3$ (c) $10/(-2) = -5$ □

The three real numbers 1, -1 and 0 have special properties. If we multiply a real number by 1, we get the real number itself. If we multiply a real number by -1, we get the opposite of the real number. If we multiply a real number by 0, we get 0.

> **Multiplication With 1, -1 and 0**
>
> If p is any real number,
>
> $$1 \cdot p = p \cdot 1 = p$$
>
> $$-1 \cdot p = p(-1) = -p$$
>
> $$0 \cdot p = p \cdot 0 = 0$$

Using 0 in quotients can be troublesome. Remember, we know that $\frac{6}{2} = 3$ because $3 \cdot 2 = 6$. In the same manner, $\frac{0}{6} = 0$ because $0 \cdot 6 = 0$. What about $\frac{6}{0}$? Is $\frac{6}{0} = 0$? This would mean that $0 \cdot 0 = 6$, which is not true. So, it is not 0. If we try any other real number, we see that $\frac{6}{0} = ?$ would mean that $? \cdot 0$ must be 6. This is not possible. Thus, we cannot divide 0 into a real number.

> **Quotients With Zero**
>
> If p and q are real numbers and p is not zero,
>
> **1.** $\frac{0}{p} = 0$
>
> **2.** $\frac{q}{0}$ is undefined.

EXAMPLE 3. Perform the operations indicated.

(a) $-1 \cdot 7$ (b) $6 \cdot 0$ (c) $1 \cdot x$

(d) $-1(-4)$ (e) $-1(-x)$ (f) $\frac{-1}{0}$

Solutions:

(a) $-1 \cdot 7 = -7$

(b) $6 \cdot 0 = 0$

(c) $1 \cdot x = x$

(d) $-1(-4) = 4$ Notice 4 is the opposite of -4.

(e) $-1(-x) = x$ Notice x is the opposite of $-x$.

(f) $\frac{-1}{0}$ is undefined. □

An important property of fractions concerns signs. Notice the signs in the following computations.

1	$\frac{-6}{2} = -3$	(A negative number divided by a positive number)
2	$\frac{6}{-2} = -3$	(A positive number divided by a negative number)
3	$-\frac{6}{2} = -3$	(The opposite of a positive number)

That is, because of the sign rules for multiplication and division, we have,

$$\frac{-6}{2} = \frac{6}{-2} = -\frac{6}{2}$$

This idea holds true in general.

A Sign Property of Fractions

For p and q integers, $q \neq 0$,

$$\frac{-p}{q} = \frac{p}{-q} = -\frac{p}{q}$$

The order of operations still applies as the next example illustrates.

EXAMPLE 4. Perform the operations indicated.

(a) $-3(-2)^2$ (b) $\frac{-4(5)}{10}$ (c) $8/4(-3)$

Solutions:

(a) We must square -2 first.

$$-3(-2)^2 = -3(4)$$
$$= -12$$

(b) We must multiply in the numerator first.

$$\frac{-4(5)}{10} = \frac{-20}{10}$$
$$= -2$$

(c) We perform multiplications and divisions in order from left to right. So, first we divide 8 by 4.

$$8/4(-3) = 2(-3)$$
$$= -6$$

□

EXAMPLE 5. A 22-gallon (gal) solution is 10% molasses by volume. How many gallons of molasses does it contain?

Solution:

The solution is 10%, or $\frac{10}{100}$, molasses. Thus it contains $\frac{10}{100} \cdot 22$ gal of molasses.

$$\begin{aligned} \frac{10}{100} \cdot 22 &= \frac{1}{10} \cdot 22 \\ &= \frac{22}{10} = 2.2 \end{aligned}$$

There are 2.2 gal of molasses in the solution. □

EXAMPLE 6. Eighteen-carat gold is 75% pure gold by weight. How much pure gold is contained in an 18-carat ring that weighs 120 grams (g)?

Solution:

75% of 120 is

$$\begin{aligned} \frac{75}{100} \cdot 120 &= \frac{3}{4} \cdot 120 \\ &= \frac{3 \cdot 120}{4} \\ &= 3 \cdot 30 = 90 \end{aligned}$$

The ring contains 90 g of pure gold. □

PROBLEM SET 1.4

Warm-ups

Perform the operations indicated.
In Problems 1 through 12, find each product. See Examples 1 and 3.

1. $(-6)(-1)$
2. $3(0)$
3. $\left(\frac{1}{2}\right)\left(-\frac{1}{2}\right)$
4. $(-5)(7)$
5. $(1.1)(-10)$
6. $-1(-3)$
7. $(-5)^2$
8. $(-3)^2$
9. $(-5)(-4)$
10. $4(-7)$
11. $0(-4)$
12. $\left(-\frac{3}{4}\right)^2$

In Problems 13 through 28, find each quotient. See Examples 2 and 3.

13. $\frac{-5}{-1}$
14. $9/(-3)$
15. $8 \div 0$
16. $\frac{18}{-9}$
17. $\frac{-35}{7}$
18. $-28/(-4)$
19. $0/(-1)$
20. $12 \div (-4)$
21. $\frac{14}{-7}$
22. $-6/(-6)$
23. $\frac{15}{-1}$
24. $\frac{-63}{-7}$
25. $25 \div (-5)$
26. $\frac{3}{5} \div \left(-\frac{9}{10}\right)$
27. $5/0$
28. $-48 \div 8$

In Problems 29 through 39, see Example 4.

29. $\frac{2(-3)}{3}$

30. $2(3)^2$

31. $(-2)^2(3)$

32. $(-4)^2/(-8)$

33. $\frac{(-7)^2}{7^2}$

34. $\frac{(-4)(3)}{2^2}$

35. $(-3)\cdot 2^2$

36. $(-1)^2(-6)^2$

37. $16(-2)/8$

38. $12/(-2)(3)$

39. $(-3)^2/3(2)$

In Problems 40 and 41, answer the question. See Example 5.

40. A gold coin is 95% pure gold by weight. If the coin weighs 2 oz, how much pure gold does it contain?

41. A solution contains 20% acid by volume. How much acid is there in 485 L of the solution?

Practice Exercises

Perform the operations indicated.

42. $(-8)(-1)$

43. $7(0)$

44. $\left(\frac{2}{3}\right)\left(-\frac{2}{3}\right)$

45. $(-3)(8)$

46. $(-2.2)(10)$

47. $(-7)^2$

48. $-1(-9)$

49. $(-6)^2$

50. $1(-2)$

51. $(-8)(-4)$

52. $-1\cdot z$

53. $(-4)(6)$

54. $(-9)(-8)$

55. $-1(-d)$

56. $(-8)(8)$

57. $(-10)^2$

58. $(-6)(-3)$

59. $5(-7)$

60. $0(-8)$

61. $\left(-\frac{5}{6}\right)^2$

62. $\frac{-7}{-1}$

63. $8/(-4)$

64. $7\div 0$

65. $\frac{28}{-7}$

66. $22\div(-2)$

67. $\frac{-45}{5}$

68. $-36/(-4)$

69. $0/(-7)$

70. $\frac{42}{-7}$

71. $-4/(-4)$

72. $\frac{17}{-1}$

73. $\frac{-56}{-7}$

74. $36\div(-6)$

75. $\frac{0}{9}$

76. $4/0$

77. $-72\div 8$

78. $\frac{-2}{2}$

79. $\frac{3}{-3}$

80. $-64/(-8)$

81. $\frac{2}{0}$

82. $\frac{2(-5)}{5}$

83. $2(6)^2$

84. $(-3)^2(3)$

85. $(-8)^2/(-4)$

86. $\frac{(-9)^2}{9^2}$

87. $\frac{(-4)(5)}{2^2}$

88. $(-4)\cdot 2^2$

89. $(-1)^2(-5)^2$

90. $(-2)(-9)/(-3)$

91. $18/(-6)(4)$

92. $45/(-3)/5$

93. If Ivory soap is $99\frac{44}{100}$% pure, how much impurity is there in a ton of it? (A ton is 2000 lb.)

Challenge Problems

94. Are $12/3^2$ and $12/3 \cdot 3$ the same?

95. Write $\frac{14}{5 \cdot 11}$ using the $\div$ symbol.

IN YOUR OWN WORDS...

96. State the rules that govern the *signs* in multiplication and division.

Write in words the meaning of the following expressions.

97. $3(-5)$ **98.** $\frac{-7}{3}$ **99.** $6 \cdot \frac{10}{11} + 5$ **100.** $6\left(\frac{10}{11} + 5\right)$

1.5 EXPRESSIONS

In this section we focus on combining the operations of addition, subtraction, multiplication, and division. Such combinations are called **numerical expressions,** or just **expressions.** We simplify such expressions with the order of operations that we have already learned.

Order of Operations

If grouping symbols are present, perform operations inside them, starting with the innermost grouping symbol, in the following order.

1. Perform any exponentiations.
2. Perform all multiplications and divisions in order from left to right.
3. Perform all additions and subtractions in order from left to right.

If grouping symbols are *not* present, perform operations in the above order.

EXAMPLE 1. Simplify the following expressions.

(a) $(-4)(3) - (5)(-6)$ (b) $-6/(-2) + 2(-3)^2 - 2^2$

Solutions:

(a) We must perform the multiplications first.

$$(-4)(3) - (5)(-6) = -12 - (-30)$$

Next, we do the subtraction.

$$\begin{aligned} -12 - (-30) &= -12 + 30 \\ &= 18 \end{aligned}$$

(b) We must perform the division and multiplications before the addition.

$$\begin{aligned} -6/(-2) + 2(-3)^2 - 2^2 &= 3 + 2(9) - 4 \\ &= 3 + 18 - 4 \\ &= 17 \end{aligned}$$

□

We must begin inside grouping symbols, if they are present. Remember that fraction bars, absolute value bars, and square root symbols are grouping symbols.

EXAMPLE 2. Simplify each expression.

(a) $\dfrac{8(-3) - 6^2}{8 - 3(-1) + (-6)}$

(b) $-6 - (5 - \sqrt{25 - 16} + 1)$

(c) $|5 - 8| \div (-3) - 4(-2)$

Solutions:

(a) We must simplify the numerator and the denominator before we divide.

$$\begin{aligned}\frac{8(-3) - 6^2}{8 - 3(-1) + (-6)} &= \frac{-24 - 36}{8 - (-3) + (-6)} \\ &= \frac{-60}{8 + 3 + (-6)} \\ &= \frac{-60}{5} \\ &= -12\end{aligned}$$

(b) We start with the innermost grouping symbol.

$$\begin{aligned}-6 - (5 - \sqrt{25 - 16} + 1) &= -6 - (5 - \sqrt{9} + 1) \\ &= -6 - (5 - 3 + 1) \\ &= -6 - 3 \\ &= -9\end{aligned}$$

(c) Remember that absolute value bars are grouping symbols.

$$\begin{aligned}|5 - 8| \div (-3) - 4(-2) &= |-3| \div (-3) - 4(-2) \\ &= 3 \div (-3) - 4(-2) \\ &= -1 - (-8) \\ &= -1 + 8 \\ &= 7\end{aligned}$$

□

It is common in algebra to use letters to name numbers. We write such things as

$$2x + 3$$

to mean

2 times *some number* plus 3

The value of $2x + 3$ depends upon the value of x. This provides us with a way to write many statements with just one expression.

In algebra, as in arithmetic, we perform four operations (addition, subtraction, multiplication, and division) with numbers. The chart below indicates how we read and write these operations with the numbers x and y.

OPERATION	SYMBOL(S)	READ
Addition	$x + y$	x plus y
Subtraction	$x - y$	x minus y
Multiplication	xy	x times y
	$x \cdot y$	
	$x(y)$	
	$(x)y$	
	$(x)(y)$	
Division	$x \div y$	x divided by y
	$\frac{x}{y}$	
	x/y	
	$y\overline{)x}$	

Meaningful collections of sums, differences, products, quotients, exponents, and grouping symbols that contain letters are called **algebraic expressions** or simply **expressions.** For example,

$$6x + 2(x - 3)$$

and

$$\frac{3x}{5 + 4x}$$

are algebraic expressions.

In a sum or difference, the numbers being added or subtracted are called **terms.** In a product the numbers being multiplied are called **factors.** These are *important* words in algebra as they are used very often, and their meanings are quite different.

EXAMPLE 3. In each of the following expressions state if 2 is a *term* or a *factor.*

(a) $5x - 2$ (b) $2x$ (c) $2(x - 5)$

Solutions:

(a) The expression $5x - 2$ is a difference. It has two *terms,* $5x$ and 2. Therefore 2 is a term of this expression.

(b) The expression $2x$ is a product. It has one *term,* namely, $2x$. This term has two *factors,* 2 and x. Thus, 2 is a factor of this expression.

(c) The expression $2(x - 5)$ is a product. It has one *term* containing two *factors,* 2 and $(x - 5)$. Therefore 2 is a factor. □

The order of operations comes up in algebra when working with expressions. Consider the expression $2x - 1$. It has many different values, depending on the value of x. If x is 2,

$$\begin{aligned} 2x - 1 &= 2(2) - 1 \\ &= 4 - 1 \\ &= 3 \end{aligned}$$

So, $2x - 1$ has a value of 3 when x is 2.
If x is 3,

$$\begin{aligned} 2x - 1 &= 2(3) - 1 \\ &= 6 - 1 \\ &= 5 \end{aligned}$$

Thus, $2x - 1$ has a value of 5 when x is 3.

When we find the value of an expression, we say that we **evaluate** the expression.

EXAMPLE 4. Evaluate the following expressions when x is -2.

(a) $10 + x$ (b) $10 - x$ (c) $\dfrac{10 - 3x}{x}$

(d) $5 + 3x + x^2$ (e) $x\sqrt{7 - x}$ (f) $|3 + 7x| + 21$

Solutions:

(a) It is convenient to think of the expression as

$$10 + x = 10 + (\ )$$

and then fill the parentheses with the number.

$$\begin{aligned} 10 + x &= 10 + (-2) \\ &= 8 \end{aligned}$$

(b) $10 - x$

$$\begin{aligned} 10 - x &= 10 - (\ ) \\ &= 10 - (-2) \\ &= 10 + 2 \\ &= 12 \end{aligned}$$

(c) When replacing a variable in an expression with a number, we must be sure to replace the variable everywhere it occurs.

$$\begin{aligned} \frac{10 - 3x}{x} &= \frac{10 - 3(\ )}{(\ )} \\ &= \frac{10 - 3(-2)}{(-2)} \\ &= \frac{10 + 6}{-2} \end{aligned}$$

(continued)

$$= \frac{16}{-2}$$

$$= -8$$

(d) $5 + 3x + x^2 = 5 + 3(\ ) + (\ )^2$

$$= 5 + 3(-2) + (-2)^2$$

The order of operations tells us we must evaluate the exponent first.

$$5 + 3x + x^2 = 5 + 3(-2) + 4$$

Then we do multiplication.

$$5 + 3x + x^2 = 5 + (-6) + 4$$

$$= 3$$

(e) Square root symbols are grouping symbols. We must perform operations inside them first.

$$x\sqrt{7 - x} = (-2)\sqrt{7 - (-2)}$$

$$= (-2)\sqrt{7 + 2}$$

$$= (-2)\sqrt{9}$$

$$= (-2)3$$

$$= -6$$

(f) Absolute value bars are also grouping symbols.

$$|3 + 7x| + 21 = |3 + 7(-2)| + 21$$

$$= |3 - 14| + 21$$

$$= |-11| + 21$$

$$= 11 + 21$$

$$= 32$$

□

CALCULATOR BOX

Computations with a Calculator

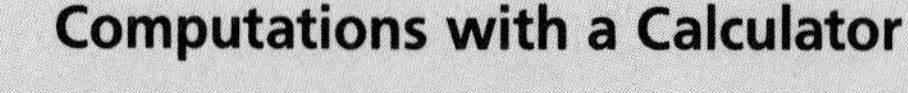

Most scientific calculators have the order of operations built in. That is, they will calculate $3 \cdot 5 - 2$ correctly by pressing the keys

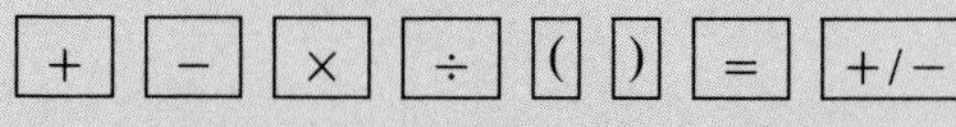

. If the calculator displays the correct result, [13], it has the order of operations built in. However, the grouping keys, [(] and [)] are needed to preserve the order of operations in a calculation like $\frac{9 - 5}{3 + 1}$. Press [(] [9] [−] [5] [)] [÷] [(] [3] [+] [1] [)] [=] and read [1] on the display.

We must be careful when entering a negative number on a calculator. The $\boxed{-}$ key is for *subtraction*. Look for a key marked $\boxed{+/-}$ or $\boxed{\text{CHS}}$.

To calculate $-6 \cdot 8$, press $\boxed{6}$ $\boxed{+/-}$ $\boxed{\times}$ $\boxed{8}$ $\boxed{=}$ and read **−48** on the display.

Calculator Exercises

Perform the following computations on a calculator and check the result with pencil and paper.

1. $71 - (8 - 3)$

2. $\dfrac{45 - 9}{3 \cdot 4}$

3. $5(18 - 11) - 2(234 - 222)$

4. $1 \div (3 + 9) \cdot 24$

5. $-11(41 - 50)$

6. $\dfrac{-7 - 4}{-2(-3 - 8)}$

Answers:

1. 66 **2.** 3 **3.** 11 **4.** 2 **5.** 99 **6.** −0.5

PROBLEM SET 1.5

Warm-ups

In Problems 1 through 8, simplify each expression. See Example 1.

1. $3(-2) + (-3)(-1)$

2. $\dfrac{-18}{-2} + \dfrac{6}{-3}$

3. $2^2 - 6/(6 - 3)$

4. $-7 + 3(-3)^2 + 5$

5. $|-3| + (-6)/(-2) - 3$

6. $2(-3)^2 - \dfrac{-60}{15} + (-8)$

7. $-\dfrac{8}{9} \div \dfrac{2}{3} + \left(-\dfrac{5}{6}\right)(-5)(-2)$

8. $(-3)(-2)(-1) - (-8) \div (-8)$

In Problems 9 through 19, simplify each expression. See Example 2.

9. $\dfrac{2(-3) + 4(-2)}{5 \cdot 2 - 3}$

10. $-5 - 2[8 - 2(4 - 6) - 7]$

11. $2|7 - 5| \div (-2 - 2)$

12. $\dfrac{3 - |0 - 3|}{4^2}$

13. $8 - 6 - (6 - 4)$

14. $\dfrac{3(-3) - 3^2 + (-12)}{(-3)^2 - 2(-3)}$

15. $\dfrac{8 - (-5) - (7 - 9)}{3(2^2 - 3^2)}$

16. $\dfrac{\sqrt{64 + 36}}{4}$

17. $14 - [2(4 - 7) - 3(8) - (5 - 11)]$

18. $10 - \sqrt{|16 - 25| + 16}$

19. $5 + 3^2 \div \sqrt{4 + 5}$

In Problems 20 through 25, state whether 7 is a term *or a* factor. *See Example 3.*

20. $7 + 3x - 5x^3$ **21.** $8x - 7$ **22.** $(1 - 3s)7$

23. $x^2 + 13x - 7$ **24.** $\frac{ax}{by} \cdot 7$ **25.** $7\sqrt{x + 3}$

In Problems 26 through 31, evaluate each expression when x has the value 3. See Example 4.

26. $15x$ **27.** $15 + x$ **28.** $\frac{15}{x}$

29. $\frac{5 + x}{5 - x}$ **30.** $\sqrt{82 + 6x}$ **31.** $x - |3x - 15|$

In Problems 32 through 37, evaluate each expression when x has the value of −3. See Example 4.

32. $15x$ **33.** $15 + x$ **34.** $\frac{15}{x}$

35. $\frac{5 + x}{5 - x}$ **36.** $\sqrt{82 + 6x}$ **37.** $x - |3x - 15|$

Practice Exercises

In Problems 38 through 59, simplify each expression.

38. $3(-5) - (-3)(-1)$

39. $\frac{-28}{4} + \frac{-9}{-3}$

40. $3^2 - 8/(6 - 3)$

41. $-8 + 2(-2)^2 + 7$

42. $|-7| + (-14)/(-2) - 14$

43. $3(-3)^2 - \frac{-30}{-6} + (-8)$

44. $3(-2) - (-8)(-4) - (-3)(-2)$

45. $3|-5| \div (-5) + (-3)^2$

46. $-\frac{6}{7} \div \frac{2}{7} + \left(-\frac{7}{9}\right)(-5)(-9)$

47. $(-4)(-3)(-1) - [(-6) \div (-6)]$

48. $\frac{2[(-3) + 4(2)]}{5(2) - 5}$

49. $-1 - 2[1 - 2(4 - 7) - 8]$

50. $3|5 - 7| \div (-3 - 3)$

51. $\frac{3 + |0 - 3|}{4^2 - 4(5) + 1}$

52. $9 - 4 - (4 - 2)$

53. $\frac{5(-6) + 2 \cdot 3^2 + (-12)}{(-3)^2 - 2 - 3}$

54. $\frac{3}{4}\left(-\frac{4}{3}\right) - \frac{1}{5}(-5) + \frac{5}{7}\left(-\frac{7}{15}\right)$

55. $|6 - 9| - 2(-3) \div (-3)$

56. $\frac{9 - (-7) - (4 - 9)}{3(4^2 - 3^2)}$

57. $4 - [2(4 - 8) - 3(4) - (3 - 7)]$

58. $\sqrt{169 - 25} + 11^2$

59. $\frac{2 + |2 - 8|}{2\sqrt{5^2 - 3^2}}$

In Problems 60 through 65, state whether x is a term *or a* factor.

60. $7 + x$ **61.** $x - 7y$ **62.** $x(1 - 3s)$

63. $x - 7$ **64.** $7ax$ **65.** $7\sqrt{y + 3} - x$

In Problems 66 through 74, evaluate each expression when x has the value 2 and again when x has the value −2.

66. $21x$ 42; **67.** $21 + x$ 23; **68.** $\frac{144}{x}$ 72;

69. $5x^2$ 20; **70.** $(5x)^2$ 100; **71.** $-x^2$ −4;

72. $\dfrac{1 - x}{1 + x}$

73. $6x|x - 6|$

74. $\sqrt{4x + 17} - 5$

Challenge Problems

Problems 75 through 78 each contain two expressions. Decide whether they are equal or not; then test by replacing x with -6.

75. $|x - 3|$ and $|x| - |3|$

76. $|x - 3|$ and $|3 - x|$

77. $|2x|$ and $|2||x|$

78. $\left|\dfrac{x}{3}\right|$ and $\dfrac{|x|}{|3|}$

IN YOUR OWN WORDS...

79. What do the expressions $20 \div 4 + x$ and $20 \div (4 + x)$ mean?

1.6 PROPERTIES OF REAL NUMBERS

In our work with real numbers in this chapter, we have not discussed their properties. Many of them are so natural to us that we completely overlook them. For example, we know that

$$2 + 3 = 3 + 2$$

and that

$$2 \cdot 3 = 3 \cdot 2$$

This illustrates what we call the **Commutative Properties.** If we add or multiply two real numbers, the *order* in which we do the operations does not matter.

> **Commutative Properties**
>
> If p and q are real numbers,
>
> Addition: $p + q = q + p$
>
> Multiplication: $pq = qp$

We also know that

$$(2 + 3) + 4 = 2 + (3 + 4)$$

and that

$$(2 \cdot 3) \cdot 4 = 2 \cdot (3 \cdot 4)$$

This illustrates the **Associative Properties.** If we add or multiply three numbers, we may group the numbers in different ways to add or multiply.

> **Associative Properties**
>
> If p, q, and r are real numbers,
>
> Addition: $(p + q) + r = p + (q + r)$
>
> Multiplication: $(pq)r = p(qr)$

EXAMPLE 1. Name the property used in each of the following.

(a) $3(4) = 4(3)$ (b) $(x + 3) + 4 = x + (3 + 4)$

(c) $x(3 + y) = x(y + 3)$ (d) $5 \cdot (y + z) = (y + z) \cdot 5$

Solutions:

(a) $3(4) = 4(3)$

This is the Commutative Property of Multiplication.

(b) $(x + 3) + 4 = x + (3 + 4)$

This is the Associative Property of Addition.

(c) $x(3 + y) = x(y + 3)$

Notice that the order of 3 and y is all that has changed. This is the Commutative Property of Addition.

(d) $5 \cdot (y + z) = (y + z) \cdot 5$

All that is changed is the order in which the two numbers 5 and $(y + z)$ are written. This is the Commutative Property of Multiplication. □

If we are given more than two numbers to add with no symbols of grouping, the commutative and associative properties allow us to choose which two numbers to add first. Suppose we want to add $2 + 3 + 4$. We may choose which two numbers to add first.

$$2 + 3 + 4 = (2 + 3) + 4 \qquad \text{Associative Property}$$
$$= \quad 5 \quad + 4 = 9$$

or

$$2 + 3 + 4 = 2 + (3 + 4) \qquad \text{Associative Property}$$
$$= 2 + \quad 7 \quad = 9$$

or

$$2 + 3 + 4 = 2 + 4 + 3 \qquad \text{Commutative Property}$$
$$= (2 + 4) + 3 \qquad \text{Associative Property}$$
$$= \quad 6 \quad + 3 = 9$$

The same idea applies to multiplication.

The Commutative and Associative Properties can be useful for working problems efficiently. Notice how they work in the next example.

EXAMPLE 2. Perform the operations indicated.

(a) $-1345 + 23{,}456 + 1345$ (b) $\frac{1}{8} \cdot 15 \cdot 16$

Solutions:

(a) First we apply the Commutative Property to the last two terms.

$$-1345 + 23{,}456 + 1345 = -1345 + 1345 + 23{,}456$$

Now we use the Associative Property.

$$-1345 + 1345 + 23{,}456 = (-1345 + 1345) + 23{,}456$$
$$= 0 + 23{,}456$$
$$= 23{,}456$$

(b) $\frac{1}{8} \cdot 15 \cdot 16 = \frac{1}{8} \cdot 16 \cdot 15$ Commutative Property

$= \left(\frac{1}{8} \cdot 16\right) \cdot 15$ Associative Property

$= 2 \cdot 15$

$= 30$ □

The Distributive Property combines addition and multiplication. For example,

$$2(3 + 4) = 2 \cdot 7$$
$$= 14$$

Notice also that

$$2 \cdot 3 + 2 \cdot 4 = 6 + 8$$
$$= 14$$

So,

$$2(3 + 4) = 2 \cdot 3 + 2 \cdot 4$$

Distributive Properties

If p, q, and r are real numbers,

$$p(q + r) = p \cdot q + p \cdot r$$
$$p(q - r) = p \cdot q - p \cdot r$$

Example 3 shows how to use the Distributive Property.

EXAMPLE 3. Multiply each of the following using the Distributive Property.

(a) $3(5 + 8)$ (b) $2(x - 3)$ (c) $3\left(\frac{2}{3} + y\right)$

Solutions:

(a) $3(5 + 8) = 3 \cdot 5 + 3 \cdot 8$

$= 15 + 24$

$= 39$

(b) $2(x - 3) = 2 \cdot x - 2 \cdot 3$

$= 2x - 6$

(c) $3\left(\frac{2}{3} + y\right) = 3 \cdot \frac{2}{3} + 3 \cdot y$

$= 2 + 3y$ □

There are two real numbers, 0 and 1, that have special properties. Since the sum of 0 and any real number is the real number itself, we call 0 the identity for addition. Likewise, since 1 times any real number is the real number itself, we call 1 the multiplicative identity.

Identities

If p is a real number, then

1. $p + 0 = 0 + p = p$ Identity for addition
2. $p \cdot 1 = 1 \cdot p = p$ Identity for multiplication

We say that 0 is the **additive identity** and that 1 is the **multiplicative identity.**

If we add opposites, we get the additive identity.

$$3 + (-3) = 0$$

The product of 2 and $\frac{1}{2}$ is the multiplicative identity.

$$2 \cdot \frac{1}{2} = 1$$

Inverses

1. ***Additive Inverse (opposite)***
   If p is a real number, then there is a real number $-p$ such that
   $$p + (-p) = (-p) + p = 0.$$
   $-p$ is called the **additive inverse,** or **opposite,** of p. (We read $-p$ as "negative p" or "the opposite of p.")
2. ***Multiplicative Inverse (reciprocal)***
   If p is a nonzero real number, then there is a real number, $\frac{1}{p}$, such that
   $$p \cdot \frac{1}{p} = \frac{1}{p} \cdot p = 1.$$
   $\frac{1}{p}$ is called the **multiplicative inverse,** or **reciprocal,** of p.

EXAMPLE 4. Give the additive inverse (opposite) and the multiplicative inverse (reciprocal) for each.

(a) -10 (b) $-\frac{3}{4}$ (c) $\frac{2}{x}$

Solutions:

(a) The additive inverse is 10 because $-10 + 10 = 0$.

The multiplicative inverse is $-\frac{1}{10}$ because

$$(-10)\left(-\frac{1}{10}\right) = 1$$

(b) The additive inverse is $\frac{3}{4}$ because $-\frac{3}{4}+\frac{3}{4}=0$.

The multiplicative inverse is $-\frac{4}{3}$ because

$$\left(-\frac{3}{4}\right)\left(-\frac{4}{3}\right)=1$$

(c) The additive inverse is $-\frac{2}{x}$ because

$$\frac{2}{x}+\left(-\frac{2}{x}\right)=0$$

The multiplicative inverse is $\frac{x}{2}$ because

$$\frac{2}{x}\cdot\frac{x}{2}=1$$

□

EXAMPLE 5. Name the property that is illustrated in each.

(a) $4+0=4$ (b) $5+(-5)=0$ (c) $1\cdot x=x$ (d) $\frac{7}{x}\cdot\frac{x}{7}=1$

Solutions:

(a) $4+0=4$	Additive identity
(b) $5+(-5)=0$	Additive inverse
(c) $1\cdot x=x$	Multiplicative identity
(d) $\frac{7}{x}\cdot\frac{x}{7}=1$	Multiplicative inverse

□

PROBLEM SET 1.6

Warm-ups

In Problems 1 through 15, name the property illustrated by each statement. See Examples 1 and 5.

1. $4\cdot 7=7\cdot 4$

2. $(3+7)+4=(7+3)+4$

3. $(x+7)+4=x+(7+4)$

4. $8+n=n+8$

5. $2(5+y)=2\cdot 5+2y$

6. $x+0=x$

7. $-\frac{1}{3}(-3)=1$

8. $-\sqrt{6}+\sqrt{6}=0$

9. $-x+y=y+(-x)$

10. $\frac{1}{t}\cdot 1=\frac{1}{t}$

11. $1\cdot 14=14$

12. $9(7-\sqrt{3})=9\cdot 7-9\cdot\sqrt{3}$

13. $-\frac{1}{2}+\frac{1}{2}=0$

14. $(1+x)+z=1+(x+z)$

15. $\frac{2}{3}\cdot\frac{3}{2}=1$

16. Match expressions that have the same value and name the property that justifies the match.

1. $2 \cdot 5 - 2 \cdot 8$	**a.** $3 + 7$
2. $7 + 3$	**b.** 9
3. $(4 + 5) + 6$	**c.** -5
4. $-5 + 0$	**d.** $2(5 - 8)$
5. $1 \cdot 9$	**e.** $(5 + 4) + 6$

In Problems 17 through 20, simplify using the Associative Property.

17. $5\left(\frac{1}{5}x\right)$ **18.** $\frac{1}{7}(7x)$ **19.** $-3\left(-\frac{1}{3}y\right)$ **20.** $\frac{3}{4}\left(\frac{4}{3}z\right)$

In Problems 21 through 24, use the Commutative and Associative Properties to perform the operations indicated efficiently. See Example 2.

21. $367 + 874 + (-367)$ **22.** $-12 + 48 + 8 + 4$

23. $-6(4)(1/6)$ **24.** $\left(\frac{1}{3}\right)\left(\frac{2}{5}\right)(6)(10)$

In Problems 25 through 29, give the additive inverse (opposite) and the multiplicative inverse (reciprocal) of each number. Assume x is not zero. See Example 4.

25. x **26.** $-\frac{2}{7}$ 27. π **28.** $-5x$ 29. 1

In Problems 30 through 35, use the Distributive Property to multiply each. See Example 3.

30. $2(1 + x)$ **31.** $3(x - 4)$ **32.** $-2(x + 9)$

33. $-4(x - 2)$ **34.** $x(y + z)$ **35.** $x(y - z)$

Practice Exercises

In Problems 36 through 49, name the property illustrated by each statement.

36. $2 \cdot 3 \cdot 4 = 3 \cdot 2 \cdot 4$ 37. $(3 + 7) + 2 = 3 + (7 + 2)$

38. $(g + 7) + 2 = (7 + g) + 2$ **39.** $5(6x) = (5 \cdot 6)x$

40. $9(r + 2) = 9r + 9 \cdot 2$ 41. $y \cdot 1 = y$

42. $-\frac{1}{3} + \frac{1}{3} = 0$ **43.** $x \cdot \frac{1}{x} = 1$

44. $-p + p = 0$ **45.** $\frac{1}{j} + 0 = \frac{1}{j}$

46. $4q + 4 \cdot 7 = 4(q + 7)$ **47.** $0 + \sqrt{7u} = \sqrt{7u}$

48. $\frac{x}{y} \cdot 1 = \frac{x}{y}$ **49.** $\frac{1}{2} \cdot 2 = 1$

In Problems 50 through 57, simplify each expression.

50. $9\left(\frac{1}{3}x\right)$ **51.** $\frac{1}{12}(4x)$ **52.** $367 + 874 + (-367)$

53. $-52 + 148 + 52$ **54.** $-14(4)\left(\frac{1}{7}\right)$ **55.** $\left(\frac{1}{3}\right)\left(\frac{2}{7}\right)(9)(14)$

56. $-5\left(-\frac{1}{15}y\right)$

57. $\frac{2}{5}\left(\frac{15}{8}z\right)$

In Problems 58 through 62, give the additive inverse (opposite) and the multiplicative inverse (reciprocal) of each number.

58. 14

59. $\frac{5}{3}$

60. $-\pi$

61. 0

62. -1

In Problems 63 through 76, use the distributive property to multiply.

63. $2(4 + x)$

64. $3(x - 8)$

65. $-2(6 + y)$

66. $-4(y - 1)$

67. $x(y + 3)$

68. $x(3 + z)$

69. $x(y - 5)$

70. $x(4 - y)$

71. $t(3 - x)$

72. $x(y + 2z)$

73. $n(3 + 2x)$

74. $k(x - 9)$

75. $c(-9 + x)$

76. $-5(t + x)$

Challenge Problems

77. Is subtraction commutative? (Is $p - q$ the same as $q - p$?)

78. Is division commutative? (Is $p \div q$ the same as $q \div p$?)

In Problems 79 through 82, use the properties to carry out the multiplications indicated.

79. $x(x + 3)$

80. $5(x + y + z)$

81. $2x \cdot 3x^2$

82. $2x(3x^2 + z)$

IN YOUR OWN WORDS...

State the following properties.

83. Commutative Properties

84. Associative Properties

85. Distributive Property

1.7 THE LANGUAGE OF ALGEBRA

We have noticed that in algebra letters are used to name numbers. We call such letters constants or variables.

A **constant** is a number whose value does not change. For example, 2 and π are constants. They always have the same fixed value. When we use letters to name constants, they have one fixed value.

A **variable** is a letter that may have many different values. For example, the circumference of a circle is given by the product

$$2\pi r$$

The letter r stands for the radius of the circle and can be *any* positive number. It is a variable. The numeral 2 and the Greek letter π stand for fixed numbers and so they are constants. (The constant π is an irrational number whose approximate value is 3.1416.)

Most problems from the fields of engineering and the sciences are solved by the use of mathematics. Real problems are usually modeled with a mathematical statement; then the laws and properties of mathematics are used to find a solution to the problem. An important step in this direction is translating from the English language to the language of mathematics.

In this section we will examine the problem of translating verbal expressions into algebraic expressions. We must learn to recognize certain English language expressions that are commonly used to indicate mathematical operations.

A list of some common phrases that indicate the various operations is given below.

PHRASE	EXAMPLE	ALGEBRAIC EXPRESSION
Addition:		
Added to	6 *added to* 9	$9 + 6$
The sum of	*The sum of* 5 and x	$5 + x$
More than	9 *more than* t	$t + 9$
Greater than	15 *greater than* y	$y + 15$
Increased by	31 *increased by* x	$31 + x$
The total of	*The total of* 3 and t	$3 + t$
Subtraction:		
Subtracted from	11 *subtracted from* 100	$100 - 11$
Minus	x *minus* 22	$x - 22$
Less than	y *less than* 10	$10 - y$
Fewer than	17 *fewer than* t	$t - 17$
Decreased by	7 *decreased by* x	$7 - x$
The difference	*The difference* when k is subtracted from 5	$5 - k$
Multiplication:		
Multiplied by	3 *multiplied by* y	$3y$
The product of	*The product of* x and y	xy
Times	4 *times* 7	$4 \cdot 7$
Twice	*Twice* x	$2x$
Of	$\frac{1}{2}$ *of* x	$\frac{1}{2}x$
Division:		
Divided by	16 *divided by* 23	$\frac{16}{23}$
Quotient of	*The quotient of* 9 and 5	$\frac{9}{5}$
The ratio of	*The ratio of* 7 and 11	$\frac{7}{11}$

EXAMPLE 1. Translate each phrase into an algebraic expression.

(a) The sum of $3x$ and 16

(b) 4 less than twice y

(c) The product of 17 and x, divided by the sum of 17 and x

(d) 17 decreased by the quotient when x is divided by the difference when 5 is subtracted from x

Solutions:

(a) The word *sum* implies addition:

$$3x + 16$$

(b) Twice y is $2y$.

$$2y - 4$$

(c) $\dfrac{17x}{17 + x}$

(d) $17 - \dfrac{x}{x - 5}$ ☐

It is often necessary to develop an expression in mathematics to represent, or model, a real-life situation.

EXAMPLE 2. Answer each question.

(a) If 5 ft are cut off the end of a board that is 16 ft long, how much is left?

(b) If x feet are cut off the end of a board that is 16 ft long, how much is left?

Solutions:

(a) To find out how much of the board remains, we subtract 5 from 16.

$$16 - 5 = 11$$

There are 11 ft of board left.

(b) We do this exactly like we did part (a). We subtract x from 16.

There are $16 - x$ feet of board left. ☐

EXAMPLE 3. Answer each question.

(a) If Harold has 20 nickels and dimes in his pocket and 7 of them are dimes, how many nickels does Harold have in his pocket?

(b) If Harold has 20 nickels and dimes in his pocket and x of them are dimes, how many nickels does Harold have in his pocket?

Solutions:

(a) If we take away the number of dimes from the total number of coins, we have the number of nickels.

$$20 - 7 = 13$$

He has 13 nickels in his pocket.

(b) In the same manner, we subtract the number of dimes, x, from the total.

He has $20 - x$ nickels in his pocket. ☐

EXAMPLE 4. Answer each question.

(a) If Harold has 7 dimes in his pocket, how much are they worth?

(b) If Harold has x dimes in his pocket, how much are they worth?

(c) If Harold has $20 - x$ nickels in his pocket, how much are they worth?

(d) Harold has x dimes and $20 - x$ nickels in his pocket. What is the value of the money Harold has in his pocket?

Solutions:

(a) Dimes are worth 10¢ each, and Harold has 7. Their value is $10 \cdot 7$ or 70 cents.

(b) By the same reasoning, x dimes are worth 10 times x cents. The x dimes are worth $10x$ cents.

(c) As nickels are worth 5¢ each, the nickels in his pocket are worth $5(20 - x)$ cents.

(d) The total value of the collection of coins is the value of the dimes plus the value of the nickels.

Harold has $10x + 5(20 - x)$ cents in his pocket. □

EXAMPLE 5. Answer each question.

(a) If Sadie is 34 years (yr) old now and she is three years older than her sister, how old will each of them be in 4 years?

(b) If Sadie is x years old now and she is 3 yr older than her sister, how old will each of them be in 4 years?

(c) If Sadie is x years old now and she is 3 yr older than her sister, what will be the sum of their ages 4 years from now?

Solutions:

(a) Since Sadie is 3 yr older than her sister, her sister must be $34 - 3 = 31$ yr old *now.* Thus in 4 yr, Sadie will be $34 + 4$ yr old and her sister will be $31 + 4$ yr old.

	NOW	IN 4 YEARS
Sadie's age	34	$34 + 4$
Sister's age	$34 - 3$	$31 + 4$

In 4 yr Sadie will be 38 yr old and her sister will be 35 yr old.

(b) We compute exactly as we did in part (a). Sadie is x years old now and her sister is $x - 3$. After they each age 4 years, Sadie will be $x + 4$ and her sister will be $(x - 3) + 4$ years old.

	NOW	IN 4 YEARS
Sadie's age	x	$x + 4$
Sister's age	$x - 3$	$(x - 3) + 4$

In 4 yr, Sadie will be $x + 4$ years old and her sister will be $(x - 3) + 4$ years old.

(c) From part (b) we see that the sum of their ages will be $(x + 4) + [(x - 3) + 4)]$ years. □

The properties of real numbers allow us to simplify certain sums and differences, such as the sum that occurs in the answer to Example 5(c).

The expression, $3x + 5x$ is a sum. In it, $3x$ and $5x$ are *terms.* We call them x-terms and we also say they are *like terms.* In the same manner, in the difference,

$$5x^2 - 7x^2$$

we say the x^2-terms, $5x^2$ and $7x^2$ are *like terms.* In the term $3x$, 3 is called the **coefficient** of x, or the numerical coefficient. In $5x^2$, the numerical coefficient is 5.

The sum, $6x + 5y$, contains an x-term and a y-term. They are *not* like terms. Likewise, $9x^2 - 7x^3$ contains two terms that are *not* alike.

If terms are the same everywhere except for the numerical coefficient, we call them **like terms.**

EXAMPLE 6. Determine if each given pair of expressions is a pair of like terms and give the coefficient for each term.

(a) $7y$ and $18y$ (b) x and $-x$ (c) $2xy^2$ and $3x^2y$

(d) $3x^3y$ and $-5x^3y$ (e) 11 and -6 (f) $15\sqrt{x}$ and $\frac{2}{3}\sqrt{x}$

Solutions:

(a) $7y$ and $18y$ are both y-terms. They *are* like terms. The coefficients are 7 and 18.

(b) x and $-x$ are both x-terms. They *are* like terms. The coefficient in the term x is 1, as x can be written as $1 \cdot x$. We call 1 an *understood* coefficient. The coefficient in the term $-x$ is -1. It is also an understood coefficient.

(c) $2xy^2$ and $3x^2y$ are *not* like terms. The coefficients are 2 and 3.

(d) $3x^3y$ and $-5x^3y$ are both x^3y-terms. They *are* like terms. The coefficients are 3 and -5.

(e) 11 and -6 *are* like terms because they are both constants. The coefficients are 11 and -6.

(f) $15\sqrt{x}$ and $\frac{2}{3}\sqrt{x}$ are both $\sqrt{x}$-terms. They *are* like terms. The coefficients are 15 and $\frac{2}{3}$. □

The Commutative and Distributive Properties allow us to simplify sums and differences which contain like terms.

$3x + 5x$	$= x \cdot 3 + x \cdot 5$	Commutative
	$= x(3 + 5)$	Distributive
	$= x \cdot 8$	
	$= 8x$	Commutative

So, $3x + 5x = 8x$. *Notice that we just added the coefficients.*

$$\begin{aligned} 5x^2 - 7x^2 &= x^2 \cdot 5 - x^2 \cdot 7 && \text{Commutative} \\ &= x^2(5 - 7) && \text{Distributive} \\ &= x^2(-2) && \\ &= -2x^2 && \text{Commutative} \end{aligned}$$

Thus, $5x^2 - 7x^2 = -2x^2$. *We just combined the coefficients.*

We call such operations **combining like terms.**

EXAMPLE 7. In each of the following, combine like terms if possible.

(a) $8x^4 + 9x^4$ (b) $6x^2y - x^2y$

(c) $3kx + 4kx - 2kx$ (d) $14x^3y - 11x^3y + 2x^3z$

Solutions:

(a) $8x^4 + 9x^4 = 17x^4$

(b) $6x^2y - x^2y = 5x^2y$

(c) $3kx + 4kx - 2kx = 5kx$

(d) $14x^3y - 11x^3y + 2x^3z = 3x^3y + 2x^3z$

Note that the last term is *not* like the other two. □

EXAMPLE 8. Simplify each expression.

(a) $6x + 1 + 5x - 3$ (b) $4x^2 - 3x + x^2 - 2x$

Solutions:

(a) First we rearrange the terms using the commutative property.

$$6x + 1 + 5x - 3 = 6x + 5x + 1 - 3$$

Then we combine like terms.

$$6x + 5x + 1 + 3 = 11x - 2$$

(b) $$\begin{aligned} 4x^2 - 3x + x^2 - 2x &= 4x^2 + x^2 - 3x - 2x \\ &= 5x^2 - 5x \end{aligned}$$ □

EXAMPLE 9. Answer each of the following questions.

(a) 26 is an even integer. What are the next two *consecutive even* integers?

(b) If x is an even integer. What are the next two *consecutive even* integers?

(c) Suppose x is the smallest of three consecutive even integers. What is their sum?

Solutions:

(a) 28 and 30 are the next two consecutive even integers following 26. Notice that 28 is $26 + 2$ and 30 is $26 + 4$.

The integers are 28 and 30.

(b) Using the same reasoning that we did in part (a), we get the next even integer after x by adding 2 to x and the next after that by adding 4.

The integers are $x + 2$ and $x + 4$.

(c) If x is the smallest of three consecutive even integers, then as we saw in part (b), they must be x, $x + 2$ and $x + 4$, and their sum is

$$x + (x + 2) + (x + 4)$$

However, we can simplify this expression by combining like terms.

$$\begin{aligned} x + (x + 2) + (x + 4) &= x + x + 2 + x + 4 \\ &= x + x + x + 2 + 4 \\ &= 3x + 6 \end{aligned}$$

The sum of the integers is $3x + 6$. □

PROBLEM SET 1.7

Warm-ups

In Problems 1 through 9, translate the given phrases into algebraic expressions. See Example 1.

1. The sum of 41 and $6y^2$.

2. 15 fewer than $2x$.

3. x times $8y$.

4. The ratio of y and twice x.

5. 10 divided by x squared.

6. $3t$ increased by 19.

7. Twice x decreased by twice y.

8. One-fifth of the difference when 5 is subtracted from x.

9. The quotient when the sum of x and 4 is divided by the difference when 4 is subtracted from x.

In Problems 10 through 19, answer each question with an English language phrase. In Problems 10 and 11, see Example 2.

10. A 4-in. piece is cut from the end of a 31-in. length of string. How much is left?

11. A 4-in. piece is cut from the end of a length of string that is x inches long. How much is left?

In Problems 12 and 13, see Example 3.

12. There are 17 red marbles in a package of 56 red and green marbles. How many green marbles are there in the package?

13. There are 17 red marbles in a package of x red and green marbles. How many green marbles are there in the package?

In Problems 14 through 16, see Example 4.

14. Perry's piggy bank contains 81 nickels and quarters. If it has 70 nickels, how many quarters does it have?

15. If Perry's piggy bank contains x nickels and quarters and it has 70 nickels, how many quarters does it have?

16. If Perry's piggy bank contains x nickels and quarters and it has 70 nickels, what is the *value* of the nickels in the bank? What is the value of the quarters? What is the value of the collection?

In Problems 17 through 19, see Example 5.

17. Don is 4 years younger than Barbara. If Barbara is 55, how old is Don?

18. Don is 4 years younger than Barbara. If Barbara is x years old, how old is Don?

19. Don is 4 years younger than Barbara. If Barbara is x years old now, how old will Don be in 10 years?

In Problems 20 through 39, simplify the given expressions. See Examples 7 and 8.

20. $3x + 7x$

21. $8x^2 - x^2$

22. $7x - 4x + x$

23. $4abc - 2abc - 5abc$

24. $3x + 2 + 5x + 13$

25. $4xy - 7xy - 5 + 8$

26. $4x^2 + x - 2 - 2x^2 - x - 1$

27. $9x^2y + 4x^2y - 5xy^2 - xy^2$

28. $3x^3 - x^2 + x^3 - 2x^2$

29. $5x^2z + 8z^2 + 31z^2 - 6x^2z$

30. $2x^2 - 3x + 2x - 3 + 2x^2$

31. $4\sqrt{x} - 2\sqrt{x}$

32. $2(x - 7) + 3x$

33. $14x + 5(3x - 7)$

34. $2(x + 3) + 3(x + 4)$

35. $-3(x - 1) + 2(x - 5)$

36. $3(10 - 7y) + 2(15 - 2y)$

37. $3(3b - 1) + 3(5b - 2)$

38. $4(3 - 2a) + 2(3a + 2)$

39. $3\sqrt{x - 1} + 7\sqrt{x - 1}$

In Problems 40 through 45, answer each question with an English language phrase. See Example 9.

40. 71 is an odd integer. What are the next two consecutive odd integers?

41. What is the sum of three consecutive odd integers if 71 is the smallest of the three?

42. Suppose x is an odd integer. What are the next two consecutive odd integers?

43. What is the sum of three consecutive odd integers if x is the smallest of the three?

44. What is the sum of three consecutive even integers if 92 is the largest of the three?

45. What is the sum of three consecutive even integers if x is the largest of the three?

Practice Exercises

In Problems 46 through 52, translate the given phrases into algebraic expressions.

46. 14 added to -11.

47. The total of $6x$ and 23.

48. 10 less than y.

49. One-third of twice t.

50. The ratio of 17 and the difference when 5 is subtracted from x.

51. The product of x cubed and -4.

52. Twice x increased by twice y.

In Problems 53 through 86, simplify the given expressions.

53. $9z + z$

54. $3y^2 - 2y^2$

55. $7x + 4x - x$

56. $3xyz - 7xyz + 5xyz$

57. $x^2 - 5x^2 - 11x^2$

58. $6xy^2y^3 - 11xy^2z^3 + 4xy^2z^3$

59. $5 + 3v + 19 + 8v$

60. $2x^3 + x^3 + y^3 + 12y^3$

61. $4x^2 + x^3 - 2x^2 + 5x^2 - x^3$

62. $1 + 2x - 3y + x + y + 4$

63. $3x^3 + x^2 - x^3 + 2x^2$

64. $5x^2z + 8z^2 - 31z^2 + 6x^2z$

65. $2x^2 - 3x - 2x - 3 - 2x^2$

66. $6\sqrt{y} + 7\sqrt{y}$

67. $3(x + 1) + 3x$

68. $18x + 3(3x - 5)$

69. $x - 4 + 2(x + 3)$

70. $1 + x + 3(x + 2)$

71. $\sqrt{2x} - 6\sqrt{2x} + 16\sqrt{2x}$

72. $2(2 - x) + 3(4 + x)$

73. $3(2x - 7) + 4(x - 1)$

74. $4(6 + x) + 5(x - 3)$

75. $(3y^2 - 4)3 + 8(y^2 - 1)$

76. $2(3x + 1) - 2x - 8$

77. $3x + 3(4x - 4) + 8$

78. $7z + 2(z - 2) - 3$

79. $2(x - 3) + 3(x - 4)$

80. $-2(x - 1) + 3(x + 5)$

81. $4(3 + 2a) + 2(3a - 2)$

82. $6\sqrt{k+2} + \sqrt{k+2}$

83. $-5(11 - 2y) + 3(17 - 2y)$

84. $3(3b - a) + 3(5b - a)$

85. $4(3 - 2x) + 2(-1 - 2x)$

86. $23(2x - 3) + (18 - 21x)$

In Problems 87 through 106, answer each question with an English language phrase.

87. Ken has \$7700 to invest. If he buys a \$2000 certificate of deposit, how much does he have left to invest?

88. Bill has \$7700 to invest. If he buys a certificate of deposit for x dollars, how much does he have left to invest?

89. In a given triangle, $\angle A$ has measure 37° and $\angle B$ is twice $\angle A$.

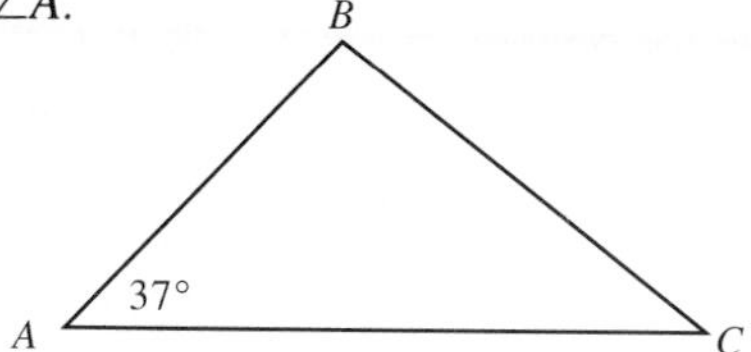

What are the measures of $\angle B$ and $\angle C$? (*Hint:* The sum of the angles in a triangle is 180°.)

90. In a given triangle, $\angle A$ has measure $x°$ and $\angle B$ is twice $\angle A$. What are the measures of $\angle B$ and $\angle C$?

91. It is 71 miles (mi) from Sharon's house to Linda's. If Sharon drives 34 mi toward Linda's house then stops for tea, how much further must she drive?

92. It is 85 mi from Leonard's house to Glenn's. If Leonard drives x miles toward Glenn's house then stops for coffee, how much further must he drive?

93. A parking meter contains 102 dimes and quarters. If it has 42 dimes, how many quarters does it contain?

94. A parking meter contains 102 dimes and quarters. If it has 42 dimes, what is the *value* of the dimes it contains? What is the *value* of the quarters it contains?

95. A parking meter contains 102 dimes and quarters. If it has x dimes, how many quarters does it contain?

96. A parking meter contains 102 dimes and quarters. If it has x dimes, what is the *value* of the dimes it contains? What is the *value* of the quarters it contains? What is the *value* of the contents of the meter?

97. If the measure of a certain angle is 56°, what is the measure of its supplement?

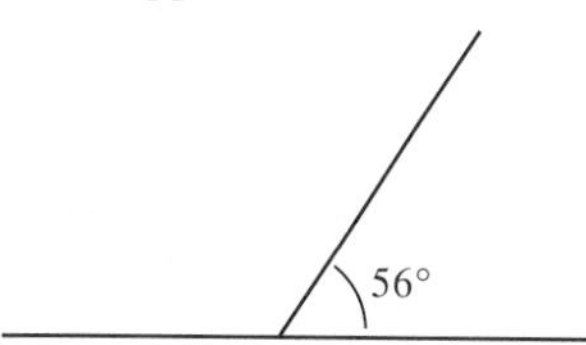

(*Hint:* An angle and its supplement have a sum of 180°.)

98. If the measure of a certain angle is $x°$, what is the measure of its supplement? What is the measure of its complement? (*Hint:* An angle and its complement have a sum of 90°.)

99. Som is 3 years older than Val. If Val is 22 yr old now, how old is Som?

100. Steve is 3 years older than Angela. If Angela is x years old now, how old is Steve?

101. Steve is 3 years older than Angela. If Angela is x years old now, how old will Steve be in 10 yr?

102. 18 is an even integer. What are the next three consecutive even integers?

103. If x is an even integer, what are the next three consecutive even integers?

104. If x is an odd integer, what are the next three consecutive odd integers?

105. If 18 is the smallest of four consecutive even integers, what is their sum?

106. If x is the smallest of four consecutive even integers, what is their sum?

Challenge Problems

Simplify each of the following.

107. $\sqrt{x^2+3} - 3\sqrt{x^2-5} - 2\sqrt{x^2-5} + 11\sqrt{x^2+3}$

108. $\sqrt{x+1} - 2\sqrt{y+1} - 2\sqrt{x+1} + 10\sqrt{y+1}$

109. $3\sqrt{x+4} - 2(3\sqrt{x+4} - \sqrt{2y}) - 4\sqrt{2y}$

110. $13\sqrt{x} - 4(3\sqrt{x} - \sqrt{y}) - 4\sqrt{y}$

IN YOUR OWN WORDS...

111. What are constants and variables?

112. Why do we have to use letters in place of numbers in algebra?

113. Make up a question that might have this statement as its answer:
There are 22 orange marbles in the bag.

1.8 EQUATIONS

Equations are a central idea of algebra. We define an **equation** to be a statement that two numbers are equal. We use the symbol = to indicate equality.

An equation may be a true statement or it may be a false statement. Consider the statements,

$$3 + 14 = 17$$

$$3 + 11 = 17$$

They are both statements that two numbers are equal. The first is *true,* and the second is *false.*

Equations are much more interesting when they contain variables. Consider the equation

$$3 + x = 17$$

Notice that if x has the value 14, the statement is *true,* but if x has the value 11, (or any number other than 14), the statement is *false.* We call 14 a **solution,** or **root,** of the equation. The set of all solutions of an equation is called its **solution set.** Since 14 is the only number that makes the statement true, we write the solution set

$$\{14\}$$

An equation has a **left side,** an equality symbol, and a **right side.**

$$\underbrace{3x - 1}_{\text{Left side}} = \underbrace{4(2 - x)}_{\text{Right side}}$$

EXAMPLE 1. Is the given number a solution of the equation?

(a) Equation: $3x + 1 = 5$
Given number: 2

(b) Equation: $t^2 - t - 6 = 0$
Given number: -2

Solutions:

(a) We must decide if replacing x with 2 makes the equation a true or false statement. Since $3(2) + 1$ is 7, the statement is false. So, 2 is not a solution of $3x + 1 = 5$.

(b) We replace t with -2 in the left side of the equation.

$$t^2 - t - 6 = (-2)^2 - (-2) - 6 \qquad \text{(Notice the use of parentheses)}$$
$$= 4 + 2 - 6$$
$$= 0$$

Since the right side is also 0, the statement is true. Thus, -2 is a solution of $t^2 - t - 6 = 0$. □

To **solve** an equation means to find its solution set. That is, we find the set of numbers that make the statement true. To do this, we have two important tools. The first tool says we can add the *same* number to *both sides* of an equation without changing the solution set. The second tool says we can multiply *both sides* of an equation by the *same* number (as long as it is not zero) without changing the solution set.

Two Tools for Solving Equations

The solution set of an equation will not change if we use either of the following two tools.

Addition Property of Equality

Add (or subtract) the same number on both sides of the equation.

Multiplication Property of Equality

Multiply (or divide) both sides by the same nonzero number.

Our general approach is to use these tools to simplify an equation to the point where it is of the form: the variable equal to a constant.

EXAMPLE 2. Solve the equation $x - 7 = 11$.

Solution:

It may seem that the solution set is obvious as the equation stands. However, as a general practice, we always simplify equations as far as we can before writing the solution set. In this case, we see that if we add 7 to the left side of the equation, the 7 and the -7 will have a sum of zero, leaving the x alone. So we add 7 to both sides of the equation.

$$x - 7 = 11$$
$$x - 7 + 7 = 11 + 7 \qquad \text{Add 7 to both sides.}$$
$$x + 0 = 18 \qquad \text{Simplify.}$$
$$x = 18$$

This statement is *true* if x has the value 18 and false otherwise. Thus the solution set contains only the number 18, which we write

$$\{18\}$$

□

A solution of an equation is a number that makes the equation a *true* statement. We can decide if a number is in fact a solution by **checking.** The type of equations we are solving now do not *require* checking as a part of the solving procedure. However, they can be checked for arithmetic errors. In checking a possible solution, *be sure to use a correct procedure.*

We must be careful that we don't just stick the number we are checking back into the equation and see if it "works out." Such a step involves a subtle, but *important,* error in logic. We recommend a checking procedure that evaluates both the left and right sides of the original equation separately. If the value obtained from the left side is the same as the value obtained from the right side, then the number being checked is in the solution set. Otherwise, it is not.

EXAMPLE 3. Solve the equation $x + 4 = 19$ and check the solution.

Solution:

We can add *or subtract* any number on both sides of an equation. In this equation, it is convenient to subtract 4 from both sides.

$$x + 4 = 19$$

$$x + 4 - 4 = 19 - 4 \qquad \text{Subtract 4 from both sides.}$$

$$x + 0 = 15 \qquad \text{Simplify.}$$

$$x = 15$$

The instructions require checking the solution. We do this *before* writing the solution set. To check, we replace x by 15 in each side of the original equation and evaluate them *separately.* We show one recommended format.

LS (left side): $x + 4 = 15 + 4 = 19$

RS (right side): 19

As the left side and the right side yield the same number when x is 15, 15 is the solution, and we write the solution set:

$$\{15\}$$

□

EXAMPLE 4. Solve the equation $3x = 21$ and check the solution.

Solution:

Adding a number to both sides of this equation will not aid in solving it. However, we can *multiply* both sides by $\frac{1}{3}$, or we can *divide* both sides by 3.

$$3x = 21$$

$$\frac{3x}{3} = \frac{21}{3} \qquad \text{Divide both sides by 3.}$$

$$x = 7$$

Again the instructions call for checking.

LS: $3x = 3(7) = 21$

RS: 21

Thus 7 checks. The solution set is

$$\{7\}$$

□

EXAMPLE 5. Solve $-\frac{1}{2}x = 4$ and check the solution.

Solution:

$$-2\left(-\frac{1}{2}x\right) = -2(4) \qquad \text{Multiply both sides by } -2.$$

$$\left[(-2)\left(-\frac{1}{2}\right)\right]x = -8 \qquad \text{Associative property}$$

$$1 \cdot x = -8$$

$$x = -8$$

To check -8, we replace x with -8.

$$\text{LS:} \quad -\frac{1}{2}x = -\frac{1}{2}(-8) = 4$$

$$\text{RS:} \quad 4$$

Since the value of the left side and right side is the same, -8 is a solution.

$$\{-8\}$$

□

EXAMPLE 6. Solve $\frac{3}{4}x = -1$

Solution:

We can solve this equation by two methods. We can multiply both sides by the reciprocal of $\frac{3}{4}$, which is $\frac{4}{3}$, or we can multiply both sides by 4 and then divide both sides by 3. We first solve this equation by multiplying both sides by $\frac{4}{3}$, the reciprocal of $\frac{3}{4}$.

1

$$\frac{3}{4}x = -1$$

$$\frac{4}{3}\left(\frac{3}{4}x\right) = -1\left(\frac{4}{3}\right)$$

$$\left(\frac{4}{3} \cdot \frac{3}{4}\right)x = -1\left(\frac{4}{3}\right) \qquad \text{Associative property}$$

$$x = -\frac{4}{3}$$

$$\left\{-\frac{4}{3}\right\}$$

Compare Method 1 with the following method.

2

$$\frac{3}{4}x = -1$$

$$4\left(\frac{3}{4}x\right) = 4(-1) \qquad \text{Multiply both sides by 4.}$$

$$\left(4 \cdot \frac{3}{4}\right)x = 4(-1) \qquad \text{Associative property}$$

$$3x = -4$$

Notice that the equation now contains no fractions. We often call this step **clearing fractions.**

$$\frac{3x}{3} = \frac{-4}{3} \qquad \text{Divide both sides by 3.}$$

$$x = -\frac{4}{3}$$

$$\left\{-\frac{4}{3}\right\}$$

The second method is longer, but it is good to see how to clear an equation of fractions. This will be useful in Chapter 3. □

The next example illustrates a situation that arises often in solving equations.

EXAMPLE 7. Solve $-x = 3$.

Solution:

If we multiply (or divide) both sides by -1, we can write the solution set.

$$-x = 3$$

$$(-1)(-x) = (-1)3 \qquad \text{Multiply both sides by } -1.$$

$$1x = -3$$

$$x = -3$$

$$\{-3\}$$

□

Equations are used in the real world to solve problems as diverse as finding interest and designing rocket boosters. They are the heart of engineering and the sciences. In an algebra course we use word problems to imitate such real-world applications.

We solve a word problem by first translating it into an equation. This equation is often called a **mathematical model** of the problem. Then we solve the equation and answer the question.

EXAMPLE 8. The sum of a number x and 8 is 12. Find the number.

(a) Translate the statement into an equation.

(b) Solve the equation and find the number.

(c) Answer the question.

Solution:

(a) $x + 8 = 12$

(b) $x + 8 - 8 = 12 - 8$ Subtract 8 from both sides.

$x = 4$

(c) The number is 4.

We can check our answer by reading the problem and replacing x with 4. The sum of a number 4 and 8 is 12. □

PROBLEM SET 1.8

Warm-ups

In Problems 1 through 8, decide if the given number is a solution of the equation. See Example 1.

1. Equation: $2x - 5 = -8$
   Given number: -1
2. Equation: $6x + 3 = 5$
   Given number: $\frac{1}{3}$
3. Equation: $1.5x - 1.25 = 0.5$
   Given number: 0.5
4. Equation: $-3 - t = -2$
   Given number: -1
5. Equation: $\frac{w-1}{w-4} = 2$
   Given number: 6
6. Equation: $|x| = 3$
   Given number: -3
7. Equation: $x^2 - 3x - 10 = 0$
   Given number: -2
8. Equation: $\sqrt{x} - 3 = -2$
   Given number: 25

In Problems 9 through 18, solve each equation. See Examples 2 and 3.

9. $x - 3 = 2$
10. $x - 7 = 1$
11. $t + 3 = 8$
12. $x + 1 = 0$
13. $15 + x = 22$
14. $-11 + z = 5$
15. $x + 3 = -2$
16. $x + 21 = -4$
17. $x - 15 = -15$
18. $y - 7 = -11$

In Problems 19 through 32, solve each equation.
In Problems 19 through 22, see Example 4.

19. $2z = 12$
20. $16 = 4x$
21. $7x = 14$
22. $3x = 0$

In Problems 23 through 26, see Examples 5 and 6.

23. $\frac{1}{2}x = 13$
24. $\frac{1}{7}x - 3 = 0$
25. $\frac{2}{3}x = 9$
26. $\frac{1}{5}t = \frac{1}{6}$

In Problems 27 through 32, see Example 7.

27. $-y = 5$
28. $-x = -7$
29. $-x = \frac{1}{2}$
30. $-\frac{2}{5} = -x$
31. $-\frac{3}{2}s = \frac{5}{2}$
32. $-\frac{2}{7}x = -11$

In Problems 33 through 40, translate each sentence into an equation, solve the equation, and find the number. See Example 8.

33. The sum of a number y and 6 is -12.
34. A number w decreased by 1 is -3.
35. A number x increased by 5 is 0.
36. One-half of a number t is -12.
37. If -7 is subtracted from a number y, the result is 13.
38. The product of a number s and 5 is -35.
39. If a number t is divided by 5, the result is 15.
40. If the reciprocal of $\frac{2}{3}$ is multiplied by a number z, the result is -12.

Practice Exercises

In Problems 41 through 48, decide if the given number is a solution of the equation.

41. Equation: $2x - 5 = x - 8$

    Given number: -3

42. Equation: $5x - 3 = 5$

    Given number: $\frac{1}{5}$

43. Equation: $6 - x = 0$

    Given number: -6

44. Equation: $0.8x - 1.0 = -0.4$

    Given number: 1.2

45. Equation: $\frac{w-1}{w+8} = \frac{1}{2}$

    Given number: -2

46. Equation: $\frac{1}{2}y - 2 = \frac{1}{3}y + 4$

    Given number: 12

47. Equation: $|x| = -3$

    Given number: -3

48. Equation: $\sqrt{x} - 3 = 1$

    Given number: 16

In Problems 49 through 62, solve the equations.

49. $x + 3 = 2$
50. $x + 7 = -1$
51. $12 + x = 27$
52. $17 + x = 4$
53. $8x = 72$
54. $11x = 121$
55. $x - 14 = -8$
56. $x - 9 = -31$
57. $-z = -20$
58. $-x = \frac{2}{3}$
59. $\frac{2}{5}w = 1$
60. $-\frac{5}{4}s = \frac{3}{4}$
61. $\frac{1}{7}x = \frac{1}{4}$
62. $-\frac{3}{2}x = -18$

In Problems 63 through 70, translate each sentence into an equation, solve the equation, and find the number.

63. The product of a number y and 6 is -12.
64. A number w decreased by 3 is -3.
65. A number x increased by 7 is 0.
66. One-fourth of a number t is -12.
67. If -5 is subtracted from a number y, the result is 18.
68. The sum of a number s and 5 is -35.
69. If a number t is divided by 7, the result is -5.
70. If the reciprocal of $\frac{4}{3}$ is multiplied by a number z, the result is -12.

Challenge Problems

In Problems 71 through 75, x is the variable and p is a constant. Solve each equation.

71. $x - 2 = p$
72. $x + p = 2$
73. $5x = p$
74. $\frac{2}{3}x = p$
75. $px = 6$ (What restriction must we place on the value of p?)

IN YOUR OWN WORDS . . .

76. What is an equation?
77. What determines whether a number is a solution of an equation?
78. What does it mean to solve an equation?
79. Make up a word problem that has the equation $2x = 8$ as its mathematical model.

CHAPTER SUMMARY

GLOSSARY

Set: A collection of objects.

Member, or **element:** One of the objects in a set.

Empty, or **null set:** A set with no members.

Subset: A set whose members are also members of a given set.

Union of two sets: The set of all elements that belong to *either* of the two sets.

Intersection of two sets: The set of all elements that belong to *both* of the two sets.

Natural numbers: 1, 2, 3, . . .

Whole numbers: 0, 1, 2, 3, . . .

Integers: . . . , −3, −2, −1, 0, 1, 2, 3, . . .

Rational numbers: All possible quotients of integers. Numbers that can be written in the form $\frac{p}{q}$, where p and q are integers and $q \neq 0$.

Irrational numbers: Numbers on the number line that are not rational, such as $\sqrt{2}$ and π.

Real numbers: The rational and irrational numbers.

Positive real numbers: All real numbers greater than zero.

Negative real numbers: All real numbers less than zero.

Absolute value: The distance of a number from zero on the number line.

Opposite: The number on the opposite side of zero on the number line from a given number, but the same distance from zero.

Reciprocal: If p is a nonzero number, then $\frac{1}{p}$ is its reciprocal.

Equation: A statement that two numbers are equal.

Variable: A letter that stands for a number.

Solution: A number that when substituted for the variable makes an equation a true statement.

THE SET OF REAL NUMBERS

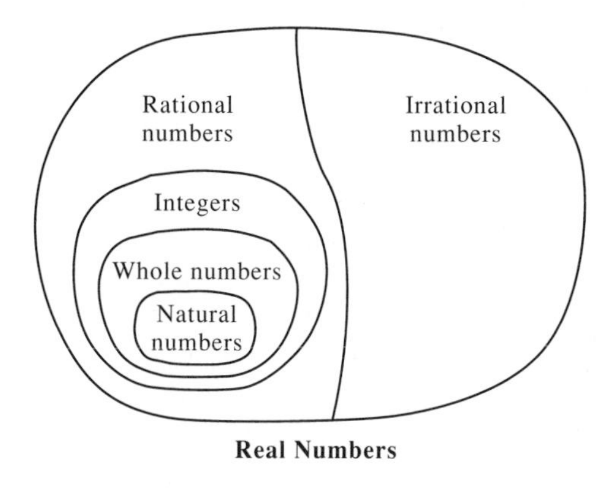

ADDITION OF REAL NUMBERS

1. To add two numbers with *like signs* (both positive or both negative), add their absolute values and keep the common sign.
2. To add two numbers with *unlike signs* (one positive and one negative), subtract the smaller absolute value from the larger absolute value and keep the sign of the number with the greater absolute value.

SUBTRACTION OF REAL NUMBERS

1. To subtract two real numbers, rewrite the problem as an addition problem.

$$p - q = p + (-q)$$

2. Follow the rules for addition.

MULTIPLICATION AND DIVISION OF REAL NUMBERS

1. To multiply or divide two numbers with *like signs* (both positive or both negative), multiply or divide their absolute values. The sign will be positive.
2. To multiply or divide two numbers with *unlike signs* (one positive and one negative), multiply or divide their absolute values. The sign will be negative.

DIVISION INVOLVING ZERO

If p and q are real numbers with $p \neq 0$, then

1. $\frac{0}{p} = 0$
2. $\frac{q}{0}$ is undefined.

PROPERTIES OF REAL NUMBERS

Addition

1. $p + 0 = 0 + p = p$	Identity
2. $p + (-p) = (-p) + p = 0$	Inverse
3. $p + q = q + p$	Commutative
4. $(p + q) + r = p + (q + r)$	Associative

Multiplication

1. $p(1) = 1(p) = p$	Identity	
2. $p\left(\frac{1}{p}\right) = \frac{1}{p}(p) = 1$	Inverse	$(p \neq 0)$
3. $pq = qp$	Commutative	
4. $(pq)r = p(qr)$	Associative	
5. $0(p) = p(0) = 0$		
6. $-1(p) = p(-1) = -p$		

Distributive Properties

1. $p(q + r) = pq + pr$
2. $p(q - r) = pq - pr$

EQUATIONS

Two tools for solving equations

1. Add or subtract the same number on both sides of the equation.
2. Multiply or divide both sides by the same nonzero number.

CHECKUPS

1. Classify each as rational or irrational:

   (a) $\frac{3}{7}$ (b) $0.\overline{78}$ (c) $\sqrt{7}$ (d) 1.7 Section 1.1; Example 3

2. Simplify each of the following: Section 1.1; Example 8

(a) $|-5|$ (b) $2|17|$ (c) $-|-7|$ (d) $|-\pi|$

3. Perform the operations indicated.

(a) $4 + (-8)$ Section 1.2; Example 3a

(b) $-7 - (-2)$ Section 1.3; Example 1c

(c) $\frac{1}{3}\left(-\frac{3}{2}\right)$ Section 1.4; Example 1c

(d) $10/(-2)$ Section 1.4; Example 2c

(e) $-6 - (5 - \sqrt{25 - 16} + 1)$ Section 1.5; Example 2b

4. Simplify the following expression.

$-6/(-2) + 2(-3)^2 - 2^2$ Section 1.5; Example 1b

5. Evaluate the following expression when x is -1.

$|3 + 7x| + 21$ Section 1.5; Example 4f

6. Name the property that is illustrated in each of the following.

(a) $4 + 0 = 4$ (b) $5 + (-5) = 0$ Section 1.6; Example 5

7. Write the opposite and reciprocal of each of the following.

(a) -10 (b) $-\frac{3}{4}$ (c) $\frac{2}{x}$ Section 1.6; Example 4

8. If Harold has 20 nickels and dimes in his pocket and x of them are dimes, how many nickels does Harold have in his pocket? Section 1.7; Example 3b

9. Simplify each expression. Section 1.7; Example 8

(a) $6x + 1 + 5x - 3$ (b) $4x^2 - 3x + x^2 - 2x$

10. Solve the equation, $x + 4 = 19$. Section 1.8; Example 3

11. Solve the equation $-\frac{1}{2}x = 4$ and check the solution. Section 1.8; Example 5

12. The sum of a number and 8 is 12. Find the number. Section 1.8; Example 8

REVIEW PROBLEMS

In Problems 1 through 20, perform the operations indicated.

1. $-5 + 3$

2. $-\frac{3}{2} - \frac{1}{3} \div \frac{1}{6}$

3. $-8.1 + (-3.2 + 2.1)$

4. $5 + \{2(6 - 8) + 3[5 + (-7)]\}$

5. $\frac{3^2(4 - 6) - 6}{2^3}$

6. $-(-5)$

7. $|6 - 11| - 3[4 + (-8)]$

8. $(-5 - 6) - 4$

9. $16/4(-2)$

10. $[-9 - (-5)] - (-7 - 2)$

11. $2|-3| \div (-6)$

12. $8 \div 4 + (-4)(-4)$

13. $\dfrac{0}{-4}$

14. $\dfrac{-4}{0}$

15. $2|0 - 5| \div (-1 - 1)$

16. $\dfrac{3(-1) - 3 \cdot 2^2 + (-9)}{2^3 - 3^2}$

17. $-\frac{1}{3}\left(-\frac{1}{5}\right)(18)(-15)$

18. $487 + 28 - 487$

19. $-3(-8) - 2(4 - 8) + (3 - 6)$

20. $8.1/(0.09) - (-10)^2$

In Problems 21 through 25, answer each question.

21. Subtract -3 from the sum of -8 and -2.
22. Divide the product of -8 and -1 by 2.
23. Subtract the sum of -15 and 10 from 25.
24. The temperature at 5 A.M. was $-10°$. If the temperature rose 2° per hour, what was the temperature at noon?
25. On 3 consecutive plays, the Atlanta Falcons football team gained 8 yd, lost 5 yd, and gained 6 yd. What was their total gain for the 3 plays?

In Problems 26 through 30, write an algebraic expression for each phrase.

26. The sum of 2 and 6 multiplied by the opposite of 5.
27. The absolute value of the sum of 8 and 11.
28. One less than 9 divided by one more than 9.
29. The cube of twice 7.
30. One-half of 11 times one-third of 11.

In Problems 31 through 35, find the solution set for each equation.

31. $x - 4 = 6$

32. $x + 8 = 11$

33. $9 + x = 24$

34. $3x = 12$

35. $-\frac{3}{4}y = 6$

In Problems 36 through 45, indicate whether each statement is true or false.

36. Every whole number is an integer.

37. $\frac{2}{3} \le 0.\overline{6}$

38. $\sqrt{7}$ is a real number.

39. $\{2, 3\} \subseteq \{1, 2, 3\}$

40. $3 \in \{1, 2, 3\}$

41. $\frac{2}{3}$ is an irrational number.

42. An integer is a real number.

43. $-5.4 > -5.3$

44. $|-3| \le 0$

45. $2.\overline{3}$ is a rational number.

...LET'S NOT FORGET...

In Problems 46 through 48, find each sum or difference.

46. $5 + (-4)$

47. $5 - (-4)$

48. $-5 - (-4)$

In Problems 49 through 52, find each product or quotient.

49. $\dfrac{0}{3}$

50. $(-1)(5)$

51. $(-1)(0)$

52. $(-1)(-6)$

In Problems 53 through 55, simplify each expression. Watch the role of the parentheses!

53. $(-3 - 5)(3 + 7)$

54. $-3 - 5(3 + 7)$

55. $(-3 - 5)3 + 7$

In Problems 56 through 58, multiply each expression by 2.

56. $\frac{1}{2} \cdot 7$

57. $\frac{1}{2}(\pi - 2)$

58. $\pi - \frac{1}{2}$

In Problems 59 through 64, label each as an expression *or as an* equation. *Solve the equations.*

59. $x + 6$

60. $x + 6 = 0$

61. $2x = 10$

62. $2x - 10$

63. $x - 4 + 7$

64. $x - 4 = 7$

CHAPTER 1 TEST

In Problems 1 through 5, choose the correct letter.

1. $\frac{4(-1) + 3(-2)}{-2 - 3} = (?)$

A. -10 B. -4

C. 4 D. 2

2. $10 - (4)(8) + 6 = (?)$

A. 52 B. -32

C. 84 D. -16

3. When expressed as a decimal, $\frac{11}{6} = (?)$.

A. $1.8\overline{3}$ B. 1.83

C. $.18\overline{3}$ D. $1.\overline{83}$

4. $2\{3 + [8 - (6 - 8)]\} = (?)$

A. 18 B. 26

C. 9 D. 13

5. $2(-3)^2 = (?)$

A. -18 B. -12

C. 12 D. 18

In Problems 6 through 10, indicate whether each statement is true or false.

6. $-\frac{3}{5} \leq -\frac{1}{2}$

7. $-(-5) > 0$

8. All rational numbers are real numbers.

9. The reciprocal of -3 is $\frac{1}{3}$.

10. $\frac{-15}{0} = 0$.

In Problems 11 and 12, find the value of each expression when x is 2 and y is -1.

11. $2x^2 - y$

12. $\frac{5x + 3y + 1}{2x}$

In Problems 13 and 14, find each quantity.

13. The product of 3 and -4 is added to 5.

14. The sum of 3 cubed and 10.

In Problems 15 through 20, perform the operations indicated.

15. $7 + (-8)$

16. $(-4)(-3)$

17. $|11| - |-5|$

18. $6 + 6 \cdot 6 - \frac{6}{6}$

19. $\frac{(-4)(-9)}{(-3)^2}$

20. $-5 + [6 + (-8)]$

21. Match each statement with the property that justifies it.

____	a. $3(8 + 1) = 24 + 3$	1. Commutative
____	b. $(-5)\left(-\frac{1}{5}\right) = 1$	2. Associative
____	c. $-3 + 0 = -3$	3. Identity
____	d. $3 + (-3) = 0$	4. Inverse
____	e. $7 + (2 + 3) = (7 + 2) + 3$	5. Distributive
____	f. $(4 + 3) + 1 = 1 + (4 + 3)$	

22. Using the following set of numbers,

$$\left\{-5, -\frac{3}{5}, 0, \sqrt{3}, \frac{17}{3}, 12.7, 102\right\}$$

list the numbers that belong to each set.

(a) Integers

(b) Rational numbers

(c) Irrational numbers

In Problems 23 through 25, find the solution set.

23. $x - 12 = 13$

24. $-5x = 20$

25. $16 + z = 22$

CHAPTER 2

See Problem Set 2.2, Exercise 128.

Polynomials

CONNECTIONS

A cardboard box manufacturer is making open boxes from a 12 in. square piece of cardboard by cutting equal squares from each of the corners and turning up the sides. She wishes to make a box of the largest possible volume. To figure out the size of the squares, she would use the expression, $4x^3 - 48x^2 + 144x$. Such expressions are called polynomials. This chapter examines polynomials and their properties. Operations of addition, subtraction, multiplication, and division are performed with polynomials just as with real numbers.

We begin the study of polynomials by looking at the definitions and properties of different kinds of exponents. We then learn how to classify, evaluate, and perform simple algebra with polynomials. Understanding how to add, subtract, multiply, and divide polynomials is essential to the study of more advanced topics in algebra.

2.1 WHOLE NUMBER EXPONENTS

We use exponents in algebra just as we use them in arithmetic. Thus, x^3 means $x \cdot x \cdot x$ just as 2^3 means $2 \cdot 2 \cdot 2$. We read x^3 as "x to the third power," or "x cubed." We now look at a definition of exponents that are natural numbers.

Natural Number Exponent

If x is a real number and n is a natural number, then

$$x^n = \underbrace{x \cdot x \cdot x \cdots x}_{n \text{ factors of } x}$$

So we see that,

$$x^1 = x$$

$$x^2 = x \cdot x$$

$$x^{10} = x \cdot x \cdot x \cdot x \cdot x \cdot x \cdot x \cdot x \cdot x \cdot x$$

In writing x^n, we call x the **base** and n the **exponent.** The exponent gives the number of times the base is used as a factor. Sometimes the exponent is called the **power.** We read x^n as x to the nth power.

Be Careful!

NOTE: x^n does *not* mean n times x. Don't confuse $x^2 = x \cdot x$ with $2x = x + x$.

EXAMPLE 1. Identify the base and simplify each of the following.

(a) 3^2 (b) $(-3)^2$ (c) -3^2 (d) $(-2)^3$ (e) -2^3

Solutions:

(a) 3^2 is 3 squared. The base is 3.

$$3^2 = 3 \cdot 3 = 9$$

(b) $(-3)^2$ is -3 squared. The base is -3.

$$(-3)^2 = (-3)(-3) = 9$$

(c) -3^2 is the opposite of 3 squared. The base is 3.

$$-3^2 = -3 \cdot 3 = -9$$

(d) $(-2)^3$ is -2 cubed. The base is -2.

$$(-2)^3 = (-2)(-2)(-2) = -8$$

(e) -2^3 is the opposite of 2 cubed. The base is 2.

$$-2^3 = -2 \cdot 2 \cdot 2 = -8$$

□

As Example 1 indicates, $(-p)^n$ and $-p^n$ are *not* the same. The difference lies in the base.

Identifying the Base

$(-p)^n$	$-p^n$
The base is $-p$.	The base is p.
$(-4)^2 = (-4)(-4) = 16$	$-4^2 = -4 \cdot 4 = -16$

Looking closer at $(-p)^n$, we can see that if n is *odd,* then we have

$$(-p)^n = (-p)(-p)\cdots(-p) \qquad (n \text{ factors of } -p\text{, and } n \text{ is odd})$$
$$= -p^n$$

However, if n is *even,* we have

$$(-p)^n = (-p)(-p)\cdots(-p) \qquad (n \text{ factors of } -p\text{, and } n \text{ is even})$$
$$= p^n$$

Simplifying $(-p)^n$

If p is a real number and n is a natural number, then

$$(-p)^n = p^n \qquad \text{if } n \text{ is even}$$
$$(-p)^n = -p^n \qquad \text{if } n \text{ is odd}$$

EXAMPLE 2. Simplify each of the following.

(a) $(-x)^4$ (b) $(-y)^7$ (c) $-z^6$

Solutions:

(a) The base is $-x$ and the exponent is *even.*

$$(-x)^4 = x^4$$

(b) The base is $-y$ and the exponent is *odd.*

$$(-y)^7 = -y^7$$

(c) The base is z, and this expression *cannot* be simplified. *Be sure to notice* the difference between this expression and the expression in part (a). □

Not only is exponential notation convenient to use, but it has some very nice properties.

Suppose we want to multiply $x^3 \cdot x^2$. We use the definition of exponent and write

$$x^3 \cdot x^2 = (x \cdot x \cdot x)(x \cdot x) = x \cdot x \cdot x \cdot x \cdot x = x^5$$

This leads us to the first property of exponents:

Product with Same Base

$$x^m \cdot x^n = x^{m+n}$$

To multiply powers of the *same* base, keep the base and *add* the exponents.

EXAMPLE 3. Simplify each expression.

(a) $\left(\frac{1}{2}\right)^2 \cdot \left(\frac{1}{2}\right)^3$ (b) $x^4 \cdot x^7$ (c) x^2y^3

Solutions:

(a) $\left(\frac{1}{2}\right)^2 \cdot \left(\frac{1}{2}\right)^3 = \left(\frac{1}{2}\right)^{2+3}$ Keep the base and add exponents.

$$= \left(\frac{1}{2}\right)^5$$

We can leave this answer or we can write

$$\left(\frac{1}{2}\right)^5 = \frac{1}{32}$$

(b) $x^4 \cdot x^7 = x^{4+7}$ Product with same base.

$$= x^{11}$$

(c) x^2y^3 cannot be simplified because the bases are not the same. □

It is *important* to note that to use this property, both *bases* must be the same.

Let's look at $(x^2)^3$. The exponent 3 has a base of x^2. We use the definition of exponents and write

$$(x^2)^3 = (x^2)(x^2)(x^2)$$

Next we use the first property and add exponents:

$$(x^2)(x^2)(x^2) = x^{2+2+2}$$
$$= x^6$$

This gives the second property of exponents:

Power to a Power

$$(x^m)^n = x^{mn}$$

To raise a power to a power, keep the base and *multiply* exponents.

EXAMPLE 4. Simplify each.

(a) $(2^2)^6$ (b) $(y^3)^5$

Solutions:

(a) $(2^2)^6 = 2^{2\cdot 6}$ Keep the base and multiply exponents.
$= 2^{12}$

(b) $(y^3)^5 = y^{3\cdot 5}$ Power to a power.
$= y^{15}$ □

Now let's consider a product raised to a power, such as $(xy)^2$. The base is xy, so we use the definition of exponent and write

$$(xy)^2 = (xy)(xy)$$

Using the Commutative and Associative Properties we can write

$$(xy)(xy) = x \cdot x \cdot y \cdot y$$
$$= x^{1+1}y^{1+1} \quad \text{Add exponents.}$$
$$= x^2y^2$$

Now we have the third property of exponents:

Product to a Power

$$(xy)^n = x^n y^n$$

To raise a product to a power, raise each factor in the product to the power.

EXAMPLE 5. Simplify each.

(a) $(3x)^4$ (b) $(-xy^2)^2$

Solutions:

(a) $(3x)^4 = 3^4x^4$ Raise each factor to the fourth power.

$= 81x^4$ $(3^4 = 3 \cdot 3 \cdot 3 \cdot 3)$

(b) $(-xy^2)^2 = (-x)^2(y^2)^2$ Product to a power

$= x^2y^4$ □

The next property is just like the previous one, except that we use a quotient instead of a product. Consider $\left(\frac{x}{y}\right)^3$. The base is $\frac{x}{y}$. We use the definition of exponent and write

$$\left(\frac{x}{y}\right)^3 = \frac{x}{y} \cdot \frac{x}{y} \cdot \frac{x}{y}$$

$$= \frac{x \cdot x \cdot x}{y \cdot y \cdot y}$$

$$= \frac{x^3}{y^3} \quad \text{Product with same base}$$

This leads to the fourth property of exponents:

Quotient to a Power

$$\left(\frac{x}{y}\right)^n = \frac{x^n}{y^n}; \quad y \neq 0$$

To raise a quotient to a power, raise the numerator and the denominator to the power.

EXAMPLE 6. Simplify each.

(a) $\left(\frac{2}{k}\right)^6$ (b) $\left(\frac{-1}{x}\right)^3$

Solutions:

(a) $\left(\frac{2}{k}\right)^6 = \frac{2^6}{k^6}$ Raise the numerator and the denominator to the sixth power.

$= \frac{64}{k^6}$

(b) $\left(\frac{-1}{x}\right)^3 = \frac{(-1)^3}{x^3}$ Quotient to a power

$= \frac{-1}{x^3}$ □

The final property of exponents deals with qoutients with the *same* base. Let's look at three examples of this situation: $\frac{x^5}{x^3}$, $\frac{x^3}{x^5}$, and $\frac{x^4}{x^4}$.

$$\frac{x^5}{x^3} = \frac{x \cdot x \cdot x \cdot x \cdot x}{x \cdot x \cdot x} = \frac{x \cdot x}{1} = x^2, \quad \text{so} \quad \frac{x^5}{x^3} = x^{5-3} = x^2$$

$$\frac{x^3}{x^5} = \frac{x \cdot x \cdot x}{x \cdot x \cdot x \cdot x \cdot x} = \frac{1}{x \cdot x} = \frac{1}{x^2}, \quad \text{so} \quad \frac{x^3}{x^5} = \frac{1}{x^{5-3}} = \frac{1}{x^2}$$

$$\frac{x^4}{x^4} = \frac{x \cdot x \cdot x \cdot x}{x \cdot x \cdot x \cdot x} = 1$$

Putting these ideas together gives us the fifth property of exponents.

Quotient with Same Base

$$\frac{x^m}{x^n} = \begin{cases} x^{m-n} & \text{if } m > n \\ 1 & \text{if } m = n \qquad x \neq 0 \text{ for each case} \\ \frac{1}{x^{n-m}} & \text{if } n > m \end{cases}$$

EXAMPLE 7. Simplify each of the following.

(a) $\frac{x^{11}}{x^8}$ (b) $\frac{(x+y)^3}{(x+y)^{17}}$ (c) $\frac{r^9}{r^9}$

Solutions:

(a) Since $11 > 8$, we have

$\frac{x^{11}}{x^8} = x^{11-8}$ Quotient with same base

$= x^3$

(b) Since $17 > 3$, we have

$\frac{(x+y)^3}{(x+y)^{17}} = \frac{1}{(x+y)^{17-3}}$ Quotient with same base

$= \frac{1}{(x+y)^{14}}$

(c) Since $9 = 9$, we have,

$\frac{r^9}{r^9} = 1$ Quotient with same base □

The properties of exponents are summarized in the following box.

Properties of Exponents

If m and n are natural numbers and x and y are real numbers,

1. $x^m x^n = x^{m+n}$ — Product with same base

2. $(x^m)^n = x^{mn}$ — Power to a power

3. $(xy)^n = x^n y^n$ — Product to a power

4. $\left(\frac{x}{y}\right)^n = \frac{x^n}{y^n}; \quad y \neq 0$ — Quotient to a power

5. $\frac{x^m}{x^n} = \begin{cases} x^{m-n} & \text{if } m > n \\ 1 & \text{if } m = n \\ \frac{1}{x^{n-m}} & \text{if } n > m \end{cases}$ $\quad x \neq 0$ for each case — Quotient with same base

EXAMPLE 8. Use the properties of exponents to simplify.

(a) $x^4(y^3)^5$ (b) $(3x)^2(3x)^2$

(c) $\left(\frac{2x}{k^2}\right)^3$ (d) $\frac{(3b)^5}{(3b)^3}$

Solutions:

(a) We begin with the parentheses, a power to a power.

$$\begin{aligned} x^4(y^3)^5 &= x^4 y^{3 \cdot 5} && \text{Power to a power} \\ &= x^4 y^{15} \end{aligned}$$

This will not simplify further as x and y are *different* bases.

(b) We notice that the base is $3x$ in both expressions. So, we add exponents.

$$\begin{aligned} (3x)^2(3x)^2 &= (3x)^{2+2} && \text{Product with same base} \\ &= (3x)^4 \end{aligned}$$

Now we have a product to a power. So we raise each factor in the product to the fourth power.

$$\begin{aligned} (3x)^4 &= 3^4 x^4 && \text{Power to a power} \\ &= 81x^4 \end{aligned}$$

(c) This is a quotient to a power.

$$\left(\frac{2x}{k^2}\right)^3 = \frac{(2x)^3}{(k^2)^3} \qquad \text{Quotient to a power}$$

This will simplify further. The numerator is a product to a power, and the denominator is a power to a power.

$$\begin{aligned} \frac{(2x)^3}{(k^2)^3} &= \frac{2^3 \cdot x^3}{k^{2 \cdot 3}} \\ &= \frac{8x^3}{k^6} \end{aligned}$$

(d) The base is $3b$ in both the numerator and the denominator. We subtract the exponents.

$$\frac{(3b)^5}{(3b)^3} = (3b)^{5-3} \qquad \text{Quotient with same base}$$

$$= (3b)^2$$

$$= 3^2b^2 \qquad \text{Product to a power}$$

$$= 9b^2$$

□

Properties 2, 3, and 4 tell us when to *multiply* exponents. They are sometimes called power rules and are used in problems in the form,

$$(\text{Expression})^{\text{power}}$$

We must be very careful in simplifying problems of this type. The properties of exponents allow products and quotients that are raised to a power to be simplified. However, the properties *do not* allow us to simplify sums and differences that are raised to powers.

EXAMPLE 9. Use the properties of exponents to simplify each expression to a power if possible.

(a) $(xy^5)^2$ (b) $(x + y)^2$

Solutions:

(a) This is a product to a power. The properties of exponents can be used to simplify.

$$(xy^5)^2 = x^2(y^5)^2 \qquad \text{Product to a power}$$

$$= x^2y^{10} \qquad \text{Power to a power}$$

(b) This is a *sum* to a power. It cannot be simplified using the properties of exponents. $(x + y)^2 = (x + y)(x + y)$, which is *not* $x^2 + y^2$. We will study how to find this product in Section 2.5. □

Be Careful!

EXAMPLE 10. Use the properties of exponents to simplify each expression to a power if possible.

(a) $\left(\frac{xy^4}{z}\right)^3$ (b) $(x - 2y)^3$

Solutions:

(a) This is a quotient to a power. The properties of exponents allow us to simplify.

$$\left(\frac{xy^4}{z}\right)^3 = \frac{x^3(y^4)^3}{z^3} \qquad \text{Quotient to a power}$$

$$= \frac{x^3y^{12}}{z^3} \qquad \text{Power to a power}$$

(b) This is a *difference* to a power. It *cannot* be simplified using properties of exponents. $(x - 2y)^3 = (x - 2y)(x - 2y)(x - 2y)$, which is *not* $x^3 - 8y^3$. □

Be Careful!

EXAMPLE 11. Simplify each expression.

(a) $(-3x^3y)^2$ (b) $\left(\frac{-2r^3s}{t^2}\right)^3$

(c) $(3mn^3)^2(2m)^3$ (d) $\frac{(-xy^3)^2}{(-x^3y)^3}$

Solutions:

(a) We notice that this is an expression in parentheses to a power and that the expression is a *product.*

$$\begin{aligned}(-3x^3y)^2 &= (-3)^2(x^3)^2y^2 && \text{Product to a power}\\ &= (-3)^2x^{3\cdot 2}y^2 && \text{Power to a power}\\ &= 9x^6y^2\end{aligned}$$

(b) This is an expression to a power and the expression is a *quotient.*

$$\begin{aligned}\left(\frac{-2r^3s}{t^2}\right)^3 &= \frac{(-2r^3s)^3}{(t^2)^3} && \text{Quotient to a power}\\ &= \frac{(-2)^3(r^3)^3s^3}{(t^2)^3} && \text{Product to a power}\\ &= \frac{-8r^9s^3}{t^6} && \text{Power to a power}\end{aligned}$$

(c) We use Property 3 to simplify each expression to a power.

$$\begin{aligned}(3mn^3)^2(2m)^3 &= 3^2m^2(n^3)^2\cdot 2^3m^3 && \text{Product to a power}\\ &= 9m^2n^6\cdot 8m^3 && \text{Power to a power}\\ &= 9\cdot 8m^2m^3n^6 && \text{Commutative Property}\\ &= 72m^{2+3}n^6 && \text{Product with same base}\\ &= 72m^5n^6\end{aligned}$$

(d) Both the numerator and the denominator are expressions raised to powers which can be simplified by using properties 2, 3, and 5.

$$\begin{aligned}\frac{(-xy^3)^2}{(-x^3y)^3} &= \frac{(-x)^2(y^3)^2}{(-x^3)^3y^3} && \text{Product to a power}\\ &= \frac{x^2y^6}{-x^9y^3} && \text{Power to a power}\\ &= \frac{y^3}{-x^7} && \text{Quotient with same base}\end{aligned}$$

□

Let's consider the expression $\frac{2^5}{2^5}$. If we apply Property 5, we notice that $m = n$, so

$$\frac{2^5}{2^5} = 1$$

Now if we try to use the first part of Property 5 we get,

$$\frac{2^5}{2^5} = 2^{5-5}$$

$$= 2^0$$

We can replace the 2 in the expression by any nonzero number with the same result. This leads us to our second definition.

> **Zero Exponent**
>
> If x is any nonzero real number, then
>
> $$x^0 = 1$$

EXAMPLE 12. Simplify.

(a) 3^0 (b) $(23x^3y^5)^0$ (c) $(-5)^0$ (d) $(x + y)^0$

(e) $\left(\frac{s}{t}\right)^0$ (f) $5x^0$ (g) -6^0 (h) $4(2a + 3b)^0$

Solutions:

(a) $3^0 = 1$

(b) $(23x^3y^5)^0 = 1$

(c) $(-5)^0 = 1$ (The base is -5.)

(d) $(x + y)^0 = 1$

(e) $\left(\frac{s}{t}\right)^0 = 1$

(f) $5x^0 = 5(1) = 5$ (The base is x.)

(g) $-6^0 = -1 = -1$ (The base is 6.)

(h) $4(2a + 3b)^0 = 4(1) = 4$ □

CALCULATOR BOX

Raising a Number to a Power with a Calculator

$\boxed{y^x}$

To find 4^5 with a calculator, press the keys

4 y^x 5 = and read 1024 on the display. So, $4^5 = 1024$.

Calculator Exercises

Find the value of each. Approximate to two decimals if necessary.

1. 3^2 **2.** 3^3 **3.** 3^{10} **4.** $(1.24)^4$ **5.** $(\sqrt{2})^3$

Answers:

1. 9 **2.** 27 **3.** 59,049 **4.** 2.36 **5.** 2.83

PROBLEM SET 2.1

Warm-ups

In Problems 1 through 6, identify the base and simplify. See Example 1.

1. 4^2
2. -4^2
3. $(-4)^2$
4. $(-4)^3$
5. -4^3
6. $\left(\frac{2}{3}\right)^3$

In Problems 7 through 9, simplify. See Example 2.

7. $(-x)^4$
8. $(-x)^9$
9. $-x^2$

In Problems 10 through 15, simplify each if possible. See Example 3.

10. x^4x^5
11. $y^{11}y^{14}$
12. $6^9 \cdot 6^9$
13. $z \cdot z^5$
14. $2^{10} \cdot 2$
15. x^2y^3

In Problems 16 through 18, simplify each. See Example 4.

16. $(x^4)^5$
17. $(y^2)^3$
18. $(5^4)^4$

In Problems 19 through 24, simplify each. See Example 5.

19. $(xy)^7$
20. $(3s)^3$
21. $(11k)^2$
22. $(-2z^3)^2$
23. $(-3t)^3$
24. $(-xy^2)^5$

In Problems 25 through 30, simplify each. See Example 6.

25. $\left(\frac{x}{y}\right)^8$
26. $\left(\frac{2}{t}\right)^5$
27. $\left(\frac{x}{5}\right)^3$
28. $\left(\frac{-3}{x}\right)^2$
29. $\left(\frac{-5}{p}\right)^3$
30. $\left(\frac{-1}{x}\right)^4$

In Problems 31 through 36, simplify each. See Example 7.

31. $\frac{x^{16}}{x^8}$
32. $\frac{y^4}{y^7}$
33. $\frac{k^{22}}{k}$
34. $\frac{7^4}{7^4}$
35. $\frac{(x+y)^6}{(x+y)^6}$
36. $\frac{(2a+1)^7}{(2a+1)^6}$

In Problems 37 through 42, simplify each. See Example 8.

37. $x(z^7)^2$
38. $(2x^4)^2(2x^4)^3$
39. $\left(\frac{5}{2t}\right)^3$
40. $x^3(x^4)^2$
41. $\frac{(7a^3)^4}{(7a^3)^6}$
42. $\left(\frac{4v^9}{3k}\right)^2$

In Problems 43 through 48, use the properties of exponents to simplify if possible. See Examples 9 and 10.

43. $(2xy)^2$
44. $\left(\frac{2x}{y}\right)^2$
45. $(2x+y)^2$
46. $(2x-y)^2$
47. $(2x^2y^3)^2$
48. $(a-b)^3$

In Problems 49 through 54, simplify each. See Example 11.

49. $(-2x^2y^3)^2$ **50.** $(-3x^3y^2)^3$ **51.** $(2pq)^3(2p^2)^3$

52. $\frac{(x^2y^3)^2}{(xz)^3}$ **53.** $\left(\frac{-6x^7t}{s^6}\right)^2$ **54.** $\left(\frac{abc}{-2d}\right)^3$

In Problems 55 through 60, simplify each. See Example 12.

55. 9^0 **56.** $4x^0yx^2$ **57.** $\left(\frac{5x^6y^7}{-3z^8}\right)^0$

58. $-x^0$ **59.** $-12y^0$ **60.** $(-8)^0$

Practice Exercises

In Problems 61 through 66, identify the base and simplify.

61. -7^2 **62.** 7^2 **63.** $(-7)^2$

64. -7^0 **65.** $(-7)^0$ **66.** 0^7

Simplify each of the following, if possible.

67. $(-k)^3$ **68.** $(-k)^8$ **69.** $(-k)^0$

70. x^5x^4 **71.** $y^{12}y^{24}$ **72.** $4^7 \cdot 4^7$

73. $p \cdot p^8$ **74.** $3^{12} \cdot 3$ **75.** s^3t^9

76. $(h^3)^3$ **77.** $(2^3)^2$ **78.** $(9^9)^3$

79. $(st)^4$ **80.** $(4 + x)^3$ **81.** $(13x)^2$

82. $\left(\frac{y}{x}\right)^5$ **83.** $\left(\frac{r}{2}\right)^6$ **84.** $\left(\frac{-6}{x}\right)^3$

85. $\frac{x^{10}}{x^8}$ **86.** $\frac{t^5}{t^6}$ **87.** $\frac{q^2}{q}$

88. $\frac{3^5}{3^5}$ **89.** $\frac{(x-y)^4}{(x-y)^4}$ **90.** $\frac{(2a+1)^6}{(2a+1)^7}$

91. $a(b^4)^9$ **92.** $(3x^5)^5$ **93.** $\left(\frac{4r}{3}\right)^3$

94. $y^2(y^3)^5$ **95.** $(4x^3y^4)^4$ **96.** $\left(\frac{3v^7}{2k^2}\right)^3$

97. $(-2x^2y^3)^3$ **98.** $(-3x^3y^2)^2$ **99.** $\frac{(7a)^2}{(7a)^3}$

100. $\left(\frac{s^3t^4}{v^5}\right)^6$ **101.** $\left(\frac{-6x^7t}{s^6}\right)^3$ **102.** $\frac{(2xy)^2}{(x^2y)^3}$

Challenge Problems

103. a. Simplify each of the following:

$(-1)^0$; $(-1)^1$; $(-1)^2$; $(-1)^3$; $(-1)^4$;

b. Make a rule for simplifying $(-1)^n$ where n is any whole number.

104. Simplify $\frac{9^5}{3^5}$. $\left(\textit{Hint:} \text{ Use Property 4, } \left(\frac{x}{y}\right)^n = \frac{x^n}{y^n}.\right)$

105. Simplify $\frac{32^3}{8^5}$. (*Hint:* Use powers of 2.)

106. What is wrong with saying $\frac{x^8}{x^4} = x^2$?

IN YOUR OWN WORDS...

107. What does x^n mean when n is a natural number?

108. What are the five properties of exponents?

109. What do the expressions -3^2 and $(-3)^2$ mean?

110. Make up an expression raised to a power that can be simplified using the properties of exponents and one that cannot be simplified using the properties of exponents.

2.2 NEGATIVE INTEGER EXPONENTS

In the previous section we worked with exponents that were natural numbers or zero. We now examine negative integer exponents. Recall the five properties of exponents.

Properties of Exponents

For m and n natural numbers and x and y real numbers,

1. $x^m \cdot x^n = x^{m+n}$	Product with same base
2. $(x^m)^n = x^{m \cdot n}$	Power to a power
3. $(x \cdot y)^n = x^n y^n$	Product to a power
4. $\left(\frac{x}{y}\right)^n = \frac{x^n}{y^n}$; $y \neq 0$	Quotient to a power

$$\textbf{5.}\ \frac{x^m}{x^n} = \begin{cases} x^{m-n} & \text{if } m > n \\ 1 & \text{if } m = n \qquad x \neq 0 \text{ for each case} \\ \frac{1}{x^{n-m}} & \text{if } n > m \end{cases}$$ Quotient with same base

Now suppose we have a definition for 2^{-5} such that the five properties of exponents are still valid. If so, the following must be true:

$$2^5 \cdot 2^{-5} = 2^{5+(-5)} \quad \text{Add exponents.}$$
$$= 2^0$$
$$= 1 \quad \text{Definition of 0 exponent}$$

If $2^5 \cdot 2^{-5} = 1$, then 2^5 and 2^{-5} are reciprocals. Therefore,

$$2^{-5} = \frac{1}{2^5}$$

Negative Exponent

For x any nonzero real number and n a natural number,

$$x^{-n} = \frac{1}{x^n}$$

This definition is consistent with all five of the properties of exponents.

EXAMPLE 1. Rewrite each expression without negative exponents and in simplest form.

(a) 5^{-3} (b) x^{-11} (c) 2^4x^{-5} (d) $(-5)^{-2}$ (e) -3^{-2} (f) $7x^{-3}$

Solutions:

(a) $5^{-3} = \frac{1}{5^3}$ Definition

$= \frac{1}{5 \cdot 5 \cdot 5}$ Definition

$= \frac{1}{125}$

Notice that the negative exponent did not give a negative answer. 5^{-3} is the *reciprocal* of 5^3.

(b) $x^{-11} = \frac{1}{x^{11}}$ Definition

(c) $2^4x^{-5} = 2^4 \cdot \frac{1}{x^5}$ Definition

$= 16 \cdot \frac{1}{x^5}$

$= \frac{16}{x^5}$

(d) $(-5)^{-2} = \frac{1}{(-5)^2}$ Definition

$= \frac{1}{25}$

Note how the parentheses show the base for the exponent is -5 and not 5.

(e) $-3^{-2} = -\frac{1}{3^2}$ Definition

$= -\frac{1}{9}$

Notice that in this case the base for the exponent is 3 and *not* -3.

Be Careful!

(continued)

(f) $7x^{-3} = 7 \cdot \frac{1}{x^3}$ Definition

$= \frac{7}{x^3}$

Notice that the exponent -3 has a base of x, not $7x$. □

Be Careful!

Including this last definition, the properties of exponents now can be written as follows.

Properties of Exponents

For m and n any *integers* and x and y real numbers,

1. $x^m \cdot x^n = x^{m+n}$ Product with same base
2. $(x^m)^n = x^{m \cdot n}$ Power to a power
3. $(x \cdot y)^n = x^n y^n$ Product to a power
4. $\left(\frac{x}{y}\right)^n = \frac{x^n}{y^n}$; $y \neq 0$ Quotient to a power
5. $\frac{x^m}{x^n} = x^{m-n}$; $x \neq 0$ Quotient with same base

Compare this form of Property 5 with that of Section 2.1 and notice the simplification that results in Property 5 when negative exponents are allowed. Consider this example:

$$\frac{2^3}{2^5} = 2^{3-5} \quad \text{Quotient with same base}$$

$$= 2^{-2}$$

$$= \frac{1}{2^2} \quad \text{Definition}$$

$$= \frac{1}{4}$$

EXAMPLE 2. Use the properties of exponents to simplify each. The answers should not contain negative exponents.

(a) $5^5 \cdot 5^{-3}$ (b) $4^3 \cdot 4^{-7}$

Solutions:

(a) $5^5 \cdot 5^{-3} = 5^{5+(-3)}$ Add exponents. Product with same base

$= 5^2$

$= 25$

(b) $4^3 \cdot 4^{-7} = 4^{3+(-7)}$ Product with same base

$= 4^{-4}$

$= \frac{1}{4^4}$ Definition

$= \frac{1}{256}$ □

EXAMPLE 3. Use the properties of exponents to simplify each. The answers should not contain negative exponents. Assume that all variables represent nonzero real numbers.

(a) $(x^{-2})^{-3}$ (b) $(3x^{-4})^{-2}$

Solutions:

(a) $(x^{-2})^{-3} = x^{(-2)(-3)}$ Multiply exponents. Power to a power

$= x^6$

(b) $(3x^{-4})^{-2} = 3^{-2}(x^{-4})^{-2}$ Raise each factor to the -2 power. Product to a power

$= 3^{-2}x^{(-4)(-2)}$ Multiply exponents. Power to a power

$= 3^{-2}x^8$

$= \frac{1}{3^2} \cdot x^8$ Definition

$= \frac{x^8}{9}$ □

EXAMPLE 4. Simplify each and write answers without negative exponents. Assume that all variables are nonzero real numbers.

(a) $\frac{a^{-3}}{a^6}$ (b) $\frac{b^6}{b^{-2}}$ (c) $\frac{x^{-1}}{x^{-3}}$ (d) $\frac{5^{-2}x^4}{5^{-3}x^{-5}}$

Solutions:

(a) $\frac{a^{-3}}{a^6} = a^{-3-6}$ Subtract exponents. Quotient with same base

$= a^{-9}$

$= \frac{1}{a^9}$ Definition

(b) $\frac{b^6}{b^{-2}} = b^{6-(-2)}$ Quotient with same base

$= b^{6+2}$

$= b^8$

(c) $\frac{x^{-1}}{x^{-3}} = x^{-1-(-3)}$ Subtract exponents.

$= x^{-1+3}$

$= x^2$

(d) $\frac{5^{-2}x^4}{5^{-3}x^{-5}} = 5^{-2-(-3)}x^{4-(-5)}$ Subtract exponents.

$= 5^{-2+3}x^{4+5}$

$= 5x^9$ □

Scientific Notation

If we multiply 12,345 by 12,345 on a calculator it may display an answer like

1.52399⁰⁸ or **1.5239903 08**

These are short for

$$1.5239903 \times 10^8$$

a large number written in scientific notation. If we divide 3 by 12,345 on a calculator, it may display an answer like

2.4301337⁻⁰⁴ or **2.4301337 −04**

These are short for

$$2.4301337 \times 10^{-4}$$

a small number written in scientific notation. Each of the numbers is written as a number between 1 and 10 times a power of 10. As is usually the case in calculator computations, these are both approximations. Scientific notation is a convenient way to express very large and very small numbers. Two examples that occur in physics are Planck's constant,

$$h \approx 0.000000000000000000000000006625 \text{ erg-sec}$$

and the speed of light in a vacuum,

$$c \approx 29{,}900{,}000{,}000 \text{ cm/sec}$$

Written in scientific notation, these become

$$h \approx 6.625 \times 10^{-27} \text{erg-sec}$$

$$c \approx 2.99 \times 10^{10} \text{ cm/sec}$$

These are both approximations, as the symbol $\approx$ indicates.

Writing a Number in Scientific Notation

1. Starting with the number in decimal format, move the decimal point until the number is between 1 and 10.
2. Multiply the number formed in Step 1 by 10 raised to the number of decimal places moved. If the original number was less than 1, the power of 10 is negative. If the original number was greater than 10, the power is positive.
3. This procedure is for a *positive* number. If the original number is negative, perform Steps 1 and 2 on the absolute value of the original number, and then append a negative sign to the result.

EXAMPLE 5. Write each in scientific notation.

(a) 93,000,000 (b) 0.000001554 (c) −254,000

Solutions:

(a) We think of 93,000,000 as 93000000.0 and move the decimal point until we have a number between 1 and 10.

Step 1 9 3 0 0 0 0 0 0.0

We moved the decimal point 7 places. Now we write

Step 2 $$9.3 \times 10^7$$

The exponent is *positive* because 93,000,000 is greater than 10.

Step 1 (b) 0.0 0 0 0 0 1 5 5 4

We moved the decimal point 6 places. We write

Step 2 $$1.554 \times 10^{-6}$$

The exponent is *negative* because 0.000001554 is less than 1.

(c) −254,000 is negative, so we work with its absolute value.

Step 1 2 5 4 0 0 0.0

We moved the decimal point 5 places. We have now

Step 2 $$2.54 \times 10^5$$

The exponent is *positive* because 254,000 is greater than 10. As the original number was negative, we write,

Step 3 $$-2.54 \times 10^5$$

□

EXAMPLE 6. Write the following numbers without exponents.

(a) 8.771×10^{14}
(b) 3.2×10^{-13}
(c) -9.99231×10^{-3}

Solutions:

To change scientific notation to standard decimal format, we reverse the steps given earlier.

(a) 8 7 7 1 0 0 0 0 0 0 0 0 0 0 0.0

877,100,000,000,000

Note that we moved the decimal point 14 places to the right (to make a number larger than 10) because 14 is positive.

(b) 0.0 0 0 0 0 0 0 0 0 0 0 0 3 2

0.00000000000032

This time we moved the decimal point 13 places to the left (to make a number smaller than 1) because −13 is negative. *(continued)*

(c) As the given number is negative, we work with its absolute value.

0.0 0 9 9 9 2 3 1

0.00999231

We moved the decimal point 3 places to the left because the exponent of 10, -3, is negative. Now we append a negative sign because the original number was negative.

$$-0.00999231$$

□

EXAMPLE 7. Write each number without exponents.

(a) $(2.1 \times 10^{-5})(-2.0 \times 10^{-2})$ (b) $\dfrac{1.6 \times 10^{5}}{3.2 \times 10^{-3}}$

Solutions:

(a) $(2.1 \times 10^{-5})(-2.0 \times 10^{-2}) = (2.1)(-2.0)(10^{-5})(10^{-2})$ Add exponents.

$= -4.2 \times 10^{-7}$

$= -0.00000042$

(b) $\dfrac{1.6 \times 10^{5}}{3.2 \times 10^{-3}} = \dfrac{1.6}{3.2} \times 10^{5-(-3)}$ Subtract exponents.

$= 0.5 \times 10^{5+3}$

$= 0.5 \times 10^{8}$

$= 50{,}000{,}000$

□

CALCULATOR BOX

Scientific Notation on a Calculator

EXP or EE

To enter 7.2×10^{12} in a calculator, press the keys

7.2 EE 12 and read **7.2¹²**. Some calculators may have an

EXP key instead of EE

To enter 1.05×10^{-8}, press the keys

1.05 EE 8 ± ; the display should read **1.05⁻⁸**

Calculator Exercises

Enter these numbers in a calculator.

1. 3.58×10^{24}

2. 1.01×10^{-5}

3. 4.3×10^{23}

4. 9.1×10^{-19}

PROBLEM SET 2.2

Assume that all variables are nonzero real numbers.

Warm-ups

In Problems 1 through 18, rewrite each expression without negative exponents and simplify. See Example 1.

1. 2^{-4}
2. 3^{-2}
3. 2^{-1}
4. x^{-2}
5. y^{-5}
6. A^{-1}
7. $(-2)^{-4}$
8. $(-2)^{-5}$
9. -2^{-2}
10. $(-5)^{-2}$
11. $(-5)^{-3}$
12. 5^{-2}
13. -5^{-2}
14. -5^{-3}
15. $2x^{-3}$
16. $3^{-2}x$
17. $-2x^{-3}$
18. $-7v^{-2}$

In Problems 19 through 38, use the properties of exponents to simplify the expressions. The answers should not contain negative exponents.
In Problems 19 through 26, see Example 2.

19. $2^3 \cdot 2^{-2}$
20. $3^{-3} \cdot 3^5$
21. $4^3 \cdot 4^{-5}$
22. $2^5 \cdot 2^{-9}$
23. $x^6 \cdot x^{-2}$
24. $z^5 z^{-4}$
25. $x^{-4}x^2$
26. $y^6 y^{-6}$

In Problems 27 through 38, see Example 3.

27. $(2^{-3})^2$
28. $(3^2)^{-2}$
29. $(2^{-2})^{-3}$
30. $(x^3)^{-4}$
31. $(x^{-5})^2$
32. $(z^{-3})^{-8}$
33. $(xy)^{-3}$
34. $(2x)^{-4}$
35. $(-3x)^{-2}$
36. $(-2s)^{-4}$
37. $(-4t)^{-2}$
38. $-(ax)^{-6}$

In Problems 39 through 47, simplify the expressions. The answers should not contain negative exponents. See Example 4.

39. $\dfrac{10^{-3}}{10^{-1}}$
40. $\dfrac{6^{-9}}{6^{-7}}$
41. $\dfrac{5^2}{5^{-1}}$
42. $\dfrac{x^4}{x^{-3}}$
43. $\dfrac{r^{-3}}{r^4}$
44. $\dfrac{y^7}{y^{-1}}$
45. $\dfrac{k^{-5}}{k^{-3}}$
46. $\dfrac{6^{-2}x^3}{6^{-3}x^{-2}}$
47. $\dfrac{2^{-4}b^{-5}}{2^{-9}b^{-3}}$

In Problems 48 through 53, write the given numbers in scientific notation. See Example 5.

48. 5280
49. 0.0000012345
50. 0.094
51. 77,722,000,000,000
52. −32,000
53. −0.000000021367

In Problems 54 through 59, write the given numbers without exponents. See Example 6.

54. 1.609×10^5

55. 5.43×10^{-5}

56. 1.1×10^{-8}

57. 6.81×10^2

58. -8.0×10^{11}

59. -4.32101×10^{-2}

In Problems 60 through 63, write the number without exponents. See Example 7.

60. $\dfrac{2.4 \times 10^5}{0.2 \times 10^{-4}}$

61. $\dfrac{4.4 \times 10^{-6}}{2.2 \times 10^{-8}}$

62. $(2.5 \times 10^5)(2.0 \times 10^{-3})$

63. $(1.1 \times 10^{-2})(5.0 \times 10^{-5})$

Practice Exercises

In Problems 64 through 123, use the properties of exponents to simplify the expressions. The answers should not contain negative exponents.

64. 2^{-5}

65. 3^{-3}

66. 3^{-1}

67. x^{-11}

68. k^{-5}

69. T^{-1}

70. $(-3)^{-2}$

71. $(-3)^{-3}$

72. -3^{-3}

73. -2^{-2}

74. $(-6)^{-2}$

75. $(-5)^{-3}$

76. 4^{-3}

77. -4^{-3}

78. -4^{-2}

79. $4x^{-2}$

80. $5^{-3}x$

81. $6x^{-7}$

82. $2r^{-1}$

83. $-8x^{-2}$

84. $-5s^{-3}$

85. bt^{-3}

86. $-cy^{-4}$

87. k^2x^{-2}

88. $a^{-1}x^{-2}$

89. $3a^{-3}t^{-2}$

90. $-q^{-2}x^{-2}$

91. $3^3 \cdot 3^{-2}$

92. $2^{-3} \cdot 2^5$

93. $4^{-3} \cdot 4^4$

94. $5^3 \cdot 5^{-5}$

95. $3^4 \cdot 3^{-8}$

96. $6^{-6} \cdot 6^4$

97. $10^4 \cdot 10^{-2}$

98. $x^{11} \cdot x^{-4}$

99. z^4z^{-5}

100. $x^{-2}x^4$

101. v^5v^{-10}

102. u^3u^{-3}

103. $(3^{-2})^3$

104. $(3^2)^{-3}$

105. $(7^{-2})^{-1}$

106. $(5^{-5})^3$

107. $(5^3)^{-5}$

108. $(5^{-5})^{-3}$

109. $(x^2)^{-3}$

110. $(x^{-4})^3$

111. $(x^{-4})^{-9}$

112. $(ab)^{-5}$

113. $(2x)^{-5}$

114. $(3t)^{-3}$

115. $(-2r)^{-6}$

116. $(-3r)^{-2}$

117. $-(bw)^{-4}$

118. $\frac{y^2}{y^{-2}}$

119. $\frac{x^{-2}}{x^4}$

120. $\frac{p^5}{p^{-1}}$

121. $\frac{w^{-6}}{w^{-4}}$

122. $\frac{4^{-8}}{4^{-6}}$

123. $\frac{3^{-4}}{3^{-5}}$

In Problems 124 through 131, write the numbers in scientific notation.

124. McDonald's has served 55 billion hamburgers.

125. The speed of light is 186,000 mi/s.

126. The mass of an electron is 0.0000000000000000000000000000009109 kg.

127. The density of oxygen is 0.001429 units.

128. Saturn is 885,000,000 mi from the sun.

129. Mount McKinley is 20,300 ft high.

130. A virus ranges in size from 0.00001 to 0.003 mm.

131. A nanosecond is 1 billionth of a second.

In Problems 132 through 137, write the given numbers without exponents.

132. 3.101×10^4

133. 1.23×10^{-3}

134. $(5.67 \times 10^3)(2.0 \times 10^{-2})$

135. 9.1×10^{-2}

136. $(-2.0 \times 10^{-3})(1.5 \times 10^{-2})$

137. 3.45×10^{10}

Challenge Problems

In Problems 138 through 141, simplify each. Answers should not contain negative exponents.

138. $\left(\frac{1}{2^{-2}}\right)$

139. $\left(\frac{1}{2}\right)^{-2}$

140. $2^{-2} - 3^{-1}$

141. $\frac{2^{-1} + 3^{-2}}{6^{-1}}$

In Problems 142 through 144, simplify each expression. Write each without negative exponents. Use Property 4 and then the definition of negative exponent. Simplify the resulting division problem.

142. $\left(\frac{2}{x}\right)^{-4}$

143. $\left(\frac{y}{3}\right)^{-3}$

144. $\left(\frac{a}{v}\right)^{-4}$

In Problems 145 through 147, evaluate each expression when x is -2.

145. $x^2 - x^{-2}$

146. $(2x - 1)^{-2}$

147. $\frac{x^{-1} + x^{-2}}{x^{-2}}$

IN YOUR OWN WORDS...

148. What is the relationship between 3^{-7} and 3^7?

149. Explain what scientific notation is and why it is useful.

2.3 DEFINING, EVALUATING, AND SIMPLIFYING POLYNOMIALS

The expression bx^n, where b is a nonzero real number and n is a nonnegative integer, is called a **monomial** in one variable. We call b the **coefficient,** or **numerical coefficient,** x the **variable**, and n the **degree.** The following are monomials in one variable.

[1] $3x^2$ [2] $\frac{1}{2}y^3$ [3] $-5z^4$

	COEFFICIENT	VARIABLE	DEGREE
[1]	3	x	2
[2]	$\frac{1}{2}$	y	3
[3]	-5	z	4

The following expressions are also monomials.

[4] x^5 [5] $-t^7$ [6] 8

Since $x^5 = 1 \cdot x^5$, we say that x^5 has an *understood* coefficient of 1. Similarly, $-t^7$ has an understood coefficient of -1. Since $8 = 8 \cdot 1 = 8X^0$, for *any* letter X, we say 8 is a monomial of degree 0. We call such monomials, **constants.**

	COEFFICIENT	VARIABLE	DEGREE
[4]	1	x	5
[5]	-1	t	7
[6]	8	any letter	0

A **polynomial** is a sum of monomials. Its degree is the highest degree of the monomials.

Consider

$$5x^7 + 3x^3 + 2x^2 + 6$$

This is a polynomial, since it is the sum of four monomials. Its degree is seven. This polynomial is written so that the monomial of highest degree comes first and the degrees of the other monomials decrease in order. This is called **standard form.** The coefficient of the first term is called the **leading coefficient.** The monomials are most often called **terms.** The last term, 6, is called the **constant term.** Another example of a polynomial is

$$3y^5 + (-2y^2) + (-9)$$

We write such polynomials as

$$3y^5 - 2y^2 - 9$$

This is a fifth-degree polynomial with three terms, written in standard form.

EXAMPLE 1. Write each polynomial in standard form and give the degree of each.

(a) $5 + 3x^2 + 2x$ (b) $2x + 3$
(c) $x^5 - x^6 + 2$ (d) $-5 - 2y^2 - 3y + 3y^5$

Solutions:

(a) Standard form is $3x^2 + 2x + 5$.
The degree is 2.

(b) Standard form is $2x + 3$.
The degree is 1.

(c) Standard form is $-x^6 + x^5 + 2$.
The degree is 6.

(d) Standard form is $3y^5 - 2y^2 - 3y - 5$.
The degree is 5. □

Polynomials that have one term are called **monomials.**
Polynomials that have two terms are called **binomials.**
Polynomials that have three terms are called **trinomials.**

Listed next are examples of each.

MONOMIALS	BINOMIALS	TRINOMIALS
x	$x^2 + x$	$x^5 + 4x^2 + 7$
$5y^4$	$z^3 - 7$	$w^2 - 8w + 6$

Polynomials have many different values depending on the value of each variable. To **evaluate** a polynomial means to find its value, given some value of the variable. Polynomials are expressions, and we evaluate them in the same manner we evaluated expressions in Section 1.5.

EXAMPLE 2. Evaluate each polynomial for the given values of each variable.

(a) $3x^2 - x + 4$ if x is -2.
(b) $2y^3 + 3y^2 + y + 5$ if y is -2.
(c) $3z^4 - 4z^3 - 2z^2 + z$ if z is -1.

Solutions:

(a)
$$\begin{aligned} 3x^2 - x + 4 &= 3(-2)^2 - (-2) + 4 \\ &= 3(4) + 2 + 4 \\ &= 12 + 2 + 4 \\ &= 18 \end{aligned}$$

(continued)

(b) $2y^3 + 3y^2 + y + 5 = 2(-2)^3 + 3(-2)^2 + (-2) + 5$
$= 2(-8) + 3(4) - 2 + 5$
$= -16 + 12 - 2 + 5$
$= -1$

(c) $3z^4 - 4z^3 - 2z^2 + z = 3(-1)^4 - 4(-1)^3 - 2(-1)^2 + (-1)$
$= 3 \cdot 1 - 4(-1) - 2 \cdot 1 - 1$
$= 3 + 4 - 2 - 1$
$= 4$ □

Sometimes we work with polynomials that contain like terms. As we saw in Section 1.5, the commutative, associative, and distributive properties allow us to combine like terms. We often simplify polynomials by combining like terms.

EXAMPLE 3. Simplify each polynomial by combining like terms.

(a) $8x^4 + 5x^4$ (b) $2y + y$
(c) $5z - 3z + 2z$ (d) $2w^3 - w^3 + 3w^3$

Solutions:

(a) $8x^4 + 5x^4 = (8 + 5)x^4 = 13x^4$
(b) $2y + y = (2 + 1)y = 3y$
(c) $5z - 3z + 2z = (5 - 3 + 2)z = 4z$
(d) $2w^3 - w^3 + 3w^3 = (2 - 1 + 3)w^3 = 4w^3$ □

EXAMPLE 4. Combine like terms in each.

(a) $3a^2 - 2a^2 + 12a + 4a$ (b) $5x^2 - 5 + 3x^2 + 4$ (c) $3y + 7y^2$

Solutions:

(a) $3a^2$ and $-2a^2$ are like terms and $12a$ and $4a$ are also like terms.

$$3a^2 - 2a^2 + 12a + 4a = a^2 + 16a$$

(b) $5x^2$ and $3x^2$ are like terms, as are -5 and 4. We use the commutative property and rearrange the terms.

$$5x^2 - 5 + 3x^2 + 4 = 5x^2 + 3x^2 - 5 + 4$$
$$= 8x^2 - 1$$

(c) There are no like terms in this polynomial. □

The distributive property can also be used to simplify a polynomial.

EXAMPLE 5. Simplify each polynomial.

(a) $2(x + 3) + 6x$ (b) $-3(4 - 3y) - 8 + y$

Solutions:

(a) The order of operations applies. We must multiply with the distributive property first.

$$2(x + 3) + 6x = 2(x) + 2(3) + 6x \quad \text{Distributive}$$
$$= 2x + 6 + 6x \quad \text{Combine like terms.}$$
$$= 8x + 6$$

(b) $-3(4 - 3y) - 8 + y = -3(4) - (-3)(3y) - 8 + y$
$$= -12 + 9y - 8 + y$$
$$= -20 + 10y$$

We can write this in standard form as $10y - 20$. □

PROBLEM SET 2.3

Warm-ups

In Problems 1 through 8, write each polynomial in standard form. Give its degree, and identify those that are monomials, binomials, or trinomials. See Example 1.

1. $x + 6$

2. $4 - 3x + x^2$

3. $\frac{1}{2}x - 3x^3 + 4x^2 - 7$

4. $4y^3 + \frac{2}{3}y - 17y^7 + 89$

5. $x^5 - 25$

6. $3x + x^2 + x^3$

7. $-5y^5 + 6y + y^7 - 1$

8. $9 - z^3$

In Problems 9 through 14, evaluate each polynomial when x is 2. See Example 2a.

9. $3x + 7$

10. $1 - x^2$

11. $8 - x$

12. $x^2 - x$

13. $4x^2 - 2x + 1$

14. $2x^3 - x - 2$

In Problems 15 through 20, evaluate each polynomial when x is −2. See Example 2b and c.

15. $3x + 7$

16. $8 - x$

17. $x^2 - x + 1$

18. $3x^3 - 5x^2 + x$

19. $x^4 + 3x^2 + 1$

20. $x^5 + 3x^3 + 5x$

In Problems 21 through 30, simplify by combining like terms. See Example 3.

21. $3x + 7x$

22. $8x^2 - x^2$

23. $7x - 4x + x$

24. $-2t - 8t - t$

25. $-w^3 - 2w^3 + 6w^3$

26. $3y - 2y - y$

27. $4a + a - 2a$

28. $12z - 8z - z$

29. $-11x^2 - 15x^2 + 5x^2$

30. $20y^2 - 25y^2 + y^2$

In Problems 31 through 40, combine like terms. See Example 4.

31. $3x + 2 + 5x + 1$

32. $-2y - 1 - 3y - 4$

33. $4r^3 - 2 + 3r^3 - 3$

34. $4x - 7x - 5 + 3$

35. $5x^2 + x + 2 - 3x^2 - 5x + 4$

36. $x^2 - 3x + 4x + 2x^2 - 5 + 4$

37. $-3y + 2y^2 - 5 - y^2 + 6 - 2y$

38. $5y^2 + 3x^2 - 4y^2 - x^2$

39. $3x^2 + 4y^2 - x^2 - y^2$

40. $3x^3 - x^2 + x^3 - 2x^2$

In Problems 41 through 50, simplify each polynomial. See Example 5.

41. $3(2x - 1) + 2x$

42. $-2(x + 3) - 4x + 3$

43. $5(3 - 2x^3) - 16$

44. $2(3y + 4) - 6y$

45. $-7(1 - t) + 7 - 7t$

46. $\frac{1}{2}(2x^2 + 4) + x^2$

47. $\frac{2}{3}(6 - 15x) - 2$

48. $-\frac{3}{5}\left(\frac{5}{2}x + \frac{1}{3}\right) + \frac{5}{2}x$

49. $1.5(2z - 3) - 1.8$

50. $-3.0(2.1 - 1.1w) + 2.1w$

Practice Exercises

In Problems 51 through 56, write each polynomial in standard form. Give its degree, and identify those that are monomials, binomials, or trinomials.

51. $5 - x$

52. $3 - 2t + t^2$

53. $x^9 - 2x^8$

54. $9z + z^4 + z^3$

55. $5y^5 - 7y - y^7 - 1$

56. $49x$

In Problems 57 through 62, evaluate each polynomial when x is 3.

57. $3x + 7$

58. $2x^3 - 3x - 22$

59. $15 - 2x$

60. $x^3 - x^2$

61. $-5x^2 + 3x - 12$

62. $2x^2 + x - 2$

In Problems 63 through 68, evaluate each polynomial when x is -3.

63. $3x + 7$

64. $x^2 - x + 1$

65. $15 - 2x$

66. $x^3 + 4x^2 - x + 5$

67. $x^4 + 3x^2 + x$

68. $x^2 + 3x^3 + 5x - 3$

In Problems 69 through 90, simplify each polynomial.

69. $3x + x$

70. $5x^2 - x^2$

71. $6x - 8x + x$

72. $-3t - 5t - t$

73. $-w^3 + 3w^3 - 6w^3$

74. $2(5x - 7) + 3x - 9$

75. $c + 3c - 2c$

76. $-x^2 - 5x^2 + 4x^2$

77. $2z - 7z - z$

78. $10y^2 - 5y^2 - y^2$

79. $4x + 1 + x - 1$

80. $-y - 3 - 5y + 4$

81. $3r^3 - 5 + 2r^3 - 1$

82. $6(y - 7) - 8 + 9y$

83. $4(x + 7) - 2x + 3$

84. $x^2 - x + 3x + 2x^2 - 7 + 1$

85. $-2y + 6y^2 - 8 - y^2 + 3 - 7y$

86. $-7(y^2 + 3) - y^2$

87. $5x^2 + 3y^2 + x^2 + y^2$

88. $\frac{1}{2}(x^2 - 6) - \frac{3}{2}x^2$

89. $\frac{1}{5}(5x + \frac{5}{2}) + \frac{3}{2}$

90. $2.5(y - 2.5) + 5.5$

Challenge Problems

Evaluate each polynomial when x is -1. Assume that n is a natural number.

91. $x^n + 1$

92. $x^{2n} + 1$

IN YOUR OWN WORDS...

93. What is a polynomial?

94. What do we mean by standard form for a polynomial?

2.4 ADDITION AND SUBTRACTION

We have seen that the Commutative, Associative, and Distributive Properties enable us to combine like terms. Addition of polynomials is just a matter of combining like terms.

EXAMPLE 1. Perform the addition in each.

(a) $(8x^2 - 3x + 11) + (x^2 + 5x - 4)$

(b) $(4x^3 + 4x^2 - x + 2) + (-7x^2 + x - 5)$

Solutions:

(a) We gather the like terms.

$$(8x^2 - 3x + 11) + (x^2 + 5x - 4) = (8x^2 + x^2) + (-3x + 5x) + (11 - 4)$$
$$= 9x^2 + 2x + 7$$

(b) $(4x^3 + 4x^2 - x + 2) + (-7x^2 + x - 5)$

$$= 4x^3 + (4x^2 - 7x^2) + (-x + x) + (2 - 5)$$
$$= 4x^3 - 3x^2 - 3$$ □

Subtraction of polynomials is defined in terms of addition as it was for real numbers. Again we use the idea of the opposite.

> If P is a polynomial, $-P$ is called the **opposite** of P, and
> $$P + (-P) = 0$$

We must learn how to find the opposite of a polynomial before we can subtract. The opposite of $2x$ is $-2x$ because

$$2x + (-2x) = 0$$

The opposite of $x^2 - x + 1$ is $-x^2 + x - 1$ because

$$(x^2 - x + 1) + (-x^2 + x - 1) = 0$$

How do we find the opposite of a polynomial? The preceding discussion suggests that to find the opposite of a polynomial, we change the sign of each term in the polynomial.

We learned in Chapter 1 that the opposite of a real number is -1 times the real number. That is, $-x = -1x$. The same idea is true for polynomials.

> If P is a polynomial,
> $$-P = -1(P)$$

We can find the opposite of a polynomial by multiplying the polynomial by -1. The distributive property allows us to do this.

EXAMPLE 2. Find the opposite of each polynomial.

(a) $4x^2 - 3x - 1$ (b) $x - 1$ (c) $y + 2$

Solutions:

(a) $-(4x^2 - 3x - 1) = -1(4x^2 - 3x - 1)$
$= -4x^2 + 3x + 1$

(b) $-(x - 1) = -1(x - 1)$
$= -x + 1$

(c) $-(y + 2) = -1(y + 2)$
$= -y - 2$

□

We define subtraction as we did for real numbers.

> If P and Q are polynomials, then
> $$P - Q = P + (-Q)$$

To subtract Q from P, we add the opposite of Q to P.

EXAMPLE 3. Perform the subtractions indicated.

(a) $(4x^2 - 17) - (x^2 - 11)$

(b) $(3x^3 - 7x + 2) - (x^3 + x - 8)$

(c) $(9x^4 + x^2 - 6) - (-4x^2 + x + 1)$

Solutions:

(a) To subtract, we add the opposite. The opposite of $x^2 - 11$ is $-x^2 + 11$.

$$(4x^2 - 17) - (x^2 - 11) = (4x^2 - 17) + (-x^2 + 11)$$
$$= 4x^2 - 17 - x^2 + 11$$
$$= 3x^2 - 6$$

(b) The opposite of $x^3 + x - 8$ is $-x^3 - x + 8$.

$$\begin{aligned}(3x^3 - 7x + 2) - (x^3 + x - 8) &= (3x^3 - 7x + 2) + (-x^3 - x + 8)\\ &= 3x^3 - 7x + 2 - x^3 - x + 8\\ &= 2x^3 - 8x + 10\end{aligned}$$

(c) The opposite of $-4x^2 + x + 1$ is $4x^2 - x - 1$.

$$\begin{aligned}(9x^4 + x^2 - 6) - (-4x^2 + x + 1) &= (9x^4 + x^2 - 6) + (4x^2 - x - 1)\\ &= 9x^4 + x^2 - 6 + 4x^2 - x - 1\\ &= 9x^4 + 5x^2 - x - 7\end{aligned}$$ □

When we combine the operations of addition, subtraction, and multiplication, the order of operations applies.

EXAMPLE 4. Perform the operations indicated.

(a) $[(x^2 - x) - (2x^2 + 3x + 2)] + (6x^2 - x + 1)$ (b) $(x^2 - 1) - 2(x^2 + 4)$

Solutions:

(a) We start with the inside parentheses.

$$\begin{aligned}&[(x^2 - x) - (2x^2 + 3x + 2)] + (6x^2 - x + 1)\\ &\quad = [(x^2 - x) + (-2x^2 - 3x - 2)] + (6x^2 - x + 1)\\ &\quad = [-x^2 - 4x - 2] + (6x^2 - x + 1)\\ &\quad = 5x^2 - 5x - 1.\end{aligned}$$

(b) We perform the multiplication first.

$$\begin{aligned}(x^2 - 1) - 2(x^2 + 4) &= (x^2 - 1) - (2x^2 + 8)\\ &= (x^2 - 1) + (-2x^2 - 8)\\ &= -x^2 - 9\end{aligned}$$ □

Often when we add and subtract polynomials, we think of them in terms of parentheses preceded by a positive or a negative sign.

Look at an addition:

$$\begin{aligned}(5x^2 + x - 8) + (3x^2 + 4x + 3) &= 5x^2 + x - 8 + 3x^2 + 4x + 3\\ &= 8x^2 + 5x - 5\end{aligned}$$

We removed the parentheses. Parentheses preceded by a positive sign can be removed without changing the problem.

Look at a subtraction.

$$\begin{aligned}(4x^2 - 3x - 7) - (x^2 + 5x - 2) &= 4x^2 - 3x - 7 - x^2 - 5x + 2\\ &= 3x^2 - 8x - 5\end{aligned}$$

Thus parentheses preceded by a negative sign can be removed if the sign of *each* term inside the parentheses is changed.

EXAMPLE 5. Subtract $3y - 4$ from the sum of $6y + 7$ and $-4y + 2$.

Solution:

We must add $6y + 7$ and $-4y + 2$ and then subtract $3y - 4$. We use grouping symbols to write this.

$$\begin{aligned}[(6y + 7) + (-4y + 2)] - (3y - 4) &= (2y + 9) - (3y - 4) \\ &= (2y + 9) + (-3y + 4) \\ &= -y + 13\end{aligned}$$ □

In using algebra, polynomials arise in many different situations. The next examples illustrate some of them.

EXAMPLE 6. Express the perimeter of the rectangle shown as a polynomial in standard form.

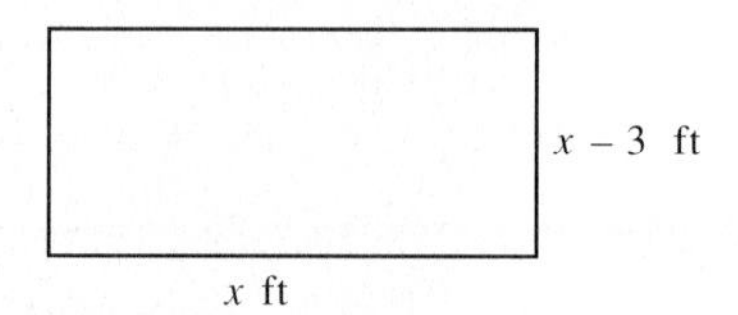

Solution:

The perimeter is the distance around the rectangle. So, it is $x + (x - 3) + x + (x - 3)$.

$$x + (x - 3) + x + (x - 3) = 4x - 6$$

The perimeter is $4x - 6$ ft. □

EXAMPLE 7. If n is the smallest of three consecutive odd integers, express the sum of the three integers as a polynomial in standard form.

Solution:

The three integers are n, $n + 2$, and $n + 4$.

$$n + (n + 2) + n + 4) = 3n + 6$$

The sum is $3n + 6$. □

PROBLEM SET 2.4

Warm-ups

In Problems 1 through 16, perform the operations indicated. See Example 1.

1. $(2x + 3) + (5x + 1)$
2. $(7x + 8) + (2x + 5)$
3. $(x^2 - 5) + (3x^2 - 2)$
4. $(5x^3 - 2) + (x^3 - 1)$
5. $(4z^2 - 6) + (3 + 4z^2)$
6. $(3 + 2x) + (5x + 7)$
7. $(5x - 8) + (7x + 4)$
8. $(11x^2 + 15) + (8x^2 - 11)$

9. $(2t^2 - 5) + (-3t^2 - 2)$

10. $(-3x + 5) + (-2x - 17)$

11. $(6y + y^2 - 5) + (2y^2 - 7)$

12. $(z^2 - 6z) + (2z^2 + 8 - z)$

13. $(x^2 + 4x + 1) + (4x^2 - 7x - 4)$

14. $(5x^3 - 6x^2 + 5x - 7) + (10x^3 + 7x^2 - x - 1)$

15. $(4x^5 - 6x^3 + 8x) + (4x^4 + 7x^3 - x^2 + 2)$

16. $(5x^6 - 3x^4 + 3) + (-4x^5 + 7x^4 + x^3 - 4)$

In Problems 17 through 26, find the opposite of each polynomial. See Example 2.

17. $3x + 1$

18. $t^2 - 3$

19. $6x - 9$

20. $1 - z$

21. $x^2 - 8x - 7$

22. $-x^3 + 2x^2 - x - 3$

23. $5 + 7x^2$

24. $x + 5$

25. $4 - x$

26. $x - 4$

In Problems 27 through 38, perform the operations indicated. See Example 3.

27. $(3x + 7) - (2x + 4)$

28. $(4x + 3) - (x - 3)$

29. $(2x^2 - 8) - (4x^2 + 5)$

30. $(-2z^3 - 3) - (-3z^3 + 1)$

31. $6x^4 - (2x^4 + 3)$

32. $33 - (x - 19)$

33. $x^3 - (2x^2 - x)$

34. $(2y - 5) - (3 + y)$

35. $(2x^2 + 3x - 4) - (x + x^2 - 3)$

36. $(y^3 - 6y - 1) - (6y + 3y^3)$

37. $(6x^3 - 5x^2 + 6x - 5) - (2x^3 - 11x^2 - x + 5)$

38. $(-y^4 - y^3 + 6y + 9) - (-5y^4 - y^2 + 9)$

In Problems 39 through 46, perform the operations indicated. See Example 4.

39. $[(x + 7) + (3x - 4)] - (2x - 3)$

40. $(2x^2 + 11) + 3(4x^2 - 1)$

41. $(4x^3 - 9) - 2(6x^3 + 8) + (8x^3 - 2x^2 + 13)$

42. $[(5x + 2) - (2x - 7)] - 2(3x - 1)$

43. $-2(4x - 1) + [(3x + 8) - (x - 1)]$

44. $(z^2 + 3) + (2z^2 - 8) - (4z^2 + 2)$

45. $2(x - 5) + 4(2 - 3x)$

46. $(5w^2 - 4w + 18) + (4w^2 - w + 4) - (2w^2 - 7w - 14)$

In Problems 47 through 52, see Example 5.

47. Add $4x^2 + 3x - 1$ to $x^2 - 1$.

48. Subtract $3s + 2$ from $s^2 + 7s - 6$.

49. Find the sum of $3x^2 - 7x + 5$ and $3x + 4$.

50. Subtract $x^2 + x + 1$ from the sum of $x^2 + 1$ and $x^2 + 2x + 1$

51. Subract $x - 3$ from $3 - x$

52. Find the sum of $4x^2 + 4x + 4$, $-x^2 - x - 1$, and $3x^2 + 3x + 3$

In Problems 53 through 58, express each quantity as a polynomial in standard form. See Examples 6 and 7.

53. The perimeter of a square with a side of length $2s + 3$ feet.

54. A board is cut into two pieces, as shown. Find the length of the board.

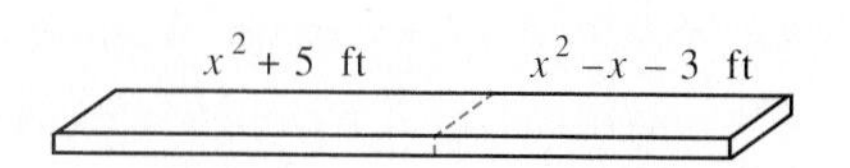

55. The circumference of a circle with radius of $r - 2$ inches.
56. If n is the largest of three consecutive integers, find the sum of the three integers.
57. John has $2x + 5$ dollars and Samuel has $5x - 7$ dollars. How many dollars do they have together?
58. Jim and Charles run from the same point in opposite directions. If Jim runs a distance of $60x$ mi and Charles runs a distance of $40x$ mi, find the distance between them.

Practice Exercises

In Problems 59 through 81, perform the operations indicated.

59. $(x - 5) + (23x + 9)$
60. $(7x + 6) - (x - 4)$
61. $(4y + 3y^2 - 8) + (3y^2 - 9)$
62. $(z^2 - 3z) - (4z^2 + z)$
63. $(x^2 - 4x - 1) - (3x^2 + 7x - 5)$
64. $(5x^2 - 2x + 4) - (7x^2 - 11 - 5x)$
65. $(3x^3 - 4x^2 + 3x - 9) + (12x^3 + 6x^2 + x + 1)$
66. $(2x^4 + x^2 + 5x^3 - x + 4) - (3x^4 - 8x^3 + 5x^2 - 2x - 1)$
67. $(8x^5 - 4x^3 + 7x) - (3x^4 - 17x^3 + x^2 - 12)$
68. $(4x^6 - 2x^4 + 13) - 2(9x^5 + 4x^4 - x^3 + 3)$
69. $(4x^3 - 2x^2 + 8x - 12) + (2x^3 - 7x^2 + 7x - 1)$
70. $(-3x^4 + 6x^3 + x^2 - 2) - (x^4 - 3x^3 + 2x^2 - 9)$
71. $(13x^5 - 4x^3 - 7x) + (11x^5 + 14x^4 - 3x)$
72. $(17x^4 - 18x^3 - 3x^2 + 19) - 3(2x^3 + 2x - 13)$
73. $(x^2 - 7x - 5) - (x^2 + 6x - 4) + (x^2 - 2x + 11)$
74. $-2(8 + 9y) + 3(10y + 6) - (y - 4)$
75. $(z^2 - 3) - (3 + z^2) + (z^2 - 2z - 1)$
76. $(2x - 3) - 3(x + 4)$
77. $[(3x^2 - 4x + 7) + (2x^2 + 10)] - (7x^2 - x)$
78. $[(2x^3 - x^2 - 3x - 7) - (5x^3 + 3x^2 - 6x)] + (4x^3 - 1)$
79. $(4v^4 - v^2 + 4) - [(2v^4 - v + 8) + (2v^3 + 3v^2 - 5v + 8)]$
80. $(t^3 + 27) - [(t^3 - 27) - (27 - t^3)]$
81. $6(x^2 - 4) - (x^2 - 20)$
82. Subtract $5s + 8$ from $2s^2 - 5s + 4$.
83. Find the sum of $6x^2 - 5x + 1$ and $x + 7$.
84. Subtract $x^2 - x - 1$ from the sum of $x^2 - 1$ and $x^2 - 2x - 1$.
85. Subtract $-7t^3 + t^2 + t - 8$ from $t^3 - 2t^2 + t + 4$.
86. Subtract $x - y$ from $y - x$.
87. Find the sum of $2x^2 + 2x + 2$, $-x^2 - 4x - 4$, and $6x^2 + 7x - 3$.

In Problems 88 through 94, express each quantity as a polynomial in standard form.

88. *Find the perimeter of a rectangle with dimensions* $5d + 3$ centimeters by $d - 7$ centimeters.
89. A ribbon is cut into three pieces, whose lengths are $5r$, $r - 1$, and $r^2 - 2r + 3$ inches. Find the length of the ribbon.

90. Find the perimeter of a square with side of length s^2 yards.

91. Two planes fly in opposite directions from Atlanta. If one plane travels a distance of $500(x + 4)$ km and the other travels a distance of $400(x - 4)$ km, find the distance between them.

92. Jason has $5d - 3$ compact disks and Steve has $8d + 5$ compact disks. How many disks do they have together?

93. Christy is $2y$ years old and Susannah is $y + 7$ years old. Find the sum of their ages.

94. The perimeter of a triangle with sides of $2s + 3$, $s - 6$, and $3s - 1$ in.

Challenge Problems

Combine like terms.

95. $2x^n + 5x^n$

96. $5x^{2n} - 3x^{2n}$

IN YOUR OWN WORDS...

97. How do we find the opposite of a polynomial?

98. How do we add polynomials?

99. How do we subtract polynomials?

2.5 MULTIPLICATION

We can multiply polynomials by using the Distributive, Commutative, and Associative Properties and the laws of exponents.

To multiply two monomials, we use the Commutative and Associative Properties to rearrange the factors.

EXAMPLE 1. Find the product of $3x^4$and $5x^2$.

Solution:

$$(3x^4) \cdot (5x^2) = 3 \cdot 5x^4x^2 \qquad \text{Associative and Commutative Properties}$$
$$= 15x^6$$

□

As illustrated by Example 1, we multiply monomials by multiplying the coefficients and multiplying the variables.

EXAMPLE 2. Find each product.

(a) $(6y^4)(-4y)$ (b) $(-2x^3)(-9x^{11})$ (c) $(-x)(-y)$

(d) $(-a)(a^2)$ (e) $\frac{1}{2}(4x)(x^2)$ (f) $\frac{2}{3}(5t^3)(6t)$

Solutions:

(a) $(6y^4)(-4y) = (6)(-4)y^4y$
$= -24y^5$

(b) $(-2x^3)(-9x^{11}) = (-2)(-9)x^3x^{11}$
$= 18x^{14}$

(c) $(-x)(-y) = (-1)(-1)xy$
$= xy$

(d) $(-a)(a^2) = (-1)(1)(a)(a^2)$
$= -1a^3$
$= -a^3$

(continued)

(e) $\frac{1}{2}(4x)(x^2) = \frac{1}{2}(4)xx^2$
$= 2x^3$

(f) $\frac{2}{3}(5t^3)(6t) = \frac{2}{3}(6)(5)\, t^3t$
$= 4(5)t^4$
$= 20t^4$

The Distributive Property allows us to multiply a monomial by a polynomial.

EXAMPLE 3. Find each product.

(a) $x^3(x - 2)$ (b) $2x^3(3x^2 - 5x + 3)$

Solutions:

(a) $x^3(x - 2) = x^3 \cdot x - x^3 \cdot 2$
$= x^4 - 2x^3$

(b) $2x^3(3x^2 - 5x + 3) = 2x^3(3x^2) - 2x^3(5x) + 2x^3(3)$
$= 6x^5 - 10x^4 + 6x^3$

The product of two binomials is one of the most important products that we learn. Let's look at $(a + b)(c + d)$, which is a product of two binomials.

Using the Distributive and Commutative Properties, we write

$$\begin{aligned}(a + b)(c + d) &= (a + b)c + (a + b)d \\ &= ac + bc + ad + bd\end{aligned}$$

If we examine this closely, we can see that there are four terms in the product and that we get them by multiplying each term in the first binomial by each term in the second binomial.

Product of Two Binomials

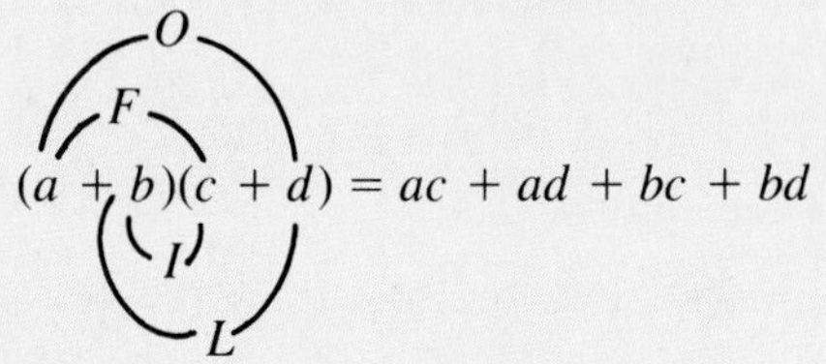

A common way to remember this is:

ac represents the product of the *f*irst terms,

ad represents the product of the *o*utside terms,

bc represents the product of the *i*nside terms, and

bd represents the product of the *l*ast terms.

This is often called the *FOIL* method.

EXAMPLE 4. Find each product.

(a) $(x+2)(x+1)$ (b) $(y-2)(y+5)$
(c) $(2x-1)(x+3)$ (d) $(z-7)(2z-1)$

Solutions:

(a) $$\begin{aligned}(x+2)(x+1) &= x(x)+x(1)+2(x)+2(1)\\ &= x^2+x+2x+2\\ &= x^2+3x+2\end{aligned}$$

(b) $$\begin{aligned}(y-2)(y+5) &= y(y)+y(5)-2(y)-2(5)\\ &= y^2+5y-2y-10\\ &= y^2+3y-10\end{aligned}$$

(c) $$\begin{aligned}(2x-1)(x+3) &= 2x(x)+2x(3)-1(x)-1(3)\\ &= 2x^2+6x-x-3\\ &= 2x^2+5x-3\end{aligned}$$

(d) $$\begin{aligned}(z-7)(2z-1) &= z(2z)+z(-1)-7(2z)-7(-1)\\ &= 2z^2-z-14z+7\\ &= 2z^2-15z+7\end{aligned}$$ □

In multiplying binomials as in Example 4, we usually compute the middle term mentally.

The Distributive Property allows us to perform other multiplications, such as

$$\begin{aligned}(a+b)(c+d+e) &= (a+b)c+(a+b)d+(a+b)e\\ &= ac+bc+ad+bd+ae+be\end{aligned}$$

Again, each term in the first polynomial is multiplied by each term in the second polynomial. It is important to note that the FOIL method works only for the product of two binomials.

Be Careful!

EXAMPLE 5. Find each product.

(a) $(x-2)(x^2-3x+4)$ (b) $(x+2)^3$
(c) $2(x+3)(x-4)$

Solutions:

(a) $$\begin{aligned}(x-2)(x^2-3x+4) &= x(x^2)-x(3x)+x(4)-2(x^2)+2(3x)-2(4)\\ &= x^3-3x^2+4x-2x^2+6x-8\\ &= x^3-5x^2+10x-8\end{aligned}$$

(b) $$\begin{aligned}(x+2)^3 &= (x+2)(x+2)(x+2)\\ &= (x^2+4x+4)(x+2)\\ &= x^2(x)+x^2(2)+4x(x)+4x(2)+4(x)+4(2)\\ &= x^3+2x^2+4x^2+8x+4x+8\\ &= x^3+6x^2+12x+8\end{aligned}$$

(continued)

(c) The Associative Property tells us that we can multiply any two of the numbers first. We will multiply the binomials first.

$$\begin{aligned} 2(x+3)(x-4) &= 2(x^2+3x-4x-12) \\ &= 2(x^2-x-12) \\ &= 2x^2-2x-24 \end{aligned}$$

□

EXAMPLE 6. Express the area of the rectangle as a polynomial in standard form.

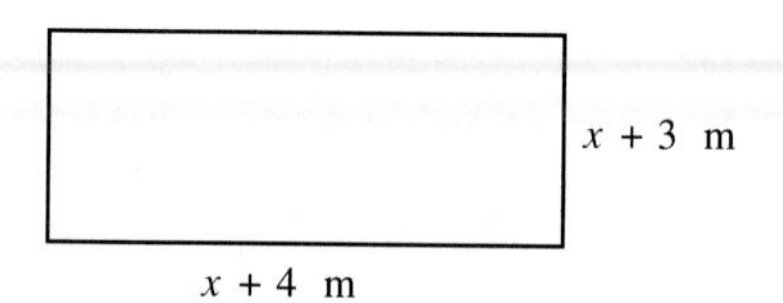

Solution:

The area of a rectangle is the product of its length and width. So, the area is $(x+4)(x+3)$. To write this in standard form, we must multiply the binomials.

$$\begin{aligned} (x+4)(x+3) &= x^2+3x+4x+12 \\ &= x^2+7x+12 \end{aligned}$$

The area of the rectangle is $x^2+7x+12$ square meters. □

EXAMPLE 7. Express the product of an integer n and the next consecutive integer as a polynomial in standard form.

Solution:

The next integer after n is $n+1$. So, the product is $n(n+1)$.

$$n(n+1) = n^2+n$$

The product is n^2+n. □

PROBLEM SET 2.5

Warm-ups

In Problems 1 through 9, find each product. See Examples 1 and 2.

1. $(2x^2)(5x^3)$
2. $(3x)(7x^8)$
3. $(-4x^3)(6x^4)$
4. $(8t^5)(-3t^2)$
5. $(-z)(-z^5)$
6. $(-9x^9)(-4x^4)$
7. $\frac{1}{2}(-8s)(3s^3)$
8. $\frac{3}{4}(7y^4)(4y^3)$
9. $-\frac{2}{3}(12z)(-z^2)$

In Problems 10 through 15, find each product. See Example 3.

10. $3(x^2+x-7)$
11. $(2x+3)4$
12. $2x(3-4x+x^3)$
13. $3x^3(3x^5-x^2-2x)$
14. $-2t(4t^3+t-6)$
15. $-5x^4(3-7x^3)$

In Problems 16 through 25, find each product. See Example 4.

16. $(x + 3)(x + 1)$
17. $(x - 5)(x + 4)$
18. $(t + 7)(t - 3)$
19. $(y - 2)(y - 6)$
20. $(2x + 3)(x + 9)$
21. $(x + 1)(5x + 8)$
22. $(3t - 5)(t + 1)$
23. $(z + 8)(5z - 9)$
24. $(6x - 3)(x - 5)$
25. $(3x + 1)(2x + 9)$

In Problems 26 through 37, find each product. See Example 5.

26. $(x + 2)(x^2 + x + 1)$
27. $(x - 3)(x^2 + 2x + 4)$
28. $(x + 4)(2x^2 - 3x + 5)$
29. $(2t - 1)(3t^2 - 2t - 6)$
30. $(3x - 2)(2x^2 + 4x - 7)$
31. $(x^2 - 3)(x^2 + 7x - 1)$
32. $(x + 1)^3$
33. $(x - 3)^3$
34. $t(3t - 2)(3t + 2)$
35. $(2y + 3)(y + 3)(y - 3)$
36. $4x(x + 3)(x - 2)$
37. $y^2(2y - 3)(y + 1)$

In Problems 38 through 45, write a polynomial in standard form to represent each quantity. See Examples 6 and 7.

38. Find the area of the square whose side is of length $x - 1$ feet.
39. Find the area of the triangle shown.

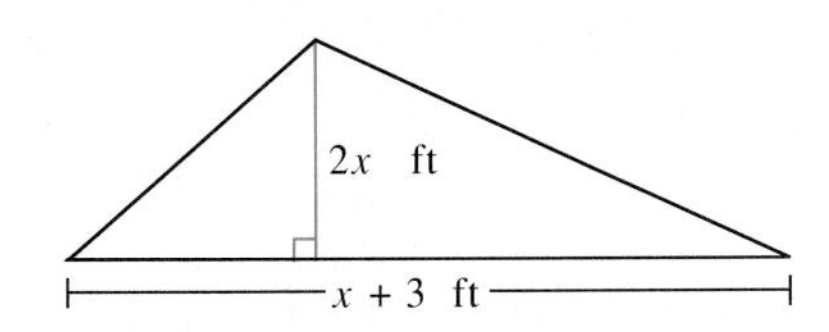

40. Find the area of the rectangle with dimensions of $3x + 2$ meters and $x + 4$ meters.

41. Find the area of the trapezoid.

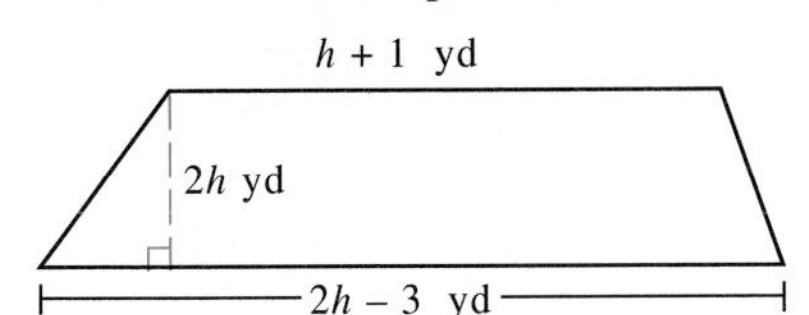

42. Find the area of a circle with radius of $r + 2$ centimeters.
43. If x is an even integer, find the product of x and the next even integer.
44. If n is an odd integer, find the product of n and the next two odd integers.
45. If y is an integer, find the product of y and the integer before y.

Practice Exercises

Find each product.

46. $(3x^2)(2x^4)$
47. $(2r)(9r^5)$
48. $(-5x^2)(7x^7)$
49. $(4w^3)(-6w^2)$
50. $(-8z^{11})(-z^{15})$
51. $(-7x^7)(-3x^3)$
52. $\frac{2}{5}(5x)(2x^2)$
53. $\frac{1}{6}(-x)(12x)$
54. $\frac{2}{7}(3w^2)(7w)$
55. $4(x - 5)$
56. $(4x - 7)5$
57. $2y(3 - 9y + y^2)$
58. $2x^2(x^3 - 3x)$

59. $-3t(7t^2 - 8t + 2)$

60. $-3x^5(3 - 2x + 5x^7)$

61. $(x + 2)(x + 5)$

62. $(x - 3)(x + 2)$

63. $(t + 3)(t - 3)$

64. $(y - 3)(y - 4)$

65. $(3t + 2)(t + 9)$

66. $(x + 2)(4x + 7)$

67. $(2x - 5)(x + 2)$

68. $(x + 6)(6x - 11)$

69. $(4x - 3)(3x + 2)$

70. $(10x - 3)(4x - 1)$

71. $(2r + 7)(3r - 5)$

72. $(3s - 5)(5s - 3)$

73. $(6 + 2x)(3 - x)$

74. $(5 - 9x)(2x + 1)$

75. $(x - 4)(x + 4)$

76. $(3x + 2)(3x - 2)$

77. $(y + 5)(y + 5)$

78. $(z - 4)^2$

79. $(5t - 3)(t - 7)$

80. $(2y + 1)(3y + 8)$

81. $(3x - 5)(5x + 2)$

82. $(12x - 1)(4x - 3)$

83. $(3s + 7)(2s - 5)$

84. $(2y - 7)(2y - 7)$

85. $(3 + 8x)(2 - x)$

86. $(2 - 7x)(3x + 1)$

87. $(x - 1)(x^2 + x + 1)$

88. $(x + 3)(x^2 + 2x + 1)$

89. $(x + 4)(x^2 - 4x + 16)$

90. $(2t - 1)(4t^2 + 2t + 1)$

91. $(2x - 3)(3x^2 + x - 4)$

92. $x(x - 1)(x^2 + 2x - 3)$

93. $(s - 1)^3$

94. $(v + 3)^3$

95. $3t(t - 5)(t + 5)$

96. $(y + 4)(y - 4)(y + 2)$

In Problems 97 through 100, express each quantity as a polynomial in standard form.

97. Find the area of the rectangle shown.

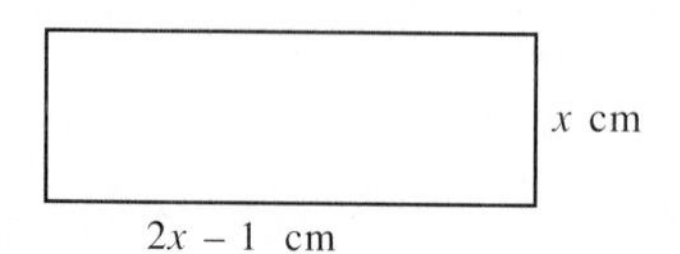

98. Find the area of a square with a side of length $y - 3$ meters.

99. If z is an even integer, find the product of the next two consecutive even integers.

100. Find the product of twice n and 3 more than n.

Challenge Problems

Find the following products.

101. $\left(\frac{1}{2}x + 1\right)\left(\frac{1}{2}x - 1\right)$

102. $\left(\frac{1}{2}y + 1\right)^2$

103. $\left(\frac{2}{3}z - 1\right)\left(\frac{1}{3}z + 2\right)$

104. $(p - q)(p^2 + pq + q^2)$

105. $(p + q)(p^2 - pq + q^2)$

IN YOUR OWN WORDS...

106. How do we multiply two binomials?

107. How do we multiply two polynomials?

2.6 SPECIAL PRODUCTS AND ORDER OF OPERATIONS

In this section we will develop some products that occur very often in algebra. They are called special products and should be memorized. The first one is called *squaring a binomial.*

EXAMPLE 1. Find each product.

(a) $(p+q)^2$ (b) $(p-q)^2$

Solutions:

(a) $(p+q)^2 = (p+q)(p+q)$
$= p^2 + pq + pq + q^2$
$= p^2 + 2pq + q^2$

(b) $(p-q)^2 = (p-q)(p-q)$
$= p^2 - pq - pq + q^2$
$= p^2 - 2pq + q^2$ □

Square of a Binomial

$$(p+q)^2 = p^2 + 2pq + q^2$$
$$(p-q)^2 = p^2 - 2pq + q^2$$

These should be memorized as a pattern and remembered in words rather than symbols. The square of a binomial is the first term squared plus (or minus) twice the product of the two terms plus the last term squared. Notice that these products are alike everywhere except for the sign in the middle term.

Notice that $(p+q)^2 \neq p^2 + q^2$ and $(p-q)^2 \neq p^2 - q^2$. To illustrate this, let's evaluate each expression when p is 4 and q is 3.

$$(p+q)^2 = (4+3)^2 = 7^2 = 49$$
$$p^2 + q^2 = 4^2 + 3^2 = 16 + 9 = 25$$

So, we see that $(p+q)^2 \neq p^2 + q^2$.

$$(p-q)^2 = (4-3)^2 = 1^2 = 1$$
$$p^2 - q^2 = 4^2 - 3^2 = 16 - 9 = 7$$

We also see that $(p-q)^2 \neq p^2 - q^2$.

$$(p+q)^2 \neq p^2 + q^2$$
$$(p-q)^2 \neq p^2 - q^2$$

Be Careful!

EXAMPLE 2. Square each binomial.

(a) $(x+4)^2$ (b) $(2x-3y)^2$

Solutions:

(a) We note this is the square of a binomial.

$(p+q)^2 = p^2 + 2pq + q^2$

$(x+4)^2 = x^2 + 2(x)(4) + 4^2$ Replace p with x and q with 4.

(the first term squared plus twice the product of the two terms plus the last term squared)

$(x+4)^2 = x^2 + 8x + 16$

(continued)

(b) $(2x - 3y)^2 = (2x)^2 - 2(2x)(3y) + (3y)^2$

(the first term squared minus twice the product of the two terms plus the last term squared)

$(2x - 3y)^2 = 4x^2 - 12xy + 9y^2$ □

Another special product comes from multiplying the sum of two numbers by the difference of the same two numbers. It is called the difference of two squares.

EXAMPLE 3. Find the product of $(p - q)$ and $(p + q)$.

Solution:

$$(p - q)(p + q) = p^2 + pq - pq - q^2$$
$$= p^2 - q^2$$

Note that $(p - q)(p + q) = (p + q)(p - q)$. □

> **Difference of Two Squares**
>
> $$(p - q)(p + q) = p^2 - q^2$$

The product of the sum of two numbers and the difference of the same two numbers is the first number squared minus the second number squared.

EXAMPLE 4. Find each product.

(a) $(x + 2)(x - 2)$ (b) $(4r + 7)(4r - 7)$

Solutions:

(a) $(p + q)(p - q) = p^2 - q^2$ Difference of two squares

$(x + 2)(x - 2) = x^2 - 2^2$ Replace p with x and q with 2.

$= x^2 - 4$

(the first number squared minus the second number squared)

(b) $(4r + 7)(4r - 7) = (4r)^2 - 7^2$

$= 16r^2 = 49$ □

Now that we have studied addition, subtraction, and multiplication of polynomials, we can combine these operations. The order of operations tells us how to simplify such problems.

EXAMPLE 5. Simplify each polynomial.

(a) $(x + 1)(x - 1) - (x^2 + 2x - 1)$ (b) $(x - 2)^2 - (x + 2)(x - 3)$

Solutions:

(a) We must multiply $(x + 1)(x - 1)$ first. It is the difference of two squares.

$$(x + 1)(x - 1) - (x^2 + 2x - 1) = x^2 - 1 - (x^2 + 2x - 1)$$

We perform the subtraction next, being *very careful* with the signs!

$$x^2 - 1 - x^2 - 2x + 1 = -2x$$

Be Careful!

(b) We perform the multiplications first.

$$(x - 2)^2 - (x + 2)(x - 3) = x^2 - 4x + 4 - (x^2 - x - 6)$$

Notice that we left the parentheses because the product of $(x + 2)$ and $(x - 3)$ is to be subtracted. Now we remove the parentheses, being careful with signs.

Be Careful!

$$x^2 - 4x + 4 - x^2 + x + 6 = -3x + 10$$ □

EXAMPLE 6. Express the area of a square with side of length $s - 4$ meters as a polynomial in standard form.

Solution:

The area of a square is the square of the length of a side. So, the area is $(s - 4)^2$.

$$(s - 4)^2 = s^2 - 2(s)(4) + 4^2$$
$$= s^2 - 8s + 16$$

The area of the square is $s^2 - 8s + 16$ square meters. □

PROBLEM SET 2.6

Warm-ups

In Problems 1 through 10, square each binomial. See Example 2.

1. $(x + 1)^2$
2. $(x - 7)^2$
3. $(x - 1)^2$
4. $(x + 2)^2$
5. $(4s + 3)^2$
6. $(5t - 4)^2$
7. $(2x - 1)^2$
8. $(3x - 2)^2$
9. $(9 + y)^2$
10. $(8 - t)^2$

In Problems 11 through 20, find each product. See Example 4.

11. $(x + 1)(x - 1)$
12. $(x - 6)(x + 6)$
13. $(2x - 3)(2x + 3)$
14. $(3x + 5)(3x - 5)$
15. $(6x + 9)(6x - 9)$
16. $(7 - 11w)(7 + 11w)$
17. $(3x + 7)(3x - 7)$
18. $(10 + 9t)(10 - 9t)$
19. $(9 - 5t)(9 + 5t)$
20. $(1 - t)(1 + t)$

In Problems 21 through 28, simplify each polynomial. See Example 5.

21. $5 - 2(x + 3)$
22. $3 - [4 - (x - 3) + 2(x + 1)]$
23. $6x[5 - 2x(x - 3)] + 4x^2$
24. $2t - 3t[t(t + 1) - t^2]$
25. $x(x - 3) + (x + 3)^2$
26. $(2x - 1)(x + 2) - x(2x - 1)$
27. $(y - 4)(y + 3) + (y - 2)(y + 2)$
28. $2x(x - 8) + (3x + 7)(3x - 7)$

In Problems 29 through 33, write a polynomial in standard form to represent each quantity. See Example 6.

29. The area of a square with side of length $3y + 2$ feet.
30. The area of a circle with radius of $r - 1$ meters.

31. The area of the rectangle shown.

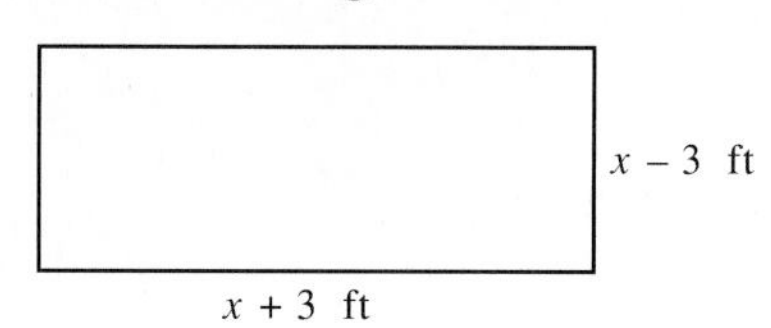

32. The square of the next integer after the integer n.

33. The square of the integer before the integer q.

Practice Exercises

In Problems 34 through 53, use special products to find each product.

34. $(x - 1)^2$

35. $(x + 3)^2$

36. $(t + 3)(t - 3)$

37. $(y - 5)(y + 5)$

38. $(3x - 4)^2$

39. $(4t + 5)^2$

40. $(3x - 2)(3x + 2)$

41. $(2t - 7)(2t + 7)$

42. $(2x + 6)^2$

43. $(2x - 3)^2$

44. $(3x + 5)^2$

45. $(2z + 1)^2$

46. $(8x + 9)(8x - 9)$

47. $(11q - 5)(11q + 5)$

48. $(7p - 1)^2$

49. $(3w - 4)^2$

50. $(7x - 8)(7x + 8)$

51. $(5 - 12k)(5 + 12k)$

52. $(5x - 1)^2$

53. $(9g + 2)^2$

In Problems 54 through 63, simplify each polynomial.

54. $(x - 7)(x + 7) + (x - 3)^2$

55. $2x - 3x[(x - 5) + 2x(x - 7)]$

56. $(s - 6)^2 - 3s(2s - 3)$

57. $3z(2 - z) - (z + 1)^2$

58. $(y - 3)^2 - (y - 1)(y + 1)$

59. $(w - 5)(w + 3) - (w + 2)(w + 4)$

60. $(z + 3)^2 - 2z(z + 1)$

61. $(t - 4)(t + 4) - (t + 2)(t + 1)$

62. $7y - y(y - 2) + (y + 2)(y - 2)$

63. $7t - t[2(t + 3) + (t + 2)(t - 3)]$

In Problems 64 through 67, express each quantity as a polynomial in standard form.

64. The area of a square with a side of length of $2y - 4$ centimeters.

65. The area of a circle with radius of $r + 7$ inches.

66. The product of the integer before and the integer after the integer p.

67. The area of the rectangle shown below.

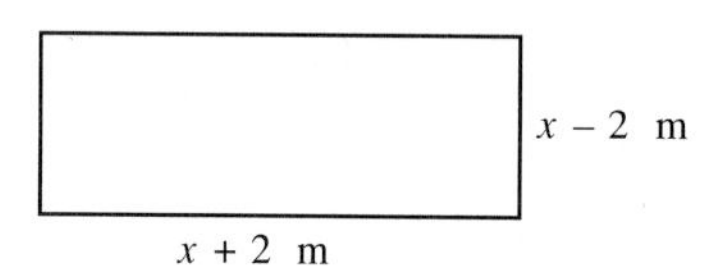

Challenge Problems

Find each product using a special product.

68. $\left(\frac{1}{2} + 3x\right)^2$

69. $\left(\frac{2}{3} - \frac{1}{2}x\right)\left(\frac{2}{3} + \frac{1}{2}x\right)$

IN YOUR OWN WORDS...

70. Explain how to square a binomial.

71. How do we recognize a product as the difference of two squares?

2.7 DIVISION

We can divide polynomials just as we can divide real numbers. There are three ways of writing a division problem. If P and D are polynomials, the following have the same meaning:

1. P divided by D or $P \div D$
2. $\dfrac{P}{D}$ or P/D
3. $D\overline{)P}$

We call P the **dividend** and D the **divisor.** The divisor *cannot* have a value of zero.

We have already discussed division of a monomial by a monomial in Sections 2.1 and 2.2 using the laws of exponents. For example,

$$\frac{x^5}{x^3} = x^2 \quad \text{and} \quad \frac{a^2}{a} = a$$

The laws of exponents do not tell us how to do other divisions, such as $\dfrac{x^4 + 3x^3}{x^2}$. We need the following result, which follows from the properties of real numbers (Section 1.5).

Division by a Monomial

If P, Q, and R are monomials, then

$$\frac{P+Q}{R} = \frac{P}{R} + \frac{Q}{R}$$

(R must not have a value of 0.)

Thus,

$$\begin{aligned} \frac{x^4 + 3x^3}{x^2} &= \frac{x^4}{x^2} + \frac{3x^3}{x^2} \\ &= x^2 + 3x \end{aligned}$$

EXAMPLE 1. Perform the divisions.

(a) $\dfrac{2x^3 + 6x^5 - 4x^2}{-2x^2}$ (b) $\dfrac{3t^2 - 5t}{2t^2}$

Solutions:

(a)
$$\begin{aligned} \frac{2x^3 + 6x^5 - 4x^2}{-2x^2} &= \frac{2x^3}{-2x^2} + \frac{6x^5}{-2x^2} - \frac{4x^2}{-2x^2} \\ &= -x + (-3x^3) - (-2) \\ &= -x - 3x^3 + 2 \end{aligned}$$

(continued)

$$\text{(b)}\quad \frac{3t^2 - 5t}{2t^2} = \frac{3t^2}{2t^2} - \frac{5t}{2t^2}$$

$$= \frac{3}{2} - \frac{5}{2t}$$

□

If the divisor is not a monomial, we use long division. It is much like long division of numbers. The steps are illustrated in this example.

Let's consider $(x^3 + 3x^2 + 5x + 4) \div (x + 1)$

1 Put each polynomial in standard form and write,

$$x + 1\overline{)x^3 + 3x^2 + 5x + 4}$$

2 The value of $\frac{x^3}{x}$ is x^2. So we write

$$\begin{array}{r} x^2 \\ x + 1\overline{)x^3 + 3x^2 + 5x + 3} \end{array}$$

3 We multiply x^2 by $x + 1$ and write:

$$\begin{array}{r} x^2 \\ x + 1\overline{)x^3 + 3x^2 + 5x + 4} \\ x^3 + x^2 \end{array}$$

Be Careful!

4 Next we *subtract* $x^3 + x^2$ from $x^3 + 3x^2$. Almost all the mistakes made in long division occur at this step, so we should be very careful to *subtract* correctly. The first few times, we might want to change the signs and add. If we do that, we should be careful to be sure the original signs are legible so we can check our work.

$$\begin{array}{r} x^2 \\ x + 1\overline{)x^3 + 3x^2 + 5x + 4} \\ \ominus \quad \ominus \\ \underline{x^3 + x^2} \\ 2x^2 \end{array}$$

5 We bring down the next term.

$$\begin{array}{r} x^2 \\ x + 1\overline{)x^3 + 3x^2 + 5x + 4} \\ \ominus \quad \ominus \\ \underline{x^3 + x^2} \\ 2x^2 + 5x \end{array}$$

Now we start the process over: $\frac{2x^2}{x}$ is $2x$, so we write

$$\begin{array}{r} x^2 + 2x \\ x + 1\overline{)x^3 + 3x^2 + 5x + 4} \\ \ominus \quad \ominus \\ \underline{x^3 + x^2} \\ 2x^2 + 5x \end{array}$$

Multiply $x + 1$ by $2x$ and enter it in the display.

$$
\begin{array}{rl}
 & x^2 + 2x \\
x + 1 & \overline{)\,x^3 + 3x^2 + 5x + 4} \\
 & \ominus \quad \ominus \\
 & \underline{x^3 + x^2} \\
 & \qquad 2x^2 + 5x \\
 & \qquad \underline{2x^2 + 2x}
\end{array}
$$

Subtract $2x^2 + 2x$ from $2x^2 + 5x$ and bring down the 4.

$$
\begin{array}{rl}
 & x^2 + 2x \\
x + 1 & \overline{)\,x^3 + 3x^2 + 5x + 4} \\
 & \ominus \quad \ominus \\
 & \underline{x^3 + x^2} \\
 & \qquad 2x^2 + 5x \\
 & \qquad \ominus \quad \ominus \\
 & \qquad \underline{2x^2 + 2x} \\
 & \qquad\qquad\quad 3x + 4
\end{array}
$$

We repeat the procedure. Since $\dfrac{3x}{x} = 3$, we multiply $x + 1$ by 3.

$$
\begin{array}{rl}
 & x^2 + 2x + 3 \\
x + 1 & \overline{)\,x^3 + 3x^2 + 5x + 4} \\
 & \ominus \quad \ominus \\
 & \underline{x^3 + x^2} \\
 & \qquad 2x^2 + 5x \\
 & \qquad \ominus \quad \ominus \\
 & \qquad \underline{2x^2 + 2x} \\
 & \qquad\qquad\quad 3x + 4 \\
 & \qquad\qquad\quad 3x + 3
\end{array}
$$

We subtract.

$$
\begin{array}{rl}
 & x^2 + 2x + 3 \\
x + 1 & \overline{)\,x^3 + 3x^2 + 5x + 4} \\
 & \ominus \quad \ominus \\
 & \underline{x^3 + x^2} \\
 & \qquad 2x^2 + 5x \\
 & \qquad \ominus \quad \ominus \\
 & \qquad \underline{2x^2 + 2x} \\
 & \qquad\qquad\quad 3x + 4 \\
 & \qquad\qquad\quad \ominus \quad \ominus \\
 & \qquad\qquad\quad \underline{3x + 3} \\
 & \qquad\qquad\qquad\quad 1
\end{array}
$$

Since the degree of 1 is less than the degree of $x + 1$, we stop the procedure. We examine the last display and conclude that

$$(x^3 + 3x^2 + 5x + 4) \div (x + 1) = x^2 + 2x + 3 + \frac{1}{x + 1}.$$

The polynomial $x + 1$ is the **divisor.**

The polynomial $x^3 + 3x^2 + 5x + 4$ is the **dividend.**

The polynomial $x^2 + 2x + 3$ is the **quotient.**

The polynomial 1 is the **remainder.**

To check our work, we multiply the divisor by the quotient and add the remainder. This should give the dividend.

Divisor · quotient + remainder = dividend

$$\begin{aligned}(x + 1)(x^2 + 2x + 3) + 1 &= x^3 + 2x^2 + 3x + x^2 + 2x + 3 + 1\\ &= x^3 + 3x^2 + 5x + 4\end{aligned}$$

Our answer checks.

EXAMPLE 2. Perform the division $\dfrac{x^2 + x - 6}{x + 3}$.

Solution:

Both polynomials are in standard form.

$$x + 3\overline{)x^2 + x - 6}$$

Since $\dfrac{x^2}{x} = x$, we write

$$\begin{array}{r} x \phantom{{}+x-6} \\ x + 3\overline{)x^2 + x - 6} \end{array}$$

Then, we multiply x by $x + 3$ and enter the product:

$$\begin{array}{r} x \phantom{{}+x-6} \\ x + 3\overline{)x^2 + x - 6} \\ x^2 + 3x \phantom{{}-6} \end{array}$$

Now we are *very careful* in subtracting.

$$\begin{array}{r} x \phantom{{}+x-6} \\ x + 3\overline{)x^2 + x - 6} \\ \ominus \quad \ominus \phantom{{}-6} \\ \underline{x^2 + 3x} \phantom{{}-6} \\ -2x \phantom{{}-6} \end{array}$$

We bring down the -6.

$$\begin{array}{r} x \phantom{{}+x-6} \\ x + 3\overline{)x^2 + x - 6} \\ \ominus \quad \ominus \phantom{{}-6} \\ \underline{x^2 + 3x} \phantom{{}-6} \\ -2x - 6 \end{array}$$

We repeat the previous steps. Since $\frac{-2x}{x} = -2$, we enter the -2 and multiply.

$$\begin{array}{r} x - 2 \\ x + 3\overline{)x^2 + x - 6} \\ \ominus \ \ominus \qquad \\ \underline{x^2 + 3x} \qquad \\ -2x - 6 \\ \underline{-2x - 6} \end{array}$$

Subtract.

$$\begin{array}{r} x - 2 \\ x + 3\overline{)x^2 + x - 6} \\ \ominus \ \ominus \qquad \\ \underline{x^2 + 3x} \qquad \\ -2x - 6 \\ \oplus \quad \oplus \\ \underline{-2x - 6} \\ 0 \end{array}$$

Since the remainder is 0, we are finished with the procedure. We still need to write the answer. Since the remainder is zero, we can write

$$\frac{x^2 + x - 6}{x + 3} = x - 2$$

Since $(x + 3)(x - 2) = x^2 + x - 6$, the answer checks. □

EXAMPLE 3. Divide $2 - 2x + x^3$ by $x + 1$.

Solution:

First, we *must* write both polynomials in standard form.

$$(x^3 - 2x + 2) \div (x + 1)$$

Notice that there is no x^2 term in the dividend. Because we need *every term* in the standard form, we insert $0x^2$ in its proper place and write the display.

Be Careful!

$$x + 1\overline{)x^3 + 0x^2 - 2x + 2}$$

We calculate $\frac{x^3}{x} = x^2$, enter it, multiply, and prepare to subtract.

$$\begin{array}{l} \qquad\quad x^2 \\ x + 1\overline{)x^3 + 0x^2 - 2x + 2} \\ \qquad\ \ \underline{x^3 + x^2} \end{array}$$

(continued)

Now we subtract.

$$
\begin{array}{rrrrr}
 & x^2 & & & \\ \cline{2-5}
x+1 \big) & x^3 & +0x^2 & -2x & +2 \\
 & x^3 & +x^2 & & \\ \cline{2-3}
 & & -x^2 & &
\end{array}
$$

We bring down the $-2x$. Since $\frac{-x^2}{x} = -x$, we enter it and multiply.

$$
\begin{array}{rrrrr}
 & x^2 & -x & & \\ \cline{2-5}
x+1 \big) & x^3 & +0x^2 & -2x & +2 \\
 & x^3 & +x^2 & & \\ \cline{2-3}
 & & -x^2 & -2x & \\
 & & -x^2 & -x & \\ \cline{3-4}
\end{array}
$$

Subtract, and bring down the 2.

$$
\begin{array}{rrrrr}
 & x^2 & -x & & \\ \cline{2-5}
x+1 \big) & x^3 & +0x^2 & -2x & +2 \\
 & x^3 & +x^2 & & \\ \cline{2-3}
 & & -x^2 & -2x & \\
 & & -x^2 & -x & \\ \cline{3-4}
 & & & -x & +2
\end{array}
$$

Since $\frac{-x}{x} = -1$, we enter -1 and multiply.

$$
\begin{array}{rrrrr}
 & x^2 & -x & -1 & \\ \cline{2-5}
x+1 \big) & x^3 & +0x^2 & -2x & +2 \\
 & x^3 & +x^2 & & \\ \cline{2-3}
 & & -x^2 & -2x & \\
 & & -x^2 & -x & \\ \cline{3-4}
 & & & -x & +2 \\
 & & & -x & -1
\end{array}
$$

We subtract, and get our final display.

$$
\begin{array}{rrrrr}
 & x^2 & -x & -1 & \\ \cline{2-5}
x+1 \big) & x^3 & +0x^2 & -2x & +2 \\
 & x^3 & +x^2 & & \\ \cline{2-3}
 & & -x^2 & -2x & \\
 & & -x^2 & -x & \\ \cline{3-4}
 & & & -x & +2 \\
 & & & -x & -1 \\ \cline{4-5}
 & & & & 3
\end{array}
$$

The degree of 3 is less than the degree of $x + 1$. We conclude that

$$(2 - 2x + x^3) \div (x + 1) = x^2 - x - 1 + \frac{3}{x + 1}. \qquad \square$$

EXAMPLE 4. Divide $(6x^3 - 13x^2 + 4x + 7)$ by $(2x - 3)$.

Solution:

$$\begin{array}{r}
3x^2 - 2x - 1 \\
2x - 3 \overline{)6x^3 - 13x^2 + 4x + 7} \\
\underline{6x^3 - 9x^2} \qquad\qquad \\
-4x^2 + 4x \qquad \\
\underline{-4x^2 + 6x} \qquad \\
-2x + 7 \\
\underline{-2x + 3} \\
4
\end{array}$$

So,

$$\frac{6x^3 - 13x^2 + 4x + 7}{2x - 3} = 3x^2 - 2x - 1 + \frac{4}{2x - 3}. \qquad \square$$

PROBLEM SET 2.7

Warm-ups

Perform the divisions indicated. In Problems 1 through 15, see Example 1.

1. $\dfrac{4a^2 - 4a + 4}{4}$
2. $\dfrac{8 - 4t^3}{-2}$
3. $\dfrac{x^2 - x + 1}{x^2}$
4. $\dfrac{3z^6 - 5z^4 - z^2}{15z^2}$
5. $(6x - 6)/(-6)$
6. $\dfrac{5 - 2y + 5y^2}{5y^2}$
7. $\dfrac{3x^6 - x^4}{x^3}$
8. $\dfrac{-4z^9 + 8z - 16}{-4z^3}$
9. $\dfrac{9x^3 - 3x^2 + 3}{-3x^2}$
10. $\dfrac{x^3 + 2x^4}{2x^4}$
11. $\dfrac{2x^3 + 3x^4}{2x^2}$
12. $\dfrac{t^5 - 7t^3}{t^3}$
13. $(3x^4 + 2x^5) \div 3x^4$
14. $\dfrac{x^2y - xy^2}{xy}$
15. $\dfrac{2x^5 - 3x^7}{6x^5}$

In Problems 16 through 39, see Examples 2, 3, and 4.

16. $\dfrac{x^2 + 3x + 2}{x + 1}$
17. $\dfrac{x^2 + 7x + 12}{x + 3}$
18. $\dfrac{x^2 - 3x - 10}{x + 2}$
19. $\dfrac{x^2 + 4x + 4}{x + 1}$
20. $\dfrac{y^2 + 6y + 11}{y + 2}$
21. $\dfrac{t^2 + t - 4}{t + 3}$
22. $\dfrac{2x^2 + x - 1}{x + 1}$
23. $\dfrac{3x^2 + 8x + 4}{x + 2}$
24. $\dfrac{2x^2 - x - 21}{x + 3}$
25. $\dfrac{s^2 + s - 1}{s - 1}$
26. $\dfrac{2x^2 + 7x - 1}{x + 4}$
27. $\dfrac{3z^2 - 11z + 2}{z - 3}$
28. $\dfrac{x^3 + 2x^2 + 2x + 1}{x + 1}$
29. $\dfrac{x^3 + 4x^2 - 3x - 10}{x + 2}$

30. $\dfrac{t^3 + 2t^2 - 4t - 3}{t + 3}$

31. $\dfrac{3x^2 + x^3 + x - 1}{x + 2}$

32. $\dfrac{x^3 - 3x^2 - x - 1}{x - 1}$

33. $\dfrac{x^3 + 4x^2 + 2x}{x + 2}$

34. $\dfrac{2x^3 + 7x^2 + 7x + 2}{2x + 1}$

35. $\dfrac{4y^3 + 4y^2 - y}{2y + 3}$

36. $\dfrac{x^3 - 26x + 7}{x - 5}$

37. $\dfrac{6v^3 - 7v^2 - 4}{2v - 1}$

38. $(5 - 7x^2 + 3x^3)/(3x + 2)$

39. $(2z^3 - 4z) \div (4 + 2z)$

Practice Exercises

Perform the divisions indicated.

40. $\dfrac{-3x^5 - 6x}{-3x}$

41. $\dfrac{3y^7 - 2y^3}{2y^3}$

42. $\dfrac{8x^4 - 16x^3 + 4x^2}{-2x^2}$

43. $\dfrac{x^4 + 4x^5}{x}$

44. $\dfrac{3x^5 + 2x^3}{x^2}$

45. $\dfrac{t^6 - 6t^4}{t^4}$

46. $\dfrac{4x^3 + 12x^4 - 4x^5}{4x^3}$

47. $\dfrac{3x^2 - 4x^5}{12x^2}$

48. $\dfrac{sx^5 - tx^7}{stx^5}$

49. $\dfrac{x^2 + 4x + 3}{x + 1}$

50. $\dfrac{x^2 + 6x + 8}{x + 2}$

51. $\dfrac{x^2 + x - 6}{x + 3}$

52. $\dfrac{x^2 + 3x + 3}{x + 1}$

53. $\dfrac{y^2 + 7y + 11}{y + 3}$

54. $\dfrac{t^2 - 3t - 7}{t + 2}$

55. $\dfrac{x^2 + x - 2}{x - 1}$

56. $\dfrac{2x^2 + 7x - 4}{x + 4}$

57. $\dfrac{3x^2 - 11x - 6}{x - 3}$

58. $\dfrac{2s^2 + s + 1}{s + 1}$

59. $\dfrac{3x^2 + 8x + 1}{x + 2}$

60. $\dfrac{2z^2 - z - 16}{z + 3}$

61. $\dfrac{x^3 + 3x^2 + 3x + 2}{x + 2}$

62. $\dfrac{x^3 + 3x^2 - 2x - 4}{x + 1}$

63. $\dfrac{t^3 + 2t^2 - t + 6}{t + 3}$

64. $\dfrac{x^3 + x + 3}{x + 1}$

65. $\dfrac{x^3 - x^2 - 9}{x - 2}$

66. $\dfrac{2x^3 + 5x^2 + 3x + 5}{x + 2}$

67. $\dfrac{3x^2 + 2x^3 - 1 - x}{2x + 1}$

68. $\frac{2y^3 - 3y^2 - 5y}{2y - 1}$

69. $\frac{x - 11 + 4x^3}{2x - 3}$

70. $\frac{6v^3 + 5v^2 - 1}{2v + 3}$

71. $(16x^3 - x) \div (4x - 3)$

72. $(3 + 5t^2 + 6t^3) \div (3t - 2)$

Challenge Problems

Perform the divisions indicated.

73. $\frac{4x^4 - 6x^3 - 4x^2 + 13x - 9}{2x - 1}$

74. $\frac{x^3 + 3x^2 + x - 1}{x^2 + x - 1}$

75. $(x^4 + 5x^3 + x^2 - 11x + 4) \div (x^2 + 2x - 1)$

76. $(x^3 - 8)/(x - 2)$

77. $\frac{x^4 + 2x^3 + 7x^2 + 2x + 5}{x^2 + 1}$

78. Find the value of k so that if $x^3 + 2x^2 + x + k$ is divided by $x + 3$, the remainder will be zero.

IN YOUR OWN WORDS...

79. When do we use long division?

80. What is the procedure for long division?

CHAPTER SUMMARY

GLOSSARY

Exponent: In the expression x^n, the exponent is n. It tells how many times to use x as a factor.

Base: In the expression x^n, the base is x. It is the number being raised to the power n.

Monomial: An expression of the form bx^n, where n is a whole number and x is a nonzero real number.

Binomial: A sum of two monomials.

Trinomial: A sum of three monomials.

Polynomial: A sum of monomials.

Coefficient: In the expression bx^n, the coefficient of x^n is b.

Degree of a monomial in one variable: The exponent of the variable.

Degree of a polynomial: Highest degree of the terms in the polynomial.

Standard form: A polynomial written so that the degree of the monomials decrease in order from left to right.

Leading coefficient: The coefficient of the first term of a polynomial.

Constant term: A term that has degree 0.

EXPONENT DEFINITIONS

1. If n is a natural number, then $x^n = x \cdot x \cdot x \cdots x$.
2. $x^0 = 1; x \neq 0$.
3. If n is a natural number and $x \neq 0$, then $x^{-n} = \frac{1}{x^n}$

PROPERTIES OF EXPONENTS

1. $x^m \cdot x^n = x^{m+n}$	Product with same base
2. $(x^m)^n = x^{mn}$	Power to a power
3. $(xy)^n = x^n y^n$	Product to a power
4. $\left(\frac{x}{y}\right)^n = \frac{x^n}{y^n}; y \neq 0$	Quotient to a power
5. $\frac{x^m}{x^n} = x^{m-n}; x \neq 0$	Quotient with same base

OPERATIONS WITH POLYNOMIALS

1. To evaluate a polynomial means to replace each variable in the polynomial with the value of the variable.
2. To add polynomials, add like terms.
3. To subtract polynomials, change the subtraction to addition of the opposite.
4. To multiply polynomials, use the distributive property and the properties of exponents.
5. To divide polynomials, use long division if the divisor is not a monomial. If the divisor is a monomial, divide each term in the dividend by the divisor.

TO WRITE A NUMBER IN SCIENTIFIC NOTATION

1. Starting with the number in decimal format, move the decimal point until the number is between 1 and 10.
2. Multiply the number formed in Step 1 by 10 raised to the number of decimal places moved. If the original number was less than 1, the power of 10 is negative. If the original number was greater than 10, the power is positive.
3. This procedure is for a *positive* number. If the original number is negative, perform steps 1 and 2 on the absolute value of the original number; then append a negative sign to the result.

SQUARE OF A BINOMIAL

1. $(p + q)^2 = p^2 + 2pq + q^2$
2. $(p - q)^2 = p^2 - 2pq + q^2$

DIFFERENCE OF TWO SQUARES

$(p - q)(p + q) = p^2 - q^2$

CHECKUPS

1. Simplify $(-xy^2)^2$	Section 2.1; Example 5b
2. Simplify $\left(\frac{2x}{k^2}\right)^3$	Section 2.1; Example 8c
3. Simplify $7x^{-3}$	Section 2.2; Example 1f
4. Simplify $(3x^{-4})^{-2}$	Section 2.2; Example 3b
5. Write 0.000001554 in scientific notation.	Section 2.2; Example 5b

6. Write $5 + 3x^2 + 2x$ in standard form and give its degree. — Section 2.3; Example 1a
7. Evaluate $3x^2 - x + 4$ if x is -2. — Section 2.3; Example 2a

In Problems 8 through 14, perform the indicated operations.

8. $(8x^2 - 3x + 11) + (x^2 + 5x - 4)$ — Section 2.4; Example 1a
9. $(3x^3 - 7x + 2) - (x^3 + x - 8)$ — Section 2.4; Example 3b
10. $[(x^2 - x) - (2x^2 + 3x + 2)] + (6x^2 - x + 1)$ — Section 2.4; Example 4a
11. $(-a)(a^2)$ — Section 2.5; Example 2d
12. $(x + 2)(x + 1)$ — Section 2.5; Example 4a
13. $\dfrac{3t^2 - 5t}{2t^2}$ — Section 2.7; Example 1b
14. $\dfrac{x^2 + x - 6}{x + 3}$ — Section 2.7; Example 2

In Problems 15 and 16, find each product using a special product.

15. $(x + 4)^2$ — Section 2.6; Example 2a
16. $(x + 2)(x - 2)$ — Section 2.6; Example 4a
17. Express the area of a square with a side of length $s - 4$ meters as a polynomial in standard form. — Section 2.6; Example 6

REVIEW PROBLEMS

In Problems 1 through 9, simplify if possible.

1. x^5x^2
2. $(x^3t^2)^3$
3. $(2z^4)^{-2}$
4. $\left(\dfrac{-5x^3}{A}\right)^2$
5. -7^{-2}
6. $\dfrac{2y^5}{y^9}$
7. $(t^{-5})^{-2}$
8. $5x^0$
9. $(\pi x^{-5}y^3)^0$
10. Write the following in scientific notation.
    a. 76,500 b. 0.00109
11. Write the following without exponents.
    a. 1.63×10^{-7} b. 8.503×10^6

In Problems 12 and 13, evaluate each polynomial when t is -2.

12. $t^2 + 3t + 4$
13. $t^3 - 3t^2 - t - 7$

In Problems 14 through 28, perform the operations indicated.

14. $(2x^3 - 3x + 4) + (x^3 - 2x^2 - 4x - 5)$
15. $(4x^5 - 3x^3 - 6x + 11) - (-x^5 + 2x^4 + 6x + 6)$
16. $(5t^2)(-3t^3)$
17. $-5x^2(1 - 5x^5)$
18. $(2x - 1)(3x + 7)$
19. $(x - 1)(x^2 - x + 1)$
20. $\dfrac{3x^4 + 5x^3}{15x^3}$
21. $\dfrac{3x^2 + x - 10}{x + 2}$

22. $\dfrac{4x^3 + 3x + 2}{2x - 1}$

23. $(x - 2)^3$

24. $x(x - 3)(2x - 5)$

25. $x(x + 5) - (x^2 - 2x + 3)$

26. $(y^2 - y + 7) + y(2y - 1)$

27. $(-xy^2)(-3xy)(-x^2y)$

28. $(t + 2)(t - 2) - (t + 1)(t + 3)$

Use special products to multiply.

29. $(x + 7)^2$

30. $(2x + 5)(2x - 5)$

31. $(2t - 3)^2$

In Problems 32 through 35, express each quantity as a polynomial in standard form.

32. The perimeter of a square with side of length $x - 2$ yards.

33. The length of the following rod.

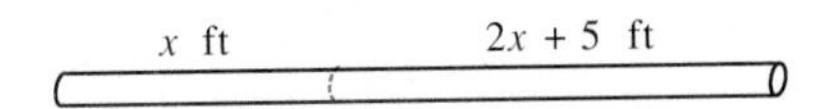

34. The sum of the angles of the triangle shown.

$x + 10°$

$x°$

$x - 7°$

35. The area of a rectangle with dimensions of $y - 3$ and $y + 3$ inches.

...LET'S NOT FORGET...

In Problems 36 through 39, find each sum or difference.

36. $-5 - 11$

37. $(3x - 5) + (6x - 7)$

38. $48 - (x - 42)$

39. $(3x^2 - x + 2) - (2x - 3)$

In Problems 40 through 49, find each product or quotient.

40. $(-x)(-2x^2)$

41. -4^2

42. -4^{-2}

43. $(-4)^{-2}$

44. $(-4)^2$

45. -4^0

46. $\dfrac{0}{-4}$

47. $\dfrac{-4}{0}$

48. $\dfrac{-2x}{x^2}$

49. $\dfrac{4x - 4}{4}$

In Problems 50 through 52, simplify each expression. Watch the role of the parentheses.

50. $(x - 3)(x + 3)$

51. $x - 3(x + 3)$

52. $(x - 3)x + 3$

In Problems 53 through 55, determine if y^2 is a factor or a term.

53. $5y^2$

54. $17 - y^2$

55. $y^2(x - 2)$

In Problems 56 through 58, multiply each expression by 3.

56. $\frac{1}{3}x$

57. $\frac{1}{3}(x - 1)$

58. $\frac{1}{3}x - 1$

In Problems 59 through 63, label each as an equation *or an* expression. *Solve the equations and perform the operations indicated with the expressions.*

59. $2x = -18$

60. $2(x + 3)$

61. $3x + 5x$

62. $x + 5 = 10$

63. $-11 + 6$

CHAPTER 2 TEST

In Problems 1 through 5, choose the correct answer.

1. $2x^3(9x^2 - x^5) = (?)$

A. $18x^5 - 2x^8$ B. $18x^6 - 2x^{15}$

C. $16x^{-6}$ D. $16x^{10}$

2. The value of the polynomial $3x^3 - 2x^2 + x$ when x is -1 is (?).

A. 2 B. 0

C. -2 D. -6

3. $\left(\frac{2a^5}{b^4}\right)^2 = (?)$

A. $\frac{2a^{10}}{b^8}$ B. $\frac{2a^7}{b^6}$

C. $\frac{4a^{10}}{b^8}$ D. $\frac{4a^7}{b^6}$

4. $\frac{(-3)^4}{(-3)^6} = (?)$

A. $\frac{1}{6}$ B. $\frac{1}{9}$

C. $-\frac{1}{6}$ D. $-\frac{1}{9}$

5. $3^{-2} = (?)$

A. -9 B. $\frac{1}{9}$

C. -6 D. $\frac{1}{6}$

6. Write the following in scientific notation.

a. 0.00233

b. 576.3

7. Write the following without exponents.

a. 9.4×10^3

b. 4.321×10^{-5}

In Problems 8 through 18, perform the operations indicated.

8. $\frac{6x^2 - 4x^3}{x^2}$

9. $(2x - 3)(x + 1)$

10. $(5x^3 - 4x^2 + 3x) + (-3x^3 + x^2 - 7x)$

11. $(3x - 2)^2$

12. $(x + 2)^3$

13. $(6x^3 - 2x^2 + x - 2) \div (x + 1)$

14. $(5x^3 - x^2 + 4) - (x^3 - 2x - 7)$

15. $(2^{-2})^{-1}$

16. $(2 - y)(2 + y)$

17. $x^3(2x^4 - 3x^3 + 3)$

18. $(x + 2)(x - 5) + 2(x - 3)$

In Problems 19 and 20, express each quantity as a polynomial in standard form.

19. The area of a rectangle with dimensions $2y - 1$ and $y + 2$ inches.

20. The length of the pipe shown.

CHAPTER 3

See Problem Set 3.4, Exercise 21.

Linear Equations and Inequalities in One Variable

CONNECTIONS

Solving equations has always been a central idea in algebra. The early Egyptians used the rule of false position to solve equations. The idea was to guess a solution and then alter the guess to make it correct. For example, to solve

$$x + \frac{x}{5} = 24,$$

guess a solution, say 5, then try it out.

$$5 + \frac{5}{5} = 6.$$

Since 24 is 4(6), the correct solution must be 4(5) or 20.

Around the year 830 A.D., a Muslim mathematician named Al-Khowarizmi published a book which gave one of the earliest descriptions on how to solve equations by using the addition and multiplication tools that we studied in Chapter 1. The title of his book contained the Arabic word al-jabr from which our word algebra is derived.

In this chapter we continue the study of equations introduced in Section 1.8. The importance of solving equations lies in solving applied problems. We state a problem in business, economics, or engineering and form an equation as a mathematical model of the real world problem. If we can solve the equation, we can solve the applied problem. Also, we will study linear inequalities and learn how to solve them.

3.1 LINEAR EQUATIONS

Equations in which the variable occurs to the first power only are called **linear equations,** or **first-degree equations.** Linear equations in one variable can be written in the form

$$Ax + B = 0, \qquad \text{where } A \text{ is not zero.}$$

Three examples of linear equations are

$$2x + 5 = 0$$

$$3(x - 7) = 11$$

$$6t = 4(t - 1) + 2$$

Three examples of **nonlinear** equations are:

$$x^2 + 2x + 5 = 0 \qquad \text{Quadratic equation}$$

$$\sqrt{x} + x = 11 \qquad \text{Equation with radicals}$$

$$\frac{1}{x} + \frac{2}{x} = x. \qquad \text{Fractional equation}$$

Remember that to **solve** an equation means to find its solution set. We use the addition and multiplication properties that we studied in Section 1.8 to solve linear equations.

Two Tools for Solving Equations

The solution set of an equation will not change if we use either of the following two tools.

Addition Property of Equality

Add (or subtract) the same number on both sides of the equation.

Multiplication Property of Equality

Multiply (or divide) both sides by the same nonzero number.

Our general approach is to use these tools to simplify an equation to the point where it is of the form

$$\text{Variable} = \text{constant}$$

EXAMPLE 1. Solve each equation. Check the solutions.

(a) $x + 8 = 2$ (b) $-\frac{1}{3}t = 6$

Solutions:

(a) We use the Addition Property and subtract 8 from both sides.

$$x + 8 = 2$$

$$x + 8 - 8 = 2 - 8$$

$$x = -6 \qquad \text{Combine like terms.}$$

To check -6, we evaluate the left side (LS) and right side (RS) of the original equation.

LS: $-6 + 8 = 2$

RS: 2

Since both the left and right sides have the value 2, we say -6 checks and is the solution.

$$\{-6\}$$

(b) We use the Multiplication Property and multiply both sides by -3.

$$-\frac{1}{3}t = 6$$

$$(-3)\left(-\frac{1}{3}t\right) = 6(-3)$$

$$\left[(-3)\left(-\frac{1}{3}\right)\right]t = 6(-3) \qquad \text{Associative Property}$$

$$t = -18$$

(continued)

We check -18.

$$\text{LS: } -\frac{1}{3}(-18) = 6$$

$$\text{RS: } 6$$

The solution is -18.

$$\{-18\}$$

□

Often we must use both the Addition and the Multiplication Properties to write an equation in the form, variable = a constant. It is usually best to use the Addition Property first.

EXAMPLE 2. Solve each equation.

(a) $2x - 3 = 9$ (b) $4 - 5t = 49$ (c) $11 = 6x + 11$ (d) $5 - x = 3$

Solutions:

(a) To write the equation in the form x = constant, we must use both the Addition and Multiplication Properties. We use the addition property first.

$$2x - 3 = 9$$

$$2x - 3 + 3 = 9 + 3 \quad \text{Add 3 to both sides.}$$

$$2x = 12 \quad \text{Combine like terms.}$$

$$\frac{2x}{2} = \frac{12}{2} \quad \text{Divide both sides by 2.}$$

$$x = 6$$

Check the solution in the original equation.

$$\{6\} \quad \text{Write the solution set.}$$

(b)

$$4 - 5t = 49$$

$$4 - 5t - 4 = 49 - 4 \quad \text{Subtract 4 from both sides.}$$

$$-5t = 45 \quad \text{Combine like terms.}$$

$$\frac{-5t}{-5} = \frac{45}{-5} \quad \text{Divide both sides by } -5.$$

$$t = -9$$

Check the solution in the original equation.

$$\{-9\} \quad \text{Write the solution set.}$$

(c) We isolate x on the right side.

$$11 = 6x + 11$$

$$11 - 11 = 6x + 11 - 11 \quad \text{Subtract 11 from both sides.}$$

$$0 = 6x \quad \text{Combine like terms.}$$

$$\frac{0}{6} = \frac{6x}{6} \quad \text{Divide both sides by 6.}$$

Be Careful!

$$0 = x \quad \text{Remember } \frac{0}{6} \text{ is 0.}$$

Check the solution in the original equation.

$\{0\}$ Write the solution set.

(d)

$$5 - x = 3$$

$$5 - x - 5 = 3 - 5 \quad \text{Subtract 5 from both sides.}$$

$$-x = -2 \quad \text{Combine like terms.}$$

$$(-1)(-x) = (-1)(-2) \quad \text{Multiply both sides by } -1.$$

$$x = 2$$

Check the solution in the original equation.

$\{2\}$ Write the solution set. □

Often we must combine like terms on the left side and on the right side before we use our addition and multiplication properties to solve an equation. Notice how this works in the next example.

EXAMPLE 3. Solve each equation and check the solutions.

(a) $3x - 5x = 12 - 8$ (b) $-5 - 5 = 6x - x$

Solutions:

(a) $3x - 5x = 12 - 8$

$$-2x = 4 \quad \text{Combine like terms.}$$

$$\frac{-2x}{-2} = \frac{4}{-2} \quad \text{Divide both sides by } -2.$$

$$x = -2$$

Check -2.

LS: $3x - 5x = 3(-2) - 5(-2) = -6 + 10 = 4$

RS: $12 - 8 = 4$

Since the left side and the right side have a value of 4, -2 checks and is in the solution set.

$$\{-2\}$$

(b) $-5 - 5 = 6x - x$

$$-10 = 5x \quad \text{Combine like terms.}$$

$$\frac{-10}{5} = \frac{5x}{5} \quad \text{Divide both sides by 5.}$$

$$-2 = x$$

Check -2.

LS: $-5 - 5 = -10$

RS: $6(-2) - (-2) = -12 + 2 = -10$

The value of both the left side and the right side is the same. Thus -2 checks.

$$\{-2\}$$ □

Sometimes the variable may appear on both sides of the equation. We must collect all terms containing the variable on one side of the equation using the addition tool.

EXAMPLE 4. Solve each equation.

(a) $6x = 5x + 7$ (b) $x + 3 = 3x - 5$

Solutions:

(a) We must collect all x-terms on the same side. We can do this if we subtract $5x$ from both sides.

$6x = 5x + 7$	
$6x - 5x = 5x + 7 - 5x$	Subtract $5x$.
$x = 7$	Combine like terms.
$\{7\}$	Write the solution set.

(b) We must collect the x-terms on one side. We can subtract $3x$ from both sides and collect them on the left side or we can subtract x from both sides and collect them on the right side. Although either would be correct, we often collect them on the side that will yield a *positive* coefficient. So, we subtract x from both sides.

$x + 3 = 3x - 5$	
$x + 3 - x = 3x - 5 - x$	Subtract x from both sides.
$3 = 2x - 5$	Combine like terms.
$3 + 5 = 2x - 5 + 5$	Add 5 to both sides.
$8 = 2x$	Combine like terms.
$\frac{8}{2} = \frac{2x}{2}$	Divide both sides by 2.
$4 = x$	
$\{4\}$	Write the solution set. ☐

Equations are used in solving many different kinds of written problems.

EXAMPLE 5. Seven more than twice a number x is -9.

(a) Translate the sentence into an equation.
(b) Solve the equation and find the number.

Solutions:

(a) Seven more than twice x is -9.

$$\begin{array}{ccccc} \text{Seven} & \text{more than} & \text{twice } x & \text{is} & -9 \\ \downarrow & \downarrow & \downarrow & \downarrow & \downarrow \\ 7 & + & 2x & = & -9 \end{array}$$

(continued)

(b) $7 + 2x - 7 = -9 - 7$ Subtract 7 from both sides.

$2x = -16$ Simplify.

$\frac{2x}{2} = \frac{-16}{2}$ Divide both sides by 2.

$x = -8$

The number is -8.
Check by replacing x with -8 in the word problem. Twice -8 is -16 and seven more than -16 is -9. ☐

Problems such as these are sometimes called applications, or word problems.

EXAMPLE 6. A 36-ft string is cut into two pieces so that the shorter piece is x feet long and the other piece is twice the shorter piece.

(a) Translate the sentence into an equation.
(b) Solve the equation and find the length of the shorter piece of string.

Solutions:

(a) A picture helps visualize the situation.

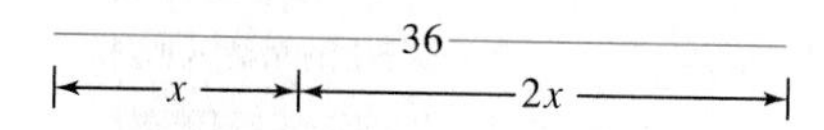

Since the string is 36 ft long, we add the lengths of the two pieces and get 36.

$$x + 2x = 36$$

(b) $3x = 36$ Combine like terms.

$\frac{3x}{3} = \frac{36}{3}$ Divide both sides by 3.

$x = 12$

The length of the shorter piece is 12 ft.
We can check by adding the lengths of the two pieces together to see if we get 36. We see that the longer piece is 24 ft long and that $12 + 24 = 36$. ☐

EXAMPLE 7. The area of the larger triangle is 44 m^2 more than the area of the smaller triangle. Find the length of the base of the larger triangle.

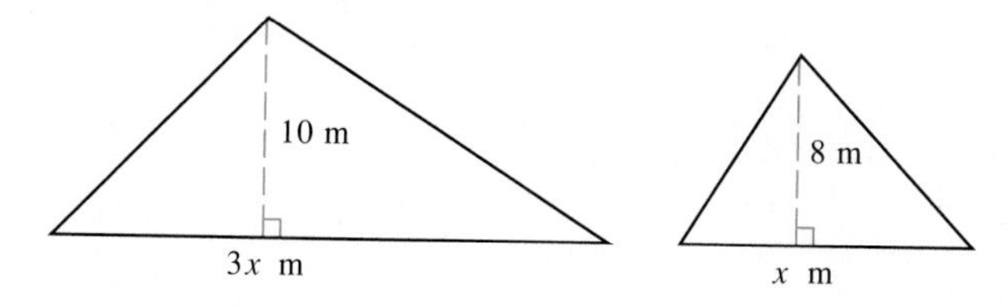

(continued)

Solution:

Remember that the area of a triangle is one-half the product of the base and the height of the triangle.

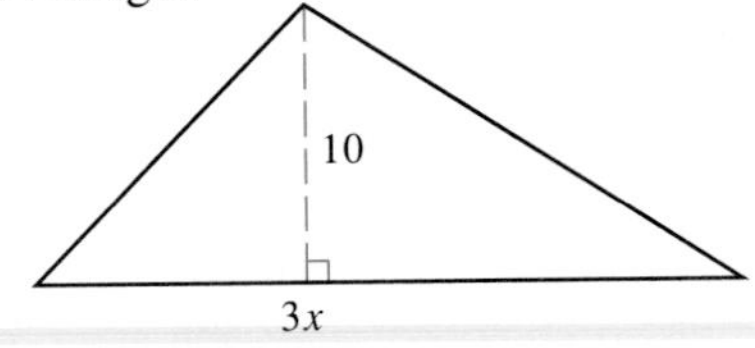

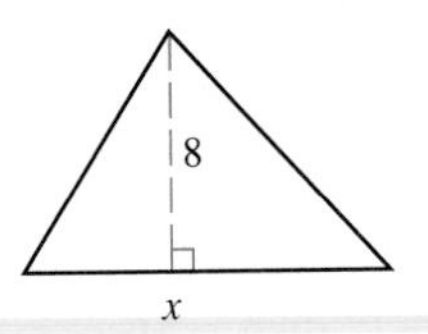

$$\text{Area} = \frac{1}{2} \cdot \text{base} \cdot \text{height} \qquad \text{Area} = \frac{1}{2} \cdot \text{base} \cdot \text{height}$$

$$\text{Area} = \frac{1}{2}(3x)(10) \qquad \text{Area} = \frac{1}{2}(x)(8)$$

$$\text{Area} = 15x \qquad \text{Area} = 4x$$

The area of the larger triangle is 44 more than the area of the smaller triangle. So, $15x$ is larger than $4x$. To make them equal, we can add 44 to the smaller number or subtract 44 from the larger number.

Add 44 to smaller number or Subtract 44 from the larger number

$$15x = 4x + 44 \qquad 15x - 44 = 4x$$

We can solve either equation.

$$15x = 4x + 44$$

$11x = 44$ Subtract $4x$ from both sides.

$x = 4$ Divide both sides by 11.

The base of the larger triangle is $3x$. So, the base of the larger triangle is $3(4) = 12$ m. □

PROBLEM SET 3.1

Warm-ups

Solve each equation.
In Problems 1 through 8, see Example 1.

1. $y + 8 = -1$ **2.** $\frac{2}{3}x = -12$ 3. $z + 8 = 8$

4. $5x = -25$ 5. $-w = \frac{1}{2}$ **6.** $-8t = -56$

7. $5 = x - 7$ **8.** $-3 = 3 + z$

In Problems 9 through 18, see Example 2.

9. $2z - 8 = 22$ **10.** $3t + 4 = 16$ 11. $5x + 2 = 22$

12. $7k - 35 = 0$ **13.** $6 + 2s = 7$ **14.** $-5 - 3x = 10$

15. $6 - 3y = 0$ **16.** $4w - 9 = -9$ **17.** $8 + 5t = 8$

18. $6 - x = 5$

In Problems 19 through 28, see Example 3.

19. $x + 4x = -35 + 20$
20. $-5x + 2x = 20 - 24$
21. $-x - 2x = -8 - 7$
22. $-2x - 7x + x = -12 - 12$
23. $t - 7t = 12 - 6$
24. $15 - 1 = 3x - 10x$
25. $4x - 5x = 2 - 3$
26. $12 = 9p - 13p$
27. $-y - 2y = -1 - 2$
28. $14 - 18 = 4x - 2x$

In Problems 29 through 42, see Example 4.

29. $5x - 3 = 6x$
30. $7x = 5 + 6x$
31. $10x - 1 = 9x$
32. $13x = 12x - 5$
33. $6x + 5 = x$
34. $3x = x - 4$
35. $4x - 1 = x + 8$
36. $2x + 5 = 3x - 7$
37. $3x + 17 = 5x + 3$
38. $x + 6 = 5 - x$
39. $2 - 3x = 2x - 13$
40. $8 - x = 4 - 3x$
41. $5x - 3 = 4x + 2$
42. $3x + 7 = 4x + 6$

In Problems 43 through 49, translate each sentence into an equation, solve the equation, and find the number. See Example 5.

43. Three times a number y decreased by 5 is 10.
44. The sum of a number z and twice z is -27.
45. Five times a number t increased by three times t is 72.
46. If 8 is added to three times a number x, the result is two more than five times x.
47. If four times a number w is subtracted from 17, the result is the sum of w and 22.
48. Twice a number z is 8 less than six times z.
49. A number y is 3 more than four times y.

In Problems 50 through 53, see Example 6.

50. A bag contains only blue and red marbles. There are b blue marbles and $b - 2$ red marbles. If there are 68 marbles in the bag, how many blue marbles are in the bag?
51. A board 48 in. long is cut into three parts, each x inches long. Find the length of each part.
52. In the following figure, the distance from A to B is $2d - 3$ miles and the distance from B to C is $d + 7$ miles. If the distance from A to C is 16 miles, how far is it from A to B?

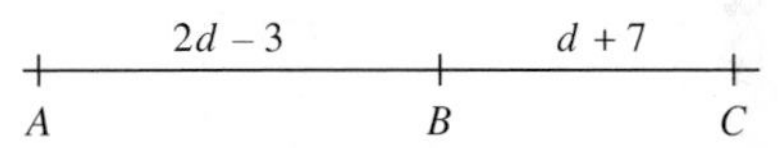

53. Mike has 27 coins, which are either nickels or dimes. If he has n nickels and five more dimes than nickels, how many of each coin does he have?

In Problems 54 through 56, see Example 7.

54. The perimeter of the larger square is 12 ft more than the perimeter of the smaller square. Find the length of a side of the larger square.

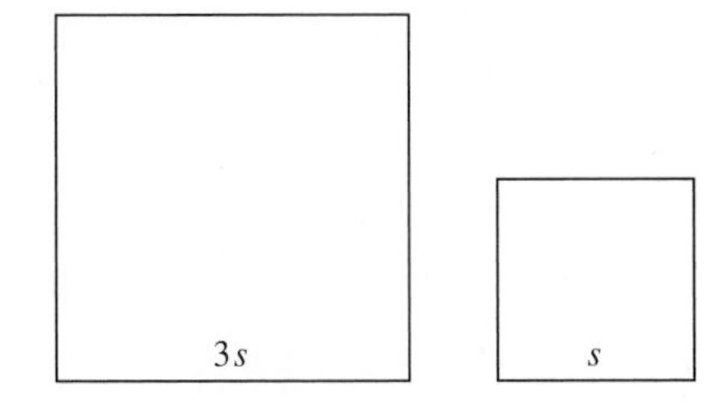

55. The area of the smaller rectangle is 20 cm^2 less than the area of the larger rectangle. Find the length of the smaller rectangle.

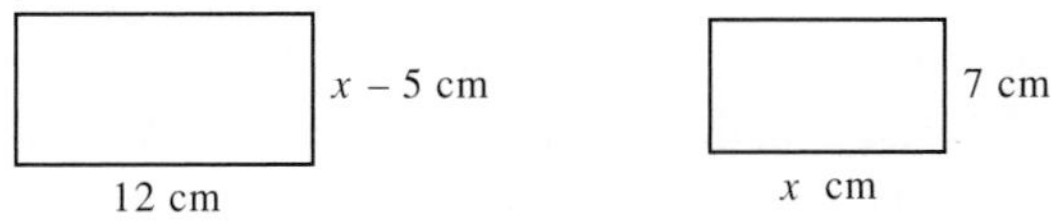

56. The circumference of the larger circle is 2π m more than the circumference of the smaller circle. Find the radius of the larger circle.

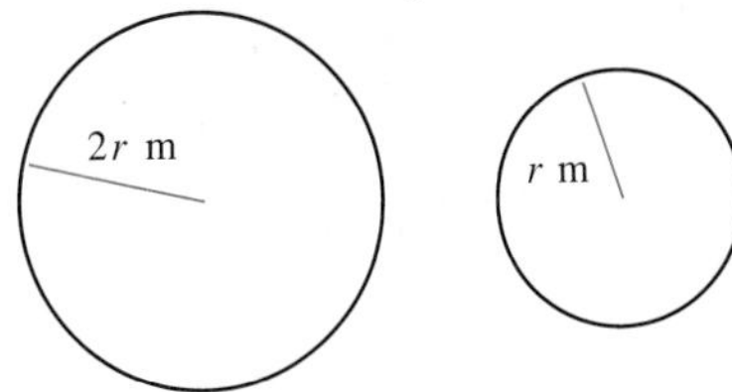

Practice Exercises

Solve the following equations.

57. $3x - 2 = -5$

58. $6 - 5t = 21$

59. $2x - 7 = x$

60. $4x = x - 12$

61. $-x = 3$

62. $1 - x = 0$

63. $3t - 1 = t + 9$

64. $3x - 7 = 4x + 5$

65. $5x - 3x = -17 + 9$

66. $-1 - 5 = -5y - y$

67. $\frac{2}{3}x = \frac{5}{3}$

68. $7 + x = 7$ {0}

69. $2s - 17 = 3s - 3$

70. $x - 9 = 5 - x$

71. $6x - 8 = 5x + 3$

72. $2x + 9 = 3x + 2$

73. $5y - 8 = 3y - 14$

74. $6x - 3x = -5 + 6$

75. $11 - y = 9 + y$

76. $10p - 12 = 7p$

77. $5y - 2y + 3 = 6y - 6$

78. $-x - 1 = x + 1$

79. $13 + 8 - 5 = -7x$

80. $16 = -8z$

81. $15 - 17 = 5x - 8x$

82. $5w + 7w - 2 = 5 + 7 - 2w$

83. $4 - 3y - 8 = 17$

84. $6x - 3 - x - 2 = 40$

85. $x - 8 = -9$

86. $3x - 7 = 7$

In Problems 87 through 91, translate each sentence into an equation, solve the equation, and find the number.

87. Twice a number z increased by 3 is -3.

88. Six times a number t decreased by twice t is 6.

89. If 7 is subtracted from three times a number y, the result is one more than twice y.

90. If seven times a number s is subtracted from 18, the result is twice s.

91. Twice a number x is 5 less than seven times x.

In Problems 92 through 96, answer each question.

92. The perimeter of the triangle on the left is 7 ft more than the perimeter of the triangle on the right. Find the lengths of all sides.

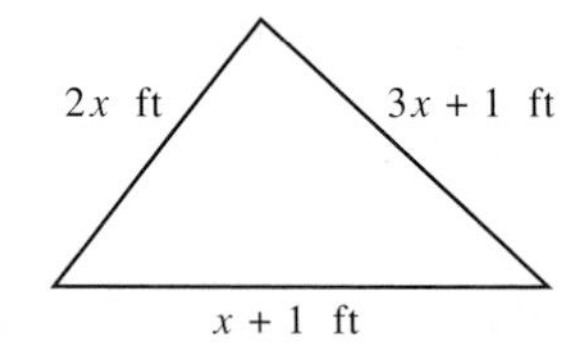

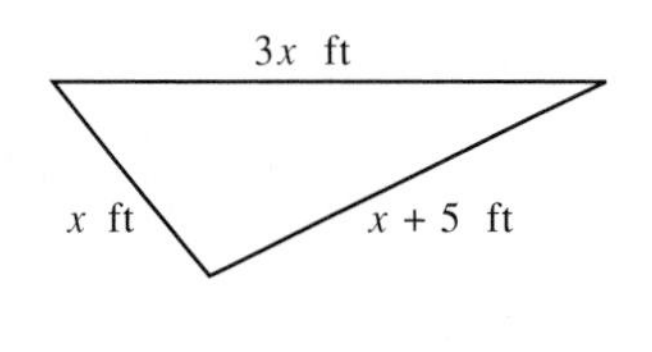

93. The area of the smaller parallelogram is 46 cm^2 less than the area of the larger parallelogram. Find the length of the base of the smaller parallelogram.

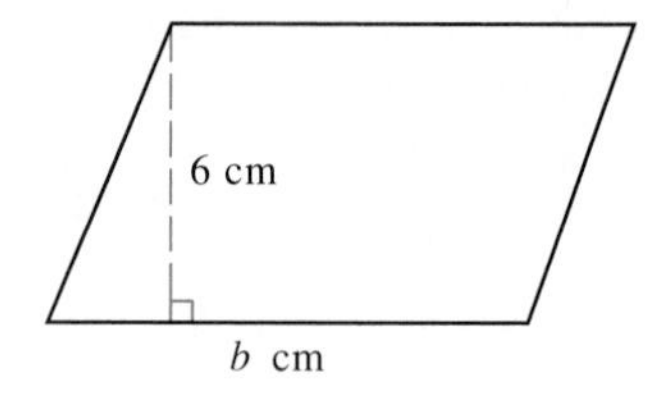

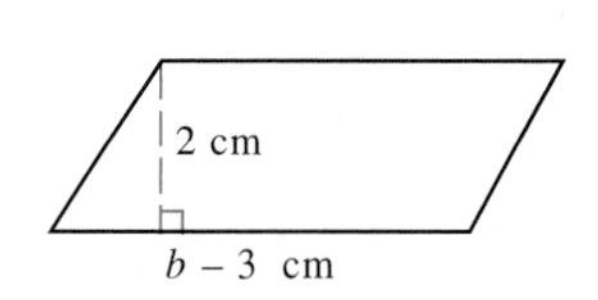

94. A rod 37 in. long is broken into two pieces so that the shorter piece is $2x - 7$ inches long and the other piece is $3x + 4$ inches long. How long is the short piece?

95. Tom has only black and blue socks. He has 12 pairs of socks. If he has s blue socks and twice that many black socks, how many black socks does he have?

96. Susie is y years old and Carol is 10 years older than Susie. If the sum of their ages is 50, how old is Susie?

Challenge Problems

Solve the following equations. The variable is x and q is a constant.

97. $2x - q = 3q$

98. $3x + q = x - 3q$

IN YOUR OWN WORDS...

99. What is a first-degree equation? What is a linear equation?

100. What is the procedure for solving a linear equation?

3.2 LINEAR EQUATIONS WITH GROUPING SYMBOLS OR FRACTIONAL COEFFICIENTS

In this section we continue our study of equations by examining linear equations that contain grouping symbols and fractional or decimal coefficients.

The properties of real numbers and the tools used in the last section provide all we need to solve these equations. However, it is often necessary to simplify as we solve equations. *Whenever possible, we should combine like terms on either side of an equation.*

A Procedure For Solving Linear Equations

1. Clear the equation of fractions (or decimals), if necessary, by multiplying by the least common denominator (LCD) of all the fractions in the equation.
2. Simplify each side by removing grouping symbols and combining like terms, where possible.
3. Collect all terms containing the variable on one side of the equation and all other terms on the other side by using the addition property.
4. Combine like terms.
5. Divide both sides by the coefficient of the variable.
6. Check if required.
7. Write the solution set.

Many times Steps 1 and 2 can be interchanged.

EXAMPLE 1. Solve each equation.

(a) $2(x + 3) = 7$ (b) $2(x - 1) = 4x - 5$

(continued)

Solutions:

(a) $2(x + 3) = 7$

Step 1 Clear fractions.

This equation has no fractions.

Step 2 Remove grouping symbols.

We must use the Distributive Property to remove the grouping symbols.

$$2(x + 3) = 7$$

$$2 \cdot x + 2 \cdot 3 = 7 \qquad \text{Distributive Property}$$

$$2x + 6 = 7$$

Step 3 Collect terms.

We subtract 6 from both sides.

$$2x + 6 - 6 = 7 - 6 \qquad \text{Subtract 6.}$$

Step 4 Combine like terms.

$$2x = 1$$

Step 5 Divide both sides by the coefficient of the variable.

We divide both sides by 2, the coefficient of x.

$$x = \frac{1}{2}$$

Step 6 Check if required.

$\frac{1}{2}$ may be checked in the original equation.

Step 7 Write the solution set.

$$\left\{\frac{1}{2}\right\}$$

(b) $2(x - 1) = 4x - 5$

Step 2 Remove grouping symbols.

$$2x - 2 = 4x - 5 \qquad \text{Distributive Property}$$

Step 3 Collect x-terms.

$$2x - 2 - 2x = 4x - 5 - 2x \qquad \text{Subtract } 2x.$$

Step 4 Combine like terms.

$$-2 = 2x - 5$$

Step 3 Collect constant terms.

$$-2 + 5 = 2x - 5 + 5 \qquad \text{Add 5.}$$

Step 4 Combine like terms.

$$3 = 2x$$

Step 5 Divide by the coefficient of the variable.

$$\frac{3}{2} = \frac{2x}{2} \qquad \text{Divide by 2.}$$

$$\frac{3}{2} = x$$

Step 6 Check if required.

$\frac{3}{2}$ may be checked in the original equation.

Step 7 Write the solution set.

$$\left\{\frac{3}{2}\right\}$$

□

EXAMPLE 2. Solve each equation, and *check* the solutions.

(a) $5(x + 2) = 2x + 19$ (b) $3t + 4 = 3(1 + 2t)$

Solutions:

(a) $5(x + 2) = 2x + 19$

Remove grouping symbols.

$$5x + 10 = 2x + 19 \qquad \text{Distributive Property}$$

Collect x-terms.

$$5x + 10 - 2x = 2x + 19 - 2x \qquad \text{Subtract 2x.}$$

Combine like terms.

$$3x + 10 = 19$$

Collect constant terms.

$$3x + 10 - 10 = 19 - 10 \qquad \text{Subtract 10.}$$

Combine like terms.

$$3x = 9$$

Divide by the coefficient of x.

$$x = 3 \qquad \text{Divide by 3.}$$

Check.

The instructions call for checking our solutions. We must do so before writing the solution set. Check 3 in the original equation.

LS: $5(x + 2) = 5(3 + 2) = 5(5) = 25$

RS: $2x + 19 = 2(3) + 19 = 6 + 19 = 25$

We get the same value in both the left side and the right side. Thus, 3 is the solution.

(continued)

Write the solution set.

$$\{3\}$$

(b)

$$3t + 4 = 3(1 + 2t)$$

$$3t + 4 = 3 + 6t \qquad \text{Distributive Property}$$

$$3t + 4 - 3t = 3 + 6t - 3t \qquad \text{Subtract } 3t.$$

$$4 = 3 + 3t \qquad \text{Combine like terms.}$$

$$4 - 3 = 3 + 3t - 3 \qquad \text{Subtract 3.}$$

$$1 = 3t \qquad \text{Combine like terms.}$$

$$\frac{1}{3} = t \qquad \text{Divide by 3.}$$

Check $\frac{1}{3}$.

$$\text{LS:} \quad 3t + 4 = 3\left(\frac{1}{3}\right) + 4 = 1 + 4 = 5$$

$$\text{RS:} \quad 3(1 + 2t) = 3\left[1 + 2\left(\frac{1}{3}\right)\right] = 3\left[1 + \left(\frac{2}{3}\right)\right] = 3\left(\frac{5}{3}\right) = 5$$

The left side has the same value as the right side, so $\frac{1}{3}$ checks.

$$\left\{\frac{1}{3}\right\} \qquad \text{Write the solution set.} \quad \square$$

Linear equations involving fractions can be solved with the addition and multiplication tools. However, arithmetic with fractions is more difficult than arithmetic with integers, and we are more likely to make mistakes. It is wise to clear fractions as a first step in such equations. If we multiply both sides of the equation by the least common denominator (LCD) of all the fractions in the equation, all the fractions will be removed. We call this **clearing fractions.**

When solving equations involving fractions, we frequently encounter expressions like

$$\frac{2}{3}x \quad \text{or} \quad \frac{2x}{3}$$

Because of the rules for multiplying fractions, these have the same value and are just two different ways of writing the same fraction. Note that the following are equal.

$$\frac{1}{5}x = \frac{x}{5}$$

$$\frac{7}{3}x = \frac{7x}{3}$$

$$\frac{3}{2}(2x - 3) = \frac{3(2x - 3)}{2}$$

EXAMPLE 3. Solve each equation.

(a) $\frac{x}{2} - 1 = \frac{2}{5}x + 3$ (b) $\frac{1}{2}(x - 2) = 4$

Solutions:

(a) $\frac{x}{2} - 1 = \frac{2}{5}x + 3$

Notice that this equation contains the fractions $\frac{x}{2}$ and $\frac{2}{5}$. The LCD of these fractions is 10. So we multiply both sides of the equation by 10.

$10 \cdot \left(\frac{x}{2} - 1\right) = 10 \cdot \left(\frac{2}{5}x + 3\right)$	Multiply by 10.
$10 \cdot \left(\frac{x}{2}\right) - 10 \cdot 1 = 10 \cdot \left(\frac{2}{5}x\right) + 10 \cdot 3$	Distributive Property
$10\left(\frac{x}{2}\right) - 10 \cdot 1 = \left(10 \cdot \frac{2}{5}\right)x + 10 \cdot 3$	Associative Property
$5x - 10 = 4x + 30$	
$5x - 10 - 4x = 4x + 30 - 4x$	Subtract $4x$.
$x - 10 = 30$	Combine like terms.
$x - 10 + 10 = 30 + 10$	Add 10.
$x = 40$	Combine like terms.
$\{40\}$.	Write the solution set.

(b)

$\frac{1}{2}(x - 2) = 4$	
$2\left[\frac{1}{2}(x - 2)\right] = 2(4)$	Multiply by 2.
$\left(2 \cdot \frac{1}{2}\right)(x - 2) = 8$	Associative Property
$x - 2 = 8$	
$x - 2 + 2 = 8 + 2$	Add 2.
$x = 10$	Combine like terms.
$\{10\}$	Write the solution set. □

If an equation contains decimals, we can work with the decimals or we can clear the decimals. Since decimals can be written as fractions, we will multiply both sides of the equation by 10, 100, 1000 or whatever power of 10 is necessary.

$$0.1 = \frac{1}{10} \qquad 0.01 = \frac{1}{100} \qquad 0.001 = \frac{1}{1000}$$

EXAMPLE 4. Solve each equation.

(a) $5.2x = 15.6$ (b) $2.3x - 2.12 = 1.2x + 0.08$

(continued)

Solutions:

(a) $5.2x = 15.6$

We clear decimals by multiplying both sides by 10.

$10(5.2x) = 10(15.6)$	Multiply by 10.
$52x = 156$	Simplify.
$\frac{52x}{52} = \frac{156}{52}$	Divide by 52.
$x = 3$	Simplify.
$\{3\}$	Write the solution set.

(b) $2.3x - 2.12 = 1.2x + 0.08$

We must multiply by 100 to clear the decimals because there are tenths and hundredths.

$100(2.3x - 2.12) = 100(1.2x + 0.08)$	
$230x - 212 = 120x + 8$	Distributive Property
$230x - 212 - 120x = 120x + 8 - 120x$	Subtract $120x$.
$110x - 212 = 8$	
$110x - 212 + 212 = 8 + 212$	Add 212.
$110x = 220$	
$x = 2$	Divide by 110.
$\{2\}$	

□

We do not always get one number in our solution set when we solve an equation. Two more equations illustrate this.

EXAMPLE 5. Solve each equation.

(a) $2x + 1 = 2(5 + x)$ (b) $3(2x + 6) = 2(9 + 3x)$

Solutions:

(a)
$$2x + 1 = 2(5 + x)$$
$$2x + 1 = 10 + 2x$$
$$2x + 1 - 2x = 10 + 2x - 2x$$
$$1 = 10$$

Oops! What has happened here? Since our work is all correct, the solution set for the original equation is the same as the solution set for the equation

$$1 = 10$$

But there is *no* value of x that will make 1 equal 10, so there are no numbers in the solution set. The solution set is the set with no elements, the empty set.

We write the empty set as

$$\{\ \} \quad \text{or} \quad \varnothing \quad \textit{Never } \{\varnothing\}$$

Be Careful!

(b)

$$3(2x + 6) = 2(9 + 3x)$$
$$6x + 18 = 18 + 6x$$
$$6x + 18 - 6x = 18 + 6x - 6x$$
$$18 = 18$$

Again, the variables disappeared from the equation. However, this time we are left with a *true* statement. That is, 18 equals 18 for *all* values of x. The solution set must be all real numbers. We use set-builder notation (Section 1.1) to write the solution set.

$$\{x \mid x \text{ is a real number}\}$$

□

EXAMPLE 6. John has x quarters and $x + 10$ nickels in his pocket. He knows that he has \$5. How many quarters does John have?

Solution:

We must find the *value* of x quarters. Since 1 quarter has a value of 25¢, x quarters have a value of $25x$ cents.

The *value* of $x + 10$ nickels is $5(x + 10)$ cents.

The *value* of the money is \$5, or 500¢.

Value of quarters in cents	+	Value of nickels in cents	=	Value of money in cents
↓		↓		↓
$25x$	+	$5(x + 10)$	=	500

We can find how many quarters John has by solving this equation.

$$25x + 5(x + 10) = 500$$
$$25x + 5x + 50 = 500 \qquad \text{Distributive Property}$$
$$30x + 50 = 500$$
$$30x + 50 - 50 = 500 - 50$$
$$30x = 450$$
$$\frac{30x}{30} = \frac{450}{30} \qquad \text{Divide by 30.}$$
$$x = 15$$

John has 15 quarters.
We check by noting that \$.25(15) is \$3.75. Also John has 25 nickels worth \$.05(25), or \$1.25. Adding the value of his quarters to the value of his nickels, we have

$$\$3.75 + \$1.25 = \$5.00$$

□

EXAMPLE 7. If x is the smallest of three consecutive integers, twice the largest integer is equal to the sum of twice the smallest integer and the middle integer. Find the integers.

Solution:

If x is the smallest integer, then $x + 1$ is the middle integer and $x + 2$ is the largest integer.

We form an equation to find the integers.

Twice the largest	is	the sum of twice smallest and middle.	
↓	↓	↓	↓
$2(x + 2)$	$=$	$2x$	$+ \quad (x + 1)$

Now we solve the equation.

$$2x + 4 = 3x + 1$$

$$2x + 4 - 2x = 3x + 1 - 2x$$

$$4 = x + 1$$

$$4 - 1 = x + 1 - 1$$

$$3 = x$$

Since the integers are x, $x + 1$, and $x + 2$, the numbers are 3, 4, and 5.

Check in the original problem. Twice the largest, 2(5), is the same as the sum of twice the smallest, 2(3), and the middle, 4. That is,

$$2(5) = 2(3) + 4$$

□

EXAMPLE 8. The formula for the volume of a pyramid is one-third the product of the area of its base and its height. If the volume of the pyramid below is 125 m^3 and the area of its base is 25 m^2, find the measure of its height.

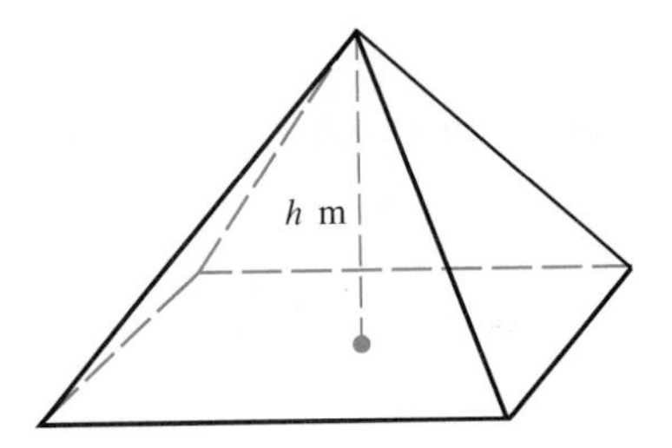

Solution:

We use the formula to write an equation.

$$\text{Volume} = \frac{1}{3} \cdot \text{area of base} \cdot \text{height}$$

$$125 = \frac{1}{3}(25)h$$

$$125 = \frac{25}{3}h$$

$$3(125) = 3\left(\frac{25}{3}\right)h \qquad \text{Multiply by 3.}$$

$$375 = 25h$$

$$\frac{375}{25} = \frac{25h}{25} \qquad \text{Divide by 25.}$$

$$15 = h$$

The height is 15 meters. □

EXAMPLE 9. If 23% of n is 99.13, find n.

Solution:

Since 23% = 0.23, we form the equation $0.23n = 99.13$.

$$0.23n = 99.13$$

$$23n = 9913 \qquad \text{Multiply by 100.}$$

$$\frac{23n}{23} = \frac{9913}{23} \qquad \text{Divide by 23.}$$

$$n = 431$$

n is 431. □

PROBLEM SET 3.2

Warm-ups

Solve each equation. In Problems 1 through 20, see Examples 1 and 2.

1. $4(x - 2) = 3$

2. $5(x + 1) = 1$

3. $6(3 + x) = 18$

4. $7(1 - x) = 3$

5. $-4(2w - 3) = 6$

6. $-5(3 + 7x) = 20$

7. $6(3 - 4x) - 22 = 0$

8. $15 - 3(2v - 9) = 0$

9. $66 - (12 - 3x) = 0$

10. $13(3 - 2x) + 1 = 1$

11. $2(y - 3) - y = 5$

12. $6x - (5x - 7) = -2$

13. $4(z + 4) - 3(z - 1) = 4$

14. $-2(w + 3) + 3(3w + 2) = -14$

15. $5x - 1 = 2(x + 3)$

16. $2(x - 1) = 6(x + 1)$

17. $2(3 + t) = 3(t - 4)$

18. $x + 3(2x - 1) = 4(2x + 1)$

19. $5(3t - 2) - (12t - 10) = 2(t + 2)$

20. $-2(1 - 3x) = 5(1 - x) - 3x$

In Problems 21 through 32, see Example 3.

21. $\frac{x}{2} + \frac{1}{2} = \frac{3}{2}$
22. $\frac{x}{3} - \frac{2}{3} = \frac{1}{3}$
23. $\frac{1}{4} - \frac{3x}{4} = 1$
24. $\frac{5}{3} - \frac{2}{3}x = 1$
25. $\frac{3}{2}x - \frac{1}{3} = \frac{1}{2}x$
26. $\frac{5}{6} - \frac{x}{2} = \frac{x}{3}$
27. $\frac{2}{3}x - \frac{3}{2}x = -\frac{10}{3}$
28. $2x + 3 = \frac{1}{2}x - 9$
29. $\frac{2}{3}x = x + \frac{5}{3}$
30. $\frac{1}{3}x = \frac{1}{2}x + 1$
31. $\frac{1}{3}(x + 2) = 1$
32. $\frac{w - 3}{3} = \frac{w}{6}$

In Problems 33 through 36, see Example 4.

33. $2.1x = 6.3$
34. $0.49x = 2.45$
35. $1.7x - 8.2 = 2.5x + 3.8$
36. $12.7z - 5.4z = z - 12.6$

In Problems 37 and 38, see Example 5.

37. $4x - 7 = 4(x - 2)$
38. $3(1 - x) = 3 - 3x$

In Problems 39 through 41, answer each question. See Example 6.

39. Tom worked x hours at K-Mart last week and earned \$4.50 an hour. How many hours did he work last week if he made \$121.50?
40. Jerry bought s 25¢ stamps at a cost of \$4.25. How many of these stamps did he buy?
41. Sharon has n nickels and $n + 8$ dimes. If she has \$2.60, how many dimes does she have?

In Problems 42 through 44, see Example 7.

42. The sum of three consecutive odd integers is 51. If n is the smallest of the integers, find the three integers.
43. The smallest of four consecutive integers is s. If the sum of the smallest one and three times the third one equals the fourth one, find the integers.
44. The smallest of three consecutive even integers is n. If twice the first integer plus three times the second is 82 more than the third integer, find the integers.

In Problems 45 through 48, see Example 8.

45. Irma deposited d dollars in her savings account at 6% for one year. If she earned \$34.20 in interest, how much did she deposit? (*Hint:* Interest = principal · rate · time.)
46. Lucille drove 385 mi into the country at a speed of r miles per hour. If the trip took 7 h, find her speed. (*Hint:* Distance = rate · time.)
47. Samson rode his bike for h hours at a speed of 12.5 miles per hour. If he traveled a distance of 100 mi, find how long he rode his bike. (*Hint:* Distance = rate · time.)
48. The volume of the following box is 300 m³. Find its height. (*Hint:* Volume = length · width · height.)

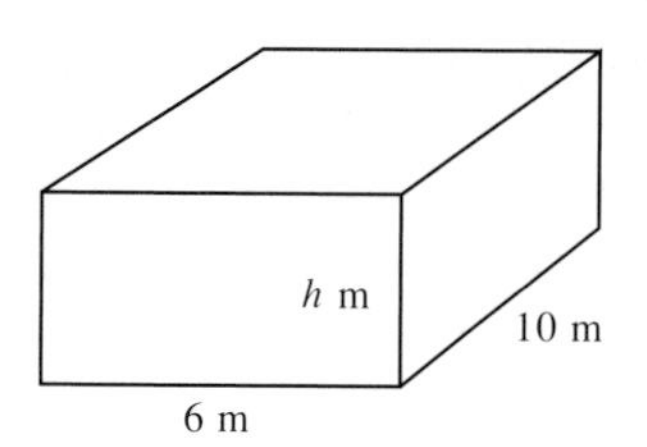

In Problems 49 through 52, see Example 9.

49. Fifteen percent of a number q is 75. What is the number?
50. Twenty-five percent of a number x is 36. What is the number?
51. A bottle contains x gallons of a 15% acid solution. If there are 3 gal of acid in the bottle, how many gallons of the solution are in the bottle?
52. A sugar solution contains 4.9 L of sugar. If the solution is 35% sugar and there are x liters of the sugar solution, how many liters of sugar solution are there?

Practice Exercises

Solve the following equations.

53. $2(x+1)=6$

54. $3(x-2)=12$

55. $\frac{2x}{3}+\frac{1}{3}=\frac{5}{3}$

56. $\frac{5}{3}x-\frac{5}{3}=\frac{4}{3}$

57. $3(9-x)=18$

58. $(5-y)7=-14$

59. $-2(3x-2)=10$

60. $-3(5+2x)=7$

61. $\frac{5}{7}-\frac{4x}{7}=-1$

62. $\frac{1}{4}-\frac{x}{8}=\frac{1}{2}$

63. $95-5(13-3x)=0$

64. $5(4-5x)+5=0$

65. $\frac{1}{2}(s+3)=1$

66. $\frac{1}{3}(x-2)=1$

67. $3(2+3x)=5$

68. $2(3-2x)=1$

69. $-2(3x+2)=10$

70. $-3(5-2x)=9$

71. $(3x+2)2=7x-6$

72. $3(3-k)=2(k+4)$

73. $7(3x+4)-2(14-3x)=54$

74. $6-2(3x-2)=4-2(3+x)$

75. $6-2(3x+2)=4-2(3-x)$

76. $3(5-x)+5(2x-3)=3(2x-1)$

77. $2[3(4-3x)-2x]-4x=-x-1$

78. $\frac{x}{3}-\frac{1}{2}=\frac{3}{2}$

79. $\frac{1}{4}-\frac{x}{2}=\frac{1}{8}$

80. $\frac{2}{3}x-\frac{3}{5}=\frac{1}{15}$

81. $\frac{3x}{2}+\frac{1}{6}=\frac{5}{3}$

82. $\frac{x}{2}-\frac{1}{2}=\frac{3}{2}$

83. $3.3t=-1.32$

84. $6(3-z)=18$

85. $\frac{1}{4}-\frac{3x}{4}=1$

86. $\frac{3}{2}-\frac{7}{2}x=3$

87. $\frac{2x}{3}+\frac{3}{5}=-\frac{1}{15}$

88. $\frac{1}{7}+\frac{5w}{3}=\frac{8}{21}$

89. $-2.7=0.3x$

90. $\frac{3}{2}t=t+\frac{5}{2}$

91. $\frac{1}{2}x=\frac{1}{3}x-1$

92. $\frac{1}{3}x+\frac{3}{4}=\frac{2}{3}x$

93. $\frac{1}{3}x+\frac{1}{4}=\frac{1}{2}x$

94. $\frac{x+3}{3}=\frac{x}{6}$

95. $\frac{1}{3}(y-2)=\frac{2}{3}-y$

96. $\frac{2}{3}(2-x)=4$

97. $\frac{1}{2}(x+5)=\frac{3}{2}-x$

98. $3[2(4+3y)-3y]-5y=4y+1$

99. $2(x-1)-3x=x-2(1+x)$

In Problems 100 through 110, answer each question.

100. James has q quarters and twice that many dimes. If he has \$4.05, how many dimes does he have?

101. Thirty percent of s is 144. What is s?

102. The volume of the cylinder below is 1440π in.^{3} If the radius of the base is 12 in. find the height. (*Hint:* Volume = $\pi \cdot (\text{radius})^2 \cdot$ height.)

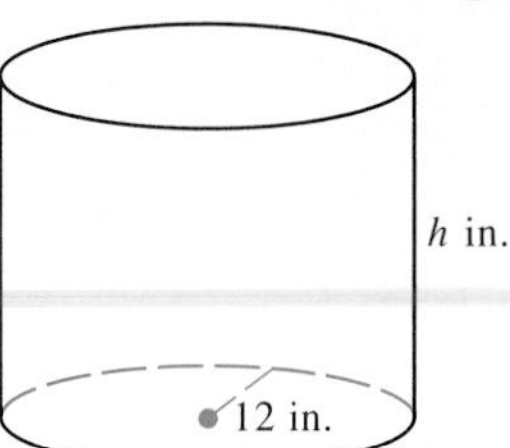

103. Ray borrowed x dollars at 14% for one year. If he paid \$105 in interest, how much did he borrow? (*Hint:* Interest = principal · rate · time.)

104. Lou bought y apples at 49¢ per apple and she bought twice that many pears at 69¢ per pear. If she spent \$9.35, how many apples did she buy?

105. Leonard has a box full of \$5 and \$10 bills worth \$140. If he has t \$5 bills and 4 less than t \$10 bills, how many \$10 does he have?

106. Jane walked x miles in the park. If she was gone $3\frac{1}{2}$ hr and she walked at a rate of 2 mph, how far did she walk? (*Hint:* Distance = rate · time.)

107. The volume of the cone is 80π ft^3 and the radius of the base is 4 ft. Find its height. (*Hint:* Volume $= \frac{1}{3}\pi \cdot (\text{radius})^2 \cdot$ height.)

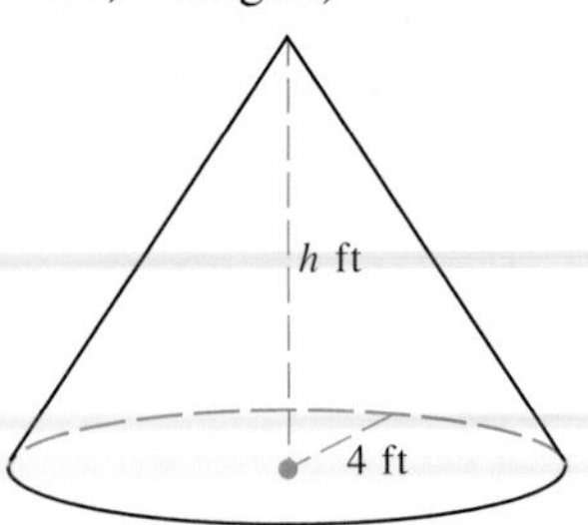

108. The smallest of three consecutive odd integers is n. If seven added to the middle integer is the same as the sum of the first and last integers decreased by 10, what are the three integers?

109. Jake bought t tapes at \$7.99 each and spent \$63.92. How many tapes did he buy?

110. Robert worked y hours at Wendy's during the week and $y + 7$ hours during the weekend. He gets paid \$3.85 per hour during the week and makes \$.45 per hour more on the weekends. If he earned \$141.50 for the week, how many hours did he work during the week?

Challenge Problems

In Problems 111 and 112, solve the equation. The variable is x and q is a constant.

111. $3(x - q) = q$

112. $\dfrac{x - q}{4} = q$

IN YOUR OWN WORDS...

113. Explain the procedure for solving a linear equation.

3.3 LITERAL EQUATIONS AND FORMULAS

Some equations contain more than one variable. For example,

$$x + y = 1$$

Notice that if we subtract y from both sides, we can write the equation in another form.

$$x + y - y = 1 - y \qquad \text{Subtract } y$$

$$x = 1 - y$$

We say we have solved the equation for x. Similarly, we can subtract x from both sides of the original equation and solve it for y.

$$x + y - x = 1 - x \qquad \text{Subtract } x$$

$$y = 1 - x$$

We have solved for y.

We sometimes call equations with more than one variable **literal equations.** When solving a literal equation for one of its variables, it's *important* to use the *same* tools we used in solving linear equations. That is, add the same number to both sides, or multiply both sides by the same nonzero number.

Formulas are literal equations that express the relationship between two or more physical conditions. For example, the area of a circle is given by πr^2, where r is the radius of the circle. We write

$$A = \pi r^2$$

Two other examples are the perimeter of a rectangle and the distance-rate-time formula.

Perimeter:	$P = 2l + 2w$	Where l is length and w is width
Distance:	$d = rt$	Distance is rate times time.

Sometimes a formula would be more convenient to use if it were solved for a different variable.

EXAMPLE 1. Solve each formula for the variable indicated.

(a) $d = rt$ for r (b) $P = 2l + 2w$ for l

Solutions:

(a) We are to solve $d = rt$ for r. Notice that if we divide the right side by t, we will isolate r. Divide both sides by t.

$$\frac{d}{t} = \frac{rt}{t}$$

$$\frac{d}{t} = r$$

Solution sets are not used when solving literal equations. Rather, the equation (or formula) is usually written with the variable being solved for on the left side.

$$r = \frac{d}{t}$$

(b) Solve $P = 2l + 2w$ for l.
We subtract $2w$ from both sides.

$$P - 2w = 2l + 2w - 2w$$

$$P - 2w = 2l$$

Now we divide both sides by 2.

$$\frac{P - 2w}{2} = \frac{2l}{2}$$

$$\frac{P - 2w}{2} = l$$

We then write the answer:

$$l = \frac{P - 2w}{2}$$

□

A Procedure for Solving Literal Equations

1. Clear the equation of fractions, if necessary.
2. Remove all symbols of grouping and combine like terms where possible.
3. Collect all terms containing the variable for which the equation is being solved on one side of the equation and all other terms on the other side of the equation.
4. Combine like terms.
5. Divide both sides of the equation by the coefficient of the variable for which the equation is being solved.

It may be convenient to interchange Steps 1 and 2.

EXAMPLE 2. The formula for the area of a triangle is one-half the base times the height. Solve this literal equation for height.

Solution:

We are to solve $A = \frac{1}{2}bh$ for h.

Step 1 Clear fractions.

First, we clear fractions by multiplying both sides by 2.

$$2 \cdot A = 2 \cdot \left(\frac{1}{2}bh\right) \qquad \text{Multiply by 2.}$$

Step 2 Remove grouping symbols.

$$2A = \left(2 \cdot \frac{1}{2}\right) \cdot bh \qquad \text{Associative Property}$$

$$2A = 1 \cdot bh$$

$$2A = bh$$

Steps 3 and 4 Collect and combine like terms.

This is unnecessary in this problem.

Step 5 Divide both sides by the coefficient of the variable for which we are solving.

Divide both sides by b.

$$\frac{2A}{b} = \frac{bh}{b}$$

$$\frac{2A}{b} = h$$

Thus the formula becomes $h = \frac{2A}{b}$ when solved for h. □

Often literal equations are simply equations containing two or more letters.

EXAMPLE 3. Solve the following literal equations for the variable indicated.

(a) $\frac{1}{2}x + s = 3t$ for x (b) $4(p + y) = 3$ for y (c) $\frac{q}{3}(w - 11b) = 4$ for w

Solutions:

(a) We have a fraction in this equation, which we can clear by multiplying both sides by 2.

$$2 \cdot \left(\frac{1}{2}x + s\right) = 2 \cdot (3t) \qquad \text{Clear fractions.}$$

We use the distributive property and the associative property for multiplication to simplify this equation.

$$2 \cdot \left(\frac{1}{2}x\right) + 2 \cdot s = 2 \cdot (3t) \qquad \text{Distributive Property}$$

$$\left(2 \cdot \frac{1}{2}\right)x + 2s = (2 \cdot 3)t \qquad \text{Associative Property}$$

$$(1)x + 2s = (6)t$$

$$x + 2s = 6t$$

Now that we have cleared the fractions, we can subtract $2s$ from both sides to solve for x.

$$x + 2s - 2s = 6t - 2s \qquad \text{Subtract } 2s.$$

$$x = 6t - 2s.$$

(b) There are no fractions in this equation, but we have a grouping symbol. We use the distributive property to remove the parentheses.

$$4p + 4y = 3 \qquad \text{Distributive Property}$$

$$4p + 4y - 4p = 3 - 4p \qquad \text{Subtract } 4p.$$

$$4y = 3 - 4p \qquad \text{Combine like terms.}$$

$$\frac{4y}{4} = \frac{3 - 4p}{4} \qquad \text{Divide by 4.}$$

$$y = \frac{3 - 4p}{4}$$

(c) $$3 \cdot \left[\frac{q}{3}(w - 11b)\right] = 3 \cdot 4 \qquad \text{Multiply by 3.}$$

$$\left(3 \cdot \frac{q}{3}\right)(w - 11b) = 3 \cdot 4 \qquad \text{Associative Property}$$

$$q(w - 11b) = 12$$

$$qw - 11bq = 12 \qquad \text{Distributive Property}$$

$$qw - 11bq + 11bq = 12 + 11bq \qquad \text{Add } 11bq.$$

$$qw = 12 + 11bq \qquad \text{Combine like terms.}$$

$$w = \frac{12 + 11bq}{q} \qquad \text{Divide by } q.$$

□

EXAMPLE 4. Solve $-3p = w$ for p.

Solution:

We divide both sides by -3.

$$\frac{-3p}{-3} = \frac{w}{-3}$$

$$p = \frac{w}{-3}$$

The fraction $\frac{w}{-3}$ can also be written as $\frac{-w}{3}$ or as $-\frac{w}{3}$. We usually write the negative sign in the numerator or in front of the fraction.

$$p = -\frac{w}{3}$$

□

PROBLEM SET 3.3

Warm-ups

In Problems 1 through 7, solve the formula for the variables indicated. See Examples 1 and 2.

1. $A = lw$ (area of rectangle); solve for w.

2. $A = \frac{1}{2}bh$ (area of a triangle); solve for b.

3. $C = 2\pi r$ (circumference of a circle); solve for r.

4. $I = prt$ (simple interest); solve for t.

5. $V = \frac{1}{3}\pi r^2 h$ (volume of a cone); solve for h.

6. $E = mc^2$ (mass-energy formula); solve for m.

7. $F = \frac{9}{5}C + 32$ (Celsius to Fahrenheit); solve for C.

Solve each for the variable indicated. See Examples 3 and 4.

8. $x + y = 1$ for x

9. $x + y = 1$ for y

10. $x + y = a$ for x

11. $x - y = a$ for x

12. $2x = q$ for x

13. $-3x = p$ for x

14. $ax + 2y = k$ for y

15. $2k - 2y = 6k$ for y

16. $ax + by = z$ for y

17. $3z - by = 4z$ for y

18. $\frac{2x}{b} + C = D$ for x

19. $\frac{ay}{4} - L = f$ for y

20. $\frac{bz}{C} - s = 14$ for z

21. $q - \frac{aw}{k} = 3q$ for w

22. $2(x + a) = 5$ for x

23. $3(y - q) = 1$ for y

24. $\frac{1}{g}(3 + w) = u$ for w

25. $6 = \frac{1}{p}(a + x)$ for x

26. $\frac{2}{b}(x + y) = 1$ for y

27. $\frac{a}{b}(k + w) = 2L$ for w

Practice Exercises

In Problems 28 through 31, solve the formulas for the variables indicated.

28. $A = lh$ (area of a parallelogram); solve for l.

29. $F = \frac{mv^2}{r}$ (centripetal force); solve for m.

30. $V = \frac{1}{3}Ah$ (volume of a pyramid); solve for A.

31. $C = \frac{5}{9}(F - 32)$ (Fahrenheit to Celsius); solve for F.

Solve each for the variable indicated.

32. $x + 2z = 5$ for x

33. $2x + z = 5$ for z

34. $x + 3z = k$ for x

35. $x - 3z = k$ for x

36. $3x = p$ for x

37. $-2x = q$ for x

38. $-qx = 4$ for x

39. $px = -7$ for x

40. $kx = a$ for x

41. $kx + a = 0$ for x

42. $3h - y = h$ for y

43. $5x + y = h$ for y

44. $2k - 3y = 5k$ for y

45. $ax + 4y = k$ for y

46. $ax + by = z$ for x

47. $ax - by = z$ for y

48. $\frac{x}{5} + \frac{b}{5} = \frac{1}{5}$ for x

49. $\frac{x}{2} - \frac{k}{2} = 3$ for x

50. $\frac{q}{3} + \frac{y}{3} = p$ for y

51. $\frac{w}{4} - \frac{y}{4} = w$ for y

52. $\frac{z}{7} + a = J$ for z

53. $\frac{z}{p} - q = T$ for z

54. $\frac{3x}{k} + M = L$ for x

55. $\frac{qy}{-2} - L = f$ for y

56. $\frac{bz}{2v} - s = 11$ for z

57. $t + \frac{aw}{k^2} = 0$ for w

58. $3(x + s) = 4$ for x

59. $2(y - p) = 1$ for y

60. $5(y + 2z) = p$ for z

61. $p(x - k) = -2$ for x

62. $j = q(s - 4)$ for s

63. $A(3x - B) = C$ for x

64. $\frac{1}{3}(5 + x) = C$ for x

65. $\frac{1}{2}(A - s) = A$ for s

66. $k^2 = \frac{1}{5}(x - y)$ for x

67. $\frac{2}{3}(x + \mathrm{d}) = r$ for x

68. $\frac{1}{C}(8 + w) = q$ for w

69. $5 = \frac{1}{r}(3r - x)$ for x

70. $\frac{2}{r}(x^2 + y) = 1$ for y

71. $\frac{c}{d}(k + w) = 5L$ for w

Challenge Problems

Many times in applications, we would like to use the same letter to name different variables. For example, suppose we have two different temperatures. We use *subscripts* to name the different temperatures. We might call one of them t_1 and the other one t_2. Each of these is a different variable. We read t_1 as "t sub-one."

72. $v = -32t + v_0$ (velocity in free fall); solve for t.

(v_0 represents initial velocity.)

73. $\frac{P_1V_1}{T_1} = \frac{P_2V_2}{T_2}$ (ideal gas laws); solve for P_2
(The variables represent different pressures, volumes, and temperatures.)

74. $F = G\frac{m_1m_2}{R^2}$ (gravitational force); solve for m_1.

(m_1 and m_2 represent different masses.)

IN YOUR OWN WORDS...

75. What is a literal equation?

3.4 APPLICATIONS

We do word problems because they are an imitation of life. They are descriptions of the various applications of mathematics in the real world.

To be successful with word problems we must resist the temptation to rush in and puzzle out the answer all at once. Instead, we must organize our efforts to solve the problem one piece at a time. We also must avoid the temptation to solve the problem by trial and error. This only works for the simplest problems and does not provide a method for the more complicated cases.

EXAMPLE 1. John wishes to cut 17 ft of cable into two pieces such that one piece is 5 ft longer than the other. How long should each piece be?

Solution:

First, we read the problem to see its general nature and to find out what it is we want to know. When we reread this problem, we see it is a relationship between the lengths of two pieces of cable and we are to find the *length of each piece.* This is one of the most important steps in solving a word problem. We let some letter (often x) stand for one of the things we are looking for. *We make this assignment in writing.*

Let x be the length of the shorter piece (in feet).

Next, we write an expression for the length of the other piece *in terms of* x. It is 5 feet longer than the piece whose length is x feet, so we can write

$x + 5$ is the length of the longer piece.

A picture, figure, or table is often useful in seeing the relationships.

$$\overbrace{\underbrace{x}\underbrace{x+5}}^{17}$$

Now that we have the lengths of the two pieces to work with (they are x feet and $x + 5$ feet), we reread the problem to see the relationship between them. From the problem we see that their lengths must add up to 17 ft. The picture also makes this clear. We write this relationship as an equation.

$$\begin{array}{ccccc} \text{Length of shorter} & + & \text{length of longer} & = & 17 \\ x & + & x + 5 & = & 17 \end{array}$$

This equation is called a **mathematical model** of the problem. Finding it is usually the hardest part of a word problem. Next, we solve the equation.

$$\begin{aligned} x + x + 5 &= 17 & & \\ 2x + 5 &= 17 & & \text{Collect like terms.} \\ 2x + 5 - 5 &= 17 - 5 & & \text{Subtract 5.} \\ 2x &= 12 & & \text{Collect like terms.} \\ x &= 6 & & \text{Divide by 2.} \end{aligned}$$

Therefore, the length of the shorter piece is 6 ft. The longer piece is 5 ft more than the shorter piece, so it must be 11 ft long. We were asked a question in writing, so we give a written reply.

The lengths of the pieces are 6 ft and 11 ft.

We should check our solution. We check word problems by taking our written solution (the lengths of the pieces are 6 ft and 11 ft) back to the *original* problem statement and make sure it does two things:

1. Makes sense
2. Answers the question

Since $6 + 11$ is 17, our written solution both answers the question and makes sense. □

Notice that we *did not* stop at the point where we wrote

$$x = 6$$

The problem asked a question and we must provide a written answer. We now give a general procedure for solving word problems.

A Procedure to Solve Word Problems

1. Read the problem and determine what quantities are to be found.
2. Assign a variable, such as x, to represent one of the quantities to be found.

(continued)

3. Express all other quantities to be found in terms of the chosen variable.
4. Draw a picture or figure to illustrate the problem, if possible. Charts or tables can also be used.
5. Reread the problem and form a mathematical model.
6. Solve the mathematical model.
7. Reread the problem and prepare a *written* answer to the question(s) asked.
8. Check the answer in the original problem statement. It should make sense and answer the question.

EXAMPLE 2. The length of a rectangle is 5 m more than its width. If the perimeter is 78 m, find the dimensions of the rectangle.

Solution:

Step 1 Determine what is to be found.

When we read the problem, we see there are two things to find, the length and the width of the rectangle. However, as the length is 5 m more than the width, we can make the following assignments.

Step 2 Assign a variable to one quantity to be found.

Let x be the width of the rectangle (meters).

Step 3 Express other quantities in terms of x.

Then $x + 5$ is the length of the rectangle.

Step 4 Draw a picture.

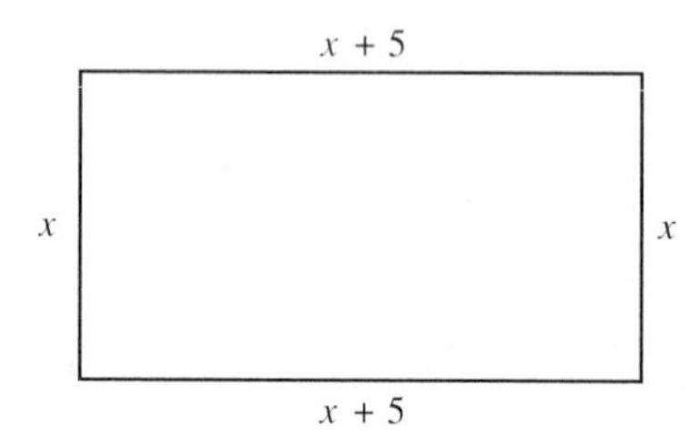

Step 5 Reread the problem and form an equation.

From the problem statement we see that the perimeter of the rectangle is 78 meters and from the picture we see it is 2 times x, plus 2 times $(x + 5)$. Therefore, the model is,

$$2 \cdot x + 2 \cdot (x + 5) = 78$$

Step 6 Solve the equation.

$$2 \cdot x + 2 \cdot (x + 5) = 78$$

$2x + 2x + 10 = 78$ Distributive Property

$4x + 10 = 78$ Collect like terms.

$$4x = 68 \qquad \text{Subtract 10.}$$

$$x = 17 \qquad \text{Divide by 4.}$$

Therefore, the width is 17 m (see Step 2) and the length is 22 m (Step 3).

Step 7 Answer the question.

The dimensions of the rectangle are 17 m by 22 m.

Step 8 Check in the *original* problem statement.

The solution certainly makes sense. And because 22 is 5 more than 17 and $2 \cdot 17 + 2 \cdot 22 = 78$, it answers the question. ☐

EXAMPLE 3. Ida is four times as old as Kathy. In 5 yr the sum of their ages will be 35 yr. How old is each person now?

Solution:

Step 1 Determine what is to be found.

We are asked to find Ida's age and Kathy's age now.

Step 2 Assign a variable to one quantity to be found.

Let x be Kathy's age now.

Step 3 Express other quantities in terms of x.

Then $4x$ is Ida's age now.

Step 4 Make a table.

When solving age problems, it is best to write down statements giving ages at other times mentioned in the problem. For example, in this problem we are given the sum of their ages 5 years *from now.* A table is convenient for recording this information.

	NOW	5 YR FROM NOW
Kathy's age	x	$x + 5$
Ida's age	$4x$	$4x + 5$

Step 5 Reread the problem and form an equation.

Now we reread the problem and see that the sum of their ages five years from now is 35 years. Or,

$$(x + 5) + (4x + 5) = 35.$$

Step 6 Solve the equation.

$$(x + 5) + (4x + 5) = 35$$

$$x + 5 + 4x + 5 = 35$$

(continued)

$$5x + 10 = 35 \qquad \text{Collect like terms.}$$
$$5x = 25 \qquad \text{Subtract 10.}$$
$$x = 5 \qquad \text{Divide by 5.}$$

Thus Kathy's age now must be 5 and Ida's, $4 \cdot 5 = 20$.

Step 7 Answer the question.

Kathy is 5 and Ida is 20.

Step 8 Check in the *original* problem statement.

This answers the question and, as $(20 + 5) + (5 + 5)$ is 35, it makes sense. □

EXAMPLE 4. Wan has 20 coins in his pocket. They are all nickels and dimes and their total value is \$1.35. How many of each type coin does Wan have?

Solution:

Determine what is to be found.

This is a problem concerning the value of a collection. We need to find out how many nickels and dimes Wan has in his pocket.

Assign a variable to a quantity to be found.

Let x be the number of nickels in Wan's pocket.

Express other quantities in terms of x.

We also need to find the number of dimes in Wan's pocket. We know that there are a total of 20 coins in his pocket and x of them are nickels. Therefore, 20 less x must be the number of dimes since there is no other type coin in his pocket.

$20 - x$ is the number of dimes in Wan's pocket.

Make a table.

The **value** of x nickels is $5x$ cents.
The **value** of $20 - x$ dimes is $10(20 - x)$ cents.
The **value** of the collection is \$1.35 or 135 cents.

	NICKELS	DIMES	TOTAL
Number of coins	x	$20 - x$	20
Their value (in cents)	$5x$	$10(20 - x)$	135

Form an equation.

$$\text{Value of nickels} + \text{value of dimes} = \text{total value}$$
$$5x + 10(20 - x) = 135$$

Solve the equation.

$$5x + 200 - 10x = 135$$

$$200 - 5x = 135 \qquad \text{Collect like terms.}$$

$$200 - 5x - 200 = 135 - 200 \qquad \text{Subtract 200.}$$

$$-5x = -65 \qquad \text{Collect like terms.}$$

$$\frac{-5x}{-5} = \frac{-65}{-5} \qquad \text{Divide by } -5.$$

$$x = 13$$

There are 13 nickels in Wan's pocket. Thus there must be 20 − 13, or 7, dimes.

Answer the question.

Wan has 13 nickels and 7 dimes in his pocket.

Check.

Since 13 nickels have a value of 65¢ and 7 dimes are worth 70¢, the value of the nickels plus the value of the dimes is 65¢ plus 70¢, or $1.35. Our answer makes sense and answers the question. □

EXAMPLE 5. How many gallons of a 60% alcohol solution should be added to 30 gal of a 10% alcohol solution to make a 20% alcohol solution?

Solution:

Determine what is to be found.

We are asked for the *number of gallons* of the 60% solution to be added.

Assign a variable.

Let x be the number of gallons of the 60% solution to be added.

Draw a figure.

A figure often helps in a mixture problem. We draw three containers; the original 30 gal of 10% solution, the x gallons of 60% solution, and the result of mixing them together.

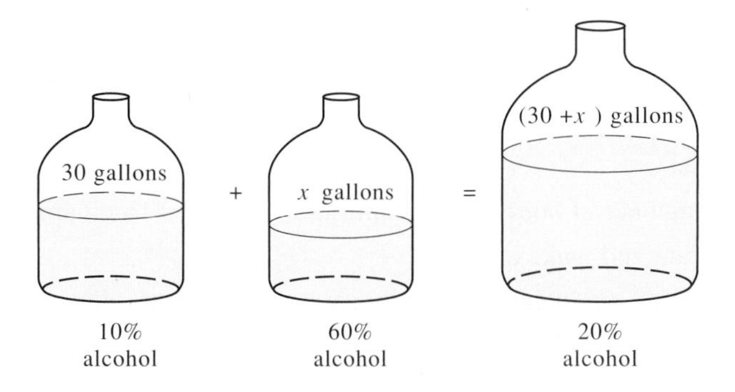

Form an equation.

We always change percents to decimals or fractions before we do arithmetic. Now, 30 gal of a 10% alcohol solution would contain $30 \cdot \frac{10}{100}$ gallons of alcohol, whereas x gallons of a 60% solution would contain $x \cdot \frac{60}{100}$ gallons of alcohol. After mixing, we have $(30 + x)$ gallons of mixture that is 20% alcohol, or $(30 + x) \cdot \frac{20}{100}$ gallons of alcohol. But, the amount of alcohol present before mixing is the same as the amount present after mixing, so we have the model

$$30 \cdot \frac{10}{100} + x \cdot \frac{60}{100} = (30 + x) \cdot \frac{20}{100}$$

Solve the equation.

$$30 \cdot \frac{1}{10} + x \cdot \frac{3}{5} = (30 + x) \cdot \frac{1}{5} \qquad \text{Reduce fractions.}$$

$$3 + x\frac{3}{5} = (30 + x)\frac{1}{5} \qquad \text{Simplify.}$$

$$5 \cdot \left(3 + x\frac{3}{5}\right) = 5 \cdot \left[(30 + x)\frac{1}{5}\right] \qquad \text{Multiply by 5 (to clear fractions).}$$

$$15 + 3x = 30 + x \qquad \text{Distributive Property}$$

$$2x = 15$$

$$x = \frac{15}{2}$$

Answer the question.

We should add 7.5 gal of the 60% solution.

Check.

This solution makes sense and answers the question. □

EXAMPLE 6. Two trains leave Atlanta at noon, traveling in opposite directions. If one train travels at a rate of 55 mph and the other at a rate of 60 mph, what time will it be when they are 414 mi apart?

Solution:

Determine what is to be found.

When we read the problem we see that we must find the number of hours it will take until the trains are 414 mi apart.

Assign a variable.

Let x be the number of hours it takes until the trains are 414 mi apart.

Draw a diagram and make a table.

Distance-rate-time problems are usually easier with a figure illustrating the problem. When we draw the figure we *look for two parts.*

Atlanta

←— Train 1 —→|←— Train 2 —→

←——— 414 mi ———→

The two parts are the train 1 portion and the train 2 portion. In distance-rate-time problems we use the basic formula distance = rate × time and the two formulas we get by solving for rate and time.

$$d = rt \qquad \text{Distance}$$

$$r = \frac{d}{t} \qquad \text{Rate}$$

$$t = \frac{d}{r} \qquad \text{Time}$$

The best approach for these problems is to identify two parts, as we did in Step 4, and then find d, r, and t for both parts. One of the three will come from the assignment statement(s) and another one should be found in the problem statement. The third can then be calculated from one of the above formulas.

For example, in this problem each train traveled x hours (time for both parts came from the assignment statement), and the rates for each train are given in the problem statement. We enter this information in a table.

	DISTANCE	RATE	TIME
Train 1		60 mph	x hours
Train 2		55 mph	x hours

We now use one of our formulas to find d for each part. The appropriate formula is $d = rt$. Inserting this calculation we get

	DISTANCE	RATE	TIME
Train 1	$60 \cdot x$	60 mph	x hours
Train 2	$55 \cdot x$	55 mph	x hours

Form an equation.
After completely filling in d, r, and t for both parts, we reread the problem. Clearly, the distance train 1 travels plus the distance train 2 travels must be 414 miles.

$$60 \cdot x + 55 \cdot x = 414$$

Solve the equation and answer the question.

$$115x = 414$$

$$x = \frac{414}{115}$$

$$x = 3.6$$

As trains are 414 miles apart after 3.6 h and 0.6 hours is 36 min, we answer the question as follows:

The trains will be 414 miles apart at 3:36 P.M.

This statement makes sense and answers the question. □

PROBLEM SET 3.4

Warm-ups

The first four word problems deal with relationships between numbers. See Example 1.

1. The sum of 18 and a number is 52. What is the number?
2. One number is 5 more than another number. The sum of the two numbers is 87. What are the numbers?
3. Margaret saws a 16-ft board into two pieces, one of which is 4 ft longer than the other. What are the lengths of the two pieces?
4. Nipa cuts a 42 in. piece of material into two pieces, one of which is twice as long as the other. What are the lengths of the two pieces?

Problems 5 through 8 are concerned with perimeters of geometric figures. See Example 2.

5. The length of a rectangle is two meters more than its width. If its perimeter is 40 m, what are its dimensions?
6. The width of a rectangle is 5 cm less than its length. What are its dimensions if its perimeter is 98 centimeters?
7. The perimeter of a rectangle is 136 yd. If the width is one-third the length, what are its dimensions?
8. The length of a rectangle is 7 m less than four times the width. If the perimeter is 126 m, what are its dimensions?

Problems 9 through 12 are age problems. See Example 3.

9. Homer is twice as old as Pete. If the sum of their ages is 96 years, how old are Homer and Pete?
10. John is 3 years older than Sam. In ten years, the sum of their ages will be 69 years. How old is each now?
11. Ellen is 4 years younger than Sarah. In 5 years, the sum of their ages will be 74 years. How old is each now?
12. Bill is 30 years older than Traci. In 10 years, he will be twice her age then. How old are Bill and Traci now?

Problems 13 through 16 are value problems. See Example 4.

13. June has \$1.70 in nickels and dimes. If she has two more dimes than nickels, how many of each kind of coin does she have?
14. Tommy has 30 bills worth \$200. If he has only fives and tens, how many of each kind does he have?
15. A collection of coins consists of 54 nickels and pennies. If the value of the collection is \$1.50, how many of each type coin are there in the collection?
16. Tickets to the ALTA Tennis Championships sold for \$5.75 for adults and \$2.50 for children under 12. If 450 tickets were sold and \$2,363.25 was collected, how many adult tickets were sold?

Problems 17 through 20 deal with mixtures. See Example 5.

17. JoAnne needs a 20% brine solution for her pickles. How many liters of a 30% brine solution should she add to 2 L of a 5% brine solution to get a 20% solution?
18. Marion wants to dilute 3 gal of a 10% soap solution to make a 3% solution. How much pure water should she add to the 3 gal of 10% solution to make the 3% solution?
19. Crude oil worth \$16 per barrel is to be mixed with crude worth \$12 per barrel to make a mixture worth \$15 a barrel. How many barrels of \$16 crude should be mixed with 210 barrels of \$12 crude?
20. Shirley wishes to mix 30 lb of cinnamon coffee worth \$5 per pound with plain coffee worth \$3 per pound in order to make a mixture worth \$3.50 per pound. How many pounds of the plain coffee should she use?

Problems 21 through 24 are distance-rate-time problems. See Example 6.

21. Rowland and Martin are enjoying the lake on a still day. They leave buoy 16 in their boats traveling in opposite directions. Martin's speed is 40 mph and Rowland's is 50 mph. In how many hours will they be 225 mi apart?

22. The slow freight and the Midnight Express are 300 mi apart but are heading toward each other on the same railroad track. If the slow freight is traveling at 40 mph and the Midnight Express at 110 mph, how much time do we have to warn them of their impending collision?

23. Federico leaves Mack's Texaco and drives toward San Diego with an average speed of 40 mph. Augusto leaves Mack's 1 hr later and drives the same route at 60 mph. How long will it take Augusto to overtake Federico?

24. A jet flew at an average speed of 1207 km/h for a period of time and then reduced its speed to 700 km/h. If it flew at 700 km/h for 1 hr longer than it flew at 1000 km/h and it flew a total distance of 4100 km, how long did it fly at 1000 km/h?

Practice Exercises

Solve the following word problems.

25. The sum of 22 and a number is 71. What is the number?

26. One number is 7 more than another number. The sum of the two numbers is 51. What are the numbers?

27. The length of a rectangle is 3 m more than its width. If its perimeter is 102 m, what are its dimensions?

28. The width of a rectangle is 11 cm less than its length. What are its dimensions if its perimeter is 106 cm?

29. Kitty is half as old as Charlie. If the sum of their ages is 48 years, how old are Kitty and Charlie?

30. Ban is 6 years older than Rieko. Ten years ago he was twice her age then. How old is each now?

31. Betty has $2.75 in nickels and dimes. If she has twice as many dimes as nickels, how many of each type coin does she have?

32. Sandi has $2.25 in nickels and quarters. If she has ten times as many nickels as quarters, how many of each type coin does she have?

33. One number is 3 more than twice another number. If their sum is 72, what are the numbers?

34. The degree measures of the angles in a triangle are three consecutive even integers. Find the number of degrees in each angle.

35. Find the length of the side of a square if its perimeter is 108 mm.

36. The perimeter of a triangle is 47 in. The length of one side is 7 in. more than the length of the shortest side, and the length of the other side is twice the length of the shortest side. What are the lengths of the three sides?

37. Tom is four times as old as Kimberly. Four years ago, Tom's age was 10 times Kimberly's age then. How old is each now?

38. Sondra is 3 yr older than Jim. One year ago, Sondra's age was four times Jim's age then. How old is each now?

39. A meter attendant emptied nickels, dimes, and pennies from a parking meter. There were 10 fewer dimes than nickels and 95 more pennies than nickels. If the total collected was $4.75, how many of each type coin were there in the meter?

40. Kaled has $2.25 in nickels, dimes, and quarters. If he has twice as many dimes as quarters and three times as many nickels as dimes, how many of each type coin does he have?

41. Pecans costing $3.50 per pound are mixed with 20 lb of pecans costing $1.70 per pound to make a mixture costing $2.30 per pound. How many pounds of the $3.50 pecans are required?

42. Kenya coffee beans selling for $8.75 per pound are to be mixed with 10 lb of Blue Mountain beans, which sell for $12.50 per pound, to make a mixture that sells for $10.00 per pound. How many pounds of the Kenya beans are required?

43. A truck and a bus leave Lansing at the same time, both traveling south. If the bus averages 60 mph and the truck averages 45 mph, how long will it be until they are 50 mi apart?

44. Two autos are racing on the Indianapolis 2.5-mi oval track. If they start together and one auto averages 150 mph while the other averages 155 mph, how long will it take for the faster car to lap the slower?
45. Jo needs a 15% brine solution. How many liters of a 20% brine solution should she add to 10 L of a 12% brine solution to get a 15% solution?
46. Joan wants to dilute 6 gal of a 5% soap solution to make a 3% solution. How much pure water should she add to the 6 gal of 5% solution to make the 3% solution?
47. Jack and Jim leave San Antonio in their cars traveling in opposite directions. Jack's speed is 62 mph and Jim's is 55 mph. In how many hours will they be 351 mi apart?
48. At noon two trains are 385 miles apart but are heading toward each other on the same railroad track. If the slow train is traveling at 50 mph and the faster train at 60 mph, at what time will they collide?
49. Henry cuts a 41-ft wire into two pieces, one of which is 11 ft longer than the other. What are the lengths of the two pieces?
50. Charlotte cuts a 87-in. piece of material into two pieces, one of which is twice as long as the other. What are the lengths of the two pieces?
51. The perimeter of a rectangle is 300 yd. If the width is one-fourth the length, what are its dimensions?
52. The length of a rectangle is 4 m less than three times the width. If the perimeter is 96 m, what are its dimensions?
53. J.R. is 6 yr younger than Kim. Three years ago she was twice J.R.'s age. How old is each now?
54. Sue is 14 yr younger than Jan. Five years ago, Jan's age was three times Sue's age then. How old is each now?
55. Jerry has 40¢ in pennies and nickels. If he has 20 coins, how many of each type does he have?
56. Tickets to the MASDA Charity Cake Walk were $1.25 for adults and 75¢ for children. If $1050 was collected and three times as many children as adults bought tickets, how many adults bought tickets?
57. Fine oil worth $35 per gallon is to be mixed with oil worth $20 per gallon to make a mixture worth $25 a gallon. How many gallons of $35 oil should be mixed with 50 gal of $20 oil?
58. Nguyen wishes to mix 30 lb of nuts worth $4 a pound with nuts worth $2 a pound in order to make a mixture worth $3.50 a pound. How many pounds of the $2 nuts should he use?
59. Barney leaves Minneapolis and drives toward St. Cloud with an average speed of 50 mph. Gerald leaves Minneapolis $\frac{1}{2}$ hr later and drives the same route at 60 mph. How long will it take Gerald to overtake Barney?
60. An airplane flew at an average speed of 600 km/h for a period of time then reduced its speed to 400 km/h. If it flew at 400 km/h for 2 hr longer than it flew at 600 km/h and it flew a total distance of 2800 km, how long did it fly at 600 km/h?
61. One number is 5 more than twice a certain number. If their sum is 134, what is the certain number?
62. The length of Rika's camper is 8 ft less than twice the length of Kim's camper. If together they are 40 ft long, what is the length of each of their campers?
63. Find the length of the side of a square if its perimeter is 76 mm.
64. The perimeter of a triangle is 68 in. The length of one side is 8 in. more than the length of the shortest side and the length of another side is twice the length of the shortest side. What are the lengths of the three sides?
65. Louise is three times as old as her daughter. In 7 yr the sum of their ages will be 70. How old is each now?
66. Dan's age is twice Jonathan's age now. Five years ago, Dan's age was seven times Jonathan's age then. How old is each now?
67. A toll box on the George Washington Bridge contains $450 in quarters, dimes, and nickels. If there are twice as many dimes as nickels and five times as many quarters as nickels, how many of each type coin are in the box?
68. Amanda has $4.50 in nickels and dimes. If she has twice as many dimes as nickels, how many dimes does she have?
69. Cashew clusters costing $4.50 per pound are mixed with 30 lb of peanut clusters costing $2.20 per pound to make a mixture costing $4.00 per pound. How many pounds of the cashew clusters are required?

70. Amaretta coffee beans selling for \$6.50 per pound are to be mixed with 10 lb of Kona beans that sell for \$11.50 per pound to make a mixture that sells for \$8.50 per pound. How many pounds of the \$6.50 beans are required?

71. A dogsled and a snowmobile leave Calgary at the same time, both traveling south. If the snowmobile averages 40 mph and the dogsled averages 8 mph, how long will it be until they are 48 mi apart?

72. Two autos are racing on a 1.5-mi oval track. If they start together and one auto averages 175 mph while the other averages 160 mph, how long will it take for the faster to lap the slower?

Challenge Problems

73. Omar drove from home to a place in the country at an average rate of 50 mph and then returned at an average rate of 60 mph. If the total trip took 2 hr, how far is the place in the country from Omar's home?

74. Lois wants to make 10 L of 29% sugar syrup by mixing syrup containing 50% sugar and syrup containing 20% sugar. How much of each should she mix?

IN YOUR OWN WORDS...

75. What is the procedure for solving a word problem?

76. Make up a word problem with an answer of 10 nickels.

3.5 LINEAR INEQUALITIES

Just as an equation is a statement that two numbers are equal, an **inequality** is a statement that one number is greater than another. We use the symbols $>$, $<$, $\geq$, and $\leq$.

$p > q$	p is **greater than** q
$p < q$	p is **less than** q
$p \geq q$	p is **greater than or equal to** q
$p \leq q$	p is **less than or equal to** q

Some *true* inequalities are

$$10 > 7$$
$$4 > -8$$
$$2 < 17$$
$$-5 < 3$$
$$6 \geq 1$$
$$6 \geq 6$$

Some *false* inequalities are,

$$10 < 7$$
$$-7 > -1$$
$$4 \leq 3.$$

Like equations, inequalities are interesting when they contain variables. For example, the inequality

$$x + 5 > 7$$

is *true* if x is 11 and is *false* if x is 1. In fact this statement is *true* if x is *any* number greater than 2 and *false* if x is 2 or *any* number less than 2. The collection of numbers that make the inequality true is called the **solution set.** To **solve** an inequality is to find its solution set.

We cannot write the solution set for an inequality, such as the one just given, by listing its elements. We must use the set-builder notation introduced in Section 1.1. As the solution set is the collection of *all* numbers greater than 2, it can be written

$$\{x \mid x > 2\}$$

Inequalities have a **left side,** an inequality symbol, and a **right side.**

Inequality symbol
↓
$x + 5 > 7$
↑ ↑
Left side Right side

EXAMPLE 1. For each inequality, determine which of the given numbers are in the solution set.

(a) $4 + 2x < 13$; $-5, 0, 5$ (b) $16 - 5x \leq -4$; $-10, 4, 10$

Solutions:

(a) If x is -5, then

$$4 + 2x = 4 + 2(-5)$$
$$= -6$$

Since $-6 < 13$ is *true*, -5 *is* in the solution set of $4 + 2x < 13$.

If x is 0, then

$$4 + 2x = 4 + 2(0)$$
$$= 4$$

Since $4 < 13$ is true, 0 *is* in the solution set of $4 + 2x < 13$.

If x is 5, then

$$4 + 2x = 4 + 2(5)$$
$$= 14$$

Because $14 < 13$ is *false*, 5 *is not* in the solution set of $4 + 2x < 13$.

(b) If x is -10, then

$$16 - 5x = 16 - 5(-10)$$
$$= 66$$

Since $66 \leq -4$ is *false*, -10 is *not* in the solution set of $16 - 5x \leq -4$.

If x is 4, then

$$16 - 5x = 16 - 5(4)$$
$$= -4$$

Because $-4 \leq -4$ is *true*, (-4 *is* less than *or equal* to -4), 4 *is* in the solution set of $16 - 5x \leq -4$.

If x is 10, then

$$16 - 5x = 16 - 5(10)$$
$$= -34$$

Because $-34 \leq -4$ is *true*, 10 *is* in the solution set of $16 - 5x \leq -4$. □

We can sometimes understand an idea much better if we can draw a picture of it. Such is the case with inequalities. We have seen that the solution set of the inequality $x < 1$ consists of all numbers less than 1 — that is, all numbers that lie to the left of 1 on a number line. We draw a picture of this set of numbers by shading a number line to the left of 1.

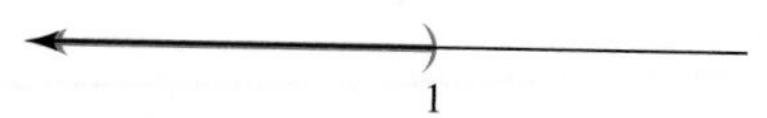

We indicate that the set continues indefinitely by an arrowhead. We indicate that the number 1 *does not* belong to the solution set with a *parenthesis*. This picture of the solution set is called the **graph** of the inequality.

Another notation is to use an open circle to indicate that a number is not included.

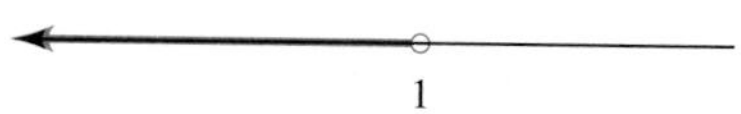

To graph the inequality $x \geq -\frac{1}{2}$, we draw a number line and shade the numbers to the right of $-\frac{1}{2}$. We use a *bracket* to indicate that $-\frac{1}{2}$ *does* belong to the solution set.

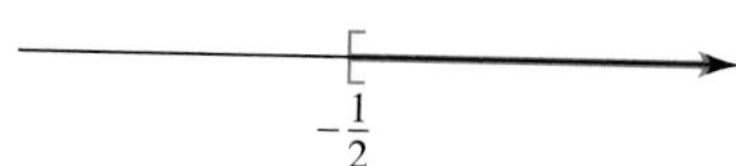

Other notations use a darkened circle to show that $-\frac{1}{2}$ is included.

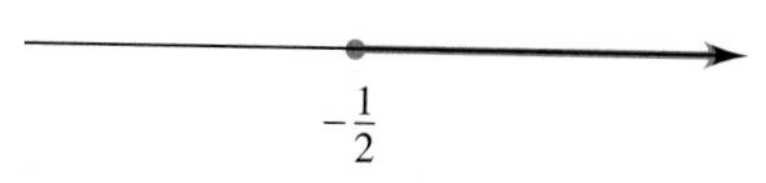

EXAMPLE 2. Graph the solution set for each inequality.

(a) $x > 0$ (b) $x \geq -2$ (c) $x < 4$ (d) $3 < x$

Solutions:

(a) We shade a number line to the right of 0. The parenthesis indicates that 0 is not included.

$x > 0$:

0

(b) We shade a number line to the right of -2. The bracket shows that -2 is included.

$x \geq 2$:

–2

(continued)

(c) We shade a number line to the left of 4. The parenthesis indicates that 4 is not included.

$x < 4$:

(d) Here $3 < x$ means the same as $x > 3$. We shade a number line to the right of 3. The parenthesis shows that 3 is not included.

$3 < x$:

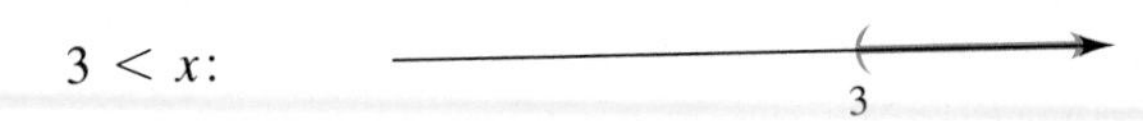

Notice that we wrote the inequality with the variable on the left side. This makes it easier to graph. □

Linear inequalities in one variable contain a variable to the first power only. We can solve linear inequalities in a manner very similar to the method we used to solve linear equations. The idea behind solving linear equations was to use the addition and multiplication tools to change the equation to a form such as

$$x = 4$$

and then write the solution set, $\{4\}$. We change an inequality to a form such as

$$x < 4 \quad \text{or} \quad 4 > x$$

and then write the solution set, $\{x \mid x < 4\}$.

Tools for Solving Inequalities

The solution set of an inequality will not change if we perform any of the following:

Addition

Add (or *subtract*) the same number on both sides of the inequality.

Multiplication by Positive Number

Multiply (or *divide*) both sides by the same *positive* number.

Multiplication by Negative Number

Multiply (or *divide*) both sides by the same *negative* number *and reverse* the direction of the inequality symbol.

To help us understand why we need two multiplication tools, consider the following example. The statement $-4 < 8$ is a true statement. What happens if we multiply both sides by -2 and do not reverse the direction of the inequality symbol?

$-4 < 8$	True statement
$-4(-2) < 8(-2)$	Multiply both sides by -2.

$$8 < -16 \qquad \textit{False statement}$$

So, when we multiply (or divide) both sides of an inequality by a negative number, we must change the direction of the inequality symbol to maintain a true statement.

The inequalities

$$x - 3 < 7$$

and

$$x < 10$$

have the same solution set, since the second one was obtained from the first by adding 3 to both sides.

The inequalities

$$5x \leq 45$$

and

$$x \leq 9$$

have the same solution set, since the second was obtained from the first by dividing both sides by the *positive* number 5.

The inequalities

$$-\frac{1}{2}z > 11$$

and

$$z < -22$$

have the same solution set, since the second was obtained from the first by multiplying both sides by -2 *and reversing the direction of the inequality symbol.*

EXAMPLE 3. Solve each inequality and graph each solution set.

(a) $x - 4 > 2$ (b) $-\frac{1}{3}z \leq 5$ (c) $-x < 2$

Solutions:

(a) We add 4 to both sides and simplify.

$$x - 4 > 2$$
$$x - 4 + 4 > 2 + 4$$
$$x > 6$$

The solution set is

$$\{x \mid x > 6\}$$

The graph of this set is

(continued)

Be Careful!

(b) We multiply both sides by -3. Since -3 is a *negative* number, we must remember to reverse the direction of the inequality symbol.

$$-\frac{1}{3}z \leq 5$$

$$(-3)\left(-\frac{1}{3}z\right) \geq (-3)\cdot 5$$

$$z \geq -15$$

The solution set is

$$\{z \mid z \geq -15\}$$

The graph is

Be Careful!

(c) We need to do something about the negative sign in order to solve this inequality. If we multiply both sides by -1, we will have x alone on the left. Because -1 is negative, *we must reverse the direction of the inequality symbol.*

$$-x < 2$$

$$(-1)\cdot(-x) > (-1)\cdot 2$$

$$x > -2$$

$$\{x \mid x > -2\}$$

–2

□

EXAMPLE 4. Find the solution set for each inequality and graph each solution set.

(a) $2x - 3 < 0$ (b) $16 - 3y \geq 22$

Solutions:

(a)

$2x - 3 < 0$	
$2x - 3 + 3 < 0 + 3$	Add 3.
$2x < 3$	Combine like terms.
$\frac{2x}{2} < \frac{3}{2}$	Divide by 2.
$x < \frac{3}{2}$	

Since 2 is positive, the inequality symbol did not change direction. We write the solution set:

$$\left\{x \mid x < \frac{3}{2}\right\}$$

The graph of this set is

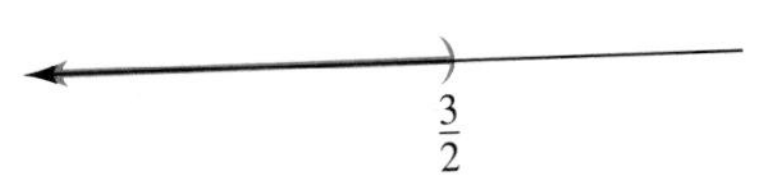

(b) $$16 - 3y \geq 22$$

$$16 - 3y - 16 \geq 22 - 16 \qquad \text{Subtract 16.}$$

$$-3y \geq 6 \qquad \text{Combine like terms.}$$

$$\frac{-3y}{-3} \leq \frac{6}{-3} \qquad \text{Divide by } -3.$$

Notice that when we divided by the *negative* number -3, we had to reverse the direction of the inequality symbol. Now we simplify, write the solution set, and draw the graph.

$$y \leq -2$$

$$\{y \mid y \leq -2\}$$

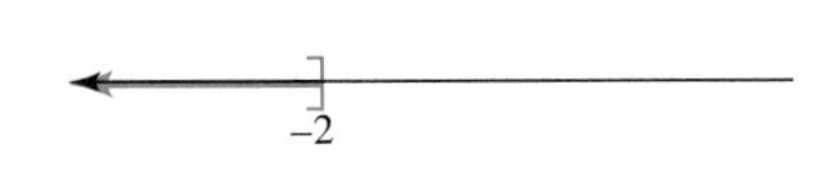

□

A Procedure for Solving Linear Inequalities

1. Clear the inequality of fractions (or decimals), if necessary, by multiplying by the least common denominator (LCD) of all the fractions in the equation. Be careful to reverse the inequality symbol when multiplying by a negative number.
2. Simplify each side by removing grouping symbols and combining like terms, where possible.
3. Collect all terms containing the variable on one side of the inequality and all other terms on the other side by using the addition tool.
4. Combine like terms.
5. Divide both sides by the coefficient of the variable. Be careful to reverse the inequality symbol when dividing by a negative number.
6. Write the solution set.

EXAMPLE 5. Solve each inequality and graph the solution set of each.

(a) $3x + 2 < 2x - 1$ (b) $2(x - 1) \geq 5x$ (c) $\frac{1}{2}x < \frac{2}{3}x + 3$

Solutions:

(a) $3x + 2 < 2x - 1$

We first collect the x-terms on one side of the inequality. We can do this if we subtract $2x$ from both sides.

$$3x + 2 - 2x < 2x - 1 - 2x \qquad \text{Subtract } 2x.$$

$$x + 2 < -1 \qquad \text{Combine like terms.}$$

(continued)

$x + 2 - 2 < -1 - 2$	Subtract 2.
$x < -3$	Combine like terms.
$\{x \mid x < -3\}$	

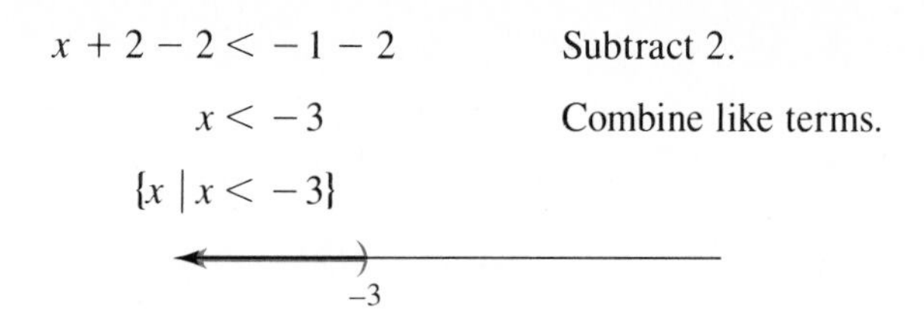

(b) We remove the parentheses first.

$2(x - 1) \geq 5x$	
$2x - 2 \geq 5x$	Distributive Property

We can collect the x-terms on either side. We subtract $2x$ from both sides to get them on the right side.

$2x - 2 - 2x \geq 5x - 2x$	Subtract $2x$.
$-2 \geq 3x$	Combine like terms.
$\frac{-2}{3} \geq \frac{3x}{3}$	Divide by 3.
$\frac{-2}{3} \geq x$	

This inequality can be written as $x \leq \frac{-2}{3}$.

$$\left\{x \mid x \leq \frac{-2}{3}\right\}$$

$-\frac{2}{3}$

(c) We clear fractions first by multiplying both sides by 6.

$\frac{1}{2}x < \frac{2}{3}x + 3$	
$6\left(\frac{1}{2}x\right) < 6\left(\frac{2}{3}x + 3\right)$	Multiply by 6.
$6\left(\frac{1}{2}x\right) < 6\left(\frac{2}{3}x\right) + 6(3)$	Distributive Property
$3x < 4x + 18$	

We collect the x-terms on the left side.

$3x - 4x < 4x + 18 - 4x$	Subtract $4x$.
$-x < 18$	
$(-1)(-x) > (-1)(18)$	Multiply by -1. (Reverse the inequality symbol.)
$x > -18$	
$\{x \mid x > -18\}$	

−18

□

The graphs that we have seen so far consist of numbers to the right or to the left of a particular number. There are many times when we want to graph solution sets that are *between* two numbers. Such a graph would be like

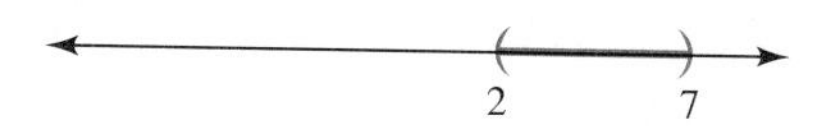

This is the graph of the numbers between 2 and 7. We write the inequality that this set represents as $2 < x < 7$. This is read "two is less than x and x is less than seven." It can also be read as "x is greater than two and x is less than seven." This type inequality is called a **compound inequality.**

EXAMPLE 6. Graph the compound inequalities.

(a) $-3 < x < 0$ (b) $5 \leq y < 10$
(c) $-5 < x \leq -1$ (d) $-4 \leq x \leq 4$

Solutions:

(a) The graph of $-3 < x < 0$ is all numbers between -3 and 0.

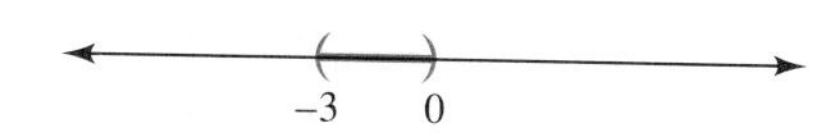

(b) The graph of $5 \leq y < 10$ consists of the numbers between 5 and 10 as well as 5. We use a bracket to show that 5 is included.

(c) The graph of $-5 < x \leq -1$ is all numbers between -5 and -1 as well as -1. We use a bracket to show -1 is included.

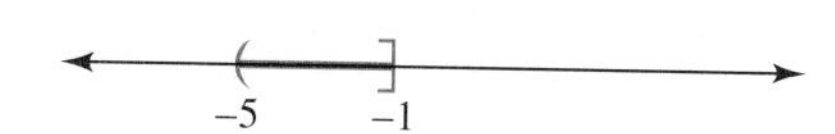

(d) The graph of $-4 \leq x \leq 4$ contains all numbers between -4 and 4. We use brackets to include both -4 and 4.

□

PROBLEM SET 3.5

Warm-ups

For each of the inequalities in Problems 1 through 5, determine which of the given numbers are in the solution set. See Example 1.

1. $6 + x < 11$; $-8, -2, 0, 2, 8$

2. $22 - x < 10$; $-13, 0, 13, 40$

3. $2x + 3 > 11$; $-20, 0, 4, 40, 400$

4. $16 - 3x < 22$; $-10, -2, 0, 2, 10$

5. $2(x + 3) > 7$; $-6, -5, -4, -3, -2, -1, 0, 1, 2, 3, 4, 5, 6$

In Problems 6 through 9, graph the inequality. See Example 2.

6. $x < 9$

7. $y \leq -1$

8. $5 < z$

9. $-3 \geq x$

In Problems 10 through 23, solve the inequalities and graph each solution set. See Example 3.

10. $x - 2 > 11$

11. $x + 6 < 24$

12. $x + 8 \leq 7$

13. $x - 23 \geq -3$

14. $2x < 18$

15. $3x \geq 18$

16. $-6x \geq 24$

17. $-11x \geq -33$

18. $\frac{1}{2}y > 5$

19. $\frac{3}{4}t \leq -2$

20. $-s \leq 14$

21. $-\frac{1}{2}x < 2$

22. $-\frac{2}{3}x > -8$

23. $\frac{6}{5}w \geq -30$

In Problems 24 through 42, find the solution set and graph each inequality. See Example 4.

24. $3 + 2x \leq -17$

25. $6 + \frac{1}{3}x < 9$

26. $7x - 4 \geq 10$

27. $4x + 5 > 1$

28. $7 - 2x > 11$

29. $4 - 5t \leq 14$

30. $3 + 7s \leq 5$

31. $13 - 6k < 11$

32. $2 - 2x > 1$

33. $7 \geq 5 - y$

34. $0 \leq 6x - 18$

35. $2x + 5 < -5$

36. $3 - 3y < 6$

37. $1 - w > 0$

38. $10x + 8 \geq -2$

39. $15 > 6 - 3z$

In Problems 40 through 51, solve each inequality and graph each solution set. See Example 5.

40. $6x < 5x - 2$

41. $11x - 10 > 7x + 2$

42. $3y - 8 \geq 5y + 2$

43. $z - 1 \leq 3z + 1$

44. $\frac{1}{2}x + \frac{2}{3} \leq 1$

45. $\frac{3}{4} - \frac{1}{3}x < \frac{1}{2}x$

46. $\frac{x}{2} > \frac{x}{4} - \frac{1}{3}$

47. $\frac{5}{3} < \frac{x}{2} - 1$

48. $2(x + 3) < 3$

49. $3(4 - x) < -12$

50. $-2(1 - x) \geq 4x$

51. $-3(x - 2) \leq 9x$

In Problems 52 through 55, graph each inequality. See Example 6.

52. $-1 < x < 3$

53. $0 \leq y < 4$

54. $\frac{1}{2} < x \leq 3$

55. $-\frac{3}{2} \leq z \leq \frac{5}{3}$

Practice Exercises

Solve each inequality and graph each solution set.

56. $x + 2 > 11$

57. $x - 6 < 24$

58. $x - 8 \leq 7$

59. $x + 23 \geq -3$

60. $3x \leq 18$

61. $2x > 16$

62. $\frac{1}{5}t \geq -8$

63. $-\frac{1}{4}z < 2$

64. $-x \leq -4$

65. $-\frac{2}{7}x < 1$

66. $-\frac{3}{2}x > 1$

67. $\frac{3}{7}s \geq -21$

68. $2 + 3x \leq -16$

69. $9 + \frac{1}{2}x < 9$

70. $\frac{1}{3}t - 3 < 1$

71. $\frac{1}{3}s + 2 \leq \frac{1}{3}$

72. $-4x \leq 24$

73. $-3x < -33$

74. $5 - 3x > 11$

75. $5 - 4t \leq -5$

76. $7 + 3s \leq 5$

77. $23 - 5k < 43$

78. $3 - \frac{1}{3}x > 1$

79. $5 \geq 7 - y$

80. $\frac{2}{3}w + \frac{1}{2} > 1$

81. $\frac{2}{3} - \frac{3}{5}v \geq \frac{1}{10}$

82. $3(x + 2) < 10$

83. $18 > -3(4 - x)$

84. $6x + 5 < 3x + 2$

85. $x - 1 < 2x$

In Problems 86 through 89, graph the compound inequality.

86. $0 < x < 1$

87. $-2 \leq t \leq 2$

88. $-3 < y \leq -1$

89. $-7 \leq x < -5$

Challenge Problems

Solve the following inequalities. The variable is x and p is a constant.

90. $x - p < 3p$

91. $2x > p$

92. $px > 2 \quad (p > 0)$

93. $px > 2 \quad (p < 0)$

IN YOUR OWN WORDS...

94. How is solving an inequality different from solving an equation?

3.6 APPLICATIONS OF INEQUALITIES

Inequalities arise in applications just as equations do. The following phrases are frequently used.

STATEMENT	SYMBOLS
p is more than q	$p > q$
p is less than q	$p < q$
p is not less than q	$p \geq q$
p is not more than q	$p \leq q$
p is at least q	$p \geq q$
p is at most q	$p \leq q$

EXAMPLE 1. Translate each statement into an inequality.

(a) s is at most 6.

(b) -3 is less than z.

(c) Five more than x is at least 10.

(d) Twice n is greater than y.

Solutions:

(a) s is at most 6

$$s \leq 6$$

(b) -3 is less than z.

$$-3 < z$$

(c) Five more than x is at least 10.

$$x + 5 \geq 10$$

(d) Twice n is greater than y.

$$2n > y$$

□

EXAMPLE 2. The area of a triangle is to be no more than 48 m^2. If the base is 8 m, find the largest possible height.

(continued)

Solution:

We let h be the height and sketch the triangle.

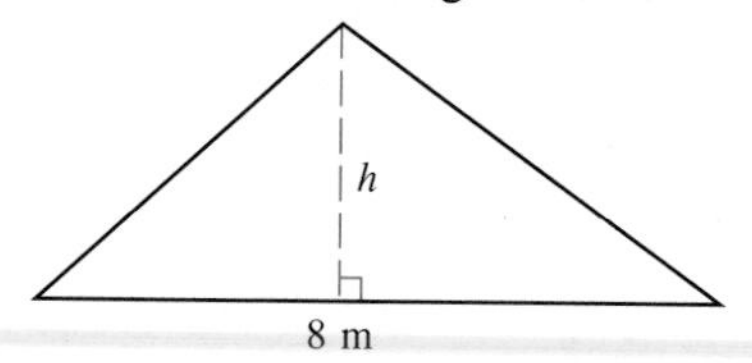

Since the area of a triangle is one-half the product of the base and the height, the area of this triangle is

$$\frac{1}{2}(8)h = 4h$$

The problem states that the area is no more than 48, so we form the inequality

$$4h \le 48.$$

$$h \le 12 \qquad \text{Divide by 4.}$$

Thus, we see the largest possible height is 12 m. □

EXAMPLE 3. Sandra has grades of 68 and 72 on her history tests. What grade must she make on her third test to have an average of at least 75?

Solution:

Let t be her third test grade. The sum of her three grades divided by three is her average. This average is to be *at least* 75. So we have the inequality,

$$\frac{68 + 72 + t}{3} \ge 75$$

$$\frac{140 + t}{3} \ge 75$$

$$3 \cdot \frac{140 + t}{3} \ge 3 \cdot 75 \qquad \text{Multiply by 3 (to clear fractions).}$$

$$140 + t \ge 225$$

$$140 + t - 140 \ge 225 - 140 \qquad \text{Subtract 140.}$$

$$t \ge 85$$

She must make at least 85 on her third test. □

PROBLEM SET 3.6

Warm-ups

In Problems 1 through 8, translate each statement into an inequality. See Example 1.

1. y is less than -2.
2. Five more than t is at most 15.
3. The sum of s and $s + 1$ is greater than 0.
4. The absolute value of y is at least 5.
5. The product of z and -3 is at most 30.

6. The square of m is greater than 0.
7. Three more than twice x is greater than -5.
8. The absolute value of the opposite of k is greater than 0.

In Exercises 9 through 14, solve each problem. In 9 through 11, see Example 2.

9. The perimeter of a square must be greater than or equal to 42 ft. How long must a side be?
10. The length of Lisa's rectangular dining room is 12 ft. If the area of the room is at least 96 ft^2, what is the smallest width the room could have?
11. Teresa cuts a ribbon into two pieces so that one piece is 6 in. longer than the other. How long can the shorter piece be if the ribbon is no longer than 30 in.?

In Problems 12 through 14, see Example 3.

12. Sally has grades of 82 and 72 in algebra. What grade must she make on her third test to have an average of at least 80?
13. In a gymnastics meet Jennifer has a score of 8.1 on vault, 7.5 on beam, and 7.8 on bars. What score must she make on floor to have an all around score of at least 30?
14. The sum of two consecutive positive integers must be no more than 5. What are the integers?

Practice Exercises

In Problems 15 through 28, translate each statement into an inequality.

15. The product of x and y is greater than or equal to -3.
16. The reciprocal of t is less than 0.
17. The sum of one-third of w and 5 is at least 9.
18. The cube of y is at most 1.
19. The opposite of x is less than 0.
20. The reciprocal of b is greater than b.
21. One-half of p is less than or equal to -4.
22. The square root of z is greater than 5.
23. The square root of p is greater than 7.
24. The opposite of z is less than w.
25. The reciprocal of $\frac{3}{5}$ is less than x.
26. Five less than three d is greater than -6.
27. The absolute value of k is at least 4.
28. The product of t and -5 is at most -6.

In Problems 29 through 40, answer the question.

29. Kristen has a score of 7.85 on her first vault. What score must she make on her second vault to have an average score of at least 8.45?
30. The width of a rectangular rug is 4 ft. If the perimeter of the rug must be at most 56 ft, what is the largest possible length it could have?
31. The sum of twice a number and one-half the number is less than 15. Find all possible numbers for which this is true.
32. The sum of two positive consecutive even integers is less than or equal to 6. What are the integers?
33. Jason sells popcorn at the ball game for 75¢ per bag. How many bags must he sell to collect at least \$100?
34. Chuck is stacking square boxes in the stock room. Each box is 16 cm high and the room is 300 cm high. What is the largest number of boxes that he can put in a vertical stack?
35. Steve wants to buy as many compact discs as he can afford. Turtle's has them on sale for \$11.99 each. If Steve has \$140, how many can he buy?
36. Robert makes \$3.50 per hour working at Wal-Mart. If he gets a bonus of \$25 this week, how many hours must he work to make at least \$165?
37. Joe works at B. Dalton bookstore and makes \$4.25 per hour. How many hours must he work to make at least \$190?

38. The low temperatures for the last two days were 28° and 15°. What must the low temperature for the next day be in order for the average temperature for the 3-day period to be less than 19°?

39. Find all numbers such that twice the number is less than 5 more than the number.

40. Lucy earned $400 and $550 in interest the last 2 yr. How much interest must she earn the next year to earn an average of $600 over the 3 yr?

Challenge Problems

41. Keith has scores of 76, 81, and 85 on three tests in English. What score must he make on his fourth test to have an average of at least 90? (The highest possible score on a test is 100.)

IN YOUR OWN WORDS...

42. Make up a word problem that could be solved using the inequality, $\frac{1}{2}x - \frac{1}{3}x \leq 1$.

CHAPTER SUMMARY

GLOSSARY

Linear Equation in one variable: An equation that can be written in the form $Ax + B = 0$ with $A \neq 0$.

Literal equation: Equation with more than one letter.

Inequality: Statement that one number is greater than another.

Solution of an inequality: A number that makes an inequality a true statement.

Solution set of the inequality: The set of all solutions of the inequality.

Linear inequality in one variable: Inequality of the form $Ax + B < 0$, where A is not zero.

SOLVING LINEAR EQUATIONS

1. Add or subtract the same number to or from both sides.
2. Multiply or divide both sides by the same nonzero number.

PROCEDURE TO SOLVE WORD PROBLEMS

1. Determine what is to be found.
2. Assign a variable, such as x, to represent one of the quantities to be found.
3. Express all other quantities to be found in terms of x (or the chosen variable).
4. Draw a figure or picture, if possible. Label it.
5. Reread the problem and form an equation.
6. Solve the equation and find the values of all the quantities to be found.
7. Check the values in the original word problem. They should answer the question and make sense.
8. Write an answer to the original question.

SOLVING LINEAR INEQUALITIES

1. Add or subtract the same number to or from both sides.
2. Multiply or divide both sides by the same *positive* number.
3. Multiply or divide both sides by the same *negative* number *and* reverse the direction of the inequality symbol.

CHECKUPS

1. Solve the following equations.
   a. $x + 3 = 3x - 5$ — Section 3.1; Example 4b
   b. $\frac{x}{2} - 1 = \frac{2}{5}x + 3$ — Section 3.2; Example 3a
2. Solve for y: $4(p + y) = 3$ — Section 3.3; Example 3b
3. The length of a rectangle is 5 m more than its width. If the perimeter is 78 m, find the dimensions of the rectangle. — Section 3.4; Example 2
4. Solve the inequality $16 - 3y \geq 22$. — Section 3.5; Example 4b
5. Graph the inequality $x \geq -2$. — Section 3.5; Example 2b
6. The area of a triangle is to be no more than 48 m^2. If the base is 8 m, find the largest possible height. — Section 3.6; Example 2

REVIEW PROBLEMS

In Problems 1 through 18, solve the given equations and inequalities.

1. $x - 6 = 14$
2. $x + 11 = 4$
3. $4 - x = 13$
4. $6 + x = 5$
5. $2x + 13 = 17$
6. $3x - 8 = -2$
7. $x - 6 < 14$
8. $x + 11 \geq 4$
9. $7 - 4x = -9$
10. $2(x - 2) = 12$
11. $4y - 11 > 9$
12. $x + 4 = 3x - 16$
13. $3(t + 2) = 10 - t$
14. $-s > 3$
15. $-x \leq -6$
16. $\frac{1}{2}x - 11 = \frac{1}{2}$
17. $5\left(x + \frac{1}{3}\right) = 3x + \frac{2}{3}$
18. $\frac{2}{3}(3 - y) > 2y$

In Problems 19 through 26, solve for the variable indicated.

19. $s + x = t$ for x
20. $u - v = 2w$ for v
21. $ax - b = c$ for x
22. $a + Bz = c$ for z
23. $2(r - at) = K$ for r
24. $2(r + at) = K$ for t
25. $\frac{a}{3}x - bh = 0$ for x
26. $\frac{abt}{c} = \pi v^2$ for t
27. One number is 3 less than twice another number. What are the numbers if their sum is 126?
28. The width of a rectangle is 3 ft more than half the length. If the perimeter is 78 ft, what are its dimensions?
29. Carol is 4 years younger than Russ. Thirty-four years ago, he was half as old as she is now. How old are Russ and Carol today?
30. A collection of nickels, dimes, and quarters has a value of \$3.60. If there are twice as many quarters as dimes and two more nickels than quarters, how many of each kind of coin are there in the collection?
31. How many grams of 50% gold should be added to 30 g of 15% gold to produce a 20% gold alloy?
32. Lee leaves Monterey walking 3 mph and Artie leaves Carmel Valley walking 2 mph. If they are initially 20 mi apart and walk toward each other, in how many hours will they meet?

33. Twice a number added to one-half the number is at most -3. Find all possible numbers.

34. Jim earns \$25 per day giving tours of a park. How many days must he work to earn at least \$500?

35. If the perimeter of a square photograph must be at most 60 in.^{2}, what is the largest possible size of the photograph?

. . . LET'S NOT FORGET . . .

In Problems 41 through 46, perform the operations indicated.

36. $(-2x)(3x)$

37. $-2x(3 + x)$

38. $\dfrac{6}{0}$

39. $(x - 9)^2$

40. -6^2

41. $\dfrac{-18z^4}{-6z^3}$

In Problems 42 through 44, simplify each expression. Watch the role of the parentheses.

42. $2 + x(x - 1)$

43. $(2 + x)(x - 1)$

44. $(2 + x)x - 1$

In Problems 45 through 48, determine if x is a factor *or a* term.

45. $x - ab$

46. $3ax$

47. $x(a + b)$

48. $x + b$

In Problems 49 through 51, multiply each expression by $\frac{1}{2}$.

49. $-6x$

50. $6(x + 7)$

51. $6x - 4$

In Problems 52 through 57, identify each of the following as an expression, equation, *or* inequality. *Solve the equations and inequalities and simplify the expressions, if possible.*

52. $2x + 3(x - 5) - 4$

53. $3x - 7 > x + 3$

54. $2x - 18 = 5(x - 3)$

55. $5 - 2(x + 6)$

56. $5 - 2x \leq 6$

57. $\frac{3}{5}x + \frac{1}{2} = \frac{1}{10}x$

CHAPTER 3 TEST

In Problems 1 through 6, choose the correct answer.

1. The solution set for the equation $6x - 3 = 4 - x$ is (?).

   A. $\left\{\frac{5}{7}\right\}$ B. $\left\{\frac{7}{5}\right\}$

   C. $\left\{\frac{1}{5}\right\}$ D. $\{1\}$

2. The solution set for the equation $3x - 5 = 5(x - 1)$ is (?).

   A. $\{-2\}$ B. $\left\{\frac{5}{4}\right\}$

   C. $\{0\}$ D. $\varnothing$

3. When solved for x, the literal equation $6 + Abx = 2A$ becomes (?).

   A. $x = \dfrac{2A - 6}{Ab}$ B. $x = \dfrac{-4}{b}$

   C. $x = 2A - 6 - Ab$ D. $x = 6 - 2A - Ab$

4. The solution set for $4x - 4 > 2x - 8$ is (?).

   A. $\{x \mid x > -2\}$ B. $\{-2\}$

   C. $\left\{x \mid x > -\frac{7}{2}\right\}$ D. $\left\{-\frac{7}{2}\right\}$

5. A board 19 ft long is cut into two pieces such that one is 5 ft longer than the other. If x is the length of the shorter piece, which of the following is true?

A. $x + 5 = 19$ B. $x + x + 5 = 19$
C. $x(x + 5) = 19$ D. $x + x - 5 = 19$

6. The solution set for the equation $\frac{1}{2}x + \frac{2}{3} = 1$ is (?).

A. $\left\{\frac{3}{2}\right\}$ B. $\left\{\frac{2}{3}\right\}$
C. $\{-1\}$ D. $\left\{\frac{4}{3}\right\}$

In Problems 7 through 10, find the solution set.

7. $3 - \frac{1}{2}x = \frac{5}{2}$

8. $2(2 - x) \leq 4 - 3x$

9. $5(t - 2) = 2 - t$

10. $\frac{1}{2}x > 1 + 2x$

In Problems 11 and 12, solve for the variable indicated.

11. $3t - ax = 5t$; solve for x.

12. $V = \frac{1}{3}\pi r^2 h$; solve for h.

13. A number is one less than twice another number. If their sum is 50, what are the two numbers?

14. Billy is 5 yr older than his sister. Six years ago he was twice her age then. How old is each now?

15. A collection of 20 nickels and dimes has a value of $1.35. How many of each type coin are there in the collection?

CHAPTER 4

See Problem Set 4.6, Exercise 7.

Factoring

CONNECTIONS

Factoring natural numbers has long been of utmost importance to mathematicians. The Fundamental Theorem of Arithmetic says that a natural number can be expressed as a product of prime numbers in only one way, apart from the order in which the factors are written. Because of this result, prime factors are often the key to understanding concepts of arithmetic. Likewise, in algebra factoring is often the key to solving a problem.

A contractor is building a warehouse which will store stacks of 18 in. and 24 in. boxes on the same shelf. To save space, he wants to know the shortest distance to put between shelves. The answer to his question is 72 in. To figure this out, he used the prime factorizations of 24 and 18 to find the least common multiple of 18 and 24.

Often in solving applications one of the intermediate steps is to factor a polynomial. Equations such as $r^2 - 49 = 0$ are very often the mathematical model for solving a problem. Factoring the polynomial $r^2 - 49$ will provide the solutions.

In arithmetic, factoring provides the basis for working with fractions. In Chapter 5 we will see the connection between factoring polynomials and working with algebraic fractions. Factoring is also a very useful tool in solving certain kinds of equations.

4.1 GREATEST COMMON FACTOR AND GROUPING

In this chapter we will study factoring. To **factor** a number means to write the number as a **product.** We learned how to factor natural numbers into prime factors in Section 0.2. To factor a polynomial means to write the polynomial as a product of polynomials.

First we will study **factoring out common factors.** This factoring is done by using the distributive property. When we multiplied polynomials, we used the distributive property:

$$p(q + r) = pq + pr$$

To factor we turn this around and write

$$pq + pr = p(q + r)$$

It is important to notice that the expression $pq + pr$ is a sum of two terms, while $p(q + r)$ is a *product.* That is, $pq + pr$ is not factored, but $p(q + r)$ is factored.

To factor out common factors, we must be able to identify the *factors* in each term. Consider the polynomial $5xy + 2x$. It has terms $5xy$ and $2x$. The factors in the first term are 5, x, and y. In the second term, 2 and x are factors. We can use the distributive property to factor.

$$5xy + 2x = x \cdot 5y + x \cdot 2 \qquad \text{Commutative Property}$$

$$= x(5y + 2) \qquad \text{Distributive Property}$$

Notice that the Distributive Property worked because x was a factor in both terms, $5xy$ and $2x$.

Let's factor $12x + 6$. Using the distributive property, we can write

$$\boxed{1} \quad 12x + 6 = 2(6x + 3)$$

or

$$\boxed{2} \quad 12x + 6 = 3(4x + 2)$$

or

$$\boxed{3} \quad 12x + 6 = 6(2x + 1)$$

Each of these is a correct way to factor $12x + 6$. However, we say that $\boxed{3}$ is factored completely. Notice that in $\boxed{1}$, $6x + 3$ can be factored as $3(2x + 1)$ and in $\boxed{2}$, $4x + 2$ can be factored as $2(2x + 1)$. If no more factoring can be done, we say that a polynomial is **factored completely.** If no factoring can be done, we say that the polynomial is **prime.** We consider monomials to be prime polynomials.

EXAMPLE 1. Factor each polynomial completely.

(a) $18y - 12$ (b) $15xy + 10y$

Solutions:

(a) $18y - 12 = 6(3y - 2)$

(b) $15xy + 10y = 5y(3x + 2)$ □

To factor a polynomial completely involves finding the largest factor common to all the terms in the polynomial. Suppose that we want to find the largest factor that two numbers have in common. Consider the numbers 12 and 18. The *largest* factor of both numbers is 6. If the numbers are so large that we cannot find this greatest common factor by inspection, we must rely on the prime factorization of the numbers. We call such a factor the **greatest common factor.**

To Find the Greatest Common Factor (GCF)

1. Factor each number completely.
2. List the prime factors that are common to all the numbers.
3. The GCF is the product of the factors in Step 2 with each factor raised to the smallest power of that factor in any number.

EXAMPLE 2. Find the GCF for each pair of numbers.

(a) 25 and 60 (b) 32 and 72

Solutions:

(a) $25 = 5^2$ Write each number as a product of prime factors.
$60 = 2^2 \cdot 3 \cdot 5$

There is only one prime factor that is common to both numbers. It is 5.

Since 5 occurs two times in 25 and one time in 60, the GCF is 5^1, or 5.

(b) $32 = 2^5$
$72 = 2^3 \cdot 3^2$

The common prime factor is 2.

Since 2 occurs five times in 32 and three times in 72, the GCF is 2^3, or 8. □

EXAMPLE 3. Find the GCF of $18x^3y^6$ and $12x^2y^3$.

Solution:

$$18x^3y^6 = 2 \cdot 3^2x^3y^6$$
$$12x^2y^3 = 2^2 \cdot 3x^2y^3$$

The common prime factors are 2, 3, x, and y.

The GCF must contain one factor of 2, one factor of 3, two factors of x, and three factors of y.

The GCF is $6x^2y^3$. □

Using the Distributive Property to factor a polynomial completely, we factor out the GCF of all the terms in the polynomial. *Always check by multiplying.*

EXAMPLE 4. Factor $36x^3 - 54x^2 + 18x$ completely.

Solution:

We must find the GCF for $36x^3$, $54x^2$ and $18x$.

$$36x^3 = 2^2 \cdot 3^2x^3$$
$$54x^2 = 2 \cdot 3^3x^2$$
$$18x = 2 \cdot 3^2x$$

The GCF is $2 \cdot 3^2 \cdot x$, or $18x$.

$$36x^3 - 54x^2 + 18x = 18x \cdot 2x^2 - 18x \cdot 3x + 18x \cdot 1$$
$$= 18x(2x^2 - 3x + 1)$$

Check by using the Distributive Property. □

EXAMPLE 5. Factor $-3y^2 + 6y - 3$ completely.

Solution:

If the first term has a negative coefficient, we have a choice of factoring out 3 or -3.

If we factor out 3,

$$-3y^2 + 6y - 3 = 3(-y^2 + 2y - 1)$$

Check by using the Distributive Property.

If we factor out -3, we must be very careful with signs.

$$-3y^2 + 6y - 3 = -3(y^2 - 2y + 1)$$

Be Careful!

Be sure to multiply this to check signs.

We usually prefer to factor out the negative number. □

Let's try to factor $a(m + n) + b(m + n)$. There are two terms in this expression, $a(m + n)$ and $b(m + n)$. We see that $(m + n)$ is a factor in both. So, it is a common factor.

$$a(m+n)+b(m+n)=(m+n)(a+b)$$

Notice how we treated $(m+n)$ as one quantity. It represents a number, just as x represents a number.

Also, it is very important to see that the expression $a(m+n)+b(m+n)$ is a *sum*, whereas $(m+n)(a+b)$ is a *product*.

$$a(m+n)+b(m+n) \text{ is not factored}$$

$$(m+n)(a+b) \text{ is factored}$$

Be Careful!

EXAMPLE 6. Factor each completely.

(a) $x(a-b)+y(a-b)$ (b) $w^2(2a+c)-z(2a+c)$
(c) $y(c+d)+(c+d)$

Solutions:

(a) $(a-b)$ is a factor common to both terms.

$$x(a-b)+y(a-b)=(a-b)(x+y)$$

(b) $(2a+c)$ is a factor in both terms.

$$w^2(2a+c)-z(2a+c)=(2a+c)(w^2-z)$$

(c) $(c+d)$ is a factor in both terms.

$$y(c+d)+(c+d)=(c+d)(y+1)$$

Notice the 1 in the second factor. Think of $(c+d)$ as $1(c+d)$.

If we do some grouping, we can factor expressions such as $am+an+bm+bn$. We notice that there are four terms and that there is no common factor. However, if we group the first two terms and last two terms together, we have,

$$am+an+bm+bn=(am+an)+(bm+bn)$$

Now we have a common factor in the first group and a common factor in the second group.

$$(am+an)+(bm+bn)=a(m+n)+b(m+n)$$

We factor out the common factor, $(m+n)$.

$$a(m+n)+b(m+n)=(m+n)(a+b)$$

This process is called factoring by **grouping.**

EXAMPLE 7. Factor by grouping.

(a) $xy+xt+ay+at$ (b) $x^2-xy+ax-ay$
(c) $ar^2+at^2-br^2-bt^2$ (d) $zx-zy+x-y$

Solutions:

(a)
$$\begin{aligned} xy+xt+ay+at &= (xy+xt)+(ay+at) \\ &= x(y+t)+a(y+t) \\ &= (y+t)(x+a) \end{aligned}$$

(continued)

(b) $x^2 - xy + ax - ay = (x^2 - xy) + (ax - ay)$
$$= x(x - y) + a(x - y)$$
$$= (x - y)(x + a)$$

(c) If we group the terms as they appear, we must be very careful with signs. Watch the negative sign with the parentheses.

$$ar^2 + at^2 - br^2 - bt^2 = (ar^2 + at^2) - (br^2 + bt^2)$$
$$= a(r^2 + t^2) - b(r^2 + t^2)$$
$$= (r^2 + t^2)(a - b)$$

(d) $zx - zy + x - y = (zx - zy) + (x - y)$
$$= z(x - y) + (x - y)$$
$$= (x - y)(z + 1)$$

□

There may be more than one way to group a polynomial. Let's look at Example 7c and group another way. Suppose we group the first and third terms and the second and fourth terms together.

$$ar^2 + at^2 - br^2 - bt^2 = (ar^2 - br^2) + (at^2 - bt^2)$$
$$= r^2(a - b) + t^2(a - b)$$
$$= (a - b)(r^2 + t^2)$$

Are the two answers the same? Why is $(a - b)(r^2 + t^2)$ the same as $(r^2 + t^2)(a - b)$?

PROBLEM SET 4.1

Warm-ups

In Problems 1 through 10, factor each completely. See Example 1.

1. $6x + 3$
2. $2x - 4$
3. $xy + xz$
4. $mn - mp$
5. $5x - 15y$
6. $18a + 9b$
7. $9ab - 6ac$
8. $7rs + 14r$
9. $6x + 6$
10. $3z - 3$

In Problems 11 through 20, find the GCF for each set of numbers. See Examples 2 and 3.

11. 12 and 18
12. 27 and 36
13. 54, 81 and 36
14. x^2 and x^3
15. xy^2 and xy
16. ab^2c^2 and a^2b^3c
17. $12x^3$, $18x^2$, and $21x$
18. $27p^2q^3$, $36pq^2$, and $45q^2$
19. $18xy^4$, $27x^2y^3$, and $36xy$
20. $20rst^2$, $30r^2st$, and $45rs^2t$

In Problems 21 through 31, factor each completely. See Example 4.

21. $5x^2 - 15x$
22. $18a^2b + 27ab^2$
23. $25x^2yz^2 - 75x^2yz$
24. $30z^3 + 45z^2$
25. $6r^2s^2t^2 + 3rst$
26. $21y^5 - 14y^7$
27. $26x^6 - 13x^4 + 39x^2$
28. $32y^8 + 36y^7 - 48y^5$
29. $7m^2n - 21m^3n^2 + 7m^2$
30. $a^2b^4c^7 - a^3b^5c^8 - a^5b^3c^5$
31. $56r^4s^2t + 64r^2s^2t^2 - 24rst$

In Problems 32 through 36, factor each completely by factoring out a negative coefficient. See Example 5.

32. $-5y - 10$

33. $-4x + 4$

34. $-2z - 2$

35. $-15y^5 + 10y^4 - 20y^2$

36. $-12a^2b - 15ab^2 + 3abc$

In Problems 37 through 44, factor completely. See Example 6.

37. $r(s + t) + 2(s + t)$

38. $z(a - b) + w(a - b)$

39. $x^2(y - z) + y^2(y - z)$

40. $x(c - d) - y(c - d)$

41. $q(p + r) - (p + r)$

42. $m^2(x + y^2) + (x + y^2)$

43. $3r^3(s^2 - 2) - 2w(s^2 - 2)$

44. $x^4(x^2 + y^2) - z^7(x^2 + y^2)$

In Problems 45 through 56, factor by grouping. See Example 7.

45. $ms + mt + ns + nt$

46. $xz - yz + xw - yw$

47. $r^2y - r^2z + s^2y - s^2z$

48. $ap - ar + bp - br$

49. $rpx + rqz - 3px - 3qz$

50. $x^2s^2 + x^2t^2 - 7s^2 - 7t^2$

51. $2z^5 + 2x - yz^5 - xy$

52. $2x^2 + 4 - wx^2 - 2w$

53. $3xy^2 - 6x - 2wy^2 + 4w$

54. $m^6 - m^3y^2 - m^3z + y^2z$

55. $ax^2 + bx^2 + a + b$

56. $as - at - s + t$

Practice Exercises

Factor each completely.

57. $5x + 10$

58. $2x - 4$

59. $rw + rz$

60. $st - sp$

61. $6x - 15y$

62. $15c + 5d$

63. $-9ab + 6ac$

64. $7r^2s + 14rs^2$

65. $2x^3 + 2$

66. $4t - 4$

67. $3x^2y - 15xy$

68. $8a^2b^2 + 20a^3b^2$

69. $15x^3yz^2 - 30x^2yz^2$

70. $35z^6 + 42z^4$

71. $-6r^3st^3 + 3r^3st$

72. $16xy^5 - 24x^2y^7$

73. $6x^2 + y^2$

74. $36p^3q^4 - 48p^3q^5$

75. $72r^4s^2 + 48r^3s^2$

76. $x^2y + 5x^2$

77. $14x^6 - 21x^4 + 28x^3$

78. $18y^6 + 27y^4 - 45y^3$

79. $21a^2b - 14ab^2 + 7ab$

80. $a^3b^2c^6 - a^2b^4c^7 - a^5b^3c^5$

81. $108r^4s^2t + 96r^2s^2t^2 - 84rst$

82. $w(x + y) + (x + y)$

83. $r(a - b) + 2t(a - b)$

84. $2b(c + d) - 3d(c + d)$

85. $x(x^2 + 1) - y(x^2 + 1)$

86. $x^2(b - 2d) + 4(b - 2d)$

87. $2r^2(a^2 - b) - (a^2 - b)$

88. $5(b + 1) - x(b + 1)$

89. $as - bs + 2at - 2bt$

90. $3bc + 3bd - 2cd - 2d^2$

91. $y^3 + y - xy^2 - x$

92. $bt^2 - 2dt^2 + 4b - 8d$

93. $pq + pr + 3q + 3r$

94. $2a - 2b + am - bm$

95. $2a^2 - 2b - a^2z + bz$

96. $ab + at - b - t$

97. $3a + 6 + ay + 2y$

98. $a^2r^2 - br^2 - a^2z + bz$

99. $x^3 + 3x^2 + x + 3$

100. $a^3 - 2a^2 - 3a + 6$

Challenge Problems

Factor each completely.

101. $ams + amt + ans + ant$

102. $2xz - 2yz + 2xw - 2yw$

103. $6bc + 6bd - 6cd - 6d^2$

104. $a^4 - 2a^3 - 3a^2 + 6a$

IN YOUR OWN WORDS...

105. What is factoring?

106. What do we mean when we say a polynomial is factored completely?

107. What is factoring out the greatest common factor?

108. What is factoring by grouping?

4.2 FACTORING BINOMIALS

We saw that the product $(p + q)(p - q)$ was $p^2 - q^2$. If asked whether $p^2 - q^2$ will factor, we know that it will. This type of factoring is called the **difference of two squares.**

Difference of Two Squares

$$p^2 - q^2 = (p + q)(p - q)$$

Notice that there are two terms in $p^2 - q^2$ and that each of them is a square. The factors are the sum and difference of the same two numbers. After we have decided that we are factoring the difference of squares, the factors will always look like

$$(\quad + \quad)(\quad - \quad)$$

EXAMPLE 1. Factor each completely.

(a) $x^2 - 4$ (b) $16x^2 - 25$

Solutions:

(a) This is the difference of two terms and both terms are squares. So, this is the difference of two squares.

$$x^2 - 4 = x^2 - 2^2$$
$$= (x + 2)(x - 2)$$

(b) $16x^2 - 25 = (4x)^2 - 5^2$

$$= (4x + 5)(4x - 5)$$

□

We can combine taking out common factors with the difference of squares.

EXAMPLE 2. Factor each completely.

(a) $x^3 - 4x$ (b) $4x^2 - 4$

Solutions:

(a) This is a difference of two terms, but the terms are not squares. However, if we first factor out a common factor of x, it will be the difference of squares.

$$\begin{aligned} x^3 - 4x &= x(x^2 - 4) \\ &= x(x + 2)(x - 2) \end{aligned}$$

(b) This is the difference of squares. What is wrong in factoring it as follows?

$$4x^2 - 4 = (2x - 2)(2x + 2)$$

It is not factored completely. To avoid this, it is best to take out common factors first.

$$\begin{aligned} 4x^2 - 4 &= 4(x^2 - 1) \\ &= 4(x + 1)(x - 1) \end{aligned}$$

□

Does the sum of squares factor? Can we factor $p^2 + q^2$? The answer is no. It is *not* $(p + q)^2$! We say that $p^2 + q^2$ is *prime*.

Sum of Squares

$p^2 + q^2$ is prime.

We can factor the sum and difference of two cubes, as the following products show.

$$\begin{aligned} (p + q)(p^2 - pq + q^2) &= p^3 - p^2q + pq^2 + p^2q - pq^2 + q^3 \\ &= p^3 + q^3 \\ (p - q)(p^2 + pq + q^2) &= p^3 + p^2q + pq^2 - p^2q - pq^2 - q^3 \\ &= p^3 - q^3 \end{aligned}$$

Sum and Difference of Cubes

$$p^3 + q^3 = (p + q)(p^2 - pq + q^2)$$
$$p^3 - q^3 = (p - q)(p^2 + pq + q^2)$$

To recognize that we have the sum or difference of cubes, we see that there are two terms and that each term is a cube. Notice the similarity in the two formulas. The second factor is obtained from the first factor. Notice that the middle term is the product of p and q and that the sign is opposite to the sign in the first factor.

Let's factor $x^3 + 8$. We see that it is the sum of cubes.

$$x^3 + 8 = x^3 + 2^3$$

We use the formula and replace p with x and q with 2.

$$p^3 + q^3 = (p + q)(p^2 - pq + q^2)$$
$$x^3 + 8 = x^3 + 2^3 = (x + 2)(x^2 - 2x + 4)$$

To factor $x^3 - 8$, we use the other formula and replace p with x and q with 2.

$$p^3 - q^3 = (p - q)(p^2 + pq + q^2)$$
$$x^3 - 8 = x^3 - 2^3 = (x - 2)(x^2 + 2x + 4)$$

EXAMPLE 3. Factor each.

(a) $w^3 + 27$ (b) $8c^3 - 27$

Solutions:

(a) This is the sum of cubes. We use the formula and replace p with w and q with 3.

$$p^3 + q^3 = (p + q)(p^2 - pq + q^2)$$
$$w^3 + 27 = w^3 + 3^3 = (w + 3)(w^2 - 3w + 9)$$

(b) This is the difference of cubes.

$$p^3 - q^3 = (p - q)(p^2 + pq + q^2)$$
$$8c^3 - 27 = (2c)^3 - 3^3 = (2c - 3)[(2c)^2 + (2c)(3) + 3^2]$$
$$= (2c - 3)(4c^2 + 6c + 9)$$

□

We can also have common factors along with the sum or difference of cubes. Common factors should be taken out first.

EXAMPLE 4. Factor $2x^3 - 16$ completely.

Solution:

Notice that there are two terms and that we have a difference. However, 2 and 16 are not cubes. We must first factor out the common factor of 2.

$$2x^3 - 16 = 2(x^3 - 8)$$
$$= 2(x - 2)(x^2 + 2x + 4)$$

□

Factoring Binomials

$p^2 - q^2 = (p - q)(p + q)$	Difference of squares
$p^2 + q^2$ is prime	Sum of squares
$p^3 - q^3 = (p - q)(p^2 + pq + q^2)$	Difference of cubes
$p^3 + q^3 = (p + q)(p^2 - pq + q^2)$	Sum of cubes

PROBLEM SET 4.2

Warm-ups

In Problems 1 through 15, factor each completely if possible. See Example 1.

1. $x^2 - 9$
2. $y^2 - 81$
3. $4x^2 - 9$
4. $x^2 - 49$
5. $16x^2 - 25$
6. $x^2 - y^2$
7. $a^2 - 4b^2$
8. $4b^2 - 1$
9. $a^2b^2 - 36$
10. $4t^2 - x^2y^2$
11. $r^2 - s^2t^2$
12. $x^2 + 4$
13. $4z^2 - 49$
14. $9x^2 - m^2$
15. $100c^2 - 9d^2$

In Problems 16 through 26, factor each completely if possible. See Example 2.

16. $2x^2 - 50$
17. $x^3 -$
18. $4x^2 - 16$
19. $8x^2 - 18$
20. $a^3 - ab^2$
21. $12b^2 - 3$
22. $5z^2 + 20$
23. $2a^3 - 8a$
24. $3ab^2 - 27a$
25. $4c^3 - 16c$
26. $x^2 + x^4$

In Problems 27 through 38, factor each completely. See Example 3.

27. $y^3 - x^3$
28. $c^3 + d^3$
29. $x^3 + 125$
30. $x^3 + 64$
31. $z^3 - 1$
32. $d^3 - 8c^3$
33. $8x^3 + 27$
34. $27k^3 + 64$
35. $8p^3 - 1$
36. $8x^3 - y^3$
37. $27t^3 + 8s^3$
38. $125z^3 - 27w^3$

In Problems 39 through 47, factor each completely. See Example 4.

39. $2y^3 + 2x^3$
40. $3z^3 - 24$
41. $8p^3q - q$
42. $2x^3 + 128$
43. $8z^3 - 8$
44. $8a^3b - 27b$
45. $64x^3 - 8y^3$
46. $16p^3 + 2q^3$
47. $-5z^3 - 5$

Practice Exercises

Factor each completely if possible.

48. $x^2 - 25$
49. $y^2 - 64$
50. $9x^2 - 4$
51. $x^2 - 121$
52. $9x^2 - 25$
53. $4x^2 - y^2$
54. $9a^2 - 4b^2$
55. $4b^2 - 81$
56. $9a^2b^2 - 1$
57. $t^2 - 4x^2y^2$
58. $25r^2 - s^2t^2$
59. $4x^2 + 4$
60. $4z^2 - 36$
61. $9x^2 - 9m^2$
62. $64c^2 - 9d^2$
63. $2x^2 - 18$

64. $x^3 - 9x$

65. $64x^2 - 16$

66. $5x^2 - 20$

67. $s^3 - st^2$

68. $6b^2 - 54$

69. $8z^2 + 72$

70. $3a^3 - 3a$

71. $50xy^2 - 2xz^2$

72. $4bc^2d^2 - 36b$

73. $c^3 - d^3$

74. $y^3 + x^3$

75. $x^3 - 64$

76. $z^3 + 1$

77. $d^3 + 8c^3$

78. $8x^3 - 27$

79. $27k^3 - 64$

80. $8p^3 + 1$

81. $8x^3 + y^3$

82. $27t^3 + 8s^3$

83. $125z^3 + 27w^3$

84. $2y^3 + 2x^3$

85. $3z^3 + 24$

86. $8p^3q + q$

87. $2x^3 - 128$

88. $8z^3 + 8$

89. $8a^3b + 27b$

90. $64x^3 + 8y^3$

91. $16p^3 - 2q^3$

92. $-8z^3 - 8$

Challenge Problems

Factor each completely.

93. $x^6 - 4$

94. $x^8 - 25$

95. $x^4 - 16$

96. $x^9 - 27$

97. $x^9 - 1$

98. $(a + b)^2 - 4$

99. $(a + b)^3 + 8$

100. $(a + b)^3 - c^3$

101. $16 - (a + b)^2$

IN YOUR OWN WORDS . . .

102. What kinds of binomials can we factor?

103. Give the rule for factoring the difference of two squares without using a formula.

104. Give the rule for factoring the sum or difference of two cubes without using a formula.

4.3 FACTORING TRINOMIALS

In this section we examine factoring trinomials. We learned that the product of two binomials is sometimes a trinomial. So, we look for two binomial factors.

We begin by looking at how $x^2 + 8x + 15$ can be factored. Notice that the last term, 15, is *positive*. We look for two binomials whose product is the given trinomial. Since the first term is x^2, the first terms in the factors must be x.

$$\square \quad \square \quad x^2 + 8x + 15 = (x \quad\quad)(x \quad\quad)$$

The numbers in the boxes must have a product of 15. There are four pairs of integers whose product is 15.

$$(-15)(-1) = 15$$

$$(-5)(-3) = 15$$

$$(15)(1) = 15$$

$$(5)(3) = 15$$

Not only must the product of the numbers be 15, but the sum of the outside and inside products must be $8x$.

$$x^2 + 8x + 15 = (x \square)(x \square)$$

Sum must be $8x$

So, we look for a pair of integers whose product is 15 and whose sum is 8.

INTEGERS	PRODUCT	SUM
−15, −1	15	−16
−5, −3	15	−8
15, 1	15	16
5, 3	15	8

Thus

$$x^2 + 8x + 15 = (x + 5)(x + 3)$$

Now, let's factor $x^2 - 8x + 15$. Notice again that the last term, 15, is *positive*. The factors must have x as the first term.

$$x^2 - 8x + 15 = (x \square)(x \square)$$

The numbers that go in the boxes must have a product of 15. To make a middle term of $-8x$, we must have two numbers whose sum is -8. So, we must use -3 and -5. Thus, $x^2 - 8x + 15 = (x - 3)(x - 5)$.

Comparing both situations, the factors look alike except for the signs.

$$x^2 + 8x + 15 = (x + 5)(x + 3) \qquad x^2 - 8x + 15 = (x - 5)(x - 3)$$

If the *last* term of a trinomial is *positive*, then the signs in the factors are *alike* and can be determined by the *middle sign*.

EXAMPLE 1. Factor each trinomial.

(a) $x^2 + 4x + 3$ (b) $x^2 - 3x + 2$

(c) $x^2 + 7x + 12$ (d) $x^2 - 9x + 18$

Solutions:

(a) $x^2 + 4x + 3$

Since the last term is positive and the middle term is also positive, the factors must look like

$$(\ \ +\ \ )(\ \ +\ \ )$$

We are looking for integers with a product of 3 and a sum of 4. They are 1 and 3.

$$x^2 + 4x + 3 = (x + 3)(x + 1)$$

(continued)

(b) $x^2 - 3x + 2$

Since the last term is positive and the middle term is negative, the factors must look like

$$(\quad - \quad)(\quad - \quad)$$

We are looking for integers with a product of 2 and a sum of 3. They are 2 and 1.

$$x^2 - 3x + 2 = (x - 2)(x - 1)$$

(c) $x^2 + 7x + 12$

The factors must look like

$$(\quad + \quad)(\quad + \quad)$$

We are looking for integers with a sum of 7 and a product of 12. There are several factors of 12.

PRODUCT OF 12	SUM
12, 1	13
6, 2	8
3, 4	7

$$x^2 + 7x + 12 = (x + 3)(x + 4)$$

(d) $x^2 - 9x + 18$

The factors must look like

$$(\quad - \quad)(\quad - \quad)$$

We look for a product of 18 and a sum of 9.

PRODUCT OF 18	SUM
18, 1	19
9, 2	11
6, 3	9

$$x^2 - 9x + 18 = (x - 6)(x - 3)$$ □

Be Careful!

After factoring any polynomial, it is *always* a good idea to *check by multiplying the factors to see if the product is the original polynomial.*

Now let's look at two examples of a second situation, where the last term is negative.

$$x^2 - 2x - 15 \qquad\qquad x^2 + 2x - 15$$

The first terms in the factors must be x.

$$(x \; \square)(x \; \square) \qquad (x \; \square)(x \; \square)$$

The product of the numbers in the boxes must be -15 and the middle terms must be $-2x$ or $2x$.

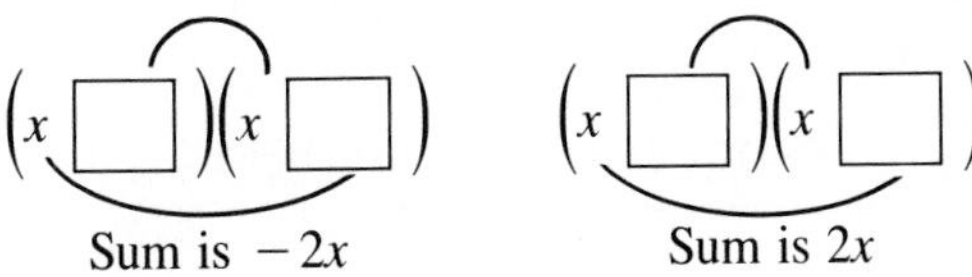

We look for possibilities in the table.

INTEGERS	PRODUCT	SUM
$-15, 1$	-15	-14
$-1, 15$	-15	14
$-5, 3$	-15	-2
$-3, 5$	-15	2

Thus we see that in the first example,

$$x^2 - 2x - 15 = (x - 5)(x + 3)$$

and in the second example

$$x^2 + 2x - 15 = (x - 3)(x + 5).$$

Notice that the factors look like

$$(\quad - \quad)(\quad + \quad)$$

That is, the signs are different in the factors. We made the middle term $2x$ or $-2x$ by where we placed the negative and positive signs.

EXAMPLE 2. Factor each trinomial.

(a) $x^2 - 2x - 3$ (b) $x^2 + 3x - 4$

Solutions:

(a) $x^2 - 2x - 3$

Since the last term is negative, the factors must look like

$$(\quad - \quad)(\quad + \quad)$$

We are looking for integers with a product of 3 and a difference of 2. They must be 1 and 3. We must be careful where we put them. We want the middle term to be $-2x$. So, the 3 must be placed with the negative sign.

$$x^2 - 2x - 3 = (x - 3)(x + 1)$$

(b) $x^2 + 3x - 4$

Again, the factors must look like

$$(\quad - \quad)(\quad + \quad)$$

(continued)

We need a product of 4 and a difference of 3.

PRODUCT OF 4	DIFFERENCE
1, 4	3
2, 2	0

We must place the 4 with the positive sign, since the middle term is $+3x$.

$$x^2 + 3x - 4 = (x - 1)(x + 4)$$ □

When factoring trinomials that are in standard form with a positive leading coefficient, the following guidelines are useful.

Guidelines for Factoring Trinomials

1. If the last term is *positive,* then the factors will look like

$$(\ + \ )(\ + \ ) \quad \text{or} \quad (\ - \ )(\ - \ )$$

The + or − sign is determined by the coefficient of the middle term.

2. If the last term is *negative,* then the factors will look like

$$(\ + \ )(\ - \ ) \quad \text{or} \quad (\ - \ )(\ + \ )$$

These guidelines place the proper signs in the factors for us. In looking at the sums and products that determine the numbers to put in, we do not have to worry about signs.

In the discussion of factoring trinomials, we have looked only at trinomials that have a leading coefficient of 1. That is, the first term was x^2. Let's consider other possibilities.

EXAMPLE 3. Factor each trinomial.

(a) $2x^2 - 5x + 3$ (b) $6x^2 + x - 2$

Solutions:

(a) $2x^2 - 5x + 3$

Since the last term is positive and the middle term is negative, the factors must look like

$$(\ - \ )(\ - \ )$$

The first term must be $2x^2$, which means that we must use $2x$ and x as first terms in the factors.

$$(2x - \ )(x - \ )$$

We are looking for a product of 3. The only possibilities are 1 and 3. We must be very careful where we place the 1 and 3. We must have a middle term of $-5x$. There are two choices.

$$(2x - 3)(x - 1) \quad \text{or} \quad (2x - 1)(x - 3)$$

Multiplying tells us that $(2x - 3)(x - 1)$ is correct.

$$2x^2 - 5x + 3 = (2x - 3)(x - 1)$$

(b) $6x^2 + x - 2$

Since the last term is negative, the factors must look like

$$(\quad - \quad)(\quad + \quad)$$

There are many possibilities for this product. The first term of $6x^2$ could give $3x$ and $2x$ or $6x$ and x as first terms in the factors. The last term of 2 gives 2 and 1 for the last terms in the factors. We list all the possible factors.

1	$(6x - 2)(x + 1)$	2	$(6x - 1)(x + 2)$
3	$(6x + 2)(x - 1)$	4	$(6x + 1)(x - 2)$
5	$(3x - 2)(2x + 1)$	6	$(3x - 1)(2x + 2)$
7	$(3x + 1)(2x - 2)$	8	$(3x + 2)(2x - 1)$

This seems to be a difficult job. However, we can eliminate 1, 3, 6, 7, because the terms in one factor of each have a common factor. These are not the factors because the original trinomial does not have a common factor. Multiplying tells us that 8 is correct.

$$6x^2 + x - 2 = (3x + 2)(2x - 1)$$ □

Any time that we factor we must always look for a common factor first. The next example shows a common factor in a trinomial.

Be Careful!

EXAMPLE 4. Factor $2a^3 - 4a^2b - 6ab^2$ completely.

Solution:

There is a common factor of $2a$. We factor it out first.

$$\begin{aligned} 2a^3 - 4a^2b - 6ab^2 &= 2a(a^2 - 2ab - 3b^2) \\ &= 2a(a + b)(a - 3b) \end{aligned}$$ □

If a trinomial is the square of a binomial, we call the trinomial a **perfect trinomial square.** The special products that we learned in Section 2.6 can be used as factoring formulas. For example, $x^2 + 6x + 9$ is $(x + 3)^2$. Perfect trinomial squares can be factored just like any other trinomial. Recognizing that the trinomial is a perfect square may save some time.

Factoring Perfect Trinomial Squares

$$p^2 + 2pq + q^2 = (p + q)^2$$
$$p^2 - 2pq + q^2 = (p - q)^2$$

PROBLEM SET 4.3

Warm-ups

In Problems 1 through 15, factor each trinomial completely. See Example 1.

1. $x^2 + 3x + 2$

2. $x^2 + 4x + 3$

3. $x^2 + 5x + 4$

4. $x^2 + 6xy + 5y^2$

5. $x^2 + 7x + 6$

6. $x^2 + 4x + 4$

7. $x^2 + 5x + 6$

8. $x^2 + 6x + 8$

9. $x^2 - 8x + 15$

10. $x^2 - 7x + 12$

11. $x^2 - 3xy + 2y^2$

12. $x^2 - 4x + 3$

13. $x^2 - 8x + 16$

14. $x^2 - 6x + 5$

15. $x^2 - 7x + 6$

In Problems 16 through 25, factor each trinomial completely. See Example 2.

16. $x^2 - x - 2$

17. $a^2 - 2ab - 3b^2$

18. $z^2 + 3z - 4$

19. $x^2 + 4x - 5$

20. $y^2 - 5y - 6$

21. $x^2 - x - 6$

22. $x^2 - 2x - 8$

23. $x^2 - 3x - 10$

24. $t^2 + t - 12$

25. $x^2 - x - 20$

In Problems 26 through 46, factor each trinomial completely. See Example 3.

26. $2x^2 + 3x + 1$

27. $3b^2 + 4b + 1$

28. $2x^2 - 5x + 2$

29. $3x^2 - 7xz + 2z^2$

30. $5w^2 + 7w + 2$

31. $3x^2 + 10x + 3$

32. $9x^2 - 12x + 4$

33. $5x^2 - 16x + 3$

34. $2x^2 - x - 1$

35. $3c^2 - 2c - 1$

36. $2x^2 + 3x - 2$

37. $3x^2 - 5x + 2$

38. $6x^2 + 5x + 1$

39. $2x^2 + 7x + 6$

40. $10x^2 - 7x + 1$

41. $3a^2 - 14a + 8$

42. $6x^2 - 7xy - 3y^2$

43. $7x^2 + 33x - 10$

44. $6t^2 + 23t + 20$

45. $8x^2 - 26x + 21$

46. $12p^2 - 5p - 25$

In Problems 47 through 50, factor each trinomial completely. See Example 4.

47. $2x^2 - 2x - 12$

48. $3x^2 - 15x + 12$

49. $4x^2 - 14x - 8$

50. $2ax^2 + 9ax + 9a$

Practice Exercises

Factor each trinomial completely.

51. $x^2 + 8x + 7$

52. $x^2 + 9x + 8$

53. $x^2 + 10x + 9$

54. $x^2 + 7x + 10$

55. $x^2 + 8x + 12$

56. $x^2 - 8x + 7$

57. $x^2 - 2x + 1$

58. $x^2 - 10x + 9$

59. $x^2 - 9x + 18$

60. $a^2 - 5ab + 6b^2$

61. $y^2 + y - 2$

62. $x^2 + 2x - 3$

63. $x^2 - 3x + 4$

64. $x^2 - 4x - 5$

65. $x^2 + 5x - 6$

66. $x^2 + xy - 6y^2$

67. $a^2 - 2a - 8$

68. $z^2 + 3z - 10$

69. $x^2 - x - 12$

70. $16y^2 + 8y + 1$

71. $5x^2 + 6x + 1$

72. $4x^2 - 6x + 1$

73. $5a^2 - 6ab + b^2$

74. $5t^2 + 8t + 3$

75. $3x^2 + 7x + 2$

76. $3x^2 - 7x + 2$

77. $49b^2 - 28b + 4$

78. $3x^2 - 8x - 3$

79. $2a^2 + ab - b^2$

80. $2w^2 - 3w - 2$

81. $3x^2 + 5x - 2$

82. $4x^2 + 8x + 3$

83. $12x^2 + 11x - 2$

84. $2x^2 - 7x + 6$

85. $6c^2 - 5c + 1$

86. $14x^2 + 15x - 9$

87. $6x^2 + 5x - 6$

88. $4x^2 - 12x + 9$

89. $6x^2 - 23x + 20$

90. $6x^2 - x - 12$

91. $16a^2 - 10ab - 9b^2$

92. $24x^2 - 38x + 15$

93. $24x^2 - 25x + 6$

94. $8t^2 + 22t + 15$

95. $6x^2 - 25x - 25$

96. $2a^2x^2 - ax^2 - 5x^2$

97. $4ax^2 + 16ax + 16a$

98. $10m^4 + 3m^3n - m^2n^2$

99. $a^4 + a^3b - 2a^2b^2$

100. $12p^3 - 9p^2q - 3pq^2$

Challenge Problems

Factor each trinomial completely.

101. $x^4 - 5x^2 + 6$

102. $x^4 + x^2 - 2$

103. $x^4 - 5x^2 + 4$

IN YOUR OWN WORDS...

104. What are the rules that determine the signs in the factors of a trinomial?

105. How do we recognize a perfect square trinomial?

4.4 SUMMARY OF FACTORING

After studying several types of factoring, we can summarize what we have learned by formulating a procedure to factor a polynomial completely.

Procedure to Factor a Polynomial Completely

1. Factor out the greatest common factor (GCF) if there is one. This should be done *first.*
2. Count the number of terms.
3. If there are two terms, check for a difference of squares or cubes or a sum of cubes. Remember, $p^2 + q^2$ *is prime.*
4. If there are three terms, look for two binomial factors.
5. If there are four or more terms, try grouping.
6. Make sure each factor is prime.
7. Check to see if the product of the factors is the original polynomial.

EXAMPLE 1. Factor $50 - 2x^2$ completely.

Solution:

There is a common factor.

$$50 - 2x^2 = 2(25 - x^2)$$

There are two terms in $25 - x^2$. This is the difference of squares.

$$\begin{aligned} 50 - 2x^2 &= 2(25 - x^2) \\ &= 2(5 - x)(5 + x) \end{aligned}$$

Check by multiplying. □

EXAMPLE 2. Factor $t^3 - 1$ completely.

Solution:

There is no common factor. There are two terms, which are the difference of cubes.

$$\begin{aligned} t^3 - 1 &= t^3 - 1^3 \\ &= (t - 1)(t^2 + t + 1) \end{aligned}$$

Check by multiplying. □

EXAMPLE 3. Factor $5(a - b) - t(a - b)$ completely.

Solution:

There are two terms and a common factor of $(a - b)$.

$$5(a - b) - t(a - b) = (a - b)(5 - t)$$

□

EXAMPLE 4. Factor $2x^2 - x - 1$ completely.

Solution:

There is no common factor. There are three terms.

$$2x^2 - x - 1 = (2x + 1)(x - 1)$$

Check by multiplying. □

EXAMPLE 5. Factor $x^4 + x^3 - 2x^2$ completely.

Solution:

There is a common factor.

$$\begin{aligned} x^4 + x^3 - 2x^2 &= x^2(x^2 + x - 2) \\ &= x^2(x + 2)(x - 1) \end{aligned}$$

Check by multiplying. □

EXAMPLE 6. Factor $6r^2 - 2rt + 3rs - st$ completely.

Solution:

There is no common factor and there are four terms. We use grouping.

$$\begin{aligned} 6r^2 - 2rt + 3rs - st &= (6r^2 - 2rt) + (3rs - st) \\ &= 2r(3r - t) + s(3r - t) \\ &= (3r - t)(2r + s) \end{aligned}$$

□

Many times we need to determine if a polynomial is factored. Which of the polynomials below is factored?

$$a^3 - b^3 \qquad (a - b)^3$$

The polynomial $a^3 - b^3$ is not factored, whereas the polynomial $(a - b)^3$ is factored. To factor a polynomial means to write it as a **product** of polynomials. Thus, $a^3 - b^3$ is a difference, *not* a product, but $(a - b)^3$ *is* a product:

$$(a - b)^3 = (a - b)(a - b)(a - b)$$

FACTORED	NOT FACTORED
$(a - b)(a + b)$	$a^2 - b^2$
$(a + b)^2$	$a^2 + 2ab + b^2$
$(a - b)^2$	$a^2 - 2ab + b^2$
$(a - b)(a^2 + ab + b^2)$	$a^3 - b^3$
$(a + b)(a^2 - ab + b^2)$	$a^3 + b^3$
$(a + b)(m + n)$	$m(a + b) + n(a + b)$

PROBLEM SET 4.4

Warm-ups

Factor each polynomial completely if possible.
In Problems 1 through 18, see Examples 1, 2, and 3.

1. $25x^2 - 25$

2. $8w^3 - 27$

3. $9b^2 - x^2$

4. $225a^5b^2c - 90a^2b^2c$

5. $s^3 + 27$

6. $x^2 - x$

7. $4x^2 + 4$

8. $49 - 4s^2$

9. $y(z - 4) + (z - 4)$

10. $x^2 + 4$

11. $8 - 2b^2$

12. $a^2b + c$

13. $16 - 25x^2$

14. $w^3 + 125x^3$

15. $ab(x - y) + c(x - y)$

16. $z^3 + 1$

17. $2000 - 16y^3$

18. $z^2(1 - y) - 4(1 - y)$

In Problems 19 through 28, see Examples 4 and 5.

19. $x^2 + x - 2$

20. $x^2 + x + 1$

21. $3x^2 - 30x + 27$

22. $9r^2 - 21r - 8$

23. $x^2 + 6x + 9$

24. $18a^3 + 36b^3c^3 + 27b^2c^2$

25. $21x^2 - 17x + 2$

26. $y^2 - 6y - 7$

27. $x^2 - 5x - 6$

28. $2x^2 + x + 1$

In Problems 29 through 32, see Example 6.

29. $a^2z + a^2x - z - x$

30. $x^6 + x^5 + x^4 + x^3$

31. $y^2 + yz + az + ay$

32. $2ab + 2ax + by + xy$

Practice Exercises

Determine whether or not each polynomial is factored. If not, factor it completely if possible.

33. $x^2 + 9x + 14$

34. $9 - 16x^2$

35. $135r^3t^5 + 225r^2t^4$

36. $x^2 - 8x + 16$

37. $1 - 27a^3$

38. $x^2 - 64$

39. $ax + ay + x + y$

40. $a^3(x + 1) + b^2(1 + x)$

41. $s^3t(2st + u)$

42. $x^2 + x$

43. $32 - 2x^2$

44. $r^3t(s - 1) + rt(s - 1)$

45. $r^3(t - x) - (t - x)$

46. $6s^2 + st - 15t^2$

47. $z^2 - x^2y^2$

48. $27r^3 + 64)$

49. $a^2b + ab^2 + a + b$

50. $x(r + s) - y(s + r)$

51. $x^2 + 3x + 2$

52. $16x^2 - 16$

53. $8x^3 + y^3$

54. $ax - ay - by + bx$

55. $x^2 + 12x + 36$

56. $x^2 - 6x + 5$

57. $3x^2 + x + 1$

58. $x^4 - x^3 + x^2 - x$

59. $8y^3 + z^3$

60. $2 - 2x^2$

61. $x^2 + x - 20$

62. $x^2 + x - 6$

63. $81 - 18x + x^2$

64. $z(a + b)$

65. $a^3 - 125$

66. $(a + b)^2$

67. $a^7b^5c^4 - a^6b^6c^6 + a^5b^6c^3$

68. $4a^2 - 4b^2$

Challenge Problems

In Problems 69 through 74, factor each polynomial completely.

69. $x^4 - x^2 - 6$

70. $x^4 - 1$

71. $5x^4 + 5x^2 - 100$

72. $x^7 - 8x$

73. $x^6 - 9$

74. $x^9 - 1$

IN YOUR OWN WORDS...

75. What procedure do we follow in factoring a polynomial?

4.5 SOLVING EQUATIONS BY FACTORING

We now look at solving equations by factoring. Remember that an equation is a statement that two numbers are equal. To solve an equation means to find its solution set. The solution set consists of all numbers that will make the equation a true statement.

The two tools that we used to solve first-degree equations, adding the same number to both sides and multiplying both sides by the same nonzero number, are not enough to solve all equations. Our number system has the following property, which provides us with another tool necessary to solve many important kinds of equations.

> **Property of Zero Products**
>
> The statement $AB = 0$ is *true* if either $A = 0$ or $B = 0$ and *false* if neither A nor B is 0.

Let's solve the equation

$$x^2 - x - 6 = 0$$

The left side is a trinomial, which we learned to factor in Section 4.3. So,

$$(x + 2)(x - 3) = 0$$

This is a statement that the product of two real numbers is 0. The Property of Zero Products says that this statement will be true if *either* $(x + 2)$ or $(x - 3)$ is zero. That is,

$$x + 2 = 0 \quad \textit{or} \quad x - 3 = 0$$

$$x = -2 \quad \textit{or} \quad x = 3$$

Thus -2 and 3 are the solutions to the equation $x^2 - x - 6 = 0$. The solution set is $\{-2, 3\}$. We can check -2 and 3 in the original equation.

EXAMPLE 1. Solve $x^2 + x = 2$.

Solution:

We write the equation so that one side is 0 and factor:

$$x^2 + x - 2 = 0$$

$$(x + 2)(x - 1) = 0$$

Apply the Property of Zero Products.

$$x + 2 = 0 \quad \text{or} \quad x - 1 = 0$$

$$x = -2 \quad \text{or} \quad x = 1$$

We write the solution set.

$$\{-2, 1\}$$

□

Once we have factored, it is easy to write the solution set directly.

Solving Equations by Factoring

1. Write the equation so that one side is 0.
2. Factor completely.
3. Set each factor equal to 0 and solve each equation.
4. Check if required.
5. Write the solution set.

EXAMPLE 2. Solve each equation.

(a) $2y^2 - y = 0$ (b) $x^2 - 9 = 0$

Solutions:

(a)

$2y^2 - y = 0$	The right side is 0.
$y(2y - 1) = 0$	Factor.
$y = 0$ or $2y - 1 = 0$	Set factors equal to 0.
$2y = 1$	Solve.
$y = \frac{1}{2}$	
$\left\{0, \frac{1}{2}\right\}$	Write the solution set.

Be Careful!

Notice that 0 is a solution just as $\frac{1}{2}$ is. A common mistake is to disregard the 0.

(b) $x^2 = 9$

$x^2 - 9 = 0$	The right side is 0.
$(x + 3)(x - 3) = 0$	Factor.
$x + 3 = 0$ or $x - 3 = 0$	Set factors equal to 0.
$x = -3$ or $x = 3$	Solve.
$\{-3, 3\}$	Write the solution set.

A common abbreviation for $\{-3, 3\}$ is $\{\pm 3\}$. We read ± 3 as "plus or minus 3." Also, we often write statements such as $x = \pm 3$, which means $x = -3$ *or* $x = 3$. □

EXAMPLE 3. Solve each equation.

(a) $-2x^2 - x + 3 = 0$ (b) $t^2 + 4t + 4 = 0$

Solutions:

(a) It is much easier to factor if the leading coefficient is positive. So we multiply both sides by -1.

$$-1(-2x^2 - x + 3) = -1(0)$$

$$2x^2 + x - 3 = 0$$

$$(2x + 3)(x - 1) = 0$$

$$2x + 3 = 0 \quad \text{or} \quad x - 1 = 0$$

$$2x = -3 \quad \text{or} \quad x = 1$$

$$x = -\frac{3}{2}$$

$$\left\{-\frac{3}{2}, 1\right\}$$

(b) $t^2 + 4t + 4 = 0$

$(t + 2)(t + 2) = 0$

$$\{-2\}$$

The solution set contains just one member because both factors are the same. We call the single solution in such cases a **repeated root.** ▭

The next example points out how important it is to write the equation so that one side is 0. The Property of Zero Products applies only to a product which is equal to *zero.*

EXAMPLE 4. Solve $(x - 2)(x + 2) = 5$.

Solution:

Be careful! The equation is not in the proper form to use the Property of Zero Products. (Why?) We must make one side of the equation zero first. This means that we must multiply $(x - 2)(x + 2)$. So,

$$(x - 2)(x + 2) = 5$$

$$x^2 - 4 = 5$$

$$x^2 - 9 = 0 \qquad \text{The right side is 0.}$$

$$(x + 3)(x - 3) = 0$$

$$\{\pm 3\}, \quad \text{or} \quad \{-3, 3\}$$ ▭

The Property of Zero Products can be extended to products with more than two factors.

EXAMPLE 5. Solve each equation.

(a) $x(x - 3)(x + 5) = 0$ (b) $y^3 - 25y = 0$

Solutions:

(a) $x(x - 3)(x + 5) = 0$

We set each factor equal to 0 and solve.

$$x = 0 \quad \text{or} \quad x - 3 = 0 \quad \text{or} \quad x + 5 = 0$$

$$x = 3 \qquad\qquad x = -5$$

$$\{-5, 0, 3\}$$

(continued)

(b) $y^3 - 25y = 0$

We factor first.

$$y(y^2 - 25) = 0$$

$$y(y + 5)(y - 5) = 0$$

$$\{\pm 5, 0\}$$

□

PROBLEM SET 4.5

Warm-ups

In Problems 1 through 22, write the solution set.

1. $(x - 9)(x + 7) = 0$

2. $y(y - 6) = 0$

3. $(x + 7)(x - 1) = 0$

4. $(x - 6)(x + 6) = 0$

5. $3t(t - 8) = 0$

6. $(x + 8)(x + 8) = 0$

7. $(2x - 5)(x + 3) = 0$

8. $(3z + 7)(2z - 5) = 0$

9. $(6x + 5)(6x - 5) = 0$

10. $(5y + 3)(2y + 9) = 0$

11. $x(7x - 8)(x + 4) = 0$

12. $(5t + 9)(2t - 1) = 0$

13. $(y - 9)^2 = 0$

14. $x^2 = 0$

15. $(2w + 9)(w + 8) = 0$

16. $(x + 2)(5x - 7)(x - 5) = 0$

17. $(x + 1)(x - 1) = 0$

18. $(2x + 3)(2x + 3) = 0$

19. $(4x + 1)(3x - 5) = 0$

20. $(x - \sqrt{2})(x + \sqrt{2}) = 0$

21. $(x + \sqrt{5})(x - \sqrt{5}) = 0$

22. $(x + \sqrt{7})(x - \sqrt{7}) = 0$

In Problems 23 through 42, find the solution set by factoring.
In Problems 23 through 38, see Examples 1, 2, and 3.

23. $x^2 = 1$

24. $x^2 = 4$

25. $x^2 - 64 = 0$

26. $4x^2 - 25 = 0$

27. $9y^2 = 4$

28. $100x^2 = 81$

29. $5x = x^2$

30. $x^2 + x = 0$

31. $x^2 + 3x + 2 = 0$

32. $z^2 + 6z + 5 = 0$

33. $x^2 - 5x - 6 = 0$

34. $w^2 = 8w - 15$

35. $x^2 + 8x + 16 = 0$

36. $y^2 - 8y = 9$

37. $2x^2 - 7x + 6 = 0$

38. $x - 10 = -3x^2$

For Problems 39 through 42, see Example 4.

39. $x(x + 3) = 4$

40. $z(z - 2) = -1$

41. $(x - 6)(x - 2) = -3$

42. $(x - 3)(x + 4) = -6$

For Problems 43 through 46, see Example 5.

43. $x^3 - 4x = 0$

44. $x^3 - x^2 - 6x = 0$

45. $(x - 8)(x + 3)(x - 1) = 0$

46. $3y^3 - 27y = 0$

Practice Exercises

In Problems 47 through 74, find the solution set by factoring.

47. $x^2 = 16$

48. $x^2 = 25$

49. $x^3 - 81x = 0$

50. $4x^2 - 81 = 0$

51. $16y^2 = 25$

52. $25x^2 = 16$

53. $x^2 - 4x = 0$

54. $x^2 = 6x$

55. $3x = x^2$

56. $x^3 - x = 0$

57. $x^2 + 9x + 8 = 0$

58. $z^2 + 11z + 10 = 0$

59. $t^2 + 6t + 8 = 0$

60. $x^2 + 8x + 15 = 0$

61. $x^2 + 5x - 6 = 0$

62. $w^2 = 2w + 15$

63. $x^2 - 6x + 9 = 0$

64. $y^2 - 3y = 4$

65. $3x^2 + 13x + 4 = 0$

66. $2x^2 + 7x + 3 = 0$

67. $2x^2 - x - 15 = 0$

68. $5x + 28 = 3x^2$

69. $6y^2 + y - 2 = 0$

70. $5t^2 = t + 4$

71. $x(x - 7) = -12$

72. $z(z + 1) = 2$

73. $(x - 4)(x + 1) = 14$

74. $(x + 5)(x + 2) = 4$

Challenge Problems

Solve the following equations.

75. $\frac{1}{2}x^2 - 2 = 0$

76. $\frac{1}{4}x^2 = 16$

77. $\frac{1}{3}x^2 - 2x + 3 = 0$

IN YOUR OWN WORDS...

78. How do we solve an equation by factoring?

79. Make up an equation that can be solved by factoring.

4.6 APPLICATIONS

Word problems can lead to mathematical models involving equations that can be solved by factoring. We follow the procedure developed in Section 3.4 to solve them.

A Procedure to Solve Word Problems

1. Read the problem to determine what quantities are to be found.
2. Assign a variable, such as x, to represent one of the quantities to be found.
3. Express all other quantities to be found in terms of x (or the chosen variable).
4. Draw a figure or picture to illustrate the problem if possible. Charts are helpful.
5. Read the problem again and form a mathematical model.
6. Solve the mathematical model.
7. Write an answer to the original question.
8. Check the answer in the original word problem to see if it makes sense and answers the question.

The first type of word problem that we will study deals with *consecutive* integers.

EXAMPLE 1. The product of two consecutive positive integers is 156. Find the integers.

Solution:

Step 2 Assign a variable.

Let n be the first integer.
Then $n + 1$ is the next integer.

Step 3 Form a model.

$n(n + 1) = 156$

Step 6 Solve the equation.

$$n^2 + n = 156$$

$$n^2 + n - 156 = 0 \qquad \text{Make the right side 0.}$$

$$(n + 13)(n - 12) = 0$$

$$n = -13 \quad \text{or} \quad n = 12$$

Since the problem asked for *positive* integers, we cannot use -13.

Step 7 Write an answer to the question.

The integers are 12 and 13.

Step 8 Check the answer.

We check by noting that the product of 12 and 13 is 156. □

The Pythagorean Theorem is one of the most used results in all of mathematics. It deals with a right triangle (a triangle containing an angle of 90°). The side opposite the right angle is called the **hypotenuse.** The hypotenuse is always the longest side of the triangle. The other two sides are often called **legs.**

The Pythagorean Theorem

In a right triangle, let a and b be the lengths of the legs and c be the length of the hypotenuse.

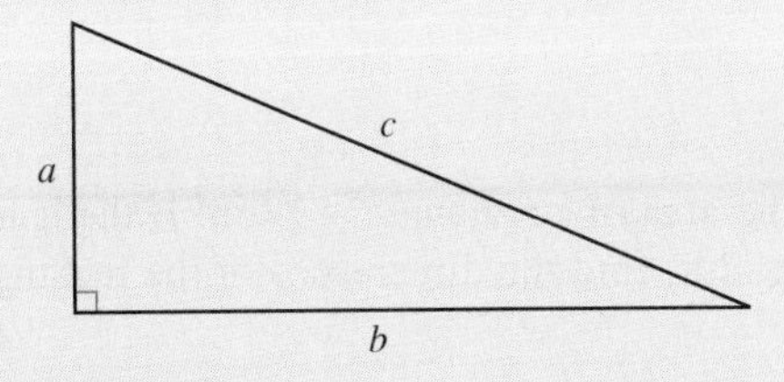

Then

$$a^2 + b^2 = c^2$$

EXAMPLE 2. The hypotenuse of a right triangle is 13 m. Find the lengths of the other two sides if one of them is 7 m longer than the other.

Solution:

Assign a variable.

Let s be the length of the shorter side (in meters).
Then $s + 7$ is the length of the other side (in meters).

Draw a figure.

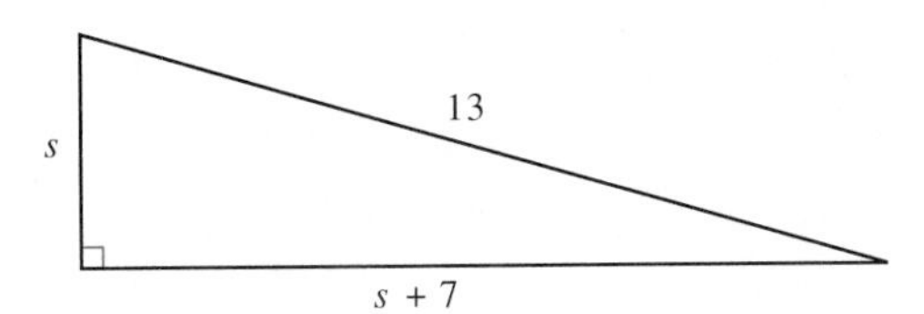

Form a mathematical model.

$$s^2 + (s + 7)^2 = 13^2 \qquad \text{Pythagorean Theorem}$$

Solve the equation.

$$s^2 + s^2 + 14s + 49 = 169$$

$$2s^2 + 14s - 120 = 0$$

$$s^2 + 7s - 60 = 0 \qquad \text{Divide by 2.}$$

$$(s + 12)(s - 5) = 0$$

$$s = -12 \quad \text{or} \quad s = 5$$

(continued)

Since s represents the length of a side, it cannot be negative. So, we cannot use -12 as an answer.

Thus $s = 5$ and $s + 7 = 12$.

Answer the question.

The sides are 5 m and 12 m.

Check the answer.

$$5^2 + 12^2 = 25 + 144 = 169$$

$$13^2 = 169$$

So, $5^2 + 12^2 = 13^2$. □

EXAMPLE 3. The area of a rectangle is 24 in.^{2} If the length of the rectangle is 2 in. more than the width, find the dimensions of the rectangle.

Solution:

Assign a variable.

Let w be the width of the rectangle (in inches).

Then $w + 2$ is the length of the rectangle (in inches).

Draw a picture.

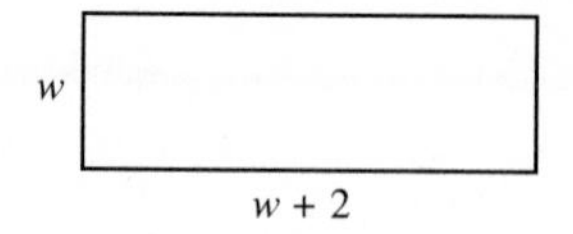

Form a model.

The area of a rectangle is the product of its length and width. So,

$$24 = w(w + 2)$$

Solve the equation.

$$24 = w^2 + 2w$$

$$0 = w^2 + 2w - 24$$

$$0 = (w + 6)(w - 4)$$

$$w = -6 \quad \text{or} \quad w = 4$$

The width of a rectangle cannot be -6. So $w = 4$ and $w + 2 = 6$.

Answer the question.

The dimensions are 6 in. by 4 in.

Check the answer.

The area of a rectangle with dimensions of 4 in. by 6 in. is 24 in.^{2} and we note that 6 is 2 more than 4. □

EXAMPLE 4. The base of a triangle is 6 less than twice the height, and its area is 54 m^2. Find the length of the base and the height of the triangle.

Solution:

Assign a variable.

Let h be the length of the height (in meters).

The problem says that the base is 6 less than twice the height.

So, $2h - 6$ is the length of the base (in meters).

Draw a picture.

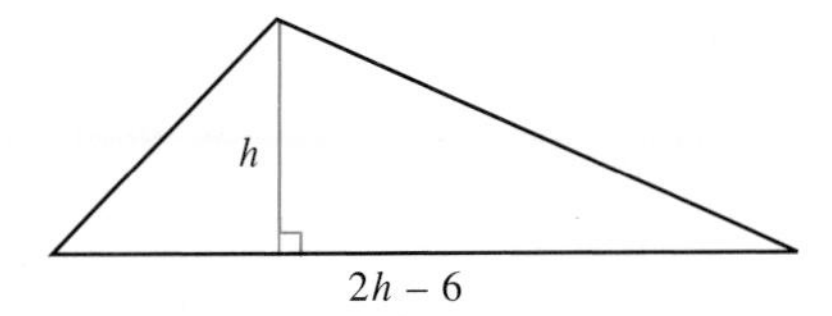

Form a model.

The formula for the area of a triangle is $A = \frac{1}{2}bh$.

$$54 = \frac{1}{2}(2h - 6)h$$

Solve the equation.

$$54 = (h - 3)h \qquad \text{Distributive Property}$$

$$54 = h^2 - 3h$$

$$0 = h^2 - 3h - 54$$

$$0 = (h - 9)(h + 6)$$

$$h = 9 \quad \text{or} \quad h = -6$$

We cannot use -6 as a length. So $h = 9$ and $2h - 6 = 12$.

Answer the question.

The base is 12 m and the height is 9 m.

Check the answer.

We check by noting that the base is 6 less than twice the height and that the area is $\frac{1}{2} \cdot 12 \cdot 9$ m^2, which is 54 m^2. □

PROBLEM SET 4.6

Warm-ups

In Problems 1 through 5, see Example 1.

1. Find two consecutive positive integers whose product is 72.
2. One positive number is 4 more than another positive number and the product of the two is 45. Find the numbers.
3. The low temperature was below zero both nights of the biology trip. Mrs. Carroll, the biology teacher, observed that the lows were consecutive odd integers and that the product of the two temperatures was 63. What were the low temperatures?

4. Lendl has 5 more aces than double faults in a tennis match. The product of the number of aces and the number of double faults is 24. How many aces does he have?

5. The product of two consecutive positive integers is 5 more than the sum of the two integers. Find the integers.

In Problems 6 through 11, see Example 2.

6. The hypotenuse of a right triangle is 10 in. If the shorter leg is two inches less than the other leg, find the lengths of both legs.
7. A hot air balloon rises 500 feet straight up from a grassy spot in the park. If the wind blows it east for 1200 feet, what is the distance between the grassy spot and the balloon?
8. The hypotenuse of a right triangle is 10 m and one of the legs is 6 m. Find the length of the other leg.
9. If one leg of a right triangle is 2 less than the hypotenuse and the other leg is one less than the hypotenuse, find the length of each side of the triangle.
10. The Ace Hardware Store is having a grand opening. John Taylor has several strings of banners which are each 17 feet long. He wants to extend them from the front of the building to the ground. The distance from the building to the point of attachment on the ground must be 7 feet less than the height of the point of attachment on the building. How high up should he attach them to the building?
11. The hypotenuse of a right triangle is 6 more than twice the shorter leg and the longer leg is 4 more than twice the shorter leg. Find the length of each side of the triangle.

In Problems 12 through 15, see Example 3.

12. The area of a rectangle is 108 m^2. If the width is 3 m less than the length, find the dimensions of the rectangle.
13. The Old Lady who lived in a shoe had her children help her make a small rectangular flower bed. The children decided that the length of the bed should be 1 yd more than twice the width of the bed. The Old Lady said the area of the bed should be 1 square yard. What are the dimensions of their flower bed?
14. The width of a rectangle is 10 cm less than its length. If the area of the rectangle is 375 cm^2, find the dimensions of the rectangle.
15. If the length of a rectangle is three times its width and its area is 48 in^2, find the dimensions of the rectangle.

In Problems 16 through 20, see Example 4.

16. The length of the base of a triangle is twice the length of the height. If the area of the triangle is 25 ft^2, find the lengths of the base and the height.
17. Jean Hess is a famous quilt maker in Louisiana. The latest design that she is using calls for triangles whose heights are four times their bases and whose areas are 32 square inches. What is the length of the base and the length of the height for her triangles?
18. One leg of a right triangle is twice the other leg and its area is 16 in^2. Find the length of each leg.
19. If the length of the base of a triangle is 2 m more than twice the length of the height and its area is 42 m^2, find the lengths of the base and the height.
20. The area of the following figure is 230 square units. Find x.

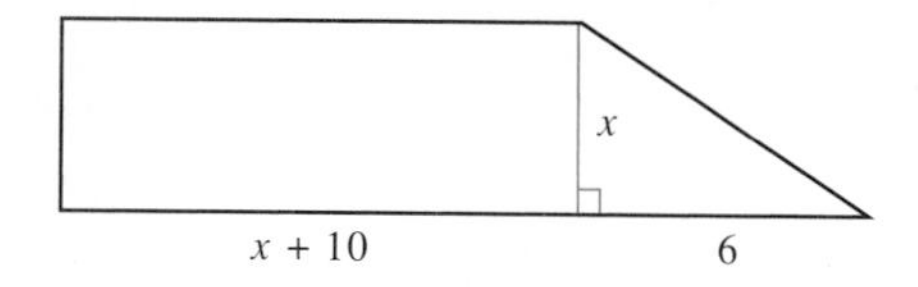

Practice Exercises

21. Dr. and Ms. Phone have two children, Mega and Xyla, born in consecutive years. If the product of the children's ages is 56, how old is each child?
22. One positive number is seven more than another positive number and the product of the two is 228. Find the numbers.

23. Today's weather summary showed that the noon temperature was below zero and was 3° higher than the 6:00 am temperature. If the product of the two temperatures is 88, what was the temperature at noon?
24. The product of two positive integers is 36. If one integer is 5 more than the other, find the integers.
25. The product of two consecutive positive integers is 1 more than the sum of the two integers. Find the integers.
26. If the hypotenuse of a right triangle is 15 in. and one leg is 12 in. find the length of the other leg.
27. Priscilla is making a string-art project. Her masterpiece is rectangular with dimensions of 8 inches by 15 inches. The design requires an expensive gold string to run across the diagonals of the rectangle. How much gold string does Priscilla need?
28. The hypotenuse of a right triangle is 20 in. If the shorter leg is 4 less than the other leg, find the lengths of both legs.
29. If one leg of a right triangle is 2 less than the hypotenuse and the other leg is 4 less than the hypotenuse, find the length of each side of the triangle.
30. The hypotenuse of a right triangle is 5 in. Find the lengths of the legs if one leg is 1 less than the other.
31. The hypotenuse of a right triangle is 2 more than twice the shorter leg and the longer leg is 2 less than twice the shorter leg. Find the length of each side of the triangle.
32. The length of the hypotenuse of a right triangle is 5 m and one leg has length 3 m. Find the length of the other leg.
33. A helicopter is searching for wreckage from a plane crash in the Atlantic Ocean. A rectangular area of 112 square miles has been designated as the main target area. If the width of the target area is 6 miles less than the length of the target area, what is the size of the target area?
34. The length of a rectangle is 1 less than three times its width. If the area of the rectangle is 2 yd^2, find the dimensions of the rectangle.
35. The width of a rectangle is 2 cm less than its length. If the area of the rectangle is 195 cm^2, find the dimensions of the rectangle.
36. If the length of a rectangle is five times its width and its area is 125 in^2, find the dimensions of the rectangle.
37. The length of the base of a triangle is four times the length of the height. If the area of the triangle is 200 ft^2, find the lengths of the base and the height.
38. If the length of the base of a triangle is 2 less than four times the length of the height and its area is 45 m^2, find the lengths of the base and the height.
39. The area of the following figure is x^2 square units. Find x.

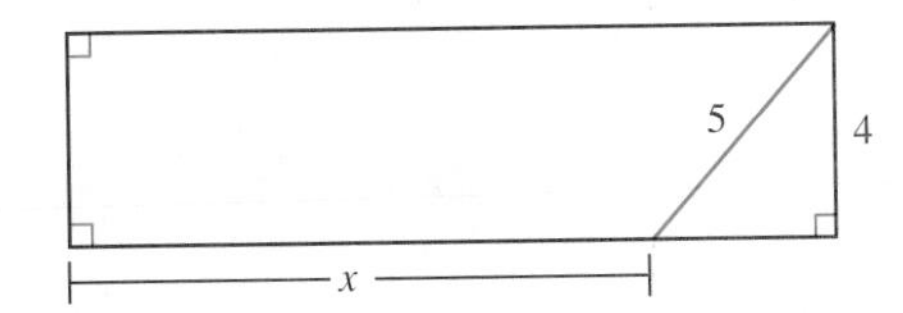

Challenge Problems

40. Find all numbers such that the cube of the number is equal to four times the number.
41. The Legg children, Beau and Peg, were born 5 years apart. Ms. Legg was born in 1942 and is 4 years older than her husband. The product of the children's age is 150. How old are Peg and Beau?

IN YOUR OWN WORDS...

42. Make up a word problem with an answer of, "The numbers are -12 and 14."

CHAPTER SUMMARY

GLOSSARY

Factor a polynomial: To write the polynomial as a product.

Prime polynomial: A polynomial that has no factors other than itself and 1.

Factor a polynomial completely: To write the polynomial as a product of polynomials which are prime.

Greatest common factor of a set of numbers: The largest number that is a factor of each number in the set.

Perfect trinomial square: A trinomial which will factor into a binomial squared.

PROCEDURE TO FACTOR A POLYNOMIAL COMPLETELY

1. Factor out the GCF if there is one. Do this *first.*
2. Count the number of terms.
3. If there are two terms, try difference of squares, difference of cubes, or sum of cubes.

   Note! $p^2 + q^2$ is prime.
4. If there are three terms, look for two binomial factors.
5. If there are four or more terms, try grouping.
6. Make sure each factor is prime.
7. Check to see if the product of the factors is the original polynomial.

FACTORING RULES

1. $p^2 - q^2 = (p - q)(p + q)$ Difference of squares
2. $p^3 - q^3 = (p - q)(p^2 + pq + q^2)$ Difference of cubes
3. $p^3 + q^3 = (p + q)(p^2 - pq + q^2)$ Sum of cubes
4. $p^2 + 2pq + q^2 = (p + q)^2$ Perfect trinomial square
5. $p^2 - 2pq + q^2 = (p - q)^2$ Perfect trinomial square

PROCEDURE TO SOLVE AN EQUATION BY FACTORING

1. Write the equation so that one side is 0.
2. Factor completely.
3. Set each factor equal to zero and solve.
4. Check, if required.
5. Write the solution set.

CHECKUPS

1. Factor $18y - 12$ completely. Section 4.1; Example 1a
2. Factor $x(a - b) + y(a - b)$. Section 4.1; Example 6a
3. Factor $x^2 - xy + ax - ay$ by grouping. Section 4.1; Example 7b
4. Factor $16x^2 - 25$. Section 4.2; Example 1b
5. Factor $8c^3 - 27$. Section 4.2; Example 3b
6. Factor $x^2 - 9x + 18$. Section 4.3; Example 1d
7. Factor $2a^3 - 2a^2b - 6ab^2$ completely. Section 4.3; Example 4

8. Solve $x^2 + x = 2$. Section 4.5; Example 1

9. Solve $(x - 2)(x + 2) = 5$. Section 4.5; Example 4

10. The product of two consecutive positive integers is 156. Find the integers. Section 4.6; Example 1

11. The hypotenuse of a right triangle is 13 m. Find the lengths of the other sides if one of them is 7 m longer than the other. Section 4.6; Example 2

REVIEW PROBLEMS

Determine whether each polynomial is factored completely. If not, factor each polynomial completely, if possible.

1. $z^2 - 16t^2$
2. $s(b + k) - t(b + k)$
3. $r^2 + 16$
4. $3(5x + 10)$
5. $2x + 2t + xt + t^2$
6. $k^2 - 11k + 10$
7. $8u^3 - v^3$
8. $(m + n)^2$
9. $a^2b^3c^4 - a^3b^2c^3 + a^2b^3c^5$
10. $27 + 8q^3$
11. $12d + 4$
12. $x^2 - 2x - 2$
13. $az^2 + bz^2 - 5a - 5b$
14. $5t^2 + 5t - 30$
15. $(2x - 6)(x + 3)$
16. $100 - d^2$
17. $x^2 + 6x + 9$
18. $27t^2 - 15t - 2$
19. $x^3y^3 - 64$
20. $36x^4 - 27x^6 + 54x^9$
21. $(a + b)^3$
22. $2a(x + y^2) + b(x + y^2)$
23. $x^2 + 9x + 14$
24. $4r^2 - 4s^2$
25. $k^3 - 125j^3$
26. $-3xy^2 + 6y^2$
27. $k^2 - 2k^2n + 4 - 8n$
28. $4x^2 + x - 14$
29. $2b^3 - 5b^2 - 3b$
30. $81x^2 - 64y^2$

In Problems 31 through 40, solve each equation.

31. $(x - 7)(x - 3) = 0$
32. $x(x + 6) = 0$
33. $(z - \sqrt{15})(z + \sqrt{15}) = 0$
34. $w^2 = 144$
35. $x^2 = 3x$
36. $x^2 - 3x - 40 = 0$
37. $x^2 - 3x = 10$
38. $y^2 + 8y + 16 = 0$
39. $3t^2 - 48 = 0$
40. $x^3 - 2x^2 - 8x = 0$

In Problems 41 through 44, solve each word problem.

41. Find two consecutive negative integers whose product is 506.
42. One leg of a right triangle is 3 units less than the length of the hypotenuse and the other leg is 24 units less than the hypotenuse. Find the length of the hypotenuse.
43. The length of a rectangle is 5 units more than the width of the rectangle. If the area of the rectangle is 176 square units, find the dimensions of the rectangle.
44. One leg of a right triangle is 6 m less than the other leg. If the area of the triangle is 108 m², find the lengths of both legs.

...LET'S NOT FORGET...

In Problems 45 through 50, perform the operations indicated.

45. $(-x)^2$

46. $\frac{0}{5}$

47. $\frac{6t^3 + t^2}{t^2}$

48. $(y - 1)^2$

49. $(x^2 - 2) - (x^2 - x + 3)$

50. -3^2

In Problems 51 through 54, simplify each expression. Watch the role of the parentheses!

51. $x + 2(x - 1)$

52. $(x + 2)(x - 1)$

53. $(x + 2)x - 1$

In Problems 54 through 56, determine if 5 is a factor or a term.

54. $\sqrt{3} + 5$

55. $5(x + 2)$

56. $\frac{1}{x + 5}$

In Problems 57 through 59, multiply each expression by $\frac{2}{3}$.

57. $\frac{3}{2}x$

58. $6(x + 7)$

59. $9y + 12$

In Problems 60 through 63, label each as an expression, equation, or inequality. Solve the equations and inequalities and simplify the expressions.

60. $(x - 2)(x + 3)$

61. $(x - 2)(x + 3) = 0$

62. $5(x + 3) - 2 < -7$

63. $5(x + 3) - 2$

In Problems 64 through 67, identify the expressions that are factored. Factor those which are not factored if possible.

64. $ay + y$

65. $x^3 + 1$

66. $(x + 1)^3$

67. $4(b + 2)$

CHAPTER 4 TEST

In Problems 1 through 5, choose the correct answer.

1. The GCF of $48x^2y$ and $80xy$ is (?).
A. 16 B. $16xy$
C. $8xy$ D. xy

2. One of the factors of $x^2 - 64$ is (?).
A. $x^2 - 8x + 16$ B. $x^2 + 8x + 16$
C. $x + 8$ D. $x + 4$

3. One of the factors of $x(a - 1) - y(a - 1)$ is (?).
A. $x + y$ B. $x - y$
C. x D. $x(a - 1)$

4. $x^2 - 12x + 36 =$ (?).
A. $(x - 9)(x - 4)$ B. $(x - 6)(x + 6)$
C. $(x + 6)^2$ D. $(x - 6)^2$

5. Which polynomial is factored?
A. $x^2 + 2x$ B. $a(r - s) + b(r - s)$
C. $x^3 - 8$ D. $(x + 2)^3$

In Problems 6 through 15, factor each polynomial completely.

6. $8a^3 - 27$

7. $2x^2 + 9x - 5$

8. $ax - bx + 2a - 2b$

9. $x(p - q) - y(p - q)$

10. $a^2 - 2a - 8$

11. $x^2 + 14x + 49$

12. $-6c^2d - 3cd^2$

13. $4y^2 - 25$

14. $27a^2b^3 - 36a^3b^2 + 45a^4b^4$

15. $t^2 + 25$

In Problems 16 through 18, solve each equation.

16. $x^2 - 81 = 0$

17. $x^2 - 4x = 12$

18. $x(x - 1)(x + 3) = 0$

19. The area of a rectangle is 234 yd^2. If the width is 5 yd less than the length, find the dimensions of the rectangle.

20. The hypotenuse of a right triangle is 2 m less than three times the shortest side. If the other side is 2 m more than twice the shortest side, what are the lengths of the three sides?

Rational Expressions

CONNECTIONS

The radio station, WKIX, often conducts unusual contests offering cash prizes. In one such contest, a secret 7 digit integer is to be guessed by listeners who call in. The contest lasts two weeks, daily guesses are taken and hints given. The prize for the first correct guess is \$1 on the first day, \$2 on the second day, \$4 on the third day, continuing to double each day that the number remains unguessed. The winner is to be awarded the sum of the daily prizes to the day of his correct guess, or \$1,000 whichever he chooses. Suppose you guess correctly on the 10th day of the contest and have 15 seconds to choose between \$1,000 or the sum of the first 10 days prizes. What do you choose?

A mathematician would think of this as a polynomial $1 + x + x^2 + x^3 + \ldots + x^9$. This polynomial can be written as the fraction $\frac{(x^{10} - 1)}{(x - 1)}$. Replace x with 2 and you can see that your prize would be \$1,023 if you select the sum of the daily prizes.

We studied polynomials in Chapter 2 and Chapter 4. In this chapter we study the quotient of two polynomials which is called a rational expression. Rational expressions are algebraic fractions. We see that performing operations with rational expressions is just like working with fractions in arithmetic. Being successful with rational expressions is strongly connected to being successful with factoring polynomials.

5.1 RATIONAL EXPRESSIONS

A **rational expression** is a fraction that has a polynomial for its numerator and a polynomial for its denominator. Three examples of rational expressions are

$$\frac{x^3 + 2x^2 - 17}{x^2 + 5x + 1} \qquad \frac{11}{x^3}, \quad \text{and} \quad \frac{16x^7}{3(x + 1)}$$

The value of a rational expression depends upon the value of the polynomials.

EXAMPLE 1. Find the value of $\frac{x + 3}{2x^2 + 1}$ when:

(a) x is 2 (b) x is -1

Solutions:

(a) Replace x with 2.

$$\frac{x + 3}{2x^2 + 1} = \frac{2 + 3}{2(2)^2 + 1}$$

$$= \frac{5}{8 + 1}$$

$$= \frac{5}{9}$$

The value of $\frac{x + 3}{2x^2 + 1}$ when x is 2 is $\frac{5}{9}$.

We also use the Fundamental Principle to build up fractions. For example, suppose we wish to find a fraction equal in value to $\frac{3}{4}$ but with 28 as its denominator. We use the Fundamental Principle to rewrite such a fraction.

$$\frac{3}{4} = \frac{3(7)}{4(7)} = \frac{21}{28}$$

In this case, we used the Fundamental Principle to multiply *both* the numerator and the denominator by the same (nonzero) number, 7.

We can reduce the rational expression $\frac{5x}{7x}$ by using the Fundamental Principle because x is a *factor* of both the numerator and the denominator of the fraction.

$$\frac{5x}{7x} = \frac{5}{7} \qquad \text{Fundamental Principle}$$

The rational expression $\frac{5+x}{7+x}$ *cannot* be reduced, since the Fundamental Principle applies only to *factors*.

$$\frac{5+x}{7+x} \qquad \text{Does } \textit{not} \text{ reduce.}$$

Be Careful!

We *must* master this idea before we continue.

EXAMPLE 1. Use the Fundamental Principle to reduce each of the following rational expressions to lowest terms.

(a) $\frac{-6x^2}{10x^3}$ (b) $\frac{x^2+2y}{x^2+4y}$

Solutions:

(a) We notice that $2x^2$ is a common *factor* of both the numerator and denominator of this fraction. We can use the Fundamental Principle.

$$\frac{-6x^2}{10x^3} = \frac{-3 \cdot 2x^2}{5x \cdot 2x^2} = \frac{-3}{5x}$$

(b) We *must* resist the temptation to reduce this fraction until we factor it. However, neither the numerator nor the denominator will factor, so the fraction *cannot be reduced.*

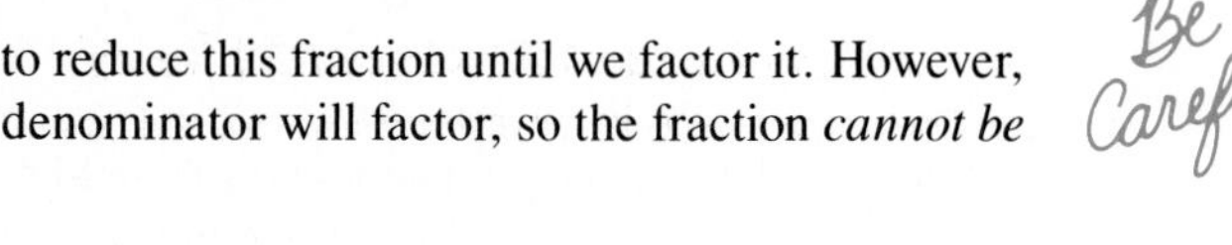

$$\frac{x^2+2y}{x^2+4y} \qquad \text{Does not reduce.}$$

□

EXAMPLE 2. Use the Fundamental Principle to reduce each of the following rational expressions to lowest terms.

(a) $\frac{3x-3y}{x-y}$ (b) $\frac{3x+9}{x^2-9}$

Solutions:

(a) Because the Fundamental Principle applies only to *factors,* we must factor the numerator and denominator completely before applying it.

$$\frac{3x-3y}{x-y} = \frac{3(x-y)}{x-y} \qquad \text{Factor the numerator.}$$

(continued)

$$= \frac{3(x-y)}{1(x-y)}$$

$$= \frac{3}{1} = 3 \qquad \text{Reduce by using the Fundamental Principle.}$$

(b) Again, the Fundamental Principle *cannot* be used until numerator and denominator are *factored*.

$$\frac{3x+9}{x^2-9} = \frac{3(x+3)}{(x-3)(x+3)} \qquad \text{Factor.}$$

$$= \frac{3}{x-3} \qquad \text{Reduce by using the Fundamental Principle.}$$

□

EXAMPLE 3. Use the Fundamental Principle to reduce $\frac{4x^2-9}{2x^2+x-6}$ to lowest terms.

Solution:

The numerator and denominator must be factored. Notice that the numerator is the difference of two squares and the denominator is a trinomial.

$$\frac{4x^2-9}{2x^2+x-6} = \frac{(2x+3)(2x-3)}{(x+2)(2x-3)} \qquad \text{Factor.}$$

$$= \frac{2x+3}{x+2} \qquad \text{Reduce by using the Fundamental Principle.}$$

□

EXAMPLE 4. Use the Fundamental Principle to reduce $\frac{x^2-1}{1-x}$ to lowest terms.

Solution:

$$\frac{x^2-1}{1-x} = \frac{(x+1)(x-1)}{1-x} \qquad \text{Factor the numerator.}$$

There would be a common factor and we could reduce this fraction if the denominator were $x - 1$ instead of $1 - x$. Is there some relationship between them? Recall that $(x - 1)$ and $(1 - x)$ are opposites. That is,

$$(1-x) = -(x-1)$$

Notice what happens if we *replace* $1 - x$ in the denominator with $-(x - 1)$.

$$\frac{(x+1)(x-1)}{1-x} = \frac{(x+1)(x-1)}{-(x-1)}$$

Now there is a common factor of $(x - 1)$ and the fraction can be reduced.

$$\frac{(x+1)(x-1)}{-(x-1)} = \frac{(x+1)}{-1} = -(x+1)$$

□

Opposites often come up in rational expressions. It is important to recognize opposites.

SOME OPPOSITES		
$x - 2$	and	$2 - x$
$6 - t$	and	$t - 6$
$3y - 8$	and	$8 - 3y$
$x + 1$	and	$-x - 1$

Don't become confused and think that $x + 2$ and $2 + x$ are opposites. The Commutative Property for Addition says that $x + 2$ and $2 + x$ are *equal*.

Reducing Opposites

$$\frac{p - q}{q - p} = \frac{p - q}{-(p - q)} = -1$$

EXAMPLE 5. Use the Fundamental Principle to reduce each of the following rational expressions to lowest terms.

(a) $\dfrac{x^2 - 4}{8 + x^3}$ (b) $\dfrac{ax - bx + 2ay - 2by}{3x + 6y}$

Solutions:

(a) We *must* factor before we give any thought to reducing this expression. We notice that the numerator is the difference of two squares and the denominator is the sum of two cubes.

$$\frac{x^2 - 4}{8 + x^3} = \frac{(x + 2)(x - 2)}{(2 + x)(4 - 2x + x^2)} \qquad \text{Factor.}$$

Now that it is factored, note the common factor, $x + 2$.

$$\frac{x^2 - 4}{8 + x^3} = \frac{x - 2}{x^2 - 2x + 4}$$

(b) The first two terms of the *numerator* have x as a common factor, whereas the last two terms have $2y$ as a common factor.

$$\frac{ax - bx + 2ay - 2by}{3x + 6y} = \frac{x(a - b) + 2y(a - b)}{3(x + 2y)}$$

$$= \frac{(a - b)(x + 2y)}{3(x + 2y)}$$

$$= \frac{a - b}{3}$$

□

When working with rational expressions, we often wish to change the denominator. If we consider the Fundamental Principle the other way around, we see how to do this.

$$\frac{A}{B} = \frac{AC}{BC}$$

Notice that we used the Fundamental Principle to *multiply* the numerator and the denominator of the fraction $\frac{A}{B}$ by the *same* number, C.

EXAMPLE 6. Write $\frac{5}{x}$ with a denominator of $2x^3$.

Solution:

We need to find a number such that,

$$\frac{5}{x} = \frac{5 \cdot ?}{x \cdot ?} = \frac{\square}{2x^3}$$

If we multiply x by $2x^2$, we will have $2x^3$, the desired denominator. So we use the Fundamental Principle to multiply both the numerator and denominator by $2x^2$.

$$\frac{5}{x} = \frac{5(2x^2)}{x(2x^2)} = \frac{10x^2}{2x^3}$$

□

EXAMPLE 7. Write each given rational expresssion as a rational expression with the indicated denominator.

(a) $\frac{x+1}{x-1}$; $x^2 - 1$ (b) $\frac{2}{x+3}$; $x^2 - x - 12$

Solutions:

(a) Since the denominator we *want*, $x^2 - 1$, can be factored as $(x - 1)(x + 1)$, we must multiply the denominator by $(x + 1)$. Of course, we also must multiply the numerator by $(x + 1)$.

$$\frac{x+1}{x-1} = \frac{(x+1)(x+1)}{(x-1)(x+1)} = \frac{x^2 + 2x + 1}{x^2 - 1}$$

(b) If we factor the desired denominator, we see that

$$x^2 - x - 12 = (x + 3)(x - 4)$$

We need to multiply the denominator by $(x - 4)$.

$$\frac{2}{x+3} = \frac{2(x-4)}{(x+3)(x-4)} = \frac{2x-8}{x^2 - x - 12}$$

□

PROBLEM SET 5.2

Warm-ups

In Problems 1 through 39, use the Fundamental Principle, when it applies, to reduce the rational expressions to lowest terms.
In Problems 1 through 15, see Example 1.

1. $\dfrac{4x}{6y}$
2. $\dfrac{6x^2}{9x^3}$
3. $\dfrac{14x^3y^2}{7x^2y^2}$
4. $\dfrac{2y^4}{3y^2}$
5. $\dfrac{3x(x+2)}{6xy}$
6. $\dfrac{-2x^2y}{xy^2}$
7. $\dfrac{(-2x)^2y}{xy^2}$
8. $\dfrac{-n^4}{(-n)^3(3+n)}$
9. $\dfrac{(x+1)^2}{(1+x)^3}$
10. $\dfrac{(y-5)^5}{(y-5)^4}$
11. $\dfrac{xy}{x+y}$
12. $\dfrac{5(2x+3y)}{15y}$
13. $\dfrac{xy^2}{xy}$
14. $\dfrac{x+y^2}{xy}$
15. $\dfrac{xy+y^2}{xy}$

In Problems 16 through 27, see Example 2.

16. $\dfrac{(x-7)(x-1)}{(x+1)(x-7)}$
17. $\dfrac{8(x-3)}{6x-18}$
18. $\dfrac{11-33y}{22(1-3y)}$
19. $\dfrac{7x^3(a+b)}{-14x^2(a+b)^4}$
20. $\dfrac{5x^2(x+2)}{20x+40}$
21. $\dfrac{x^2+3x}{6+2x}$
22. $\dfrac{3xy+9x^2y}{x^2z+3x^3z}$
23. $\dfrac{2(x+4)}{x^2-16}$
24. $\dfrac{1-x^2}{2-2x}$
25. $\dfrac{8x^2z^2-2x^2}{16z+8}$
26. $\dfrac{axyz+txyz}{a^2-t^2}$
27. $\dfrac{x^2-9}{(x-3)^2}$

In Problems 28 through 33, see Example 3.

28. $\dfrac{x^2+3x+2}{3x+6}$
29. $\dfrac{2x^2-6x}{x^2-2x-3}$
30. $\dfrac{x^2-x-2}{x^2-1}$
31. $\dfrac{27x^2-3y^2}{9x^2+6xy+y^2}$
32. $\dfrac{x^2+3x+2}{x^2-x-2}$
33. $\dfrac{x^2-x-6}{x^2+x-2}$

In Problems 34 through 39, see Example 4.

34. $\dfrac{5-7t}{7t-5}$
35. $\dfrac{3t-1}{1-3t}$
36. $\dfrac{2x-6}{3-x}$
37. $\dfrac{10x-6}{3-5x}$
38. $\dfrac{3-2x}{4x^2-9}$
39. $\dfrac{1-x}{x^2-3x+2}$

In Problems 40 through 45, use the Fundamental Principle to reduce the rational expressions to lowest terms. See Example 5.

40. $\dfrac{x+1}{x^3+1}$
41. $\dfrac{3-y}{27-y^3}$
42. $\dfrac{ax+ay+bx+by}{5a+5b}$

43. $\frac{(x-y)^3}{x^3-y^3}$

44. $\frac{3x-y}{3x+3x^2-xy-y}$

45. $\frac{1-8x^3}{1-2x-2y+4xy}$

In Problems 46 through 52, write each rational expression as a fraction with the indicated denominator.
In Problems 46 through 48, see Example 6.

46. $\frac{1}{x}$; $3x^2$

47. $\frac{7}{2x}$; $6x^3$

48. $\frac{2y}{3ax}$; $12a^2x$

In Problems 49 through 52, see Example 7.

49. $\frac{2}{x+1}$; $5(x+1)$

50. $\frac{-6}{x-9}$; x^2-81

51. $\frac{x+1}{x-2}$; x^2-4x+4

52. $\frac{-16}{2x-7}$; $6x^2-7x-49$

Practice Exercises

In Problems 53 through 94, use the Fundamental Principle, when it applies, to reduce the rational expressions to lowest terms.

53. $\frac{8x}{6y}$

54. $\frac{6x^3}{8x^4}$

55. $\frac{16x^5t^3}{8x^3t^4}$

56. $\frac{2t^6}{3t^3}$

57. $\frac{2x(x-3)}{6xy}$

58. $\frac{12x^3y}{xy^4}$

59. $\frac{-2x^2y}{x^3y}$

60. $\frac{(-n)^4}{-n^3(3+n)}$

61. $\frac{(s+5)^3}{(s+5)^5}$

62. $\frac{(y-5)^4}{(y-5)^2}$

63. $\frac{st^2}{t-s}$

64. $\frac{6(3x+2y)}{15y}$

65. $\frac{(x+7)(x+1)}{(x+1)(x-7)}$

66. $\frac{6(x+2)}{8x+16}$

67. $\frac{13+52y}{26(1+4y)}$

68. $\frac{6x^2(a+b)}{-42x^3(a+b)^2}$

69. $\frac{7t^2(t-3)}{21t-63}$

70. $\frac{z^3-3z}{2z^2-6}$

71. $\frac{2st-8s^2t}{s^2t-4s^3t}$

72. $\frac{3(x-4)}{x^2-16}$

73. $\frac{1-x^2}{3+3x}$

74. $\frac{8x^2z^2-2x^2}{8z-4}$

75. $\frac{aklm-tklm}{a^2-t^2}$

76. $\frac{x^2-4}{(x+2)^2}$

77. $\frac{x^2-3x+2}{3x-6}$

78. $\frac{2x^2+6x}{x^2+2x-3}$

79. $\frac{x^2+x-2}{x^2-1}$

80. $\frac{16x^2-4y^2}{4x^2-4xy+y^2}$

81. $\frac{x^2+3x+2}{x^2+x-2}$

82. $\frac{3x-6}{2-x}$

83. $\frac{x^2+x-6}{x^2-x-2}$

84. $\frac{2-3y}{9y^2-4}$

85. $\frac{2-t}{t^2-3t+2}$

86. $\frac{6x^2+13x-15}{2x^2+5x-3}$

87. $\frac{7-3t}{3t-7}$

88. $\frac{6x^3+7x^2y-20xy^2}{6x^2y+23xy^2+20y^3}$

89. $\frac{s-1}{s^3-1}$

90. $\frac{4+x}{64+x^3}$

91. $\frac{ax-ay+bx-by}{7a+7b}$

92. $\dfrac{(x+y)^2}{x^3+y^3}$

93. $\dfrac{25x^2-y^2}{5x+5x^2-xy-y}$

94. $\dfrac{1-4x^2}{1+2x+2y+4xy}$

In Problems 95 through 103, write each rational expression as a fraction with the indicated denominator.

95. $\dfrac{2}{y}$; $5y^2$

96. $\dfrac{16}{3x}$; $6x^3$

97. $\dfrac{2b}{3ax}$; $21a^2x^2$

98. $\dfrac{2}{x-2}$; $5(x-2)$

99. $\dfrac{-x}{x-7}$; x^2-49

100. $\dfrac{x+3}{x-3}$; x^2-9

101. $\dfrac{x-1}{x+2}$; x^2+4x+4

102. $\dfrac{x-5}{2x+7}$; $2x^2-3x-35$

103. $\dfrac{-10}{6x+5}$; $6x^2-13x-15$

Challenge Problems

In Problems 104 through 111, use the Fundamental Principle to reduce the rational expressions to lowest terms.

104. $\dfrac{x^4-1}{x^2+1}$

105. $\dfrac{x^4-16}{x+2}$

106. $\dfrac{2t^3-1}{4t^6-1}$

107. $\dfrac{1-x^4}{x-1}$

108. $\dfrac{s^6-16}{4-s^3}$

109. $\dfrac{6y^4-y^2-35}{6y^5+11y^3-7y}$

110. $\dfrac{ax+bx+ay+by}{2ax-ay+2bx-by}$

111. $\dfrac{pr+qs-qr-ps}{pr-qs-qr+ps}$

IN YOUR OWN WORDS...

112. What does the Fundamental Principle of Rational Expressions allow us to do with a fraction?

113. Explain the steps in reducing $\dfrac{x^3+4x^2}{x^2-16}$ to lowest terms.

5.3 MULTIPLICATION AND DIVISION

Because rational expressions are fractions, we multiply and divide them exactly as we multiply and divide fractions.

Let's multiply $\frac{3}{5}$ times $\frac{5}{9}$ to review the procedure for multiplication of fractions.

$$\frac{3}{5}\cdot\frac{5}{9}=\frac{3(5)}{5(9)}$$ Multiply numerators and multiply denominators.

Instead of actually multiplying, we look for common factors in both the numerator and denominator so that we can reduce. 5 is a factor of both numerator and denominator.

$$\frac{3(5)}{5(9)}=\frac{1}{3}$$ Reduce.

Multiplication of Rational Expressions

If A, B, C and D are polynomials and B and D do not have value 0,

$$\frac{A}{B}\cdot\frac{C}{D}=\frac{A\cdot C}{B\cdot D}$$

EXAMPLE 1. Perform the operations indicated.

(a) $\dfrac{2x}{3y} \cdot \dfrac{5x}{7y^2}$ (b) $\dfrac{3x^2}{10y} \cdot \dfrac{5xy^2}{6}$

Solutions:

(a) $\dfrac{2x}{3y} \cdot \dfrac{5x}{7y^2} = \dfrac{2x(5x)}{3y(7y^2)} = \dfrac{10x^2}{21y^3}$

(b) $\dfrac{3x^2}{10y} \cdot \dfrac{5xy^2}{6} = \dfrac{3x^2(5xy^2)}{10y(6)} = \dfrac{15x^3y^2}{60y}$

We notice that this fraction will reduce. Very often, multiplication of fractions results in a fraction that will reduce. We *must* always look for common factors in the numerator and denominator.

$$\frac{15x^3y^2}{60y} = \frac{15y(x^3y)}{15y(4)} = \frac{x^3y}{4}$$

□

EXAMPLE 2. Perform the operations indicated.

(a) $\dfrac{5x - 10}{x + 3} \cdot \dfrac{2x + 6}{3x - 6}$ (b) $\dfrac{x^2 - 1}{x^2 - 4} \cdot \dfrac{x + 2}{x - 1}$

Solutions:

(a) $\dfrac{5x - 10}{x + 3} \cdot \dfrac{2x + 6}{3x - 6} = \dfrac{(5x - 10)(2x + 6)}{(x + 3)(3x - 6)}$

Notice the parentheses! If we forget the parentheses in this step, we will surely make a mistake in this problem.

Now we wish to simplify the resulting fraction, if possible. The first step is to factor.

$$\frac{5(x - 2)2(x + 3)}{(x + 3)3(x - 2)} = \frac{10}{3} \quad \text{Reduce.}$$

(b) $\dfrac{x^2 - 1}{x^2 - 4} \cdot \dfrac{x + 2}{x - 1} = \dfrac{(x^2 - 1)(x + 2)}{(x^2 - 4)(x - 1)}$

Again note the parentheses. Now we factor to see if there is a common factor, and, if so, we simplify.

$$\frac{(x^2 - 1)(x + 2)}{(x^2 - 4)(x - 1)} = \frac{(x + 1)(x - 1)(x + 2)}{(x + 2)(x - 2)(x - 1)} = \frac{x + 1}{x - 2}$$

□

Procedure for Multiplying Rational Expressions

1. Multiply the numerators to get the new numerator and the denominators to get the new denominator.
2. Simplify the resulting fraction, if possible.

EXAMPLE 3. Perform the operation indicated.

$$\frac{x^2 - 2x + 1}{x^2 - x - 6} \cdot \frac{x^2 - 4}{x^2 - 3x + 2}$$

Solution:

$$\frac{x^2 - 2x + 1}{x^2 - x - 6} \cdot \frac{x^2 - 4}{x^2 - 3x + 2} = \frac{(x^2 - 2x + 1)(x^2 - 4)}{(x^2 - x - 6)(x^2 - 3x + 2)}$$

$$= \frac{(x - 1)^2(x + 2)(x - 2)}{(x - 3)(x + 2)(x - 1)(x - 2)}$$

$$= \frac{x - 1}{x - 3}$$

□

EXAMPLE 4. Perform the operation indicated.

$$\frac{3 + x}{3 - x} \cdot \frac{x^2 - 9}{x + 2}$$

Solution:

$$\frac{3 + x}{3 - x} \cdot \frac{x^2 - 9}{x + 2} = \frac{(3 + x)(x^2 - 9)}{(3 - x)(x + 2)}$$

$$= \frac{(3 + x)(x + 3)(x - 3)}{(3 - x)(x + 2)}$$

Notice that $x - 3$ and $3 - x$ are opposites.

$$= \frac{-(3 + x)(x + 3)}{(x + 2)}$$

Be Careful!

Since $(3 + x)$ *does* equal $(x + 3)$ (by the Commutative Property for Addition), we would write the answer

$$= \frac{-(x + 3)^2}{x + 2}$$

$$= -\frac{(x + 3)^2}{x + 2}$$

□

To divide fractions, multiply by the reciprocal of the divisor. For example, let's divide $\frac{4}{7}$ by $\frac{2}{21}$.

$$\frac{4}{7} \div \frac{2}{21} = \frac{4}{7} \cdot \frac{21}{2}$$ Multiply by reciprocal.

$$= \frac{4(21)}{7(2)}$$ Multiply numerators and multiply denominators.

$$= 6$$ Simplify.

> **Division of Rational Expressions**
>
> If A, B, C, and D are polynomials and B, C, and D do not have value 0,
>
> $$\frac{A}{B} \div \frac{C}{D} = \frac{A}{B} \cdot \frac{D}{C} = \frac{A \cdot D}{B \cdot C}$$

EXAMPLE 5. Perform the division indicated.

$$\frac{x + 1}{x - 1} \div \frac{x + 1}{x - 2}$$

Solution:

Multiply by the reciprocal of the divisor.

$$\frac{x + 1}{x - 1} \div \frac{x + 1}{x - 2} = \frac{x + 1}{x - 1} \cdot \frac{x - 2}{x + 1}$$

$$= \frac{(x + 1)(x - 2)}{(x - 1)(x + 1)}$$

$$= \frac{x - 2}{x - 1} \qquad \text{Simplify.}$$

□

> **Procedure for Dividing Rational Expressions**
>
> 1. Change the problem to multiplication by multiplying by the reciprocal of the divisor.
> 2. Multiply the fractions.
> 3. Simplify the resulting fraction, if possible.

EXAMPLE 6. Perform the operations indicated.

(a) $\dfrac{xy}{x^2 - 4} \div \dfrac{yz}{x^2 + 4x + 4}$ (b) $\dfrac{3x}{5} \div (3x - 6)$

Solutions:

(a) $$\frac{xy}{x^2 - 4} \div \frac{yz}{x^2 + 4x + 4} = \frac{xy}{x^2 - 4} \cdot \frac{x^2 + 4x + 4}{yz}$$

$$= \frac{xy(x^2 + 4x + 4)}{(x^2 - 4)yz}$$

Important! Note the use of parentheses.

Now, factor and simplify.

$$= \frac{xy(x + 2)^2}{(x + 2)(x - 2)yz}$$

$$= \frac{x(x + 2)}{(x - 2)z}$$

(b) The reciprocal of $(3x - 6)$ is $\frac{1}{3x - 6}$, so

$$\frac{3x}{5} \div (3x - 6) = \frac{3x}{5} \cdot \frac{1}{3x - 6}$$

$$= \frac{3x}{5(3x - 6)}$$

$$= \frac{3x}{(5)(3)(x - 2)} \qquad \text{Factor.}$$

$$= \frac{x}{5(x - 2)} \qquad \text{Simplify.}$$

□

EXAMPLE 7. Perform the operation indicated.

$$\frac{6x^3}{2x - 3} \div \frac{9x}{3 - 2x}$$

Solution:

$$\frac{6x^3}{2x - 3} \div \frac{9x}{3 - 2x} = \frac{6x^3}{2x - 3} \cdot \frac{3 - 2x}{9x}$$

$$= \frac{6x^3(3 - 2x)}{(2x - 3)9x}$$

Notice that $3 - 2x$ and $2x - 3$ are opposites.

$$= \frac{-2x^2}{3}$$

□

PROBLEM SET 5.3

Warm-ups

In Problems 1 through 42, perform the operations indicated.
In Problems 1 through 6 see Example 1.

1. $\frac{2x}{3y} \cdot \frac{4x}{7y}$ **2.** $\frac{3a}{5b^2} \cdot \frac{3x}{10b^2}$ **3.** $\frac{11m^3}{7n} \cdot \frac{2m}{13n^2}$

4. $\frac{2x}{5z} \cdot \frac{3z^2}{8}$ 5. $\frac{-2x^2y}{7m^2n^3} \cdot \frac{21mn^4}{2xy^2}$ **6.** $\frac{18}{5x^3} \cdot 25x^4$

In Problems 7 through 14, see Example 2.

7. $\frac{x + 1}{-2x} \cdot \frac{6x^2}{x + 1}$ **8.** $\frac{-4x^3}{y^2 + 2} \cdot \frac{y^2 + 2}{12x}$ **9.** $\frac{x + 5}{x - 5} \cdot \frac{x - 5}{x + 6}$

10. $\frac{4x - 2}{x + 1} \cdot \frac{x - 1}{2x - 1}$ 11. $\frac{3k^2}{2x^3 - 6x} \cdot \frac{5x^4 - 15x^2}{9k}$

12. $\frac{x^2 - 1}{3x + 1} \cdot \frac{3x + 1}{x - 1}$ 13. $\frac{2x - 3}{2x + 3} \cdot \frac{4x^2 - 9}{4x - 6}$

14. $\frac{2 + 6x^2}{3 - 6x} \cdot \frac{5 - 10x}{6 + 18x^2}$

In Problems 15 through 19, see Example 3.

15. $\dfrac{x^2 + 2x + 1}{x - 1} \cdot \dfrac{2x + 1}{x^2 - 1}$

16. $\dfrac{x^2 + 4x + 4}{4x + 2} \cdot \dfrac{4x^2 - 1}{x^2 - 4}$

17. $\dfrac{s^2 + s - 6}{s^2 - 4} \cdot \dfrac{2 + s}{s^2 - s - 6}$

18. $\dfrac{t^2 - 3t - 4}{t + 4} \cdot \dfrac{t^2 + 3t - 4}{t - 4}$

19. $\dfrac{2x^2 + x - 1}{x^2 + 2x - 3} \cdot \dfrac{x^2 + 7x + 12}{2x^2 - 9x + 4}$

In Problems 20 and 21, see Example 4.

20. $\dfrac{x}{x - y} \cdot \dfrac{y - x}{y}$

21. $\dfrac{2 - 3w}{5} \cdot \dfrac{10}{3w - 2}$

In Problems 22 through 27, see Example 5.

22. $\dfrac{2x}{3y} \div \dfrac{3y}{2x}$

23. $\dfrac{3a}{5b^2} \div \dfrac{4x}{7a^2}$

24. $\dfrac{12m^2}{5n^2} \div \dfrac{2n}{-5m}$

25. $\dfrac{2x}{5y} \div \dfrac{3x}{7y}$

26. $\dfrac{5x^2y^3}{6mn^2} \div \dfrac{10x^3y^2}{11mn^3}$

27. $\dfrac{-5x^3}{8y^2} \div 25x^2$

In Problems 28 through 40, see Example 6.

28. $\dfrac{3x^2 - 2x}{13y^3} \div \dfrac{6x - 4}{-39xy}$

29. $\dfrac{12y^2}{7y^2 + 21} \div \dfrac{30y}{5y^3 + 15y}$

30. $\dfrac{3 - 9x^2}{4x + 18x^3} \div \dfrac{4x - 12x^3}{10 + 45x^2}$

31. $\dfrac{x^2 + 2x^3}{-7} \div \dfrac{x^2}{7}$

32. $\dfrac{3x}{4y} \div (3y - 6)$

33. $(x^2 + 1) \div \dfrac{x^2 + 1}{x - 1}$

34. $\dfrac{(x + 1)^2}{x - 1} \div (2x + 2)$

35. $(3x - 9x^2) \div \dfrac{4 - 12x}{3}$

36. $\dfrac{x^2 - 4}{2x - 1} \div \dfrac{2 + x}{x - 2}$

37. $\dfrac{x^2 - 2x + 1}{x + 1} \div \dfrac{x^2 - 1}{2x + 1}$

38. $\dfrac{x^2 - 6x + 9}{3x + 9} \div \dfrac{x^2 - 9}{2x + 6}$

39. $\dfrac{y^2 - 2y - 8}{y + 1} \div \dfrac{y^2 + y - 20}{1 + y}$

40. $\dfrac{6x^2 - x - 2}{x^2 + 4x - 5} \div \dfrac{3x^2 + x - 2}{x^2 + 3x - 10}$

In Problems 41 and 42, see Example 7.

41. $\dfrac{x^2 - 1}{2x + 3} \div \dfrac{1 - x}{4x + 6}$

42. $\dfrac{q^2 - 9}{5 - q} \div \dfrac{q^2 - 2q - 15}{2q - 10}$

Practice Exercises

Perform the operations indicated.

43. $\dfrac{3x}{4y} \cdot \dfrac{5x}{7y}$

44. $\dfrac{2b}{5a^2} \cdot \dfrac{3y}{15a^3}$

45. $\dfrac{9m^2}{5n} \cdot \dfrac{3m}{11n^3}$

46. $\dfrac{3x}{4y} \cdot \dfrac{3y^3}{2}$

47. $\dfrac{2xy^2}{5m^3n^2} \cdot \dfrac{20mn^4}{2x^2y}$

48. $\dfrac{12x^3y^3}{65x^2} \cdot 65x^5$

49. $\dfrac{-3x}{2y} \div \dfrac{2y}{3x}$

50. $\dfrac{2a}{3b^2} \div \dfrac{5x}{7a}$

51. $\dfrac{18x^2}{5y^3} \div 6x$

52. $\dfrac{3x}{4y} \div \dfrac{5x}{7y}$

53. $\dfrac{6x^3y^2}{5mn^3} \div \dfrac{-12x^2y^3}{13mn^3}$

54. $-24x^4y^2 \div \dfrac{16x^2y^2}{13z}$

55. $\dfrac{y-1}{3y} \cdot \dfrac{6y^2}{y-1}$

56. $\dfrac{3x^2}{x^2-5} \cdot \dfrac{x^2-5}{12x}$

57. $\dfrac{x-5}{x-6} \cdot \dfrac{x+5}{x-5}$

58. $\dfrac{6x-3}{x-1} \cdot \dfrac{x+1}{2x-1}$

59. $\dfrac{2k^2}{9x-3x^3} \cdot \dfrac{7x^5-21x^3}{8k}$

60. $\dfrac{3-6x^2}{2+6x} \cdot \dfrac{5+15x}{6-12x^2}$

61. $\dfrac{3x-2}{4y} \div \dfrac{3x-2}{y^3}$

62. $\dfrac{-5m}{2mn-3} \div \dfrac{3mn}{2mn-3}$

63. $\dfrac{3x+2}{3x-2} \div \dfrac{2+3x}{2x+1}$

64. $\dfrac{2s^2-3s}{11t^3} \div \dfrac{4s-6}{33st}$

65. $\dfrac{16w^2}{5w^2-20} \div \dfrac{28w}{3w^3-12w}$

66. $\dfrac{2+8x^2}{3x+18x^3} \div \dfrac{6x+24x^3}{15+90x^2}$

67. $\dfrac{-2x}{5z} \cdot 4z^3$

68. $3(x+4) \cdot \dfrac{-11}{x+4}$

69. $\dfrac{x-2}{2+x} \cdot (x+2)$

70. $2(y+3y^3) \cdot \dfrac{2}{x+3xy^2}$

71. $\dfrac{2x}{5y} \div (2y-2)$

72. $(x^2-4) \div \dfrac{x^2-4}{x-2}$

73. $\dfrac{(r-2)^2}{r+2} \div (4-2r)$

74. $(2x+8x^2) \div \dfrac{4+16x}{3}$

75. $\dfrac{x^2-9}{2x-1} \cdot \dfrac{2x-1}{x+3}$

76. $\dfrac{3x+2}{2x-3} \cdot \dfrac{4x^2-9}{6x+4}$

77. $\dfrac{x^2-1}{2x+1} \div \dfrac{x+1}{x-1}$

78. $\dfrac{2x-3}{2x+3} \div \dfrac{4x^2-9}{4x+6}$

79. $\dfrac{x^2-2x+1}{x+1} \cdot \dfrac{2x+1}{x^2-1}$

80. $\dfrac{x^2-4x+4}{4x-2} \cdot \dfrac{4x^2-1}{x^2-4}$

81. $\dfrac{x^2+2x+1}{x-1} \div \dfrac{x^2-1}{2x-1}$

82. $\dfrac{x^2+6x+9}{3x-9} \div \dfrac{x^2-9}{2x-6}$

83. $\dfrac{y^2-y-6}{y^2-9} \cdot \dfrac{y+3}{y^2+y-6}$

84. $\dfrac{s^2-4s-5}{s-1} \cdot \dfrac{s^2+4s-5}{s+1}$

85. $\dfrac{x^2-4}{4-x} \div \dfrac{x^2-2x-8}{3x-12}$

86. $\dfrac{w+3}{w^2-25} \div \dfrac{w^2+6w+9}{w-5}$

87. $\dfrac{3x^2+x-2}{x^2+2x-3} \cdot \dfrac{x^2+8x+15}{3x^2+13x-10}$

88. $\dfrac{u^2}{u-2v} \cdot \dfrac{2v-u}{v^2}$

89. $\dfrac{6x^2+x-2}{x^2-4x-5} \div \dfrac{3x^2-x-2}{x^2-3x-10}$

90. $\dfrac{x^2-9}{2x+4} \div \dfrac{3-x}{4x+8}$

Challenge Problems

Perform the operations indicated.

91. $\dfrac{x^2-5x-14}{3x^2-2x-1} \div \dfrac{x^2-9x+14}{1+2x-3x^2}$

92. $\dfrac{16-x^2}{x^2+3x+2} \cdot \dfrac{x^2+x-2}{x^2-5x+4}$

93. $\dfrac{3x^2-2x-8}{2x^2-3x-5} \cdot \dfrac{2x^2-11x+15}{3x^2+10x+8} \cdot \dfrac{x^2-6x-7}{4x^2-7x-2}$

94. $\dfrac{6x^2-5x+1}{x^2+x-2} \cdot \dfrac{3x^2+8x+4}{4x^2+8x-5} \div \dfrac{3x^2-7x-6}{3x^2-4x+1}$

IN YOUR OWN WORDS...

95. Explain how to multiply two rational expressions.

96. Explain how to divide two rational expressions.

5.4 ADDITION AND SUBTRACTION

We add and subtract rational expressions exactly as we add and subtract fractions. If necessary, we use the Fundamental Principle of Fractions to adjust the expressions so that they have the same denominator; then we use that denominator and add or subtract the numerators.

Let's add $\frac{2}{7}$ to $\frac{3}{7}$ to review the procedure of adding fractions with the same denominator.

$$\frac{2}{7}+\frac{3}{7}=\frac{2+3}{7}$$
$$=\frac{5}{7}$$

Addition and Subtraction of Rational Expressions

If A, B and C are polynomials and if C does not have value 0,

$$\frac{A}{C}+\frac{B}{C}=\frac{A+B}{C} \quad \text{and} \quad \frac{A}{C}-\frac{B}{C}=\frac{A-B}{C}.$$

EXAMPLE 1. Perform the operations indicated.

(a) $\frac{5}{x+y}+\frac{3}{x+y}$ (b) $\frac{x+1}{x-3}-\frac{x-1}{x-3}$

Solutions:

(a) $\frac{5}{x+y}+\frac{3}{x+y}=\frac{5+3}{x+y}$

$$=\frac{8}{x+y}$$

Be Careful!

(b) It is *important* to understand the use of parentheses in the following example!

$$\frac{x+1}{x-3}-\frac{x-1}{x-3}=\frac{(x+1)-(x-1)}{x-3}$$

We wish to subtract the *entire numerator,* $x-1$, not just x, from $x+1$.

$$\frac{(x+1)-(x-1)}{x-3}=\frac{x+1-x+1}{x-3} \quad \text{Note signs.}$$
$$=\frac{2}{x-3}$$

□

When adding and subtracting rational expressions, we usually reduce answers to lowest terms.

EXAMPLE 2. Perform the operations indicated.

(a) $\dfrac{5}{12xy} + \dfrac{1}{12xy}$ (b) $\dfrac{x^2}{(x+1)(x-3)} - \dfrac{x+6}{(x+1)(x-3)}$

Solutions:

(a) $$\frac{5}{12xy} + \frac{1}{12xy} = \frac{5+1}{12xy}$$

$$= \frac{6}{12xy}$$

$$= \frac{1}{2xy} \qquad \text{Simplify.}$$

(b) Notice the parentheses in this example.

$$\frac{x^2}{(x+1)(x-3)} - \frac{x+6}{(x+1)(x-3)} = \frac{x^2 - (x+6)}{(x+1)(x-3)}$$

Be Careful!

Be careful with signs!

$$= \frac{x^2 - x - 6}{(x+1)(x-3)}$$

If the numerator is factored, this can be simplified.

$$= \frac{(x-3)(x+2)}{(x+1)(x-3)}$$

Now we see there is a common factor in numerator and denominator.

$$= \frac{x+2}{x+1}$$ □

Often we wish to add (or subtract) rational expressions that have different denominators. In that case, we must decide on an appropriate denominator and rewrite each fraction with that denominator. We use the Fundamental Principle of Fractions to rewrite the fractions.

Suppose we were to add the fractions $\frac{1}{12}$ and $\frac{1}{18}$. We need a common denominator. So we inspect 12 and 18 to find the smallest number that is a multiple of each of them. We select 36 because 36 is $2 \cdot 18$ and $3 \cdot 12$. There is no smaller integer that is a multiple of both. Now we can add.

$$\frac{1}{12} + \frac{1}{18} = \frac{1(3)}{12(3)} + \frac{1(2)}{18(2)} \qquad \text{Fundamental Principle of Fractions}$$

$$= \frac{3}{36} + \frac{2}{36} = \frac{5}{36} \qquad \text{Addition of Fractions}$$

How do we find the number 36? First, we factor 12 and 18 completely.

$$12 = 2 \cdot 2 \cdot 3$$

$$18 = 2 \cdot 3 \cdot 3$$

The factors of 36 are $2 \cdot 2 \cdot 3 \cdot 3$. Notice that 2 occurs twice in the factors of the least common denominator and 2 occurs twice in 12. Notice that 3 occurs twice in the least common denominator and 3 occurs twice in 18. This is the key. The least common denominator is made up of the factors of the denominators, each taken the greatest number of times it occurs *in any one denominator.*

Procedure for Finding a Least Common Denominator (LCD)

1. Factor each denominator completely.
2. List all the *different prime factors.*
3. For each factor in Step 2, find the *greatest* number of times it occurs in any *one* denominator.
4. The LCD is the product of the factors found in Step 2, each taken the number of times found in Step 3.

EXAMPLE 3. Find the LCD for the given denominators.

(a) 72 and 48 (b) x^2y^3z and x^4y^2
(c) $7y$ and y (d) $x(x + 2)$ and x
(e) x and $x + 3$ (f) $x + 4$ and $(x + 4)^2$

Solutions:

(a) 72 and 48

Step 1 Factor each denominator completely.

$$72 = 2^3 \cdot 3^2 \qquad 48 = 2^4 \cdot 3$$

Step 2 List all the different prime factors.

The different prime factors are 2 and 3.

Step 3 Determine the greatest number of times each prime factor occurs in any one denominator.

The factor 2 occurs three times in 72 and four times in 48. The greatest number of times that it occurs in either denominator is *four.*

The factor 3 occurs two times in 72 and one time in 48. The greatest number of times that it occurs in either denominator is *two.*

Step 4 Write the LCD.

The LCD contains *four* factors of 2 and *two* factors of 3. The LCD is $2^4 \cdot 3^2$, which is 144.

(b) x^2y^3z and x^4y^2

Factor each denominator.
The denominators are factored.

List the different prime factors.
The different prime factors are x, y, and z.

Count the number of times each prime factor occurs in each denominator.

x occurs as a factor twice in x^2y^3z and four times in x^4y^2. Thus, we need four factors of x.

y occurs three times in x^2y^3z and two times in x^4y^2. So, we need three factors of y.

z occurs one time in x^2y^3z and no times in x^4y^2. Thus, we need one factor of z.

The LCD is x^4y^3z.

(c) $7y$ and y

Both denominators are factored.
The different prime factors are 7 and y.
7 occurs one time in $7y$ and no times in y.
y occurs one time in $7y$ and one time in y.
Thus, we need one 7 and one y.

The LCD is $7y$.

(d) $x(x + 2)$ and x

Both are factored.
The different prime factors are x and $x + 2$.
x occurs in $x(x + 2)$ one time and one time in x. We need one x in the LCD.
$x + 2$ occurs one time in $x(x + 2)$ and no times in x. We need one $x + 2$ in the LCD.

The LCD is $x(x + 2)$.

(e) x and $x + 3$

Both denominators are prime.
The different prime factors are x and $x + 3$.
x occurs one time in x and *no* times in $x + 3$. (The x in $x + 3$ is not a factor.)
$x + 3$ occurs no times in x and one time in $x + 3$.

The LCD is $x(x + 3)$.

(f) $x + 4$ and $(x + 4)^2$

Both denominators are factored.
The prime factor is $x + 4$.
$x + 4$ occurs one time in $x + 4$ and two times in $(x + 4)^2$. So, we need two factors of $x + 4$.

The LCD is $(x + 4)^2$ ☐

EXAMPLE 4. Find the LCD for the given denominators.

(a) $x - y$ and $x^2 - y^2$ (b) $x^2 - 4x + 4$ and $x^2 - 4$

Solutions:

(a) $x - y$ and $x^2 - y^2$

We factor the two denominators completely.

$$x - y = x - y \qquad \text{(It is prime.)}$$
$$x^2 - y^2 = (x + y)(x - y)$$

(continued)

The different factors that occur are $x - y$ and $x + y$. The greatest number of times that either one occurs *in any one denominator* is once. We need one of each.

The LCD is $(x + y)(x - y)$.

(b) $x^2 - 4x + 4$ and $x^2 - 4$

$$x^2 - 4x + 4 = (x - 2)^2$$

$$x^2 - 4 = (x - 2)(x + 2)$$

The different factors are $(x - 2)$ and $(x + 2)$; we need two factors of $(x - 2)$ and one of $(x + 2)$.

The LCD is $(x - 2)^2(x + 2)$. □

To add or subtract rational expressions with different denominators, we must first find the least common denominator. We then use the Fundamental Principle of Rational Expressions to write each fraction with the common denominator.

EXAMPLE 5. Perform the operations indicated.

(a) $\frac{x}{5} - \frac{1}{x}$ (b) $\frac{1}{x} + \frac{1}{x + 3}$

Solutions:

(a) The LCD is $5x$.

$$\frac{x}{5} - \frac{1}{x} = \frac{x \cdot x}{5 \cdot x} - \frac{1 \cdot 5}{x \cdot 5} \quad \text{Fundamental Principle}$$

$$= \frac{x^2}{5x} - \frac{5}{5x}$$

$$= \frac{x^2 - 5}{5x}$$

Notice that this cannot be simplified.

(b) The denominators are x and $x + 3$. Neither factors, and the LCD is $x(x + 3)$. We use the Fundamental Principle to write each of the fractions with a denominator of $x(x + 3)$.

$$\frac{1}{x} + \frac{1}{x + 3} = \frac{1 \cdot (x + 3)}{x \cdot (x + 3)} + \frac{1 \cdot x}{(x + 3) \cdot x}$$

$$= \frac{x + 3}{x(x + 3)} + \frac{x}{x(x + 3)}$$

Now that we have a common denominator, we can add the fractions.

$$\frac{x + 3}{x(x + 3)} + \frac{x}{x(x + 3)} = \frac{x + 3 + x}{x(x + 3)}$$

$$= \frac{2x + 3}{x(x + 3)}$$

□

EXAMPLE 6. Perform the operations indicated.

(a) $\dfrac{3}{2x^2} + \dfrac{4}{x^3}$ (b) $\dfrac{1}{12x^2y} - \dfrac{1}{18xy^2}$

Solutions:

(a) The LCD is $2x^3$.

$$\frac{3}{2x^2} + \frac{4}{x^3} = \frac{3 \cdot x}{2x^2 \cdot x} + \frac{4 \cdot 2}{x^3 \cdot 2} \qquad \text{Fundamental Principle}$$

$$= \frac{3x}{2x^3} + \frac{8}{2x^3}$$

$$= \frac{3x + 8}{2x^3}$$

(b) We factor the two denominators.

$$12x^2y = 2 \cdot 2 \cdot 3 \cdot x^2y$$

$$18xy^2 = 2 \cdot 3 \cdot 3 \cdot xy^2$$

The LCD is $2 \cdot 2 \cdot 3 \cdot 3 \cdot x^2y^2$, or $36x^2y^2$.

$$\frac{1}{12x^2y} - \frac{1}{18xy^2} = \frac{1 \cdot 3y}{12x^2y \cdot 3y} - \frac{1 \cdot 2x}{18xy^2 \cdot 2x}$$

$$= \frac{3y}{36x^2y^2} - \frac{2x}{36x^2y^2}$$

$$= \frac{3y - 2x}{36x^2y^2}$$

□

EXAMPLE 7. Perform the operations indicated.

(a) $\dfrac{x}{x + 2} + \dfrac{4x}{x^2 - 4}$ (b) $\dfrac{1}{x^2 + 2x + 1} - \dfrac{1}{x^2 - 1}$

Solutions:

(a) The denominators are $x + 2$ and $x^2 - 4$. We must factor $x^2 - 4$ to find the LCD.

$$x + 2 \qquad \text{Prime}$$

$$x^2 - 4 = (x + 2)(x - 2)$$

The factors are $x + 2$ and $x - 2$, and the most either one occurs *in any one denominator* is once. Therefore, the LCD is $(x + 2)(x - 2)$.

$$\frac{x}{x + 2} + \frac{4x}{x^2 - 4} = \frac{x}{x + 2} + \frac{4x}{(x + 2)(x - 2)}$$

$$= \frac{x \cdot (x - 2)}{(x + 2)(x - 2)} + \frac{4x}{(x + 2)(x - 2)}$$

(continued)

$$= \frac{x^2 - 2x + 4x}{(x + 2)(x - 2)}$$

$$= \frac{x^2 + 2x}{(x + 2)(x - 2)}$$

$$= \frac{x(x + 2)}{(x + 2)(x - 2)} \qquad \text{Factor the numerator.}$$

$$= \frac{x}{x - 2} \qquad \text{The common factor is } x + 2.$$

(b) We factor the two denominators.

$$x^2 + 2x + 1 = (x + 1)^2$$

$$x^2 - 1 = (x + 1)(x - 1)$$

The LCD is $(x + 1)^2(x - 1)$.

$$\frac{1}{x^2 + 2x + 1} - \frac{1}{x^2 - 1} = \frac{1}{(x + 1)^2} - \frac{1}{(x + 1)(x - 1)}$$

$$= \frac{1 \cdot (x - 1)}{(x + 1)^2(x - 1)} - \frac{1 \cdot (x + 1)}{(x + 1)^2(x - 1)}$$

$$= \frac{(x - 1) - (x + 1)}{(x + 1)^2(x - 1)} \qquad \text{Note parentheses.}$$

$$= \frac{x - 1 - x - 1}{(x + 1)^2(x - 1)} \qquad \text{Note signs.}$$

$$= \frac{-2}{(x + 1)^2(x - 1)}$$

□

EXAMPLE 8. Perform the operations indicated.

(a) $3y^3 + \frac{2}{5y}$ (b) $\frac{2}{3x - 1} - 2$

Solutions:

(a) Think of $3y^3$ as $\frac{3y^3}{1}$. The LCD is $5y$.

$$3y^3 + \frac{2}{5y} = \frac{3y^3(5y)}{1(5y)} + \frac{2}{5y} \qquad \text{Fundamental Principle}$$

$$= \frac{15y^4}{5y} + \frac{2}{5y}$$

$$= \frac{15y^4 + 2}{5y}$$

(b) If we think of the number 2 as the fraction $\frac{2}{1}$, we see that the LCD is $3x - 1$.

$$\frac{2}{3x - 1} - 2 = \frac{2}{3x - 1} - \frac{2}{1}$$

$$= \frac{2}{3x - 1} - \frac{2(3x - 1)}{1(3x - 1)}$$

$$= \frac{2 - 2(3x - 1)}{3x - 1}$$

$$= \frac{2 - 6x + 2}{3x - 1}$$

$$= \frac{4 - 6x}{3x - 1}, \quad \text{or} \quad \frac{2(2 - 3x)}{3x - 1}$$

The numerator will factor, but the fraction cannot be reduced. So, it is not necessary to write the answer in factored form. □

EXAMPLE 9. Perform the operations indicated.

(a) $\frac{x}{2x - 3} + \frac{6}{3 - 2x}$ (b) $\frac{5}{x - 7} - \frac{3}{7 - x}$

Solutions:

(a) The denominators are $2x - 3$ and $3 - 2x$. We should *never forget* that $2x - 3$ and $3 - 2x$ are *opposites* of each other. That is,

$$(3 - 2x) = -(2x - 3)$$

We replace $3 - 2x$ in the second fraction with $-(2x - 3)$ and simplify.

$$\frac{x}{2x - 3} + \frac{6}{3 - 2x} = \frac{x}{2x - 3} + \frac{6}{-(2x - 3)}$$

$$= \frac{x}{2x - 3} + \frac{-6}{2x - 3} \qquad \text{Sign Property of Rational Expressions}$$

$$= \frac{x - 6}{2x - 3}$$

Be Careful!

(b) The denominators are opposites. Replace $7 - x$ with $-(x - 7)$.

$$\frac{5}{x - 7} - \frac{3}{7 - x} = \frac{5}{x - 7} - \frac{3}{-(x - 7)}$$

$$= \frac{5}{x - 7} + \frac{3}{x - 7} \qquad \text{Sign Property of Rational Expressions}$$

$$= \frac{5 + 3}{x - 7}$$

$$= \frac{8}{x - 7}$$

□

In adding and subtracting rational expressions, we must be very careful when simplifying. Consider this example.

Be Careful!

$$\frac{2}{a + b} + \frac{3}{a - b} = \frac{2(a - b)}{(a + b)(a - b)} + \frac{3(a + b)}{(a + b)(a - b)}$$

Why do we not reduce these fractions at this point? Notice that if we did reduce, we would end up where we started. Continuing,

$$\frac{2(a - b)}{(a + b)(a - b)} + \frac{3(a + b)}{(a + b)(a - b)} = \frac{2(a - b) + 3(a + b)}{(a + b)(a - b)}$$

Can we reduce this fraction now? Are there common factors in both numerator and denominator? $(a - b)$ and $(a + b)$ appear in the numerator and in the denominator. However, the numerator is *not* factored. Thus neither $(a - b)$ nor $(a + b)$ is a *factor* of the numerator. We *cannot* simplify this fraction. We continue by using the distributive property.

$$\frac{2(a-b)+3(a+b)}{(a+b)(a-b)} = \frac{2a-2b+3a+3b}{(a+b)(a-b)}$$

$$= \frac{5a+b}{(a+b)(a-b)}$$

PROBLEM SET 5.4

Warm-ups

In Problems 1 through 10, perform the operations indicated and reduce answers to lowest terms. See Examples 1 and 2.

1. $\frac{5}{x} + \frac{8}{x}$

2. $\frac{3}{x+2} + \frac{2x}{2+x}$

3. $\frac{x}{x-1} + \frac{x-2}{x-1}$

4. $\frac{3t}{2-t} + \frac{t-1}{2-t}$

5. $\frac{x}{x-1} - \frac{x+1}{x-1}$

6. $\frac{3t}{2-t} - \frac{t-1}{2-t}$

7. $\frac{x^2}{x-2} - \frac{4}{x-2}$

8. $\frac{2x+3}{x-1} - \frac{x+1}{x-1}$

9. $\frac{x^2+3}{5+x} + \frac{3x-13}{x+5}$

10. $\frac{x^2}{x^2-x-2} + \frac{2x+1}{x^2-x-2}$

In Problems 11 through 22, find the LCD of the given denominators. See Examples 3 and 4.

11. 5 and 7

12. x and $x + 2$

13. 3 and 27

14. x and x^3

15. 60 and 90

16. $6x$ and $9x$

17. x^2 and $x(x - 2)$

18. x^7y^4 and x^2y^5z

19. $5x$ and $x + 5$

20. $x^2 + 4x + 4$ and $x^2 - 4$

21. $x^2 - x - 6$ and $3x + 6$

22. $18x^3y^4$ and $27xy^3$

In Problems 23 through 64, perform the operation indicated and reduce answers to lowest terms.
In Problems 23 through 32, see Example 5.

23. $\frac{5}{x} + \frac{2}{3y}$

24. $\frac{x}{7} + \frac{3}{2x}$

25. $\frac{5}{x} - \frac{3}{2}$

26. $\frac{3}{2x} - \frac{x}{3}$

27. $\frac{3}{x+7} - \frac{1}{x}$

28. $\frac{3}{x-1} + \frac{1}{x}$

29. $\frac{2}{r-2} - \frac{1}{r+2}$

30. $\frac{3}{3z-2} - \frac{1}{z+1}$

31. $\frac{t}{t+1} + \frac{t}{t-1}$

32. $\frac{1}{x+1} + \frac{2}{2x-3}$

In Problems 33 through 38, see Example 6.

33. $\dfrac{1}{x^2} + \dfrac{1}{2x}$

34. $\dfrac{1}{3x^2} + \dfrac{2}{x^3}$

35. $\dfrac{3}{4a^2} - \dfrac{5}{2a}$

36. $\dfrac{7}{6xy} - \dfrac{1}{12y}$

37. $\dfrac{4}{15x^2y} - \dfrac{3}{10xy^2}$

38. $\dfrac{1}{3x^2} + \dfrac{5}{2xy}$

In Problems 39 through 50, see Example 7.

39. $\dfrac{3x}{1 - 5x} + \dfrac{3x}{4(1 - 5x)}$

40. $\dfrac{2z}{z(z + 9)} - \dfrac{11}{z + 9}$

41. $\dfrac{5}{4(k - 6)} - \dfrac{1 - k}{8(k - 6)}$

42. $\dfrac{2x}{(x - 7)^2} - \dfrac{1}{x - 7}$

43. $\dfrac{2y}{(y + 4)^2} - \dfrac{1}{y + 4}$

44. $\dfrac{2}{k - 7} + \dfrac{3}{2k - 14}$

45. $\dfrac{3d}{5d + 10} - \dfrac{2}{6 + 3d}$

46. $\dfrac{x}{x^2 - 1} + \dfrac{1}{1 + x}$

47. $\dfrac{3x}{x^2 - 2x + 1} - \dfrac{1}{x - 1}$

48. $\dfrac{2}{x^2 + 4x + 4} - \dfrac{3}{x^2 - 4}$

49. $\dfrac{c}{c^2 + c - 12} + \dfrac{c}{c^2 - 2c - 3}$

50. $\dfrac{t + 5}{t^2 - 2t - 15} + \dfrac{t}{t^2 - 6t + 5}$

In Problems 51 through 60, see Example 8.

51. $1 + \dfrac{x}{7}$

52. $x + \dfrac{1}{3}$

53. $y - \dfrac{2}{y}$

54. $3x^2 - \dfrac{1}{x}$

55. $1 + \dfrac{2}{x + 3}$

56. $\dfrac{1}{z + 2} + 1$

57. $\dfrac{x}{1 + 2x} - 2$

58. $\dfrac{5}{t - 1} - 3$

59. $\dfrac{x + 1}{x - 1} + 1$

60. $2 - \dfrac{x - 1}{x + 1}$

In Problems 61 through 64 see Example 9.

61. $\dfrac{x}{2x - 1} + \dfrac{1}{1 - 2x}$

62. $\dfrac{3}{t - 14} - \dfrac{3}{14 - t}$

63. $\dfrac{a}{b - 1} + \dfrac{1}{1 - b}$

64. $\dfrac{x}{x - y} + \dfrac{y}{y - x}$

Practice Exercises

In Problems 65 through 73, find the LCD of the given denominators.

65. 8 and 3

66. y and $y - 3$

67. 8 and 2

68. t and t^4

69. 126 and 84

70. $4x$ and $6x$

71. x^2 and $x(x + 3)$

72. $x^2 - 4$ and $x - 2$

73. $2x^2 - 5x - 3$ and $2x^2 + 7x + 3$

In Problems 74 through 103, perform the operation indicated and reduce answers to lowest terms.

74. $\frac{r^2}{r-3} + \frac{r-12}{r-3}$

75. $\frac{2y}{5-y} + \frac{5-3y}{5-y}$

76. $\frac{r}{r-3} - \frac{1+r}{r-3}$

77. $\frac{2y}{5-y} - \frac{1-y}{5-y}$

78. $\frac{x^2-2}{x+1} - \frac{2x+1}{1+x}$

79. $\frac{2t}{2-3t} + \frac{2t}{3(2-3t)}$

80. $\frac{7}{8(x-5)} - \frac{1+x}{6(x-5)}$

81. $\frac{2}{3x^3} + \frac{1}{2xy}$

82. $\frac{3}{16x^2y} - \frac{5}{18xy^2}$

83. $\frac{3}{x-4} - \frac{1}{x}$

84. $\frac{1}{t-1} - \frac{1}{t+1}$

85. $\frac{2}{r+2} + \frac{r}{r-2}$

86. $\frac{3}{m-2} + \frac{3}{m+2}$

87. $\frac{a}{a-3} - \frac{a}{a+2}$

88. $\frac{x}{7} - \frac{2}{28x}$

89. $\frac{1}{6t^2} - \frac{3}{4t^3}$

90. $\frac{5}{x-7} - \frac{7x}{(x-7)^2}$

91. $\frac{1}{y+5} + \frac{y}{(y+5)^2}$

92. $\frac{3}{s-5} - \frac{2}{3s-15}$

93. $\frac{7}{4d+12} - \frac{5}{3d+9}$

94. $\frac{1}{x^2-1} - \frac{x}{x+1}$

95. $\frac{2x}{x^2+2x+1} - \frac{5}{x+1}$

96. $\frac{2}{x^2-6x+9} - \frac{1}{x^2-9}$

97. $\frac{c}{c^2-c-12} - \frac{c}{c^2+2c-3}$

98. $1 - \frac{1}{x-1}$

99. $\frac{2}{x-2} + 1$

100. $\frac{9}{x} - 3$

101. $\frac{3}{x} - 4$

102. $\frac{x}{3x-2} - \frac{1}{2-3x}$

103. $\frac{2}{s-11} + \frac{1}{11-s}$

Challenge Problems

Perform the operations indicated and simplify.

104. $\frac{16}{s^2-10s+21} - \frac{40}{s^2-4s-21} + \frac{3(s+1)}{s^2-9}$

105. $\frac{6}{25+5x+x^2} + \frac{1}{x-5} + \frac{15x}{125-x^3}$

106. $\frac{1}{t-5} - \frac{1}{t^2-10t+25} + \frac{1}{5-t} - \frac{1}{25-t^2}$

IN YOUR OWN WORDS...

107. Explain how to find the least common denominator for 108 and 48.

5.5 COMPLEX FRACTIONS

A fraction that contains a fraction is called a **complex fraction.** For example,

$$\frac{\frac{1}{x+1}}{x+3}, \qquad \frac{4x^2}{\frac{1}{x}-\frac{2}{y}}, \quad \text{and} \quad \frac{\frac{1}{x}+1}{5-\frac{1}{x}}$$

are complex fractions. There are two methods generally used for simplifying complex fractions. The first is to think of the complex fraction as a division problem and then use the rules for dividing fractions. The second method employs the Fundamental Principle of Fractions.

EXAMPLE 1. Simplify $\dfrac{\frac{1}{x+1}}{\frac{2}{(x+1)^2}}$ using both methods.

Solution:

Method 1 We think of the example as a division problem.

$$\frac{\frac{1}{x+1}}{\frac{2}{(x+1)^2}} = \frac{1}{x+1} \div \frac{2}{(x+1)^2}$$

$$= \frac{1}{x+1} \cdot \frac{(x+1)^2}{2}$$

$$= \frac{1 \cdot (x+1)^2}{(x+1) \cdot 2}$$

which simplifies to

$$\frac{x+1}{2}$$

Method 2 In Method 2 we use the Fundamental Principle of Fractions and multiply both numerator and denominator of the original fraction by the LCD of the fractions *within* the numerator and denominator. This will result in a single fraction.

The LCD of $x + 1$ and $(x + 1)^2$ is $(x + 1)^2$.

$$\frac{\frac{1}{x+1}}{\frac{2}{(x+1)^2}} = \frac{\frac{1}{x+1} \cdot (x+1)^2}{\frac{2}{(x+1)^2} \cdot (x+1)^2}$$

$$= \frac{\frac{1 \cdot (x+1)^2}{x+1}}{\frac{2 \cdot (x+1)^2}{(x+1)^2}}$$

(continued)

Both the numerator fraction and the denominator fraction will simplify, yielding

$$\frac{1 \cdot (x + 1)}{2}$$

This is a single fraction, which we write as

$$\frac{x + 1}{2}$$

□

Procedure for Simplifying Complex Fractions

Method 1

1. Write the fraction as a division problem.
2. Perform the division.
3. Reduce the result to lowest terms.

Method 2

1. Find the LCD of all denominators.
2. Multiply numerator and denominator by the LCD.
3. Reduce the result to lowest terms.

EXAMPLE 2. Simplify $\dfrac{\frac{1}{x} + 2}{\frac{3}{x^2} - 1}$ using both methods.

Solution:

Method 1 To divide as we did in Example 1, we must first have a single fraction in the numerator and in the denominator. The *numerator* becomes

$$\frac{1}{x} + 2 = \frac{1}{x} + \frac{2x}{x}$$
$$= \frac{1 + 2x}{x}$$

The *denominator* becomes

$$\frac{3}{x^2} - 1 = \frac{3}{x^2} - \frac{x^2}{x^2}$$
$$= \frac{3 - x^2}{x^2}$$

Therefore, the *fraction* becomes

$$\frac{\frac{1}{x} + 2}{\frac{3}{x^2} - 1} = \frac{\frac{1 + 2x}{x}}{\frac{3 - x^2}{x^2}}$$

Notice that we have a single fraction divided by a single fraction. That is, there is only *one* fraction in the numerator and *one* fraction in the denominator. The complex fraction *must* be in that form to use Method 1. Now we divide.

$$\frac{\frac{1+2x}{x}}{\frac{3-x^2}{x^2}} = \frac{1+2x}{x} \div \frac{3-x^2}{x^2}$$

$$= \frac{1+2x}{x} \cdot \frac{x^2}{3-x^2}$$

$$= \frac{(1+2x)x^2}{x(3-x^2)} \qquad \text{Note parentheses.}$$

x is a common factor of the numerator and the denominator.

$$\frac{(1+2x)x^2}{x(3-x^2)} = \frac{(1+2x)x}{(3-x^2)}$$

Method 2 The LCD is x^2. We multiply numerator and denominator by x^2.

$$\frac{\frac{1}{x}+2}{\frac{3}{x^2}-1} = \frac{\left(\frac{1}{x}+2\right)x^2}{\left(\frac{3}{x^2}-1\right)x^2} \qquad \text{Note parentheses.}$$

We use the distributive property to multiply in the numerator and the denominator.

$$\frac{\left(\frac{1}{x}+2\right)x^2}{\left(\frac{3}{x^2}-1\right)x^2} = \frac{\left(\frac{1}{x}\right)x^2 + 2 \cdot x^2}{\left(\frac{3}{x^2}\right)x^2 - 1 \cdot x^2}$$

$$= \frac{\frac{x^2}{x} + 2x^2}{\frac{3x^2}{x^2} - x^2} = \frac{x+2x^2}{3-x^2}$$

□

A complex fraction may be simplified by either method. However, as Examples 1 and 2 indicated, one method may be better than another for a particular problem.

In general, Method 1 is best if the numerator and denominator each contain a *single* term. Method 2 is preferred if *either* the numerator or the denominator has more than one term.

EXAMPLE 3. Simplify $\dfrac{\frac{1}{x-1}+5}{\frac{1}{x-1}-1}$

Solution:

Since there is more than one term in the numerator (and in the denominator), we choose Method 2.

(continued)

The LCD is $(x - 1)$.

$$\frac{\frac{1}{x-1}+5}{\frac{1}{x-1}-1} = \frac{\left(\frac{1}{x-1}+5\right)(x-1)}{\left(\frac{1}{x-1}-1\right)(x-1)}$$

$$= \frac{\frac{1}{x-1}\cdot(x-1)+5\cdot(x-1)}{\frac{1}{x-1}\cdot(x-1)-1\cdot(x-1)} \qquad \text{Distributive Property}$$

$$= \frac{1+5x-5}{1-x+1}$$

$$= \frac{5x-4}{2-x}$$

□

EXAMPLE 4. Simplify $\dfrac{\frac{1}{x}+\frac{1}{x+1}}{\frac{1}{x}-\frac{1}{x+1}}$

Solution:

There is more than one term in the numerator, so we choose Method 2.

The LCD is $x(x + 1)$.

$$\frac{\frac{1}{x}+\frac{1}{x+1}}{\frac{1}{x}-\frac{1}{x+1}} = \frac{\left(\frac{1}{x}+\frac{1}{x+1}\right)\cdot x(x+1)}{\left(\frac{1}{x}-\frac{1}{x+1}\right)\cdot x(x+1)}$$

$$= \frac{\frac{1}{x}\cdot x(x+1)+\frac{1}{x+1}\cdot x(x+1)}{\frac{1}{x}\cdot x(x+1)-\frac{1}{x+1}\cdot x(x+1)} \qquad \text{Distributive Property}$$

$$= \frac{(x+1)+x}{(x+1)-x}$$

$$= \frac{2x+1}{1} = 2x+1$$

□

PROBLEM SET 5.5

Warm-ups

Simplify each complex fraction. Write answers in lowest terms.
In Problems 1 through 15, see Example 1.

1. $\dfrac{\frac{1}{2}}{\frac{3}{2}}$

2. $\dfrac{\frac{4}{5}}{\frac{4}{3}}$

3. $\dfrac{\frac{2}{3}}{\frac{4}{9}}$

4. $\dfrac{\frac{1}{x}}{\frac{1}{x^2}}$

5. $\dfrac{\frac{2}{y^2}}{\frac{6}{y}}$

6. $\dfrac{\frac{2x^2}{3}}{\frac{4x}{12}}$

7. $\dfrac{\frac{4}{x+1}}{\frac{18}{x+1}}$

8. $\dfrac{\frac{xy}{x+y}}{\frac{x^2}{x+y}}$

9. $\dfrac{\frac{6w^2}{xyz}}{\frac{8w^3}{xyz}}$

10. $\dfrac{\frac{x+1}{x}}{\frac{x-1}{x}}$

11. $\dfrac{\frac{3x^2}{y^2-3}}{\frac{12x}{y^2-3}}$

12. $\dfrac{\frac{2-3x}{2x}}{\frac{3-4x}{3x}}$

13. $\dfrac{\frac{35x^2}{24}}{\frac{49x}{12}}$

14. $\dfrac{\frac{1}{6z^3}}{\frac{z+1}{9z^2}}$

15. $\dfrac{\frac{2}{x-1}}{\frac{1}{x+1}}$

In Problems 16 through 21, see Example 2.

16. $\dfrac{\frac{1}{2}+\frac{5}{2}}{\frac{3}{2}+\frac{7}{2}}$

17. $\dfrac{\frac{2}{3}-\frac{1}{2}}{\frac{4}{3}+\frac{1}{2}}$

18. $\dfrac{\frac{3}{4}+1}{\frac{5}{8}-2}$

19. $\dfrac{\frac{1}{x}+1}{\frac{1}{x}-1}$

20. $\dfrac{\frac{2}{y}-3}{\frac{4}{y}+5}$

21. $\dfrac{2+\frac{3}{z}}{4-\frac{5}{z}}$

In Problems 22 through 24, see Example 3.

22. $\dfrac{\frac{1}{z-3}+1}{\frac{2}{z-3}}$

23. $\dfrac{\frac{2}{s+1}}{\frac{s}{s+1}+1}$

24. $\dfrac{\frac{5}{x+2}-1}{\frac{6}{x+2}}$

In Problems 25 through 27, see Example 4.

25. $\dfrac{\frac{1}{x}+1}{\frac{1}{y}+1}$

26. $\dfrac{\frac{1}{x}+\frac{1}{y}}{\frac{1}{x}-\frac{1}{y}}$

27. $\dfrac{\frac{3}{x-1}-\frac{2}{x}}{\frac{2}{x-1}+\frac{3}{x}}$

Practice Exercises

28. $\dfrac{\frac{1}{3}}{\frac{2}{3}}$

29. $\dfrac{\frac{3}{4}}{\frac{3}{5}}$

30. $\dfrac{\frac{3}{4}}{\frac{9}{16}}$

31. $\dfrac{\frac{1}{3}+\frac{4}{3}}{\frac{4}{3}+\frac{7}{3}}$

32. $\dfrac{\frac{3}{5}-\frac{1}{4}}{\frac{4}{5}+\frac{1}{4}}$

33. $\dfrac{\frac{2}{3}-1}{\frac{5}{6}+2}$

34. $\dfrac{\frac{1}{x^2}}{\frac{1}{x}}$

35. $\dfrac{\frac{6}{y^2}}{\frac{8}{y^3}}$

36. $\dfrac{\frac{5s}{16}}{\frac{s^3}{12}}$

37. $\dfrac{\frac{6}{x-1}}{\frac{27}{x-1}}$

38. $\dfrac{\frac{2xy^2}{x-y}}{\frac{x^2}{x-y}}$

39. $\dfrac{\frac{6w^2}{xyz}}{\frac{15w}{xyz}}$

40. $\dfrac{\frac{y-2}{xy}}{\frac{y+3}{xy}}$

41. $\dfrac{\frac{4x^3}{y^2+1}}{\frac{12x}{y^2+1}}$

42. $\dfrac{\frac{2t+5}{5t}}{\frac{3t+7}{6t}}$

43. $\dfrac{\frac{35x^3}{54}}{\frac{25x}{27}}$

44. $\dfrac{\frac{s-1}{12s^2}}{\frac{3}{18s^4}}$

45. $\dfrac{\frac{x}{x-2}}{\frac{x}{x+2}}$

46. $\dfrac{\frac{1}{x}-2}{\frac{1}{x}+3}$

47. $\dfrac{\frac{3}{y}-4}{\frac{5}{y}+6}$

48. $\dfrac{3+\frac{4}{z}}{5-\frac{6}{z}}$

49. $\dfrac{\frac{2}{z-4}-1}{\frac{1}{z-4}}$

50. $\dfrac{\frac{t}{t-1}}{\frac{t^2}{t-1}+t}$

51. $\dfrac{\frac{2}{k+3}-4}{\frac{4}{k+3}}$

52. $\dfrac{\frac{1}{x}-1}{\frac{1}{x^2}-\frac{1}{x}}$

53. $\dfrac{\frac{1}{xy}+\frac{1}{xy^2}}{\frac{1}{x}-\frac{1}{xy^2}}$

54. $\dfrac{\frac{2}{x+1}-\frac{3}{x}}{\frac{3}{x+1}+\frac{2}{x}}$

Challenge Problems

55. Is $\dfrac{\frac{a}{b}}{c}$ the same as $\dfrac{a}{\frac{b}{c}}$?

IN YOUR OWN WORDS...

56. Discuss the two methods for simplifying a complex fraction.

5.6 FRACTIONAL EQUATIONS

In Section 3.2 we solved linear equations that contained fractional coefficients. In this section we look at solving equations that contain variables in their denominators. We call such equations **fractional equations.**

Let's solve

$$\frac{x}{x-1}-\frac{2}{x-1}=3$$

Notice that there are variables in the denominators. Any time we have a variable in the denominator of a fraction, we must be *very* careful. For example, if x has the value 1, then the first fraction in the equation is of the form

$$\frac{1}{1-1}, \quad \text{or} \quad \frac{1}{0}$$

However, in Section 5.1 we learned that a fraction with a denominator that is 0 is undefined. Thus 1 cannot be in the solution set of this equation.

So how do we solve equations of this type? We solve them by multiplying both sides of the equation by the least common denominator. However, before writing the solution set, we check to make sure the possible solution does not make any denominator in the equation have the value of zero.

We multiply both sides by the least common denominator, $(x - 1)$, to clear fractions.

$$(x-1)\cdot\left(\frac{x}{x-1} - \frac{2}{x-1}\right) = (x-1)\cdot 3$$

$$(x-1)\cdot\frac{x}{x-1} - (x-1)\cdot\frac{2}{x-1} = (x-1)\cdot 3 \qquad \text{Distributive Property}$$

$$\frac{(x-1)x}{x-1} - \frac{(x-1)2}{x-1} = (x-1)3$$

$$x - 2 = (x-1)3$$

$$x - 2 = 3x - 3$$

$$x - 2 - x = 3x - 3 - x$$

$$-2 = 2x - 3$$

$$-2 + 3 = 2x - 3 + 3$$

$$1 = 2x$$

$$\frac{1}{2} = x$$

Replacing x with $\frac{1}{2}$ does *not* make any denominator in the original equation have the value of zero, so we write the solution set

$$\left\{\frac{1}{2}\right\}$$

A Procedure for Solving Fractional Equations

1. Clear the equation of fractions by multiplying both sides by the least common denominator of all denominators in the equation.
2. Solve the resulting equation for possible solutions.
3. Check to see if any of the possible solutions make a denominator of the original equation have a value of zero. If so, we *cannot* include it in the solution set.

EXAMPLE 1. Solve $\frac{1}{x} + \frac{2}{3} = \frac{3}{5}$.

Solution:

Step 1 Clear the equation of fractions and solve the resulting equation.

The least common denominator of all the fractions in the equation is $15x$. Multiply both sides by $15x$.

$$15x \cdot \left(\frac{1}{x} + \frac{2}{3}\right) = 15x \cdot \frac{3}{5}$$

$$15x \cdot \frac{1}{x} + 15x \cdot \frac{2}{3} = 15x \cdot \frac{3}{5} \qquad \text{Distributive Property}$$

$$\frac{15x}{x} + \frac{30x}{3} = \frac{45x}{5}$$

$$15 + 10x = 9x$$

$$15 + 10x - 9x = 9x - 9x$$

$$15 + x = 0$$

$$x = -15$$

Step 2 Check that -15 does not make a denominator have value 0.

Replacing x by -15 does not make any denominator in the original equation equal 0, so we may put -15 in the solution set.

Step 3 Write the solution set.

$$\{-15\}$$

□

EXAMPLE 2. Solve $\frac{x + 2}{x + 5} - \frac{2x - 1}{x + 5} = 1$.

Solution:

Multiply both sides by $(x + 5)$.

$$(x + 5)\left(\frac{x + 2}{x + 5} - \frac{2x - 1}{x + 5}\right) = (x + 5)1$$

$$(x + 5) \cdot \frac{x + 2}{x + 5} - (x + 5) \cdot \frac{2x - 1}{x + 5} = x + 5 \qquad \text{Distributive Property}$$

$$x + 2 - (2x - 1) = x + 5 \qquad \text{Note parentheses.}$$

$$x + 2 - 2x + 1 = x + 5$$

$$-x + 3 = x + 5$$

$$3 = 2x + 5$$

$$-2 = 2x$$

$$-1 = x$$

Since -1 does not make a denominator have value zero, the solution set is

$$\{-1\}$$ □

EXAMPLE 3. Solve $\dfrac{x}{x+1} + 2 = -\dfrac{1}{x+1}$.

Solution:

We multiply both sides by $(x + 1)$ and simplify.

$$(x+1)\cdot\left(\frac{x}{x+1} + 2\right) = (x+1)\cdot\left(-\frac{1}{x+1}\right)$$

$$(x+1)\cdot\frac{x}{x+1} + (x+1)\cdot 2 = (x+1)\cdot\left(-\frac{1}{x+1}\right) \qquad \text{Distributive Property}$$

$$\frac{(x+1)x}{x+1} + (x+1)2 = -\frac{x+1}{x+1}$$

$$x + 2x + 2 = -1$$

$$3x + 2 = -1$$

$$3x + 2 - 2 = -1 - 2$$

$$3x = -3$$

$$x = -1$$

Before we write the solution set we must be sure the possible solution does not make any denominator have the value zero. However, when we inspect the original equation, we find that if x has the value -1, two denominators are 0. The number -1 is not a solution. So, because -1 was our *only* possible solution, there are no numbers in the solution set. Therefore, the solution set is the empty set, which we write

Be Careful!

$$\{\ \}, \quad \text{or} \quad \varnothing$$ □

The least common denominator is used in working several different kinds of problems in this chapter. In Section 5.4 we used the LCD to add and subtract fractions; in Section 5.5 the LCD was used to simplify complex fractions; in this section we used the LCD to clear an equation of fractions.

EXAMPLE 4. Follow the directions in each and notice how the LCD is used.

(a) Perform the operation indicated: $\dfrac{3}{x} + \dfrac{1}{7}$.

(b) Solve: $\dfrac{3}{x} + \dfrac{1}{7} = \dfrac{22}{7}$.

(c) Simplify: $\dfrac{\frac{3}{x}}{\frac{1}{7}}$.

(continued)

Solutions:

(a) This is addition of fractions. The LCD is $7x$.

$$\frac{3}{x} + \frac{1}{7} = \frac{3(7)}{x(7)} + \frac{1(x)}{7(x)} \qquad \text{Fundamental Principle}$$

$$= \frac{21}{7x} + \frac{x}{7x}$$

$$= \frac{21 + x}{7x}$$

(b) This is a fractional equation. The LCD is $7x$. We clear fractions by multiplying by $7x$.

$$\frac{3}{x} + \frac{1}{7} = \frac{22}{7}$$

$$7x\left(\frac{3}{x} + \frac{1}{7}\right) = 7x \cdot \frac{22}{7} \qquad \text{Multiply by } 7x.$$

$$7x \cdot \frac{3}{x} + 7x \cdot \frac{1}{7} = 7x \cdot \frac{22}{7} \qquad \text{Distributive Property}$$

$$21 + x = 22x$$

$$21 = 21x$$

$$1 = x$$

$$\{1\}$$

(c) This is a complex fraction. The LCD is $7x$. We multiply numerator and denominator by $7x$.

$$\frac{\frac{3}{x}}{\frac{1}{7}} = \frac{\frac{3}{x} \cdot 7x}{\frac{1}{7} \cdot 7x} \qquad \text{Multiply by } 7x.$$

$$= \frac{21}{x}$$

□

Fractional equations can lead to equations other than linear equations.

EXAMPLE 5. Solve $\frac{x}{2} - \frac{4}{x} = 1$.

Solution:

We clear fractions by multiplying by $2x$.

$$2x\left(\frac{x}{2} - \frac{4}{x}\right) = 2x(1)$$

$$x^2 - 8 = 2x \qquad \text{Distributive Property}$$

$$x^2 - 2x - 8 = 0 \qquad \text{Make one side 0.}$$

$$(x - 4)(x + 2) = 0 \qquad \text{Factor.}$$

$$x = 4 \text{ or } x = -2$$

Neither -2 nor 4 make a denominator in the original equation have a value of 0.

$$\{-2, 4\}$$

□

PROBLEM SET 5.6

Warm-ups

In Problems 1 through 32, solve the equations.
In Problems 1 through 12, see Example 1.

1. $\frac{3}{y} - \frac{4}{y} = 1$

2. $\frac{x + 2}{x} + \frac{x - 1}{x} = 3$

3. $\frac{2p - 1}{p} - \frac{p + 1}{p} = 3$

4. $\frac{3t - 1}{2t} - \frac{2 - t}{2t} = 1$

5. $\frac{12}{x} = \frac{6}{5}$

6. $\frac{3}{2y} = \frac{1}{2}$

7. $\frac{1}{x} - \frac{1}{2} = -\frac{1}{6}$

8. $\frac{2}{5} - \frac{1}{x} = \frac{1}{5}$

9. $\frac{x - 4}{x} + \frac{1}{7} = \frac{4}{7}$

10. $\frac{2}{3} - \frac{3}{x} = 1$

11. $\frac{4}{3x} + \frac{1}{x} = \frac{1}{2}$

12. $\frac{7}{2y} - \frac{1}{y} = \frac{5}{4}$

In Problems 13 through 18, see Example 2.

13. $\frac{3}{x + 1} - \frac{x - 2}{x + 1} = 1$

14. $\frac{x + 1}{x - 1} - \frac{2x - 6}{x - 1} = 2$

15. $\frac{1}{w + 3} + \frac{1}{2} = \frac{3}{w + 3}$

16. $\frac{3}{v - 1} + \frac{2}{3} = 1$

17. $\frac{2}{x + 3} + \frac{10}{7} = 2$

18. $\frac{2x - 1}{x - 1} - \frac{3}{2} = 1$

In Problems 19 through 25, see Example 3.

19. $\frac{x - 3}{x - 2} + \frac{1}{x - 2} = 2$

20. $\frac{x}{x + 2} + \frac{3}{x + 2} = \frac{1}{x + 2}$

21. $\frac{3}{y + 3} - 2 = \frac{-y}{y + 3}$

22. $\frac{2}{t - 2} = \frac{t}{t - 2} - 2$

23. $\frac{2k + 3}{k} - \frac{3k + 3}{k} = 1$

24. $\frac{x + 3}{x} - \frac{3 - x}{x} = 3$

In Problems 25 through 28, perform any operations indicated and simplify any complex fractions. See Example 4.

25. $\dfrac{\dfrac{x}{x + 2}}{\dfrac{3}{x + 2}}$

26. $\frac{7}{2y} - \frac{1}{y}$

27. $\frac{3}{x + 1} - \frac{x + 2}{x + 1}$

28. $\frac{1}{w + 3} + \frac{1}{2}$

In Problems 29 through 32, solve each equation. See Example 5.

29. $x + \frac{3}{x} = -4$

30. $x - \frac{4}{x} = 3$

31. $\frac{x}{2} + \frac{5}{2} + \frac{3}{x} = 0$

32. $\frac{x}{4} - \frac{1}{4} = \frac{3}{x}$

Practice Exercises

Solve any equations, perform any additions or subtractions or simplify any complex fractions.

33. $\frac{5}{y} = \frac{1}{2}$

34. $\frac{3}{4} + \frac{1}{x} = \frac{5}{4}$

35. $\frac{4}{w} = \frac{2}{w} + 4$

36. $\frac{r+3}{r} + \frac{2r+1}{r} = 2$

37. $\frac{1}{x} + \frac{1}{3} = \frac{5}{6}$

38. $\frac{1}{x} - \frac{2}{7} = \frac{1}{21}$

39. $\frac{5-z}{z} - \frac{3-2z}{z} = -1$

40. $\frac{2t-3}{t} - \frac{5t+7}{t} = 2$

41. $\frac{6x-5}{x} - \frac{7-x}{x} = 3$

42. $\frac{x+7}{x} = \frac{2}{3}$

43. $\frac{2b-5}{b} = \frac{5}{3}$

44. $\frac{7}{5y} = \frac{2}{y}$

45. $\frac{3}{8} - \frac{1}{x} = \frac{1}{4}$

46. $\frac{5}{12} + \frac{3}{x} = \frac{2}{3}$

47. $\frac{7}{2} - \frac{x+3}{x} = 1$

48. $\frac{4}{5} - \frac{3}{x} = \frac{x-4}{5}$

49. $\frac{1}{x} - \frac{1}{3} + \frac{1}{6}$

50. $\frac{1}{x} + \frac{2}{7}$

51. $\frac{2}{x+1} + \frac{x}{x+1} = 2$

52. $\dfrac{\frac{x+1}{x-1}}{\frac{5}{x-1}}$

53. $\frac{x}{x+2} - \frac{1}{2} = \frac{1}{x+2}$

54. $\frac{y+9}{y-3} + 1 = \frac{4y}{y-3}$

55. $x + 2 = \frac{3}{x}$

56. $\frac{x-3}{x-2} - \frac{1}{x-2} = 2$

57. $\frac{x-16}{x+4} - 2 = \frac{5x}{x+4}$

58. $\frac{3x-1}{x-1} - \frac{3}{2} = 2$

59. $\frac{3}{x+2} = \frac{3}{2}$

60. $x - \frac{10}{x} = 3$

61. $\frac{5}{k+5} - 4 = \frac{-k}{k+5}$

62. $\frac{-2}{x-2} + 2 = \frac{-x}{x-2}$

Challenge Problems

In the following problems, Q, R, and S are constants, with $S \neq 0$. Find the solution set for each.

63. $\frac{x}{S} + \frac{1}{S} = \frac{3}{S}$

64. $\frac{x}{3} + \frac{Q}{3} = Q$

65. $\frac{Q}{5} - \frac{x}{5} = \frac{R}{5}$

66. $\frac{Q}{S} + \frac{x}{S} = \frac{R}{S}$

67. $\frac{1}{2}(x + R) = R$

68. $\frac{2(x-Q)}{3} = 2Q$

69. $\frac{3x}{2S} - \frac{x}{S} = \frac{5}{S}$

70. $\frac{4}{3S} + \frac{1}{2S}x = 1$

IN YOUR OWN WORDS...

71. Explain the procedure for solving a fractional equation.

5.7 RATIO AND PROPORTION, SIMILAR TRIANGLES, AND OTHER APPLICATIONS

Ratios and proportions are commonly used to express the relationship of one quantity to another. Both are used in many fields, and using them leads to fractional equations.

A church choir has 5 men and 13 women. We say the ratio of men to women in the choir is 5 to 13, which is often written 5 : 13. We also write ratios as fractions. The ratio of men to women in the choir is $\frac{5}{13}$. Because ratios are fractions, they may often be reduced. A bin of oranges at the Farmers Market contains 500 naval oranges and 20 tangerines. The ratio of naval oranges to tangerines in the bin is 500 : 20. Since this can be written as the fraction $\frac{500}{20}$, which can be reduced to $\frac{25}{1}$, we say the ratio is 25 : 1.

Ratio

If p and q are numbers with $q \neq 0$, the **ratio** of the number p to the number q is written

$$p : q, \quad \text{or} \quad \frac{p}{q}$$

EXAMPLE 1. Write a ratio for each.

(a) The ratio of 3 males to 5 females.

(b) The ratio of 7 cats to 11 dogs.

Solutions:

(a) $\dfrac{\text{Male}}{\text{Female}} = \dfrac{3}{5}$

(b) $\dfrac{\text{Cats}}{\text{Dogs}} = \dfrac{7}{11}$ □

Application problems are solved by making two ratios equal.

Proportion

A **proportion** is a statement that two ratios are equal.

We read the proportion $p : q = s : t$ as "p is to q as s is to t." The "inside" numbers in this proportion, q and s, are called the **means,** and the "outside" numbers, p and t, are called the **extremes.** This proportion can also be written

$$\frac{p}{q} = \frac{s}{t}$$

In this equation, if we multiply both sides by qt, we see that

$$qs = pt$$

That is, in a proportion, the *product* of the *means* equals the *product* of the *extremes.*

EXAMPLE 2. Find x in the proportion $7 : 30 = x : 5$.

Solution:

Writing $7 : 30 = x : 5$ is the same as writing $\frac{7}{30} = \frac{x}{5}$. To find x, we must solve this equation.

$$\frac{7}{30} = \frac{x}{5}$$

$$30 \cdot \frac{7}{30} = 30 \cdot \frac{x}{5} \qquad \text{Multiply by 30.}$$

$$7 = 6x$$

$$\frac{7}{6} = x \qquad \text{Divide by 6.}$$

In the proportion $7 : 30 = x : 5$, x has the value $\frac{7}{6}$. □

Proportion problems are often stated in words.

EXAMPLE 3. A shirt manufacturer uses 3 lb of cotton for every 2 lb of rayon in a certain grade of shirt. How many pounds of rayon should the manufacturer buy to use with 840 lb of cotton to make these shirts?

Solution:

We assume the ratio of cotton to rayon stays the same in the manufacture of these shirts. We are given the ratio

$$\frac{\text{Number of pounds of cotton}}{\text{Number of pounds of rayon}} = \frac{3}{2}$$

If we let x be the number of pounds of rayon needed to use with 840 lb of cotton, then we form the proportion

$$\frac{3}{2} = \frac{840}{x}$$

$$2x \cdot \frac{3}{2} = 2x \cdot \frac{840}{x} \qquad \text{Multiply by } 2x.$$

$$3x = 1680$$

$$x = \frac{1680}{3}$$

$$x = 560$$

The manufacturer should buy 560 lb of rayon to use with 840 lb of cotton. □

Similar triangles are triangles that are the same shape. In plane geometry we find that corresponding sides of similar triangles are proportional.

EXAMPLE 4. If a triangle has sides of lengths 18 ft, 24 ft, and 40 ft, what is the length of the longest side of a *similar* triangle whose shortest side is 63 ft?

Solution:

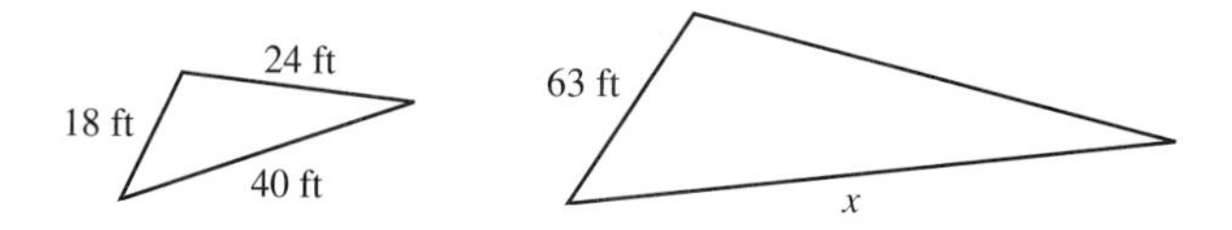

If x is the length of the longest side of the unknown triangle, then we form the proportion

$$\frac{40}{x} = \frac{18}{63}$$

$$63x \cdot \frac{40}{x} = 63x \cdot \frac{18}{63} \qquad \text{Multiply by } 63x.$$

$$63 \cdot 40 = 18x$$

$$2520 = 18x$$

$$\frac{2520}{18} = x \qquad \text{Divide by 18.}$$

$$140 = x$$

The length of the longest side is 140 ft. □

EXAMPLE 5. The ratio of number of peppermints to number of butterscotch pieces in a can of party mints is 5 : 8. If there are 26 pieces of candy in a can, how many of them are peppermints?

Solution:

Let x be the number of peppermints. Since there are 26 pieces in all, we have $26 - x$ pieces of butterscotch.

We are given the ratio

$$\frac{\text{Peppermint}}{\text{Butterscotch}} = \frac{5}{8}$$

We form a proportion.

$$\frac{5}{8} = \frac{x}{26 - x}$$

We multiply by the LCD, $8(26 - x)$.

$$8(26 - x) \cdot \frac{5}{8} = 8(26 - x) \cdot \frac{x}{26 - x}$$

$$(26 - x)5 = 8x$$

(continued)

$$130 - 5x = 8x \qquad \text{Distributive Property}$$

$$130 = 13x$$

$$10 = x$$

There are 10 peppermints in the can. □

Other applications of fractional equations include number problems and distance problems.

EXAMPLE 6. One number is 5 times another number. If the sum of their reciprocals is $\frac{3}{5}$, what are the numbers?

Solution:

Let n be the smaller number.
Then $5n$ is the other number.

We form an equation.

$$\frac{1}{n} + \frac{1}{5n} = \frac{3}{5}$$

$$5n\left(\frac{1}{n} + \frac{1}{5n}\right) = 5n \cdot \frac{3}{5} \qquad \text{Multiply by } 5n.$$

$$5 + 1 = 3n \qquad \text{Distributive Property}$$

$$6 = 3n$$

$$2 = n$$

If n is 2, then $5n$ is 10. The numbers are 2 and 10. □

EXAMPLE 7. A Delta 737 flies 50 mph faster than a British Airways DC 9. The 737 travels 500 mi in the same amount of time that the DC 9 travels 400 mi. Find the speeds of the planes.

Solution:

Let s be the speed of the DC 9 in miles per hour. Then $s + 50$ is the speed of the 737 in miles per hour.

We make a table.

	D	*R*	*T*
737	500	$s + 50$	
DC 9	400	s	

To find the time for each plane, we use the formula $d = rt$, or $t = \frac{d}{r}$.

	D	R	T
737	500	$s + 50$	$\frac{500}{(s + 50)}$
DC 9	400	s	$\frac{400}{s}$

Since the time each plane flies is the same, we form the equation

$$\frac{500}{s + 50} = \frac{400}{s}.$$

We multiply both sides by $s(s + 50)$.

$$s(s + 50)\frac{500}{s + 50} = s(s + 50)\frac{400}{s}$$

$$500s = (s + 50)\ 400$$

$$500s = 400s + 20{,}000$$

$$100s = 20{,}000$$

$$s = 200$$

If s is 200, $s + 50$ is 250.

The speed of the DC 9 is 200 mph and the speed of the 737 is 250 mph. □

PROBLEM SET 5.7

Warm-ups

In Problem 1, see Example 1.

1. This fall 28 students have enrolled in Ms. Gavant's beginning algebra class, 16 of whom are females. There are 8 students from out of state and 7 from out of the country.
   Write the following ratios.
   (a) Females to males
   (b) Males to females
   (c) Out-of-state students to out-of-country
   (d) Females to out-of-state
   (e) Males to total number of students

In Problems 2 through 5, find the value of x that makes the proportion correct. See Example 2.

2. $x : 5 = 3 : 1$

3. $8 : x = 1 : 7$

4. $2 : 3 = x : 12$

5. $3 : 5 = 15 : x$

In Problems 6 through 11, see Example 3.

6. If a person 6 ft tall casts a shadow 4 ft long, how tall is a flagpole that casts a 60-ft shadow?
7. If there are 2.54 cm in 1 in., how many inches in 508 cm?
8. If 5 lb of chicken will serve 18 people, how many pounds will it take to serve 360 people?
9. If it takes 2 hr to plant 30 tomato plants, how many tomato plants can be planted in 20 min?
10. Seven drops of special dye 3 are needed to color 400 L of oil. How many drops should be added to color 20,000 L?
11. Gloria's car averages 22 mi/gal. How many gallons of gasoline did she use during the 209,000-mi lifetime of her car?

In Problems 12 and 13 the triangles are similar. Find x in each. *See Example 4.*

12.

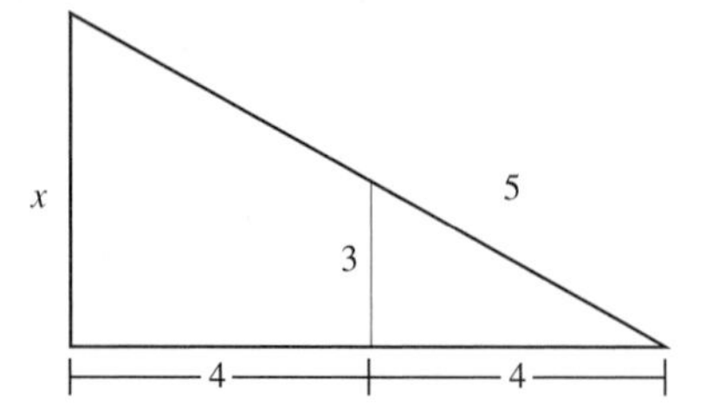

13.

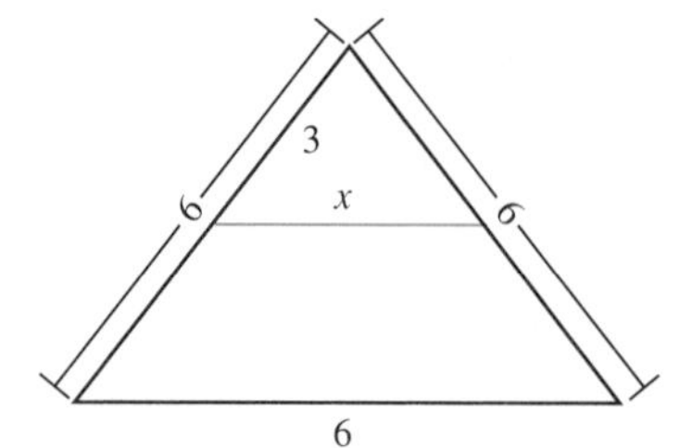

In Problems 14 and 15, see Example 5.

14. The ratio of woodpeckers to mockingbirds in a forest is 2 : 7. If a total of 27 of these two kinds of birds were counted one day, how many of them were woodpeckers?
15. A cable 30 m long is divided into two parts so that the ratio of the length of the shorter piece to that of the longer piece is 2 : 3. Find the length of each piece.

In Problems 16 and 17, see Example 6.

16. If twice the reciprocal of a number is 6, find the number.
17. One number is four times another number. If the sum of their reciprocals is $\frac{5}{8}$, what are the numbers?

In Problems 18 through 20, see Example 7.

18. Jody and Meredith live 2 and 3 mi, respectively, from the Joffrey Ballet Studio in New York. They walk to the studio from their homes each day to take lessons. The walk takes Jody 15 min less than Meredith. If they walk at the same rate, how long does it take Meredith to walk to the studio?
19. Frank and Doug ran races at a field day. Frank ran 10 km and Doug ran 12 km in the same amount of time. If Doug's rate was 2 km/h faster than Frank's rate, how fast did Frank run?
20. Two trains leave Grand Central Station at 5:00 P.M. One train travels 400 mi in the same amount of time that the other travels 600 mi. If one of the trains traveled 10 mph faster than the other, how fast did the slower train travel?

Practice Exercises

21. The triangles pictured are similar. Find x.

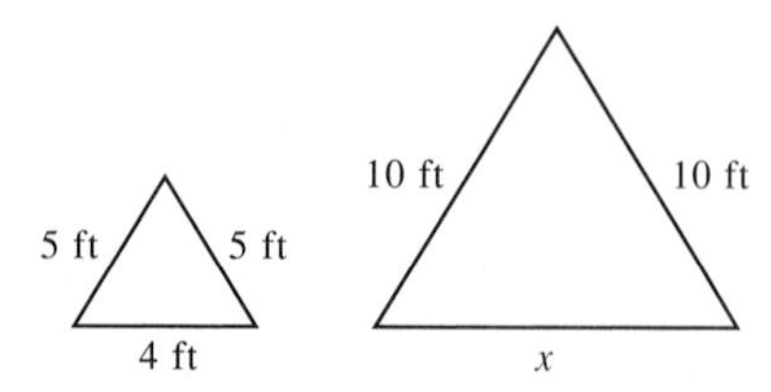

22. There are 1.6 km per mile. How many miles are there in 48.8 km?
23. A box of coffee beans weighing 40 lb is to be divided into two parts so that the ratio of the weight of the lighter part to the weight of the heavier part is 3 : 5. How many pounds are there in both parts?
24. If it takes 2 hr to work 30 problems, how many problems can be worked in 12 min?

25. An investment of 100 shares of a certain stock earned a dividend of \$23. An investment of how many shares would have received a dividend of \$29.90?
26. A sugar solution is made by dissolving 2 lb of sugar in 5 gal of water. How many pounds of sugar are needed for 30 gallons of water to make a solution of the same strength?
27. The reciprocal of 2 more than a number is twice the reciprocal of the number. Find the number.
28. One number is twice another number. If the sum of their reciprocals is $\frac{1}{2}$, find the numbers.
29. A train travels 10 mph faster than a boat. The train goes 250 mi in the same amount of time the boat goes 200 mi. What is the speed of each?
30. It takes Gail 20 min longer to walk 3 mi than it takes Carol to walk 2 mi. If they walk at the same rate, how long does it take Carol to walk the 2 mi?

Challenge Problems

31. Prove that the product of the means is equal to the product of the extremes.
32. The sum of the reciprocals of two consecutive even integers is 0. Find the integers.

IN YOUR OWN WORDS...

33. Make up a word problem that uses a ratio of 2 : 3.

CHAPTER SUMMARY

GLOSSARY

Rational expression: The quotient of two polynomials. The expression is **undefined** when the polynomial in the denominator has the value of zero.

Ratio: A quotient of two numbers, written as a fraction.

Reduce a rational expression to lowest terms: To write the expression so that there are no common factors in the numerator and the denominator.

Proportion: A statement that two ratios are equal.

FUNDAMENTAL PRINCIPLE OF FRACTIONS

$$\frac{AC}{BC} = \frac{A}{B} \quad \text{(where } B \text{ and } C \text{ do not have value 0)}$$

We use the Fundamental Principle to reduce and rewrite fractions.

MULTIPLICATION AND DIVISION OF RATIONAL EXPRESSIONS

Multiply $\frac{A}{B} \cdot \frac{C}{D} = \frac{A \cdot C}{B \cdot D} = \frac{AC}{BD}$

Divide $\frac{A}{B} \div \frac{C}{D} = \frac{A}{B} \cdot \frac{D}{C} = \frac{AD}{BC}$

ADDITION AND SUBTRACTION OF RATIONAL EXPRESSIONS

Add $\frac{A}{B} + \frac{C}{B} = \frac{A + C}{B}$

Subtract $\frac{A}{B} - \frac{C}{B} = \frac{A - C}{B}$

SIMPLIFYING COMPLEX FRACTIONS If there is only one term in the numerator and denominator, we use Method 1.

Method 1 Treat the fraction as a division problem.

If either the numerator or the denominator has more than one term we use Method 2.

Method 2 Multiply numerator and denominator by the LCD.

FRACTIONAL EQUATIONS Fractional equations can be solved by clearing the equation of fractions (multiply both sides by the LCD) and then solving the resulting equation for possible solutions. Possible solutions that make any denominator in the original equation have a value of zero *cannot* be included in the solution set.

PROPORTION The proportion $p : q = r : s$ can be written as $\frac{p}{q} = \frac{r}{s}$.

CHECKUPS

1. The rational expression $\frac{x + 3}{(x - 1)(x + 2)}$ is undefined when x has the value of (?). — Section 5.1; Example 3
2. The rational expression $\frac{3x + 9}{x^2 - 9}$ reduces to (?). — Section 5.2; Example 2b
3. $\frac{5x - 10}{x + 3} \cdot \frac{2x + 6}{3x - 6} = (?)$ — Section 5.3; Example 2a
4. $\frac{x + 1}{x - 1} \div \frac{x + 1}{x - 2} = (?)$ — Section 5.3; Example 5
5. The LCD for x^2y^3z and x^4y^2 is (?). — Section 5.4; Example 3b
6. $\frac{2}{x + 2} + \frac{x}{x^2 - 4} = (?)$. — Section 5.4; Example 7a
7. $\frac{1}{x^2 + 2x + 1} - \frac{1}{x^2 - 1} = (?)$. — Section 5.4; Example 7b
8. $\dfrac{\frac{1}{x - 1} + 5}{\frac{1}{x - 1} - 1} = (?)$. — Section 5.5; Example 3
9. Solve $\frac{1}{x} + \frac{2}{3} = \frac{3}{5}$. — Section 5.6; Example 1
10. Find x in the proportion $7 : 30 = x : 5$. — Section 5.7; Example 2
11. One number is 5 times another number. If the sum of their reciprocals is $\frac{3}{5}$, what are the numbers? — Section 5.7; Example 6

REVIEW PROBLEMS

In Problems 1 through 4, find the values of x, if any, for which the rational expression is undefined.

1. $\frac{11}{2x}$

2. $\frac{x}{x-2}$

3. $\frac{x+1}{(x+2)(x-3)}$

4. $\frac{x^2-9}{x^2-3x-10}$

In Problems 5 through 8, express each rational expression as an expression of equal value with the indicated denominator.

5. $\frac{21}{x^2}$; $4x^3$

6. $\frac{3}{x-1}$; $4(x-1)$

7. $\frac{x-1}{x+2}$; x^2+x-2

8. $\frac{4}{x+4}$; x^2-16

In Problems 9 through 12, reduce each expression to lowest terms.

9. $\frac{3t^3}{5t^2}$

10. $\frac{7-21x}{2(1-3x)}$

11. $\frac{x^2-2x}{x^2-4}$

12. $\frac{3x-6}{x^2-x-2}$

In Problems 13 through 28, perform the operations indicated. Express each number in lowest terms.

13. $\frac{2x}{x-1}\cdot\frac{x-1}{2x+6}$

14. $\frac{x^2-4}{xy^3}\cdot\frac{x^2-2x}{x+2}$

15. $\frac{x^3y^2}{6-2x}\div\frac{xy^3}{3x-9}$

16. $\frac{x^2-x-6}{x^2-25}\div\frac{x^2-9}{x^2-2x-15}$

17. $\frac{x+2}{3x-2}+\frac{x-3}{6x-4}$

18. $\frac{4}{x^2-1}+\frac{2x}{x+1}$

19. $\frac{1}{4x+6}-\frac{1}{6x+9}$

20. $\frac{3}{2x^2+3x-27}-\frac{1}{2x^2+13x+18}$

21. $\frac{4t^2-t}{-t^3}\div(16t^2-1)$

22. $\frac{1}{2-5y}+\frac{1}{5y-2}$

23. $x-\frac{3}{2x}$

24. $\frac{2-s-s^2}{s+1}\cdot\frac{s}{s^2-1}$

25. $\frac{1}{w+1}+\frac{3}{w^2-w-2}$

26. $\frac{ux+3u-2x-6}{u^2x}\cdot\frac{ux^2}{u^2+u-6}$

27. $\frac{x^2-16}{x^2+8x+16}\div\frac{4-x}{4+x}$

28. $\frac{3}{2x-1}-\frac{5}{3-6x}$

29. $\dfrac{\frac{1}{2}-\frac{1}{3}}{\frac{1}{2}+\frac{1}{3}}$

30. $\dfrac{\frac{1}{x}+\frac{1}{x^2}}{\frac{1}{x^2}}$

31. $\dfrac{1 + \dfrac{1}{y}}{1 - \dfrac{1}{y}}$

32. $\dfrac{\dfrac{3}{t-1} - 1}{\dfrac{2}{t-1}}$

In Problems 33 through 38, find the solution set.

33. $\dfrac{3}{x} - 1 = 0$

34. $\dfrac{1}{2x} + \dfrac{2}{x} = \dfrac{5}{2}$

35. $\dfrac{x-3}{x} + \dfrac{1}{4} = \dfrac{7}{x}$

36. $\dfrac{3x+4}{x} - \dfrac{2x+3}{x} = 2$

37. $\dfrac{7}{k+7} - 6 = \dfrac{-k}{k+7}$

38. $\dfrac{4}{x-3} + 1 = \dfrac{5}{x-3}$

39. $x - \dfrac{3}{x} = 2$

40. $\dfrac{x}{2} = \dfrac{8}{x}$

41. The ratio of foxes to wolves in an area is 3 : 8. If 15 foxes are in the area, how many wolves are in the area?

42. If 1 gal is 3.8 L, how many liters in 3.5 gal?

43. The sum of the reciprocals of two consecutive odd integers is 0. What are the numbers?

44. Jean Marie and Paul drive different routes from their house to the mall. Jean Marie takes the scenic route, which is 6 mi, and Paul takes the interstate highway, which is 15 mi. It takes the same amount of time, but Paul travels 45 mph faster than Jean Marie. How fast does Jean Marie drive?

...LET'S NOT FORGET...

In Problems 45 through 50, perform the operations indicated.

45. -5^2

46. $(x - 5)^2$

47. $(-5)^2$

48. $\dfrac{5}{0}$

49. $(x + 2) \cdot \dfrac{x}{x+2}$

50. $\dfrac{4}{x-1} - \dfrac{3}{1-x}$

In Problems 51 through 54, simplify each expression. Watch the role of the parentheses.

51. $x - 2(3x - 7)$

52. $(x - 2)3x - 7$

53. $(x - 2)(3x - 7)$

54. $x - 2(3x) - 7$

55. $\dfrac{x+1}{y} - \dfrac{x-1}{y}$

56. $\dfrac{x+1}{(x-1)y} - \dfrac{1}{y}$

In Problems 57 through 60, determine if 2 is a factor *or a* term.

57. $\dfrac{3x+2}{5x}$

58. $\dfrac{7x^2 - x + 1}{y+2}$

59. $\dfrac{7x^3 - 1}{2x(3x+1)}$

60. $\dfrac{5(x-1) + (1+x)}{2y}$

In Problems 61 through 63, multiply each expression by $\dfrac{2}{x}$.

61. $\dfrac{x}{2}$

62. $x^2(x + 7)$

63. $2x^2 + 3x$

In Problems 64 through 67, label each as an expression, equation, *or* inequality. *Solve the equations and inequalities and simplify the expressions.*

64. $\dfrac{x^2 - 9}{3 - x}$

65. $\dfrac{2}{x} + \dfrac{3}{4x} = \dfrac{11}{4}$

66. $5(x - 1) - 3 < -7$

67. $\dfrac{2}{x} + \dfrac{3}{4x}$

In Problems 68 through 71, identify the expressions that are factored. Factor those that are not factored, if possible.

68. $(x - 3)^2$

69. $x^2 - 9$

70. $x^2 - 6x + 9$

71. $x^3 + 27$

72. $(x + 3)^3$

73. $3x^3 - 81$

CHAPTER 5 TEST

In Problems 1 through 8, choose the correct answer.

1. The rational expression $\dfrac{x - 1}{x(x - 2)}$ is undefined if x has the value of (?).

A. 1 B. -1

C. 0 or 2 D. -2

2. If the rational expression $\dfrac{3x}{x - 5}$ is written with a denominator of $x^2 - 25$, it becomes (?).

A. $\dfrac{3x}{x^2 - 25}$ B. $\dfrac{3x^2 + 15x}{x^2 - 25}$

C. $\dfrac{9x^2}{x^2 - 25}$ D. $\dfrac{3x^2 - 15x}{x^2 - 25}$

3. When reduced to lowest terms, the rational expression $\dfrac{2x^2 - 6x}{2x^2 - 5x - 3}$ becomes (?).

A. $\dfrac{3}{4}$ B. $\dfrac{6x}{5x + 3}$

C. $\dfrac{2x}{2x + 1}$ D. None of the above

4. $\dfrac{12x^2}{7x^2 - 21} \cdot \dfrac{5x^3 - 15x}{9x} = (?)$.

A. $\dfrac{10x^3}{7}$ B. $\dfrac{20x^4 - 60x^2}{21x^2 - 63}$

C. $60x(x^3 - x)$ D. $\dfrac{20x^2}{21}$

5. $\dfrac{6x - 4}{x^2(x + 1)} \div \dfrac{x(3x - 2)}{2x + 2} = (?)$.

A. $\dfrac{4}{x^3}$ B. $\dfrac{4}{x}$

C. $\dfrac{1}{x}$ D. None of the above

6. $\dfrac{x}{x - 2} - \dfrac{2x - 3}{2x - 4} = (?)$.

A. $\dfrac{-3}{2(x - 2)}$ B. $\left\{\dfrac{1}{2}\right\}$

C. $\dfrac{-5x + 3}{(x - 2)(2x - 4)}$ D. $\dfrac{3}{2(x - 2)}$

7. $\dfrac{1}{(x - 1)^2} + \dfrac{1}{x^2 - 1} = (?)$

A. $\dfrac{2}{(x - 1)^2}$ B. $\dfrac{2x}{(x - 1)^2(x + 1)}$

C. $\dfrac{2}{x^2 - 1}$ D. $\dfrac{x^2 - (x - 1)^2}{(x - 1)^2(x^2 - 1)}$

8. When simplified, the complex fraction $\dfrac{x + \dfrac{1}{x}}{1 - \dfrac{2}{x}}$ becomes (?).

A. $\dfrac{x^2 + 1}{x - 2}$ B. $\dfrac{2x}{1 - 2x}$

C. $\dfrac{x + 1}{-2}$ D. $\dfrac{(x^2 + 1)(x - 2)}{x^2}$

9. Evaluate the rational expression $\dfrac{3x - y^2}{2x + y^3}$ when x is 3 and y is -2.

In Problems 10 through 14, perform the operations indicated. Write answers in lowest terms.

10. $\dfrac{1}{x - x^2} + \dfrac{x}{1 - x}$

11. $\dfrac{(t + 2)^2}{t^2 + 6t + 8} \cdot \dfrac{t^2 - 2t - 8}{(t + 2)^3}$

12. $\dfrac{x^3 - 3x^2}{x^3 - 6x^2 + 9x} \div \dfrac{x}{x^2 - 9}$

13. $\dfrac{x}{x - 2} - \dfrac{10}{x^2 + x - 6}$

14. $\dfrac{s^2 - s - 6}{s^2 + s - 2} \cdot \dfrac{s^2 + 3s - 4}{s^2 - 7s + 12}$

In Problems 15 through 17, find the solution set.

15. $\dfrac{1}{x} + \dfrac{1}{2x} = \dfrac{3}{2}$

16. $\dfrac{2}{x + 2} - \dfrac{1}{x} = \dfrac{-4}{x(x + 2)}$

17. $x - \dfrac{7}{x} = 6$

18. There are 1500 registered voters in a district, 900 of whom are Republicans; the remainder are Democrats. Express each of the following as a ratio:

(a) The number of Republicans to the number of Democrats

(b) The number of Democrats to the number of voters

19. On one map, 2 in. represents 75 mi. If two cities are 5 in. apart on the map, find the number of miles between the two cities.

20. One number is four times another number. If the sum of their reciprocals is $\frac{5}{12}$, find the two numbers.

CHAPTER 6

See Problem Set 6.2, Exercise 65.

Linear Equations and Inequalities in Two Variables

CONNECTIONS

The inclination of the roof on a home in Vermont determines how much snow it can safely hold. A particularly steep ski slope might be developed specifically for advanced skiers. A road which has a long, steep grade sometimes has a runaway ramp for 18-wheelers that lose their brakes driving down the incline. These are all applications of the idea of slope of a line which we will explore in this chapter.

In 1638, the French mathematician Rene Descartes published the "Discourse on Method." In one of the appendices, Descartes describes a method of drawing pictures to represent abstract mathematical ideas. These pictures, called graphs, are the basis of what we today call Analytic Geometry. Analytic Geometry shows the relationship between algebra and geometry.

In this chapter, we introduce the rectangular coordinate system popularized by Descartes and link it to the graphs of points and lines. Lines, their equations and slopes will be emphasized. Linear inequalities in two variables will be discussed and graphed. Functions, a central concept of calculus, will also be introduced.

6.1 THE GRAPH OF $Ax + By = C$

In Chapter 1 we used the number line to help us visualize the set of real numbers. Each point on the number line represents a real number, and each real number can be represented by a point on the number line. To help us visualize another set, we place two copies of the number line, one horizontally and one vertically, so that they intersect at the zero point of each number line.

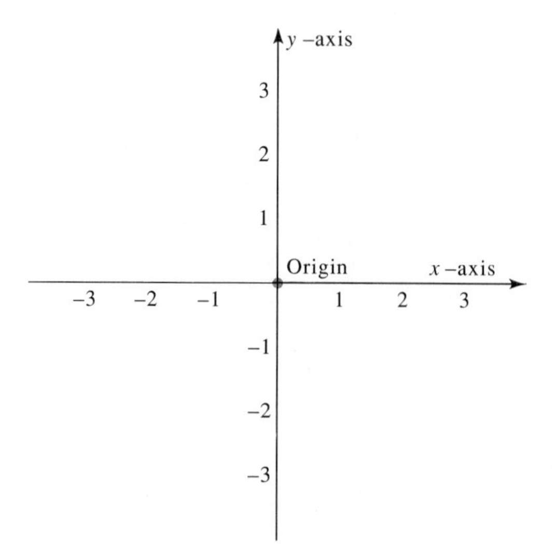

We call this a **rectangular coordinate system,** or a **Cartesian coordinate system,** named for the French mathematician René Descartes, (1596–1650). We call the horizontal number line the ***x*-axis** and the vertical number line the ***y*-axis.** The point where the x-axis and the y-axis intersect is called the **origin.**

With this coordinate system we can associate each point in the plane with two numbers, called **coordinates.** Suppose we take two numbers, say 1 and 2, and write them in the form

$$(1, 2)$$

We call this form an **ordered pair.** We associate a point with the ordered pair (1, 2) by first locating 1 on the x-axis and then locating 2 on the y-axis. The point is determined by the point of intersection of the vertical line drawn through 1 and the horizontal line drawn through 2.

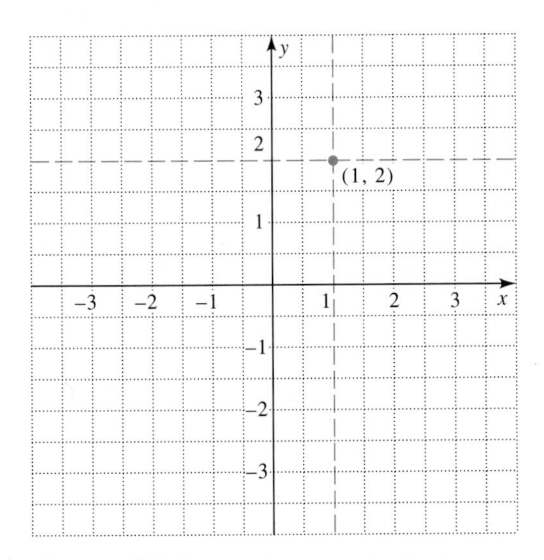

This unique point corresponds to the ordered pair (1, 2) and is called its **graph.** We call the first number in the ordered pair the ***x*-coordinate,** or **abscissa,** and the second number the ***y*-coordinate,** or **ordinate.** To **plot** an ordered pair means to graph its ordered pair.

Notice that the graphs of the ordered pairs (1, 2) and (2, 1) are different points. The order in which the numbers are written *is important.*

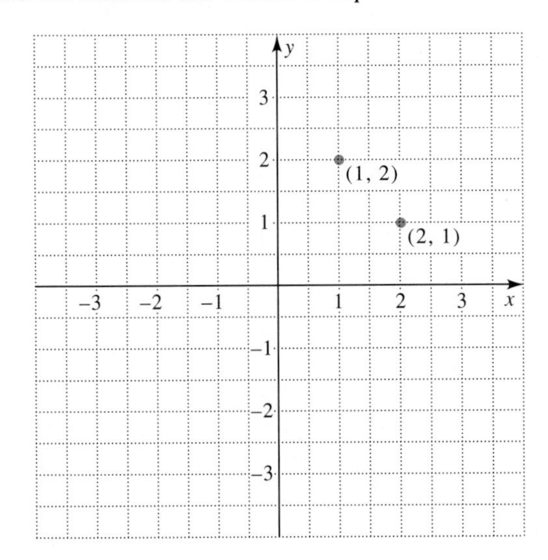

EXAMPLE 1. Graph the points $(0, 2)$, $(-1, 0)$, $(-1, 2)$, $(1, -2)$, and $(-1, -2)$.

Solution:

The five points are graphed in the following graph. Notice how the point $(-1, 0)$ lies on the x-axis and the point $(0, 2)$ lies on the y-axis.

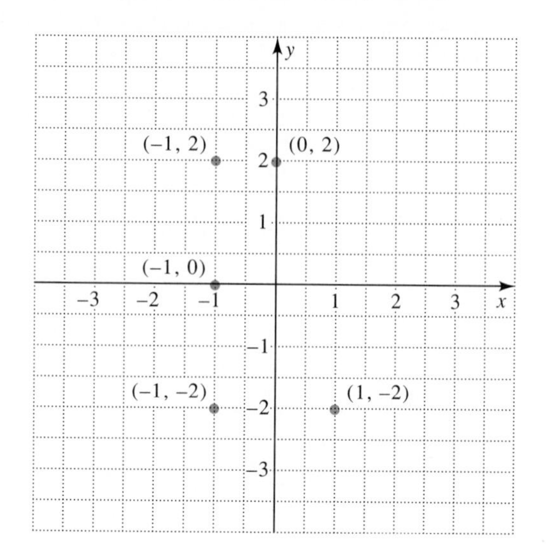

□

The plane in which the coordinate system lies is divided into four parts called **quadrants.** They are named with the Roman numerals I, II, III, and IV in a counterclockwise direction.

EXAMPLE 2. Identify the quadrants in which the following points lie.

(a) $(2, 3)$ (b) $(-3, 1)$ (c) $(2, -3)$
(d) $(-2, -1)$ (e) $(2, 0)$

Solutions:

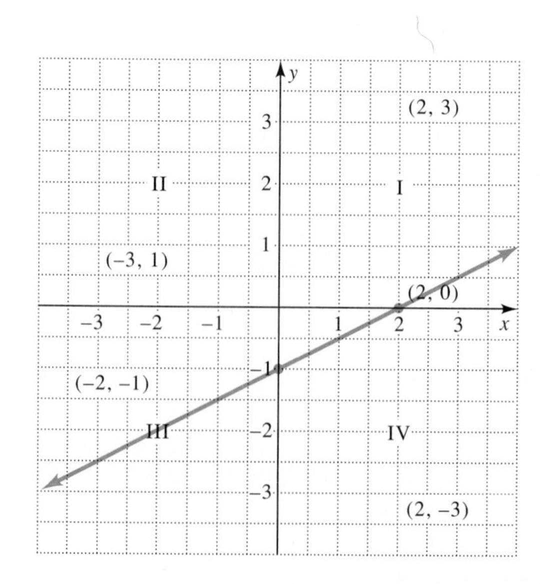

(a) (2, 3) lies in quadrant I. (b) $(-3, 1)$ lies in quadrant II.
(c) $(2, -3)$ lies in quadrant IV. (d) $(-2, -1)$ lies in quadrant III.
(e) A point on an axis does not lie in any quadrant. □

Let's consider solving the equation

$$2x - 3y = 6$$

Since there are two variables, a solution will consist of a value of x and a value of y that make the equation a *true* statement.

We can see that if x is 3 and y is 0, we have

$$2(3) - 3(0) = 6$$

For these values the equation is a true statement. We can write this solution as an ordered pair, writing the x-value first and the y-value second:

$$(3, 0)$$

Also, if x is 0 and y is -2, we have

$$2(0) - 3(-2) = 6$$

This is also a true statement. Thus we write the ordered pair

$$(0, -2)$$

Likewise, if x is 6 and y is 2, the equation is a true statement:

$$2(6) - 3(2) = 6$$

So we write the ordered pair

$$(6, 2)$$

Still another solution is the ordered pair $(-3, -4)$.

We have found four solutions, which we write as ordered pairs. They are:

$$(3, 0), \quad (0, -2), \quad (6, 2), \quad (-3, -4)$$

There are an infinite number of solutions. So, we cannot list them all. However, if we plot the ordered pairs that we have found to be in the solution set, we can make an interesting observation.

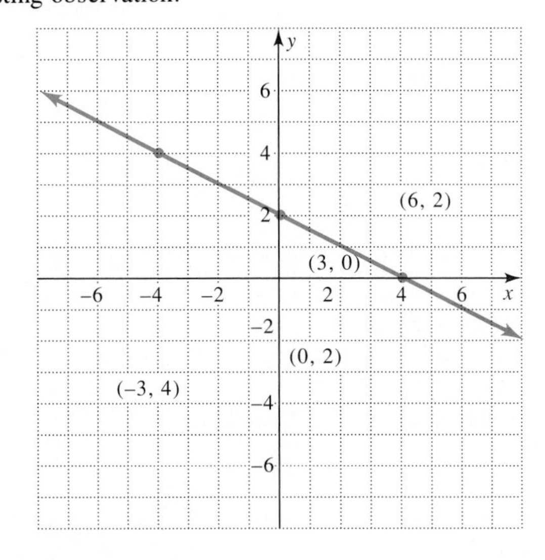

The points appear to lie on a line. In fact, every member of the solution set of the equation $2x - 3y = 6$ is an ordered pair whose graph lies on this line. Furthermore, any point on this line represents an ordered pair that belongs to the solution set. So, the line is a picture of the solution set of the equation $2x - 3y = 6$. We call this line the **graph** of the equation.

The Graph of Linear Equations in Two Variables

If A, B, and C are constants with A and B not *both* zero, the graph of

$$Ax + By = C$$

is a line.

$Ax + By = C$ is called the **standard form** of the equation.

Some examples of linear equations in two variables written in standard form are as follows:

$3x + 7y = 5$ (A is 3, B is 7, and C is 5)

$x - 2y = 0$ (A is 1, B is -2, and C is 0)

$x = 3$ (A is 1, B is 0, and C is 3)

EXAMPLE 3. Determine if the given ordered pair is in the solution set of the given equation.

(a) $5x + 3y = 8$; $(-1, 3)$ (b) $x - 2y = 0$; $(0, 0)$ (c) $x = y$; $(-1, 1)$

(d) $x = 2$; $(0, 4)$ (e) $y + 3 = 0$; $(-1, -3)$

Solutions:

(a) $5x + 3y = 8$; $(-1, 3)$
Replacing x with -1 and y with 3 gives

$$5(-1) + 3(3) = -5 + 9$$
$$= 4$$

Since 4 is not equal to 8, $(-1, 3)$ is not in the solution set.

(b) $x - 2y = 0$; $(0, 0)$
Substituting 0 for x and y gives

$$0 - 2(0) = 0$$

Because this makes a true statement, $(0, 0)$ is in the solution set.

(c) $x = y$; $(-1, 1)$
Since -1 is not equal to 1, $(-1, 1)$ is not in the solution set.

(d) $x = 2$; $(0, 4)$
This equation is the same as,

$$x + (0)y = 2$$

If we replace x with 0 and y with 4, we have

$$0 + 0(4) = 0$$

Since 0 is not equal to 2, $(0, 4)$ is not in the solution set.

(e) $y + 3 = 0$; $(-1, -3)$

If we subtract 3 from both sides, this becomes

$$y = -3$$

This is the same as,

$$(0)x + y = -3$$

We replace x with -1 and y with -3.

$$0(-1) + (-3) = -3$$

This makes a true statement, so $(-1, -3)$ is in the solution set. □

EXAMPLE 4. Find four points that are on the graph of the equation $2x + y = 4$.

Solution:

We choose a value for x, say 0, and then determine what the value of y must be to make the equation a true statement. If we replace x with 0, the equation becomes

$$2(0) + y = 4$$

Solving for y gives us

$$y = 4$$

Thus the point (0, 4) is on the graph.
It is convenient to list this information in a table with the x-coordinate on the left and the y-coordinate on the right.

x	y
0	4

Now, if we let y be 0 and solve for x,

$$2x + 0 = 4$$
$$2x = 4$$
$$x = 2$$

x	y
0	4
2	0

So the point (2, 0) is also on the graph.
Let's choose 3 as a value for x. We replace x by 3:

$$2(3) + y = 4$$
$$6 + y = 4$$
$$y = -2$$

x	y
0	4
2	0
3	−2

The point $(3, -2)$ is on the graph.

To find a fourth point, let y have a value of -3.

$$2x + (-3) = 4$$
$$2x = 7$$
$$x = \frac{7}{2}$$

x	y
0	4
2	0
3	-2
$\frac{7}{2}$	-3

This gives the point $\left(\frac{7}{2}, -3\right)$.

Four points on the graph of $2x + y = 4$ are

$$(0, 4), \quad (2, 0), \quad (3, -2), \quad \text{and} \quad \left(\frac{7}{2}, -3\right)$$ □

A common problem is to graph a linear equation. Since the graph of a linear equation is a line, two distinct points are enough to determine the line. If we look at Example 4, we can see that the easiest points to find are those where x is 0 and where y is 0. So, to graph a linear equation, we will find these two points. A third point may be found to serve as a check.

EXAMPLE 5. Draw the graph of $x + 2y = 2$.

Solution:

We let x be 0 and then solve for y.

$$x + 2y = 2$$
$$0 + 2y = 2$$
$$2y = 2$$
$$y = 1$$

x	y
0	1

We let y be 0 and then solve for x.

$$x + 2(0) = 2$$
$$x + 0 = 2$$
$$x = 2$$

x	y
0	1
2	0

To get a third point, we let y be 2 and solve for x.

$$x + 2(2) = 2$$
$$x + 4 = 2$$
$$x = -2$$

x	y
0	1
2	0
-2	2

We plot these points and use a straightedge to draw the line.

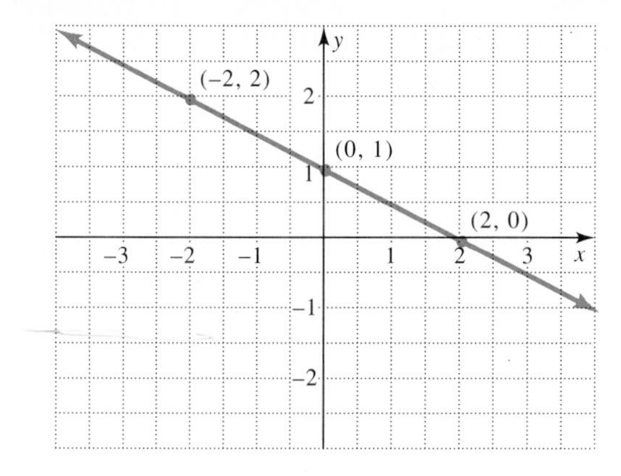

□

Notice that when we let x have the value 0 and find the y-value that makes the equation true, we have found where the line crosses the y-axis. We call this y-value the **y-intercept.** Likewise the **x-intercept** is the x-value where the line crosses the x-axis. In Example 5 the x-intercept is 2 and the y-intercept is 1.

EXAMPLE 6. Find the intercepts of the equation $2x - 4y = 4$ and draw its graph.

Solution:

To find the x-intercept, we replace y with 0 and solve for x. To find the y-intercept, we replace x with 0 and solve for y.

x	y
2	0
0	-1

To graph the line, we plot the points $(0, -1)$ and $(2, 0)$.

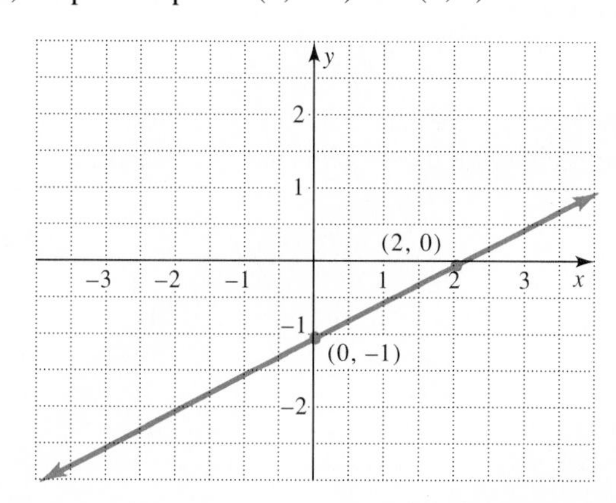

□

To Find the Intercepts of a Line

1. To find the x-intercept, replace y with 0 and solve for x.
2. To find the y-intercept, replace x with 0 and solve for y.

EXAMPLE 7. Draw the graph of $x + y = 0$.

Solution:

To find the x-intercept, we replace y with 0 and solve for x. We see that y is also 0. This gives us the point (0, 0). When we find the y-intercept, we get (0, 0) again. Both the x-intercept and the y-intercept are at the origin.

We need to find another point. If we let y be 1 and solve for x, we get -1 for x. Thus $(-1, 1)$ is a second point on the graph.

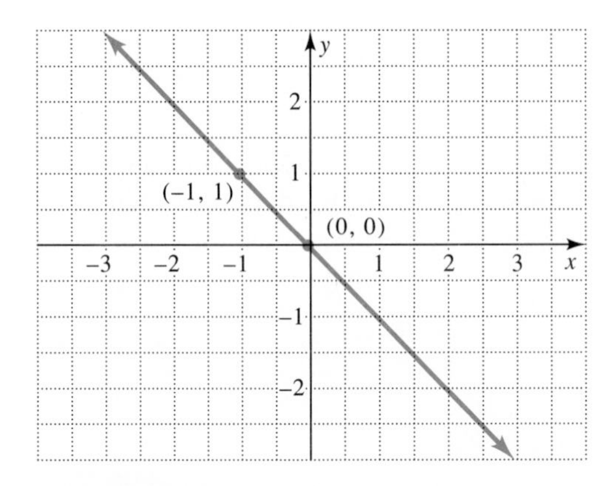

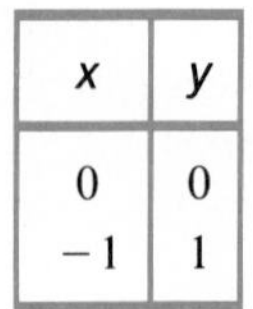

x	y
0	0
−1	1

□

Notice that the line in Example 7 goes through the origin. If C is 0 in the standard form $Ax + By = C$, then the line will go through the origin. The x-intercept as well as the y-intercept are 0.

Does every line have both an x-intercept and a y-intercept?

EXAMPLE 8. Draw the graph of $x = 3$.

Solution:

If we write this equation as

$$x + (0)y = 3$$

we see that y can be any value and x is always 3. So (3, 1) and $(3, -2)$ are two points on the graph. We plot these points and draw the line.

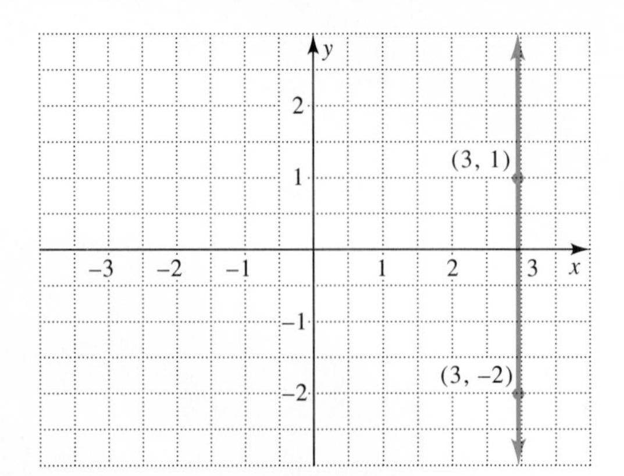

x	y
3	1
3	−2

□

The line in Example 8 is called a *vertical line*.

Vertical Lines

The graph of the equation $x = p$
is a **vertical** line with an x-intercept of p.

EXAMPLE 9. Draw the graph of $y = 2$.

Solution:

If we write this equation as

$$(0)x + y = 2$$

we can see that x can be any value and y is always 2.
The points (1, 2) and (−3, 2) are on the graph. We plot these points and draw the line.

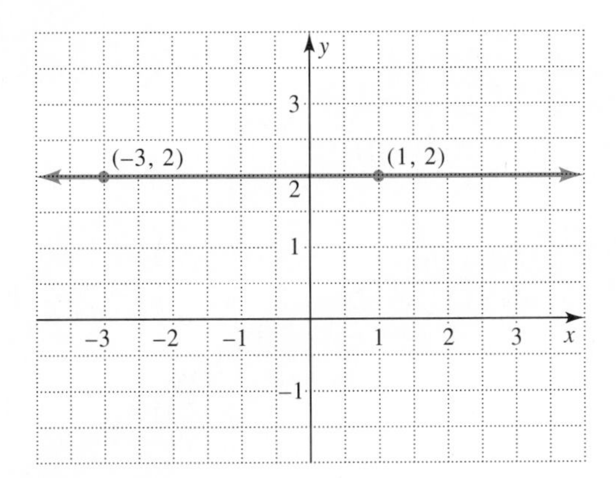

x	y
1	2
−3	2

The line in Example 9 is called a *horizontal line.*

Horizontal Lines

The graph of the equation $y = q$
is a **horizontal** line with a y-intercept of q.

PROBLEM SET 6.1

Warm-ups

Plot the following ordered pairs. See Example 1.

1. (2, 3) **2.** (1, −3) **3.** (−2, 2) **4.** (−2, −3)

Give the quadrant in which each point lies. See Example 2.

5. $(1, 9)$ **6.** $(-2, 8)$ 7. $\left(\frac{1}{2}, -6\right)$ **8.** $\left(-\frac{4}{3}, -\frac{9}{5}\right)$

Determine which of the given ordered pairs are in the solution set of the given equation. See Example 3.

9. $2x - y = 4$; $(2, 0), (0, 2), (1, 3)$

10. $x + 3y = 6$; $(0, 2), (-1, 2), (1, 3)$

11. $x + y = 0$; $(0, 0), (1, 1), (1, -1)$

12. $x = 2y$; $(1, 2), (2, 1), (-2, -1)$

Find three points on the graph of each equation. See Example 4.

13. $x + y = 4$

14. $x = 3y$

15. $\frac{1}{2}x - y = 0$

16. $x - \frac{1}{2}y = 3$

17. $y = 4$

18. $2x - 3y = 6$

19. $x = \frac{7}{3}$

20. $\frac{3}{2}x + y = 5$

Find the intercepts of each equation and graph each. See Examples 5 and 6.

21. $x + 3y = 6$ **22.** $x - y = 3$ **23.** $x + 3y = 0$ **24.** $2x - 5y = 10$

25. $\frac{1}{5}x - 1 = 0$ **26.** $\frac{1}{2}y + 1 = 0$ **27.** $x = -1$ **28.** $y = \frac{1}{2}$

Practice Exercises

Plot the following ordered pairs.

29. $(3, 2)$ **30.** $(-3, -2)$ 31. $(-3, 3)$ **32.** $(2, -2)$

33. $(0, 2)$ **34.** $(0, -1)$ 35. $\left(-\frac{7}{3}, \frac{2}{3}\right)$ **36.** $\left(\frac{1}{2}, 0\right)$

37. $(0, -3)$

38. $(-3, 0)$

Give the quadrant in which each point lies.

39. $(1, -9)$

40. $(-2, -8)$

41. $(-4, 6)$

42. $(6, 0)$

Determine which of the given ordered pairs are in the solution set of the given equation.

43. $3x - y = 4$; $(2, 0), (0, -4), (1, 3)$

44. $x + 2y = 5$; $(0, 5), (-1, -2), (3,$

45. $x - y = 0$; $(0, 0), (1, 1), (1, -1)$

46. $x = -2y$; $(1, 2), (2, 1), (2, -1)$

47. $5x - y = 3$; $(0, -3), (0, -1), (-1, -8)$

48. $x - 4y = 5$; $(3, -1), (5, 0), (0, 5)$

49. $x = 7$; $(0, 7), (7, 7), (7, 3)$

50. $y = 5$; $(0, 5), (-3, 5), (0, 0)$

Find three points on the graph of each equation.

51. $x + y = 3$

52. $x = -3y$

53. $x - 4y = 0$

54. $3x - y = 4$

55. $y = -3$

56. $2x - 4y = 12$

57. $x = 8$

58. $5x + 2y = 10$

Find the intercepts of each equation and graph each.

59. $x + 4y = 4$

60. $x - y = 2$

61. $x = -2$

62. $y = \frac{5}{2}$

63. $2x - 3y = 12$

64. $3x = y$

65. $\frac{1}{3}x - y = 3$

66. $4x + y = 4$

67. $x = \frac{-3}{2}$

68. $x = y$

69. $\frac{1}{3}x + \frac{1}{5}y = 1$

70. $\frac{2}{3}x - \frac{3}{2}y = 0$

71. $x + 5y = 0$

72. $3x - 5y = 15$

73. $x - 1 = 0$

74. $y + 3 = 0$

75. $2x - 3y = -6$

76. $4x = y$

77. $x - 4y = 4$

78. $3x + y = 3$

79. $x = \frac{-3}{4}$

80. $2x = y$

81. $5x + 4y = 20$

82. $x - 5y = 0$

Challenge Problems

83. Plot five points with the same x- and y-coordinates on the same set of axes. For example, (2, 2) and $(-3, -3)$. Connect these points. What seems to be true? Write an equation for a line whose points have the same x and y coordinates.

84. Write an equation for a line whose points have coordinates which are opposites.

85. Write an equation for a line on which each y-coordinate is twice the x-coordinate.

IN YOUR OWN WORDS...

86. What is a rectangular coordinate system?

87. What are the intercepts of a line and how are they found?

6.2 SLOPE

How are the following two lines different?

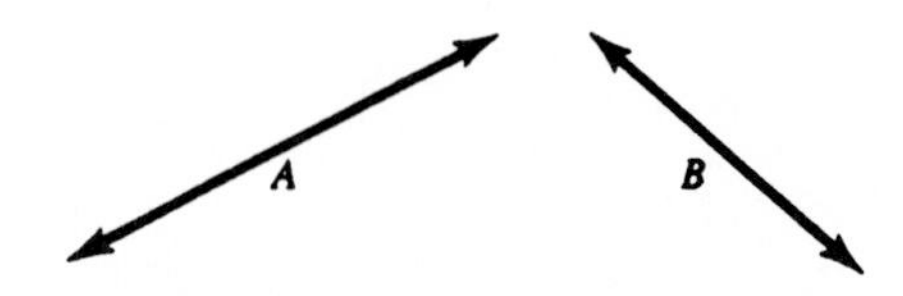

Moving along each line from left to right, we could say that line A is going up, whereas line B is going down. The rate at which a line goes up or down is the most important property of a line. We call this property the **slope** of the line. To determine the slope of a line, we start with any two points on the line, say P_1 with coordinates (x_1, y_1) and P_2 with coordinates (x_2, y_2).

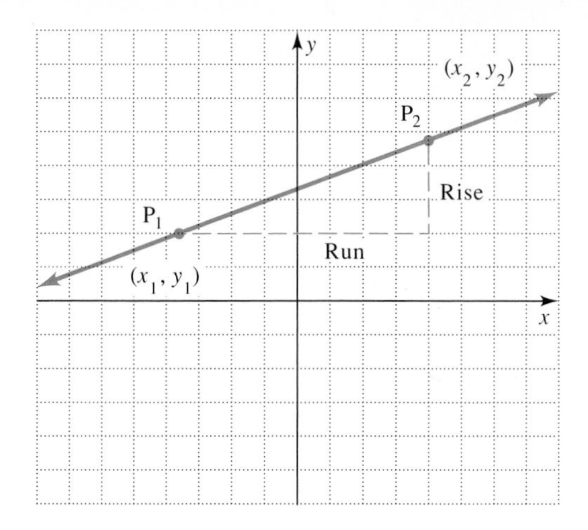

As we move from P_1 to P_2 the vertical change is called the **rise,** and the horizontal change is called the **run.** We define the slope of a line as the ratio of the rise to the run.

$$\text{Slope} = \frac{\text{rise}}{\text{run}}$$

To find the rise, we calculate the difference in the y-coordinates. The y-coordinate of P_1 is y_1 and that of P_2 is y_2.

$$\text{Rise} = y_2 - y_1$$

To find the run, we find the difference in the x-coordinates. The x-coordinate of P_1 is x_1 and that of P_2 is x_2. So,

$$\text{Run} = x_2 - x_1$$

We use the letter m to designate the slope. Thus the slope of the line containing the points (x_1, y_1) and (x_2, y_2) is given by the following formula.

The Slope Formula

The slope of the line through the points (x_1, y_1) and (x_2, y_2) is given by

$$m = \frac{y_2 - y_1}{x_2 - x_1}, \qquad x_1 \neq x_2$$

EXAMPLE 1. Find the slope of the line that contains the points (1, 5) and (4, 7).

Solution:

If we let P_1 be the point (1, 5) and P_2 be the point (4, 7), then we can see that x_1 is 1, y_1 is 5, x_2 is 4, and y_2 is 7. Substituting these values into the formula gives

$$m = \frac{y_2 - y_1}{x_2 - x_1}$$

$$= \frac{7-5}{4-1}$$

$$= \frac{2}{3}$$

The slope is $\frac{2}{3}$.

(continued)

The graph of the line is as follows. Notice that the line is going up from left to right.

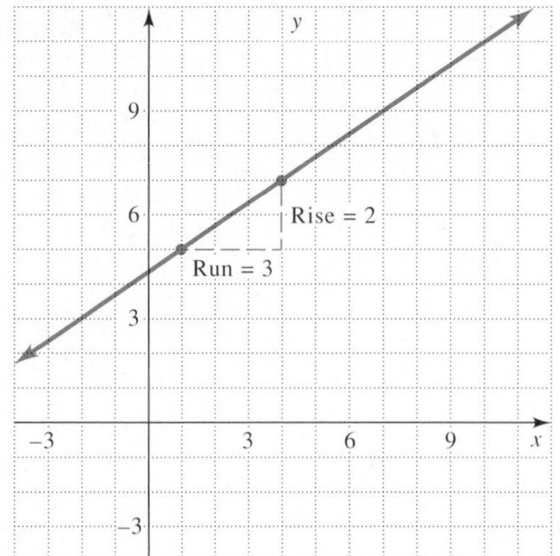

□

What would happen in Example 1 if we let P_1 be the point (4, 7) and P_2 be the point (1, 5)? It would mean that x_1 is 4, y_1 is 7, x_2 is 1, and y_2 is 5. Substituting these values into the slope formula gives

$$\begin{aligned} m &= \frac{5-7}{1-4} \\ &= \frac{-2}{-3} \\ &= \frac{2}{3} \end{aligned}$$

Notice that the slope is the same in both cases. It doesn't make any difference which point we call P_1 or P_2. However, it *is* important that the coordinates are subtracted in the same order. The order in which the x-values are subtracted *must be the same* as the order in which the y-values are subtracted.

EXAMPLE 2. Find the slope of the line that contains the points (3, 9) and (8, 5).

Solution:

If we let P_1 be (3, 9) and P_2 be (8, 5), we have $x_1 = 3$, $y_1 = 9$, $x_2 = 8$, and $y_2 = 5$. We substitute these into the slope formula.

$$\begin{aligned} m &= \frac{y_2 - y_1}{x_2 - x_1} \\ &= \frac{5-9}{8-3} \\ &= \frac{-4}{5} \\ &= -\frac{4}{5} \end{aligned}$$

The slope is $-\frac{4}{5}$.

The graph of the line is shown next. Notice that it goes down from left to right.

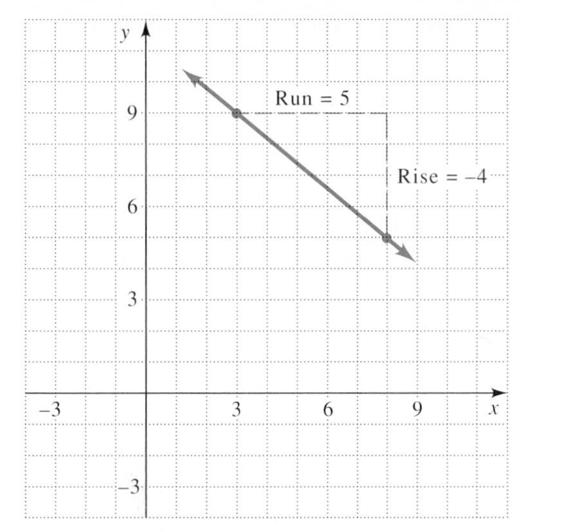

□

The next examples show that we must be careful with signs.

EXAMPLE 3. Find the slope of the line containing the points $(-2, 4)$ and $(3, -1)$.

Solution:

If P_1 is $(-2, 4)$ and P_2 $(3, -1)$, we have x_1 is -2, y_1 is 4, x_2 is 3, and y_2 is -1.

$$m = \frac{y_2 - y_1}{x_2 - x_1}$$

$$= \frac{-1 - 4}{3 - (-2)}$$

$$= \frac{-5}{5}$$

$$= -1$$

The slope is -1. □

EXAMPLE 4. Find the slope of the line that goes through the points $(-3, 0)$ and $(-1, 2)$.

Solution:

If we let P_1 be $(-3, 0)$ and P_2 be $(-1, 2)$, we have

$$m = \frac{y_2 - y_1}{x_2 - x_1}$$

$$= \frac{2 - 0}{-1 - (-3)}$$

$$= \frac{2}{2}$$

$$= 1$$

The slope is 1. □

The graphs of the lines in Examples 3 and 4 are shown next. Notice that lines going up have a *positive slope,* whereas lines going down have a *negative slope*.

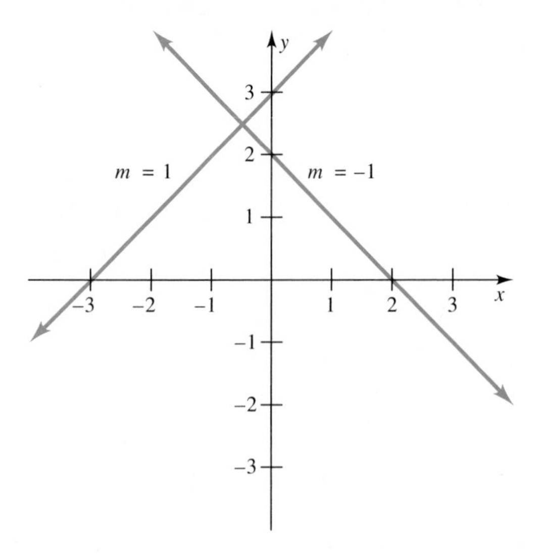

In the next example, the line determined by the given points is a horizontal line. Look carefully at the calculation of its slope.

EXAMPLE 5. Find the slope of the line that contains the points (1, 3) and (4, 3).

Solution:

If we let P_1 be (1, 3) and P_2 be (4, 3), the slope formula gives

$$m = \frac{y_2 - y_1}{x_2 - x_1} = \frac{3-3}{4-1}$$

$$= \frac{0}{3}$$

$$= 0$$

The slope is 0. □

The y-coordinates of all points on a horizontal line are the same, so the rise is 0. Thus the slope is 0. The graph is shown next.

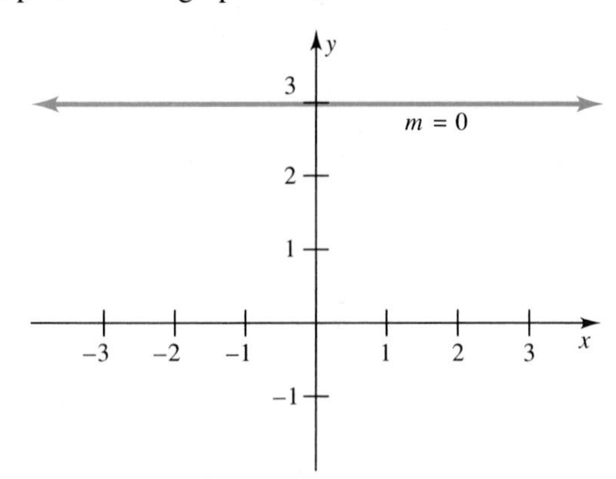

The next example illustrates a vertical line.

EXAMPLE 6. Find the slope of the line through the points (4, −1) and (4, 5).

Solution:

If P_1 is (4, −1) and P_2 is (4, 5), we have

$$m = \frac{y_2 - y_1}{x_2 - x_1} = \frac{5 - (-1)}{4 - 4}$$

$$= \frac{6}{0}$$

Since division by 0 is not allowed, the slope is undefined. We say a vertical line has no slope. ▭

The x-coordinates of all points on a vertical line are the same. The run will be 0. Thus a vertical line has no slope. The graph is shown next.

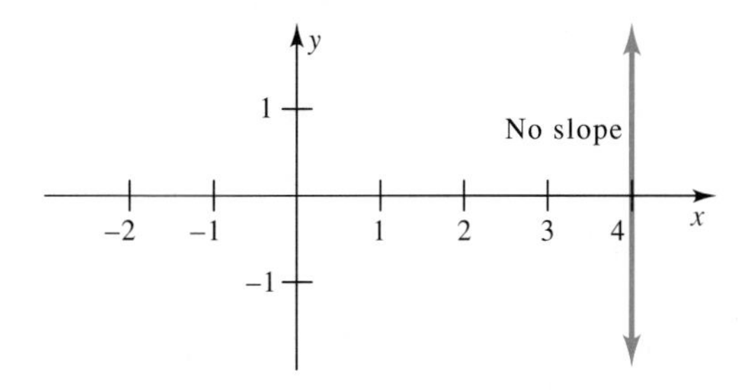

Horizontal and Vertical Lines

1. *Horizontal lines* have equations of the form

$$y = q$$

and have a slope of 0.

2. *Vertical lines* have equations of the form

$$x = p$$

and have *no* slope.

If we are given the equation of a line, then we can find its slope by finding two points on the line and using the slope formula.

EXAMPLE 7. Find the slope of the graph of $x + 2y = 4$.

Solution:

The graph is a line. We find two points on the line. The intercepts are convenient to use.

$$x + 2(0) = 4$$
$$x = 4$$
$$0 + 2y = 4$$
$$y = 2$$

x	y
4	0
0	2

(continued)

We use the points (4, 0) and (0, 2) as P_1 and P_2 in the slope formula.

$$m = \frac{y_2 - y_1}{x_2 - x_1} = \frac{2 - 0}{0 - 4}$$

$$= \frac{2}{-4}$$

$$= -\frac{1}{2}$$

The slope is $-\frac{1}{2}$ □

EXAMPLE 8. Find the slope of the graph of $x - 2y = 0$.

Solution:

We have a line that goes through the origin. So, we need one other point. If y has a value of 1, then

$$x - 2(1) = 0$$

$$x = 2$$

This gives the point (2, 1). We use the points (0, 0) and (2, 1) in the slope formula:

$$m = \frac{y_2 - y_1}{x_2 - x_1} = \frac{1 - 0}{2 - 0}$$

$$= \frac{1}{2}$$

The slope is $\frac{1}{2}$. □

EXAMPLE 9. Find the slope of the graphs of each equation.

(a) $x = 14$ (b) $y = -8$

Solutions:

(a) The graph is a vertical line. It has no slope.

(b) The graph is a horizontal line. Its slope is 0. □

Parallel and Perpendicular Lines

l_1 is a line with slope m_1 and l_2 is a line with slope m_2. Assume that neither line is vertical.

1. If l_1 and l_2 are **parallel** lines, then their slopes are the same; that is,

$$m_1 = m_2$$

2. If l_1 and l_2 are **perpendicular** lines, then the product of their slopes is -1; that is,

$$m_1 m_2 = -1, \quad \text{or} \quad m_1 = -\frac{1}{m_2}$$

EXAMPLE 10. If the slope of l_1 is $\frac{3}{4}$, find the slope of a line that is

(a) Parallel to l_1
(b) Perpendicular to l_1

Solutions:

(a) The slope of any line parallel to l_1 is $\frac{3}{4}$.
(b) The slope of any line perpendicular to l_1 is $-\frac{4}{3}$, since $\frac{3}{4} \cdot (-\frac{4}{3}) = -1$. □

EXAMPLE 11. Find the slope of a line parallel to the line containing the points $(-2, 5)$ and $(3, -1)$.

Solution:

We first find the slope of the line containing the points $(-2, 5)$ and $(3, -1)$. The slope formula gives

$$\begin{aligned} m &= \frac{y_2 - y_1}{x_2 - x_1} \\ &= \frac{5 - (-1)}{-2 - 3} \\ &= \frac{6}{-5} \\ &= -\frac{6}{5} \end{aligned}$$

The slope of a line parallel to this line is the same. So, the slope of the line is $-\frac{6}{5}$.
□

EXAMPLE 12. Find the slope of a line that is perpendicular to the line through the points $(-3, 4)$ and $(-1, -5)$.

Solution:

We use the slope formula to find the slope of the line through the points $(-3, 4)$ and $(-1, -5)$.

$$\begin{aligned} m &= \frac{y_2 - y_1}{x_2 - x_1} \\ &= \frac{-5 - 4}{-1 - (-3)} \\ &= \frac{-9}{2} \\ &= -\frac{9}{2} \end{aligned}$$

The slope of a line perpendicular to this line has slope $\frac{2}{9}$ because $-\frac{9}{2} \cdot \frac{2}{9} = -1$.

□

EXAMPLE 13. Find the slope of a line that is parallel to the graph of $x + 2y = -2$ and find the slope of a line that is perpendicular to the graph of $x + 2y = -2$.

Solution:

We can find the slope of the graph of $x + 2y = -2$ by finding two points on the line and then using the slope formula. We see that $(0, -1)$ and $(-2, 0)$ are on the line.

$$m = \frac{0 - (-1)}{-2 - 0}$$

$$= -\frac{1}{2}$$

The slope of the graph of $x + 2y = -2$ is $-\frac{1}{2}$. The slope of a line parallel to this line will also be $-\frac{1}{2}$. The slope of a line perpendicular to this line is 2 because $-\frac{1}{2}(2) = -1$. □

PROBLEM SET 6.2

Warm-ups

In Problems 1 through 15, find the slope of the line if it exists containing the given points. In Problems 1 through 6, see Examples 1 and 2.

1. $(2, 1), (3, 2)$

2. $(4, 1), (2, 5)$

3. $(0, 2), (4, 3)$

4. $(3, 4), (1, 1)$

5. $(1, 1), (2, 2)$

6. $(0, 2), (2, 0)$

In Problems 7 through 12, see Examples 3 and 4.

7. $(2, 1), (3, -6)$

8. $(-1, 2), (1, 3)$

9. $(3, -1), (-2, 2)$

10. $(2, -4), (-3, -5)$

11. $(-2, -2), (-3, 0)$

12. $(-5, -4), (-1, -3)$

In Problems 13 through 15, see Examples 5 and 6.

13. $(-1, 2), (3, 2)$

14. $(3, -2), (3, 4)$

15. $(-1, 0), (-1, 4)$

In Problems 16 through 21, find the slope of the graph of each equation. See Examples 7 through 9.

16. $x + 2y = 2$

17. $3x - y = 6$

18. $x - 4y = 0$

19. $2x + 3y = 0$

20. $x = 15$

21. $y = -5$

In Problems 22 through 25, the two given points determine a line. For each line, find (a) the slope of a line parallel to the given line, and (b) the slope of a line perpendicular to the given line. See Examples 11 and 12.

22. $(2, 3)$ and $(-1, 2)$

23. $(1, 4)$ and $(2, 0)$

24. $(2, 4)$ and $(2, 3)$

25. $(-1, -3)$ and $(-4, 2)$

In Problems 26 through 28, the equation of a line is given. For each equation, find (a) the slope of a line parallel to the given line, and (b) the slope of a line perpendicular to the given line.
See Example 13.

26. $x + y = 3$

27. $x - y = 2$

28. $2x - y = 2$

Practice Exercises

Find the slope of the line if it exists determined by the two given points.

29. (3, 2), (4, 3)

30. (5, 6), (3, 2)

31. (0, 3), (5, 4)

32. (−1, 2), (2, 0)

33. (3, −1), (−2, 2)

34. (3, 3), (−1, 9)

35. (4, 4), (2, 2)

36. (−2, −2), (3, 3)

37. (0, 3), (3, 0)

38. (−4, −3), (−2, 3)

39. (3, 3), (4, 3)

40. (0, 3), (0, 0)

41. (4, 2), (4, 7)

42. (−2, 3), (2, 4)

43. (4, −2), (3, 3)

Find the slope of the graph of each equation.

44. $x + y = 3$

45. $x - y = 2$

46. $2x + y = 6$

47. $x + 2y = 4$

48. $3x - y = 6$

49. $x - 3y = 8$

50. $x = -3$

51. $2x + 4y = 8$

52. $y = 7$

The two given points determine a line. For each line, find (a) the slope of a line parallel to the given line, and (b) the slope of a line perpendicular to the given line.

53. (1, 3) and (−2, 2)

54. (2, 5) and (1, 0)

55. (1, 4) and (3, 5)

56. (2, −3) and (1, 2)

57. (3, −3) and (2, −3)

58. (−1, −3) and (0, 0)

The equation of a line is given. For each equation, find (a) the slope of a line parallel to the given line, and (b) the slope of a line perpendicular to the given line.

59. $x + y = 1$

60. $x - y = 5$

61. $2x - y = -2$

62. $x - 3y = 8$

63. $3x - 5y = 15$

64. $x - y = 0$

65. An airplane flying from San Francisco to Boston is approaching Logan International Airport for a landing. If the airplane is 32 mi from Logan at an altitude of 2 mi, what should be the slope of its glide path to make a successful landing?

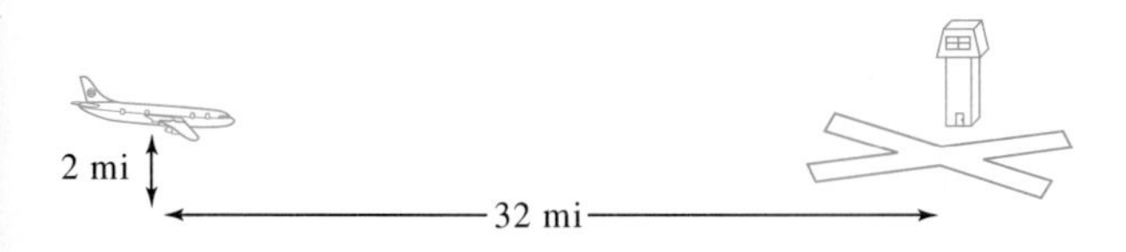

66. The pitch, or slope, of a roof gives it the ability to shed water or snow. Find the pitch of the roof shown.

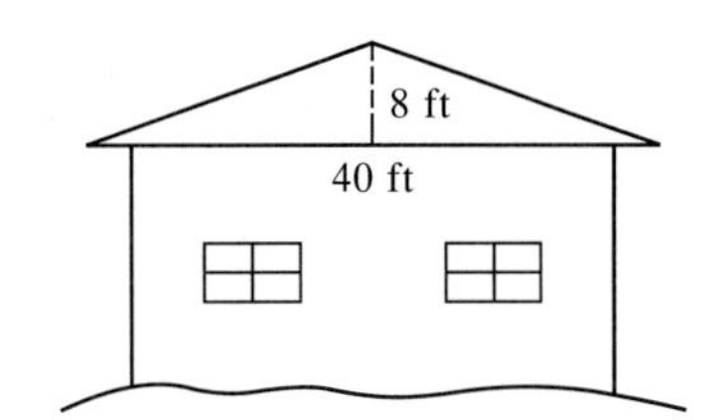

67. A ski slope descends 2200 ft in a straight line run of 16,000 feet. What is its slope?

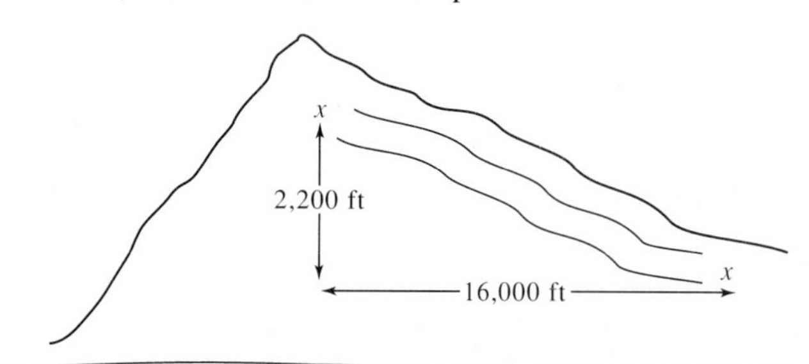

68. An airplane at an altitude of 10,000 ft is going to land at an airport 20 mi distant. What should the slope of its glide path be? (1 mi = 5280 ft).

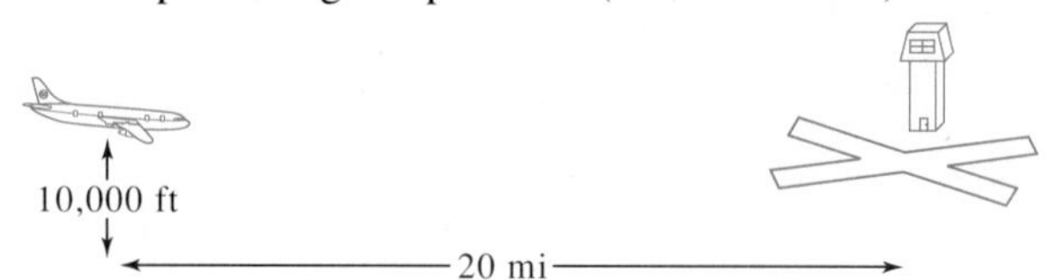

69. What is the pitch (slope) of the roof shown?

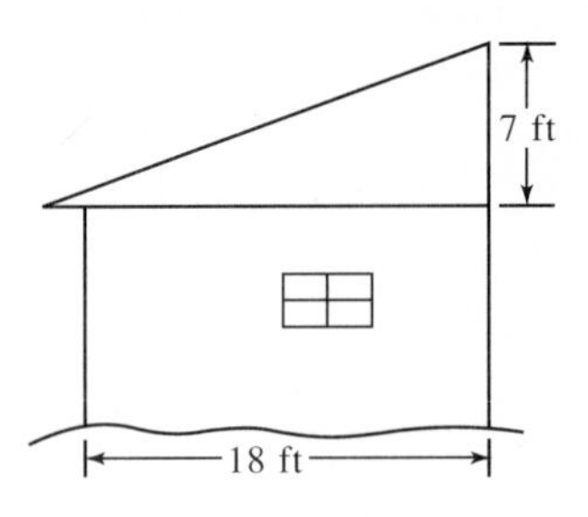

Challenge Problems

70. Find the slope of a line perpendicular to the graph of $x = 0$.

71. Find the slope of a line that is parallel to the graph of $x = 0$.

72. Find the slope of a line that is perpendicular to the graph of $y = 0$.

73. Find the slope of a line that is parallel to the graph of $y = 0$.

IN YOUR OWN WORDS...

74. What do we mean by the slope of a line?

75. How is the slope of a line calculated?

6.3 EQUATIONS OF LINES

We have been graphing linear equations written in standard form by finding the intercepts. We found the slope of the lines by using the slope formula. Consider the graph of $2x + 3y = 6$. We see that the x-intercept is 3 and the y-intercept is 2, giving the graph shown below.

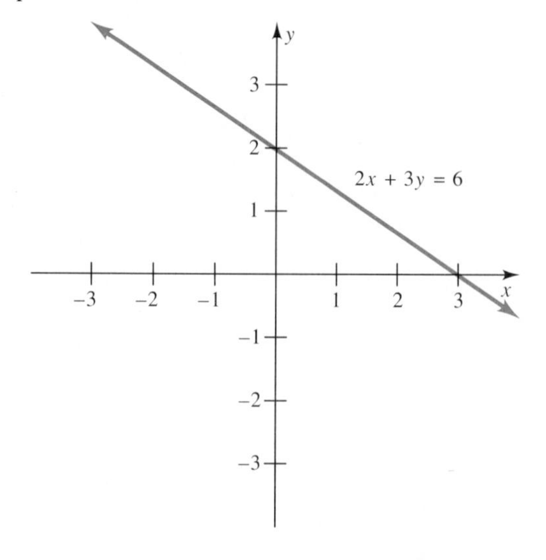

We can calculate the slope of this line by using the two points (0, 2) and (3, 0) in the slope formula.

$$m = \frac{2-0}{0-3}$$

$$= -\frac{2}{3}$$

The slope of the line is $-\frac{2}{3}$.

The equation $2x + 3y = 6$, can be written in a form other than the standard form. By solving for y, we can rewrite the equation as follows:

$$2x + 3y = 6$$

$$3y = -2x + 6 \qquad \text{Subtract } 2x.$$

$$y = -\frac{2}{3}x + 2 \qquad \text{Divide by 3.}$$

This form is called the **slope-intercept form.** Looking closely at this, we can make some interesting observations. The slope of the line, $-\frac{2}{3}$, and the y-intercept, 2, are very visible. This is not an accident. The equation of any nonvertical line can be written in this form.

Slope-Intercept Form

If a linear equation can be written in the form

$$y = mx + b$$

then the graph has slope m and y-intercept b.

Any nonvertical line can be written in this very useful form by solving for y.

EXAMPLE 1. Find the slope and y-intercept of the graph of each equation.

(a) $3x + y = 6$ (b) $5x - 2y = 4$

Solutions:

(a) We rewrite the equation in slope-intercept form by solving for y.

$$3x + y = 6$$

$$y = -3x + 6 \qquad \text{Subtract } 3x.$$

The slope is -3 and the y-intercept is 6.

(b) Writing the equation in slope-intercept form, we have

$$5x - 2y = 4$$

$$-2y = -5x + 4 \qquad \text{Subtract } 6x.$$

$$y = \frac{5}{2}x - 2 \qquad \text{Divide by } -2.$$

The slope is $\frac{5}{2}$ and the y-intercept is -2. □

It is often convenient to use the slope-intercept form when drawing the graph of a linear equation. The y-intercept gives us a point on the line, and we use the slope to find a second point.

EXAMPLE 2. Sketch the graph of the equation $y = 2x + 1$.

Solution:

As the equation is in slope-intercept form, we see that the slope is 2 and the y-intercept is 1. To draw the line, we locate the y-intercept. This gives us one point on the line. To find a second point, we use the slope, which is 2.

$$m = 2,$$

or,

$$m = \frac{2}{1} = \frac{\text{rise}}{\text{run}}$$

The rise is 2 and the run is 1. We start at the y-intercept and move *right* 1 unit and then *up* 2 units. This gives us a second point on the line.

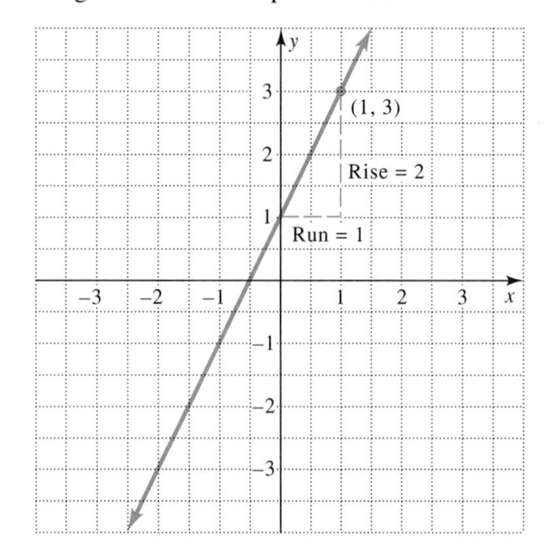

□

EXAMPLE 3. Sketch the graph of $y = -\frac{2}{3}x - 1$.

Solution:

The slope is $-\frac{2}{3}$ and the y-intercept is -1.
We locate the y-intercept and use the slope to find a second point.

$$m = -\frac{2}{3}$$

or,

$$m = \frac{-2}{3}$$

The rise is -2 and the run is 3. From the y-intercept, we go *right* 3 units and *down* 2 units.

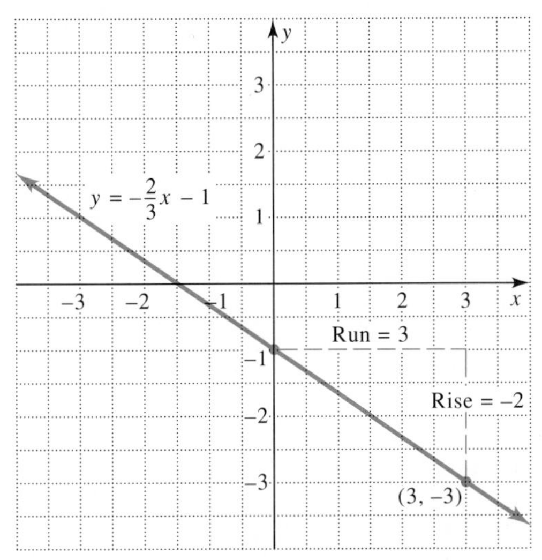

□

Notice that the run is always from left to right and that the rise is *up* if the slope is *positive* and *down* if the slope is *negative*.

Often we are given a line and asked to find an equation of that line. We need two pieces of information about the line:

1. The slope of the line.

2. The coordinates of any point on the line.

Suppose a line has slope 2 and contains the point (2, 4). Then we know that $m = 2$ and the point (2, 4) is on the line. Let (x, y) be *any* other point *on the line.*

If we substitute this information into the slope formula, we can get an equation of the line. Let P_1 be (2, 4) and P_2 be (x, y). Then

$$m = \frac{y_2 - y_1}{x_2 - x_1}$$

Slope → $2 = \dfrac{y - 4}{x - 2}$ ← From the point (2, 4)

$2(x - 2) = y - 4$ Multiply by $(x - 2)$.

$2x - 4 = y - 4$ Distributive Property

$2x - y = 0.$

Notice that we have found an equation for a line with slope 2 that contains the point (2, 4).

When using a given slope of m, a given point of (x_1, y_1), and a general point of (x, y), the slope formula becomes

$$m = \frac{y - y_1}{x - x_1}$$

If we multiply both sides by $x - x_1$, we have

$$m(x - x_1) = y - y_1$$

We usually turn this around and write,

$$y - y_1 = m(x - x_1)$$

This is called the **point-slope** form of the equation of a line.

Linear equations are often given in *standard* form. We have seen that one very useful form is the *slope-intercept* form. The *point-slope* form is convenient for finding an equation of a line given a point on the line and its slope.

Forms of the Linear Equation in Two Variables

Standard Form

$$Ax + By = C, \qquad A \text{ and } B \text{ not both } 0.$$

Slope-Intercept Form

$$y = mx + b$$

Point-Slope Form

$$y - y_1 = m(x - x_1)$$

EXAMPLE 4. Find an equation of the line with slope $\frac{1}{2}$ that contains the point (3, 1).

Solution:

We use the point-slope form, where m is $\frac{1}{2}$ and (x_1, y_1) is (3, 1). Substituting into the point-slope formula gives

$$y - y_1 = m(x - x_1)$$

$$y - 1 = \frac{1}{2}(x - 3)$$

$$2y - 2 = x - 3 \qquad \text{Multiply by 2.}$$

$$1 = x - 2y$$

or

$$x - 2y = 1$$

The desired equation is $x - 2y = 1$. □

EXAMPLE 5. Find an equation of the line with slope -4 that goes through the point $(-3, -2)$.

Solution:

In the point-slope formula, we substitute -4 for m and -3 for x_1 and -2 for y_1.

$$y - y_1 = m(x - x_1)$$

$$y - (-2) = -4[x - (-3)]$$

$$y + 2 = -4(x + 3)$$

$$y + 2 = -4x - 12$$

$$4x + y = -14$$

□

EXAMPLE 6. Find an equation of the line containing the points $(-1, 2)$ and $(3, -4)$.

Solution:

To write an equation of a line, we need to know the *slope* of the line and the *coordinates* of any *point* on the line. We have the coordinates of two points on the line, so we need the slope. We can find it with the slope formula.

$$m = \frac{y_2 - y_1}{x_2 - x_1} = \frac{2 - (-4)}{-1 - 3}$$

$$= \frac{6}{-4}$$

$$= -\frac{3}{2}$$

Now we pick *either* of the two given points to be (x_1, y_1). Let's use $(-1, 2)$ and substitute into the point-slope form.

$$y - y_1 = m(x - x_1)$$

$$y - 2 = -\frac{3}{2}[x - (-1)]$$

$$2y - 4 = -3(x + 1) \qquad \text{Multiply by 2.}$$

$$2y - 4 = -3x - 3$$

$$3x + 2y = 1$$

An equation of the line is $3x + 2y = 1$. □

If a line has a slope, we can use the point-slope form to write an equation for it.

> **To Write an Equation of a Line that has Slope**
>
> 1. Find m, the slope of the line.
> 2. Find (x_1, y_1), the coordinates of any point on the line.
> 3. Substitute into the point-slope form,
>
> $$y - y_1 = m(x - x_1)$$
>
> 4. Write the equation in whichever form is preferred.

A special case of this occurs if we are given the slope and y-intercept of a line.

EXAMPLE 7. Write an equation of the line with slope of 2 and y-intercept of 3.

Solution:

Since we know the slope and the y-intercept, we can use the *slope-intercept* form and write an equation of the line directly.

$$y = mx + b$$

$$y = 2x + 3$$

This is an equation of the line in slope-intercept form. □

Since a vertical line has no slope, we must treat it in a different way. It is usually best to treat horizontal lines in the same manner.

Equations of Horizontal and Vertical Lines

Horizontal Lines

A line with slope 0 is *horizontal* and has as its equation

$$y = q$$

where q is the y-coordinate of any point on the line.

Vertical Lines

A line with no slope is *vertical* and has as its equation,

$$x = p$$

where p is the x-coordinate of any point on the line.

EXAMPLE 8. Find an equation of the line through the points (1, 2) and (3, 2).

Solution:

First we find the slope of the line.

$$m = \frac{2 - 2}{1 - 3} = \frac{0}{-2} = 0$$

Since the slope is 0, the line is horizontal and we can write an equation directly:

$$y = 2$$

□

EXAMPLE 9. Find an equation of the vertical line through the point (3, −2).

Solution:

We cannot use the point-slope form because a vertical line has no slope. We remember that an equation of a vertical line must be of the form,

$$x = p$$

An equation of the line is $x = 3$. □

PROBLEM SET 6.3

Warm-ups

In Problems 1 through 6, write each equation in slope-intercept form. Find the slope and y-intercept of each graph and draw the graph of each. See Examples 1 through 3.

1. $2x + y = 0$
2. $2x + 3y = 6$
3. $2x - 3y = 9$

4. $x + 2y = 2$

5. $3x - y = 1$

6. $4x - 2y = 4$

In Problems 7 through 20, write an equation of the line satisfying the given conditions.
In Problems 7 through 9, see Examples 4 and 5.

7. Through the point $(2, -3)$ with slope $\frac{1}{2}$.

8. Through the point $(-1, 4)$ with slope $\frac{-2}{3}$.

9. Through the origin with slope of 1.

In Problems 10 through 13, see Example 6.

10. Containing the points $(1, 3)$ and $(-2, 1)$.

11. Containing the points $(7, -2)$ and $(3, -5)$.

12. Containing the points $(1, 3)$ and $(0, 0)$.

13. Containing the points $(0, 3)$ and $(-2, 0)$.

In Problems 14 and 15, see Example 7.

14. With y-intercept of 2 and slope of 1.

15. With y-intercept of -1 and slope of $\frac{2}{3}$.

In Problems 16 through 20, see Examples 8 and 9.

16. Containing the points $(2, 4)$ and $(2, 7)$.

17. Through $(1, 2)$ with no slope.

18. Through $(1, 2)$ with 0 slope.

19. Horizontal line through the point $(1, -2)$.

20. Vertical line through the point $(1, -2)$.

Practice Exercises

In Problems 21 through 29, write each equation in slope-intercept form. Find the slope and y-intercept of each graph and draw the graph of each.

21. $2x + y = 2$

22. $4x + y = 1$

23. $x - y = 3$

24. $5x + y = 0$

25. $2x - 3y = -6$

26. $2x + 4y = 8$

27. $x + 3y = 3$

28. $2x - y = 2$

29. $3x - 4y = 4$

In Problems 30 through 48, write an equation of the line satisfying the given conditions.

30. Through the point (4, 3) with slope 5.
31. Through the point $(-2, 0)$ with slope -1.
32. Through the point $(3, -3)$ with slope $\frac{3}{2}$.
33. Through the point $(-1, 3)$ with slope $\frac{-3}{2}$.
34. Through the origin with slope of -1.
35. With y-intercept of -2 and slope of 2.
36. With y-intercept of 4 and slope of $\frac{2}{5}$.
37. Through $(-1, 2)$ with no slope.
38. Through $(-1, 2)$ with 0 slope.
39. Vertical line through the point $(2, -2)$.
40. Horizontal line through the point $(2, -2)$.
41. Containing the points (5, 3) and $(-2, 4)$.
42. Containing the points $(-2, 5)$ and (0, 7).
43. Containing the points $(4, -6)$ and $(7, -5)$.
44. Containing the points (2, 2) and $(-1, 1)$.
45. Containing the points (1, 4) and (1, 7).
46. Containing the points $(-1, 3)$ and (0, 0).
47. Containing the points $(0, -3)$ and (2, 0).
48. x-intercept of -4 and y-intercept of 1.

Challenge Problems

49. Find an equation of the line through the origin with no slope.
50. Find an equation of the line through the origin with 0 slope.
51. Find an equation of the line through the origin parallel to the graph of $y = 2x + 4$.
52. Find an equation of the line through the origin perpendicular to the graph of $y = \frac{2}{3}x - 1$.
53. Find an equation of the line through the point (1, 3) and parallel to the line through the points (1, 4) and $(-1, 3)$.
54. Find an equation of the line through the point (2, 1) and perpendicular to the line through the points $(2, -5)$ and $(-3, 7)$.
55. Find an equation of the line through the point (1, 2) and parallel to the x-axis.
56. Find an equation of the line through the point $(-1, -3)$ and perpendicular to the y-axis.

IN YOUR OWN WORDS...

57. Compare the *standard form*, the *slope-intercept form*, and the *point-slope form* of an equation of a line.
58. Describe the general procedure for finding an equation of a line.

6.4 INEQUALITIES IN TWO VARIABLES

In this section we examine linear inequalities in two variables. Consider the inequality

$$x + y \leq 2$$

The solution set is the set of all ordered pairs that make the inequality a true statement. Because there are an infinite number of such ordered pairs, we don't try to list them. We draw a picture of the solution set and call it the **graph** of the inequality.

We begin by noting that the inequality $x + y \leq 2$ means

$$x + y = 2 \quad \text{or} \quad x + y < 2.$$

We can draw the graph of $x + y = 2$, as shown at the top of p. 353. Every point on the line will make $x + y = 2$ a true statement. This line is called a **boundary line** because it divides the plane into two regions, A and B.

Let's pick some points in each region and see if they are in the solution set of $x + y < 2$, as shown in the table under the graph.

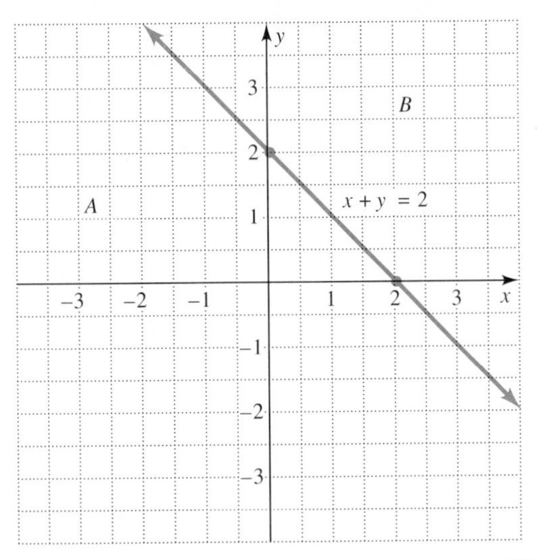

REGION	TEST POINT IN REGION	STATEMENT $x + y < 2$	TRUTH OF STATEMENT	REGION IN SOLUTION SET?
A	$(0, 0)$	$0 + 0 < 2$	True	Yes
A	$(-1, 0)$	$-1 + 0 < 2$	True	Yes
A	$(0, -1)$	$0 + (-1) < 2$	True	Yes
A	$(-1, -1)$	$-1 + -1 < 2$	True	Yes
B	$(2, 1)$	$2 + 1 < 2$	False	No
B	$(3, 0)$	$3 + 0 < 2$	False	No
B	$(0, 3)$	$0 + 3 < 2$	False	No
B	$(4, 2)$	$4 + 2 < 2$	False	No

It appears that every point in region A is in the solution set, whereas no point in region B is in the solution set. This is correct. To graph this solution set, we shade region A.

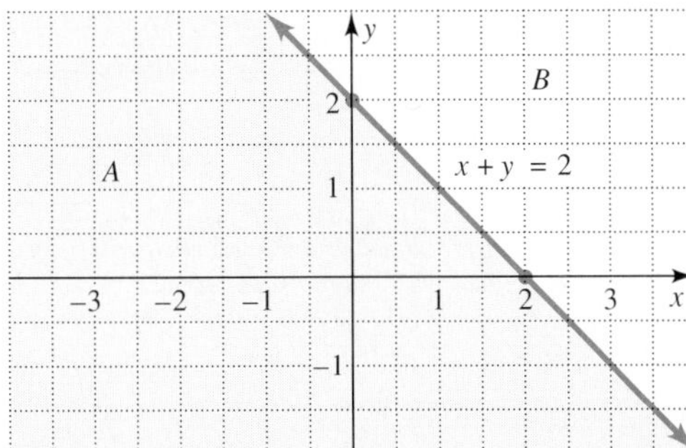

To graph a linear inequality in two variables, we first draw the boundary line. This line divides the plane into two regions. Pick one point in each region and test it in the original inequality to determine if the point is in the solution set. If one point from a region is in the solution set, then the entire region is in the solution set. If any point from a region is not in the solution set, then the entire region is not in the solution set.

If the boundary line itself is in the solution set, it is drawn as a *solid* line. If the boundary line is not in the solution set, it is drawn as a *dashed* line.

EXAMPLE 1. Graph the solution set for $x - y < 0$.

Solution:

The boundary line is $x - y = 0$. We graph it as a *dashed* line because equality is not included in $x - y < 0$. We pick a point from region A and one from region B.

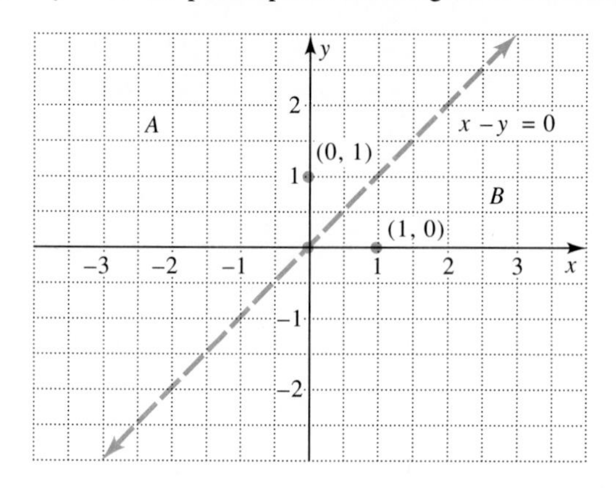

REGION	TEST POINT IN REGION	STATEMENT $x - y < 0$	TRUTH OF STATEMENT	REGION IN SOLUTION SET?
A	(0, 1)	$0 - 1 < 0$	True	Yes
B	(1, 0)	$1 - 0 < 0$	False	No

We shade region A.

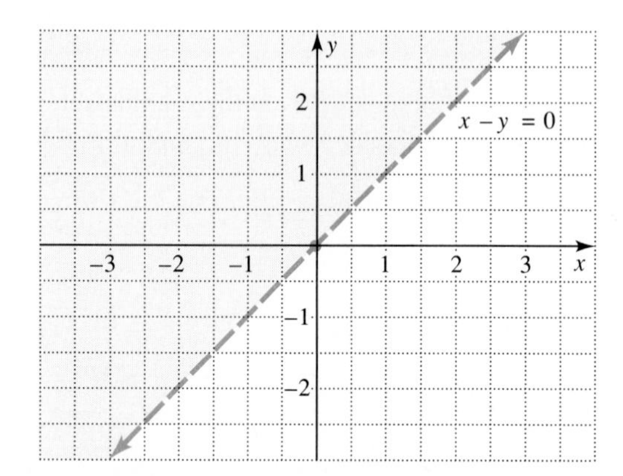

□

Procedure for Graphing Linear Inequalities in Two Variables

1. Graph the boundary line.
   (a) Draw a solid line if equality is included ($\leq$ or $\geq$).
   (b) Draw a dashed line if equality is not included ($<$ or $>$).
2. Determine which region formed by the line makes the inequality true by testing with one point from each region.
3. Shade the region(s) that make(s) the inequality true.

EXAMPLE 2. Graph the solution set for $x + 2y \geq 2$.

Solution:

The boundary line is $x + 2y = 2$.

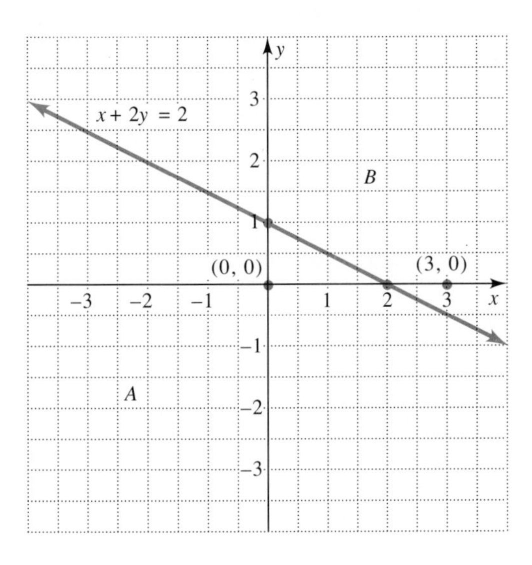

We draw a *solid* line because equality is included.

REGION	TEST POINT IN REGION	STATEMENT $x + 2y > 2$	TRUTH OF STATEMENT	REGION IN SOLUTION SET?
A	(0, 0)	$0 + 0 > 2$	False	No
B	(3, 0)	$3 + 0 > 2$	True	Yes

Notice that the origin, (0, 0), is always a good test point to use *if* it lies in one of the regions.

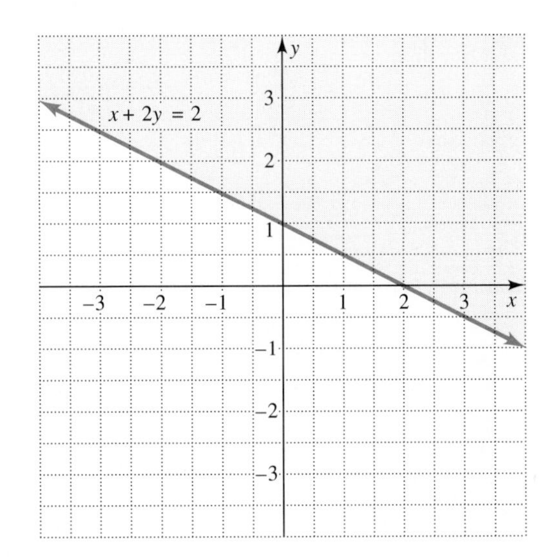

EXAMPLE 3. Graph the solution set for $y \le 1$ on the *xy-plane.*

Solution:

The boundary line is $y = 1$.
We draw a solid line because equality is included.

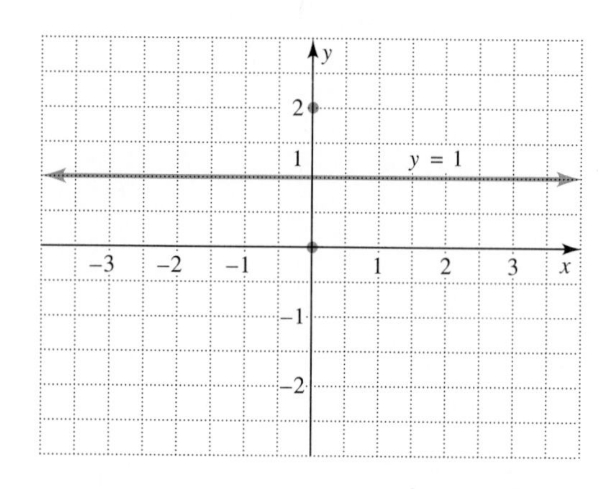

REGION	TEST POINT IN REGION	STATEMENT $y \le 1$	TRUTH OF STATEMENT	REGION IN SOLUTION SET?
A	(0, 0)	$0 \le 1$	True	Yes
B	(0, 2)	$2 \le 1$	False	No

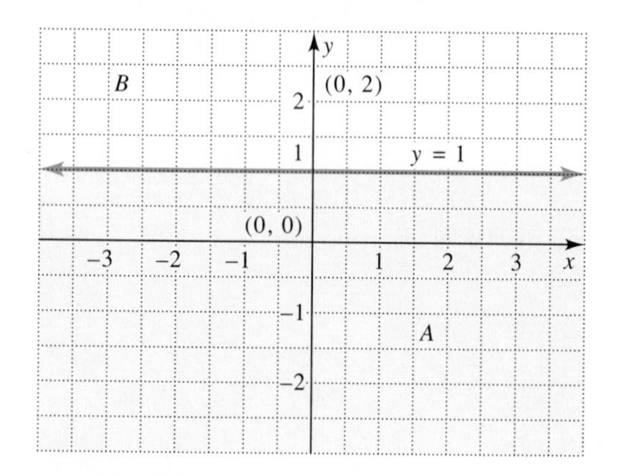

PROBLEM SET 6.4

Warm-ups

Graph the solution set for each inequality on the xy-plane.
In problems 1 through 3, see Examples 1 and 2.

1. $x + y \le 1$
2. $3x + 2y \ge 6$
3. $2x + y > 4$

In problems 4 through 6, see Example 3.

4. $x \leq 1$

5. $y > -2$

6. $2x \geq 7$

Practice Exercises

Graph the solution set for each inequality on the xy-plane.

7. $x + y \leq 2$

8. $3x + y > 1$

9. $x - y \geq -1$

10. $x \leq 7$

11. $3x + 2y < -6$

12. $4x - y > 1$

13. $4x - y \geq 0$

14. $y > -5$

15. $2x + 5y < 10$

16. $2x - 5y \geq 10$

17. $x > -1$

18. $y \leq \frac{3}{2}$

Challenge Problems

Graph the solution set for each inequality on the xy-plane.

19. $x \geq 0$

20. $x > 0$

21. $x \leq 0$

22. $x < 0$

23. $y > 0$

24. $y \geq 0$

25. $y < 0$

26. $y \leq 0$

Assume that a *is a positive real number and that* b *is a negative real number. Graph the solution set for each inequality on the xy-plane.*

27. $x > a$

28. $y \leq b$

29. $x \leq b$

IN YOUR OWN WORDS...

30. Describe how to graph a linear inequality in two variables.

6.5 FUNCTIONS (OPTIONAL)

So far, this chapter has concerned itself with pairs of numbers as coordinates of points on a plane and as solutions of equations and inequalities in two variables. It is very common to have numbers paired together in other situations. For example:

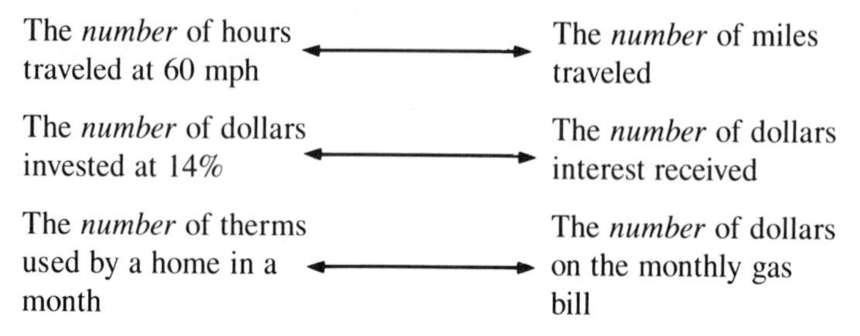

Such relationships lead us to an important mathematical concept.

> **Function**
>
> A *function* is a rule that assigns to each number in one set (the domain) exactly one number in another set (the range).

To illustrate the key ideas in this definition, we examine some functions that have domains containing just a few numbers.

Suppose we have a function named f with domain $\{2, 4, 5\}$. Further, suppose that f assigns the number 10 to 2, 20 to 4, and 25 to 5. This is conveniently illustrated by the following diagram.

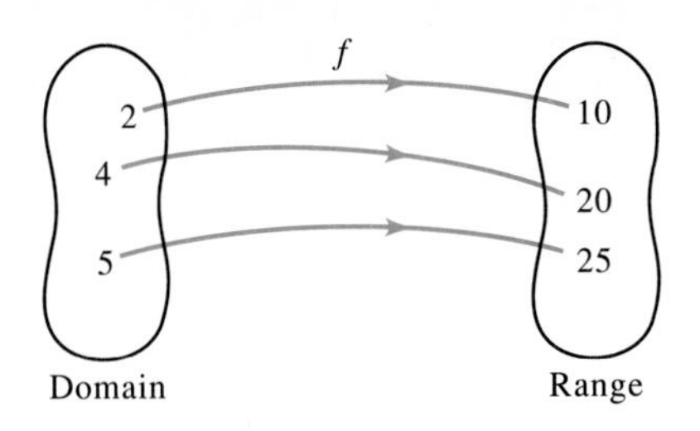

These are sometimes called "peanut drawings."

EXAMPLE 1. Consider the assignment made by each picture. Is each a function?

(a)

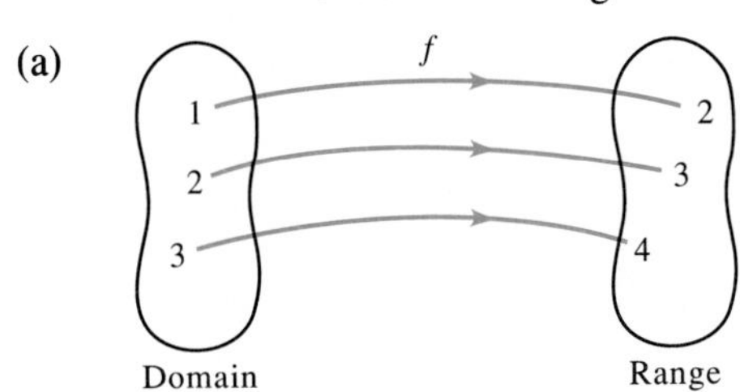

(b)

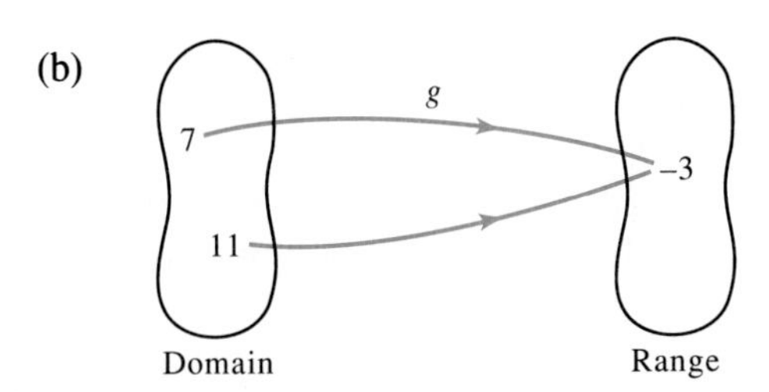

(c)

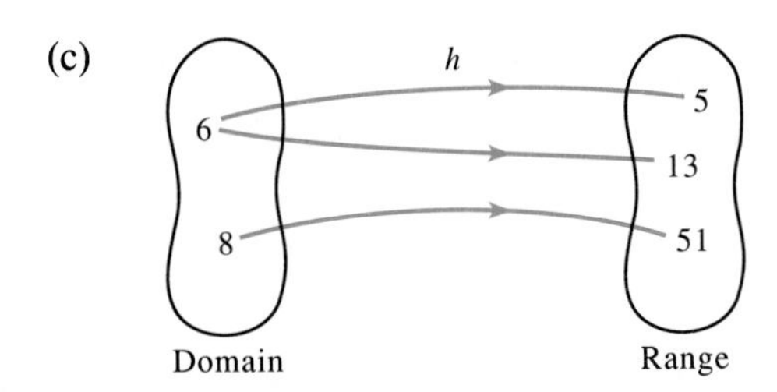

Solutions:

(a) Yes, f is a function. It assigns exactly one number in the range to each number in the domain.

(b) Yes, g is also a function. Although g assigns the same number to both 7 and 11, it assigns *exactly one* number to each.

(c) No, h is *not* a function. It assigns *two* numbers from the range to the number 6.

□

There are several ways to illustrate a function. Using the peanut drawings of Example 1 is one way. Sometimes the assignments are given as ordered pairs or in tables.

EXAMPLE 2. Suppose the rule for the function F is "add 5 to the number" and the domain is the set $\{1, 2, 3, 4\}$.

(a) Illustrate F with a drawing, as in Example 1.

(b) Illustrate F with a set of ordered pairs.

(c) What is the range of F?

Solutions:

(a)

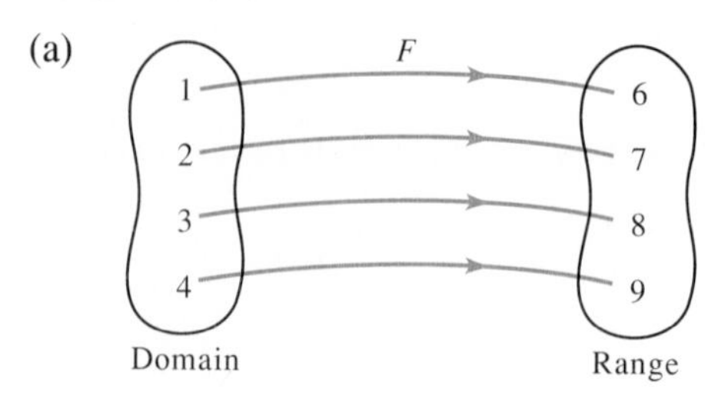

(b) $F = \{(1, 6), (2, 7), (3, 8), (4, 9)\}$

(c) $\{6, 7, 8, 9\}$ □

In mathematical applications of functions, the rule is often given using **functional notation.** The rule for the function F in Example 2 was "add 5 to the number." In functional notation this is written

$$F(x) = x + 5$$

The symbol $F(x)$ looks like F times x, but that is *not* what it means. We read $F(x)$ as "F of x" or "F evaluated at x."

Notice what happens if we replace x with a domain element in the functional notation for F.

$$F(1) = 1 + 5 = 6$$

$$F(2) = 2 + 5 = 7$$

$$F(3) = 3 + 5 = 8$$

$$F(4) = 4 + 5 = 9$$

If x is a number in the domain of F, then $F(x)$ is the corresponding number in the range.

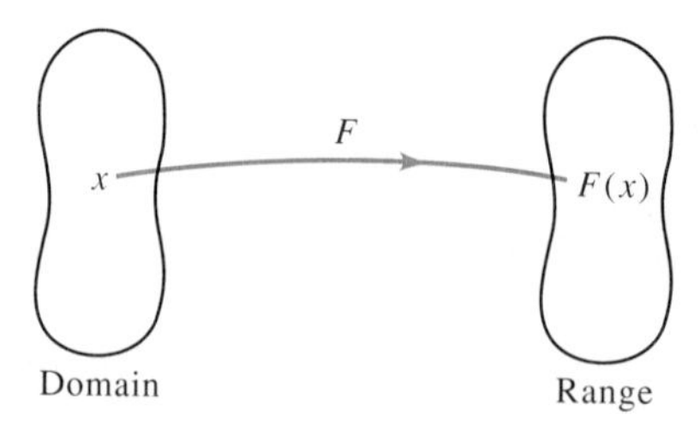

To define a function, a rule and a domain must be specified. The rule is usually specified with functional notation.

EXAMPLE 3. If $g(x) = 2x - 7$ with domain all real numbers, find each value.

(a) $g(2)$ (b) $g(0)$ (c) $g(-3)$ (d) $g(\pi)$

Solutions:

(a) $g(2) = 2(2) - 7 = -3$ (b) $g(0) = 2(0) - 7 = -7$

(c) $g(-3) = 2(-3) - 7 = -13$ (d) $g(\pi) = 2\pi - 7$ □

Although to define a function it is necessary to specify both a rule and a domain, it is common practice to state the rule with functional notation and omit the domain. *When this occurs, we assume the domain to be the* natural domain *of the function*. The **natural domain** is the largest subset of the real numbers for which the rule has meaning.

EXAMPLE 4. Give the natural domain of the following rules.

(a) $f(x) = 13x + x^2$ (b) $g(x) = \dfrac{1}{x - 6}$ (c) $h(x) = 7 + \sqrt{x}$

Solutions:

(a) We can multiply any number by 13 and we can square any number, so the natural domain of f is all real numbers.

(b) We *cannot* divide by zero, so this rule does not have meaning when x is 6. The natural domain is all real numbers *except* 6.

(c) $\sqrt{x}$ is a real number only if x is greater than or equal to 0. The natural domain of h is

$$\{x \mid x \geq 0\}$$ □

The graph of a function is a useful tool for visualizing its properties. To graph the function f, we graph the equation $y = f(x)$ for every x in the domain of f.

EXAMPLE 5. Graph the function $h(x) = 2x + 1$.

Solution:

We graph the equation $y = h(x)$. However, $h(x) = 2x + 1$, so we graph $y = 2x + 1$. The domain was not specified, so we use the natural domain, which is all real numbers.

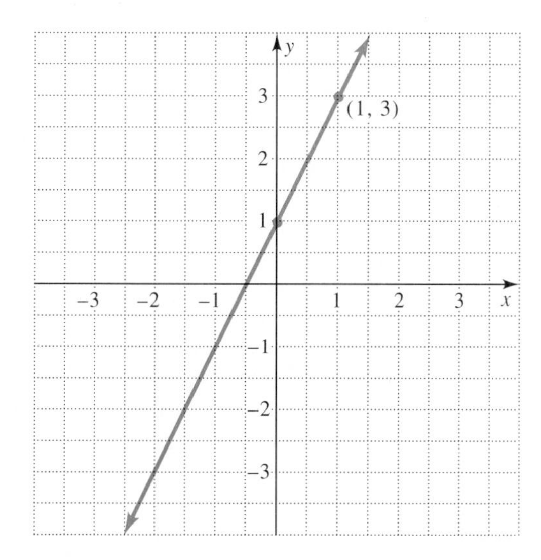

□

We can examine a graph to determine whether it can be the graph of a function or not. Consider the circle graphed next. Can it be the graph of a function?

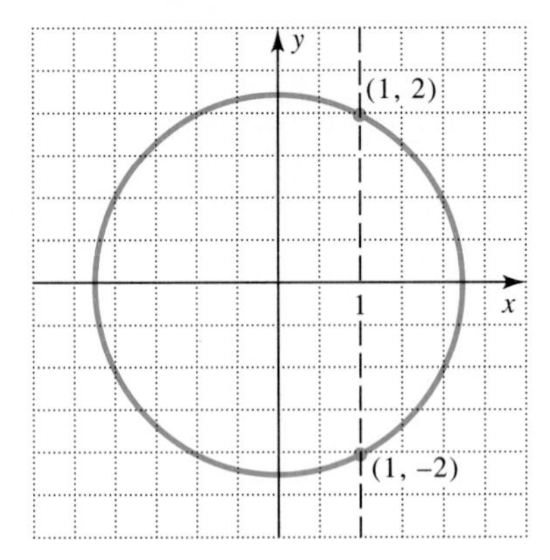

The graph contains the points (1, 2) and (1, −2). This means the domain element 1 must be assigned to *both* 2 and −2. This cannot be the graph of a function. Notice that both of the points are on the same *vertical* line.

> **Vertical Line Test**
>
> If any vertical line intersects a graph in more than one point, the graph is not the graph of a function.

EXAMPLE 6. Is each the graph of a function?

(a)

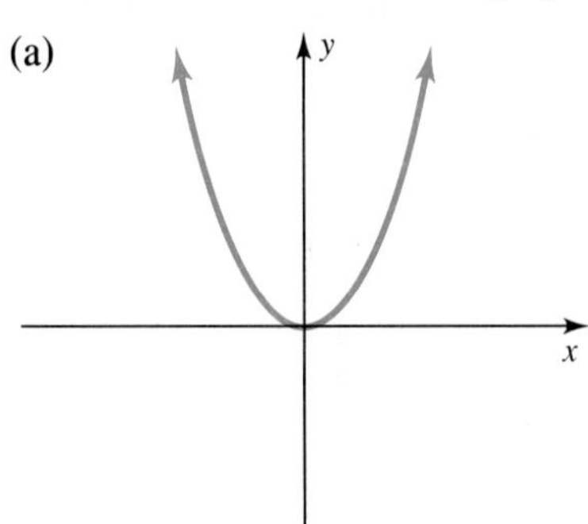

(b)

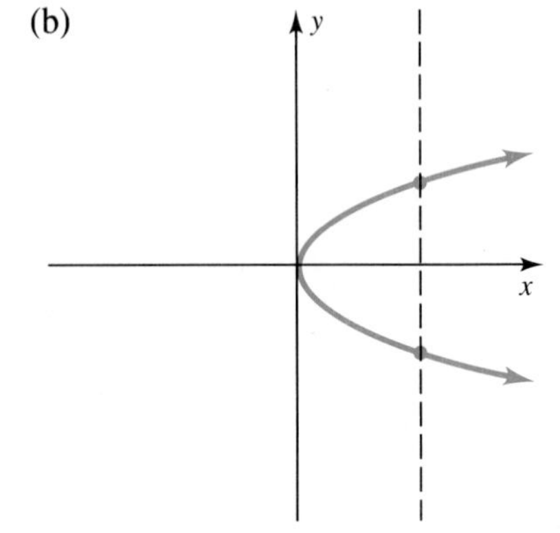

(c)

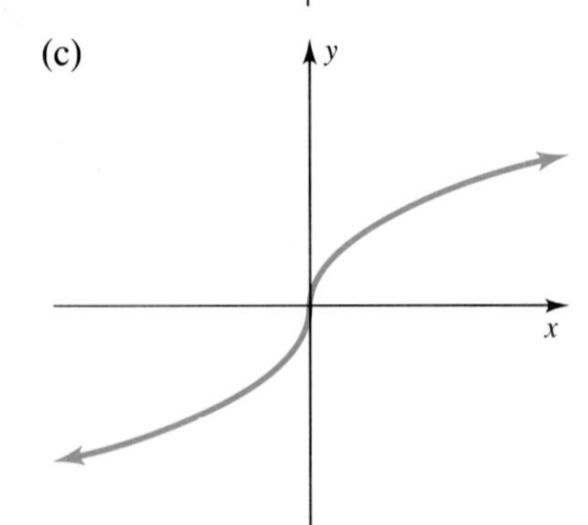

(d)

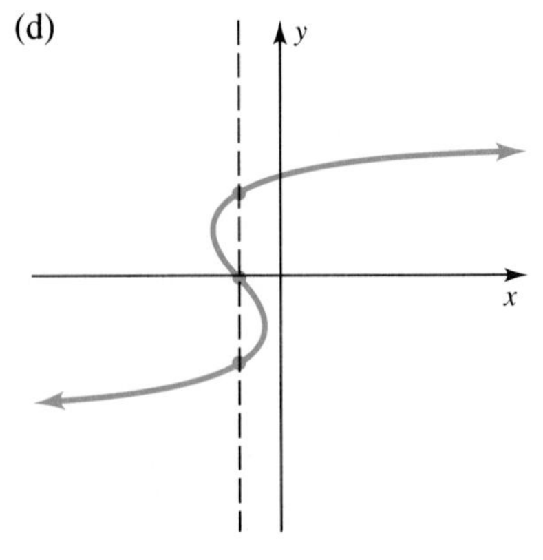

Solutions:

(a) Any vertical line strikes the graph at most once. This is the graph of a function.

(b) The dotted vertical line intersects the graph in two places. This graph cannot be the graph of a function.

(c) This graph passes the vertical line test and is the graph of a function.

(d) This graph fails the vertical line test and cannot be the graph of a function. □

Notice that the x-coordinate of any point on the graph of a function is an element of the domain of the function and the y-coordinate is an element of the range. Therefore, we can estimate the domain and range of a function by examining its graph.

EXAMPLE 7. Determine the domain and range of each function graphed below.

(a)

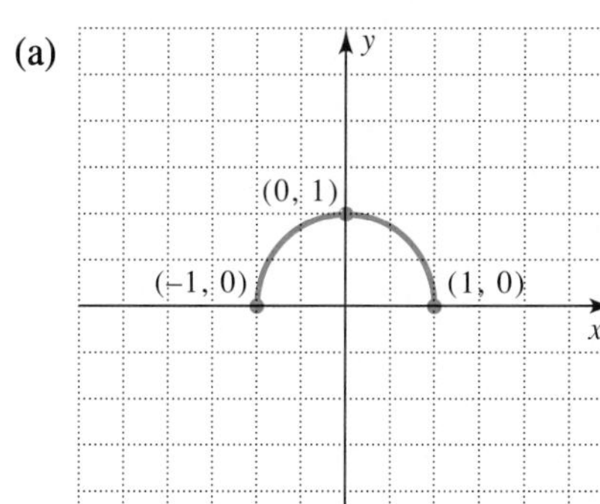

(b)

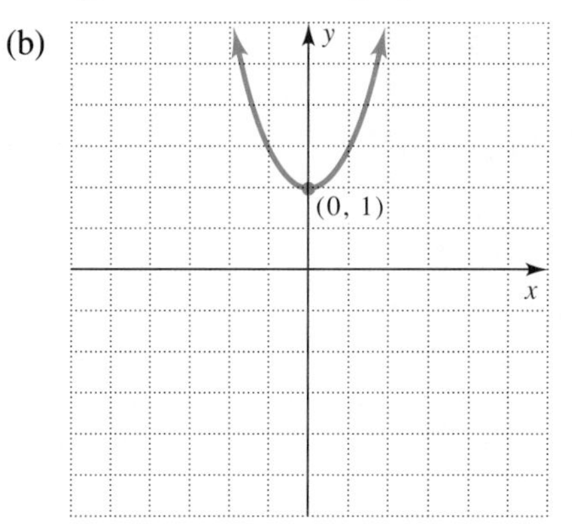

(c)

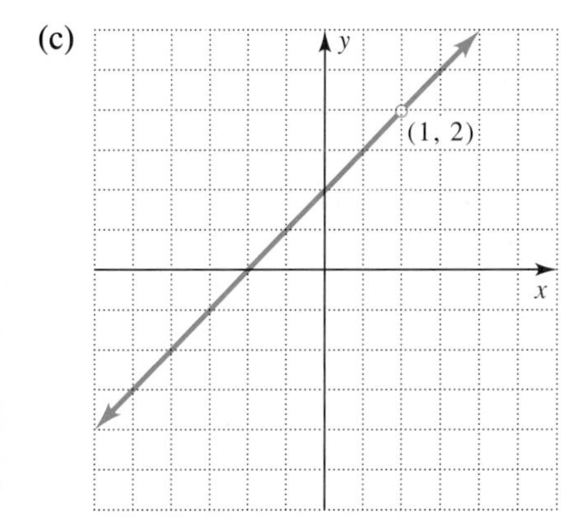

(d)

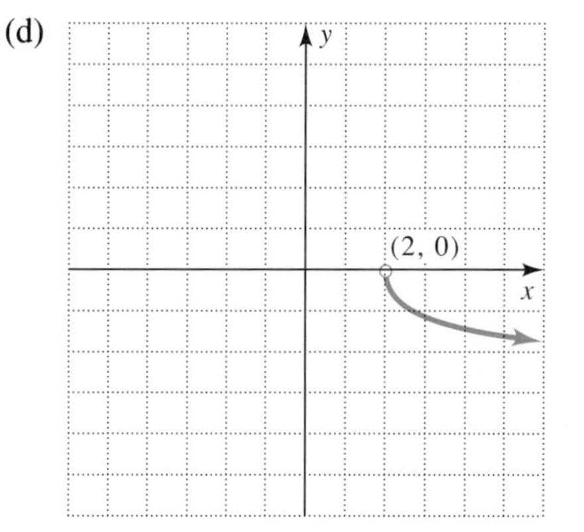

Solutions:

(a) Domain: $\{x \mid -1 \leq x \leq 1\}$
Range: $\{y \mid 0 \leq y \leq 1\}$

(b) Domain: {all real numbers}
Range: $\{y \mid y \geq 1\}$

(c) Domain: $\{x \mid x \neq 1\}$
Range: $\{y \mid y \neq 2\}$

(d) Domain: $\{x \mid x > 2\}$
Range: $\{y \mid y < 0\}$ □

PROBLEM SET 6.5

Warm-ups

In Problems 1 through 4, which assignments represent functions? See Example 1.

1.

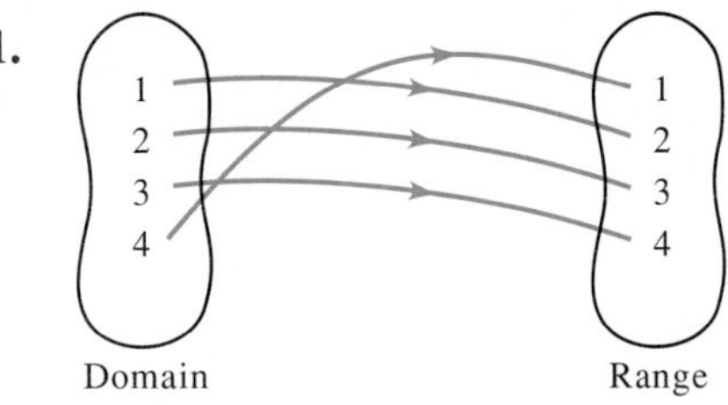

2.

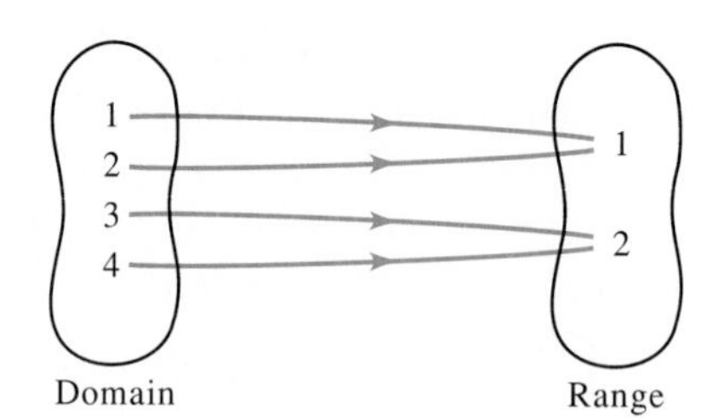

3.

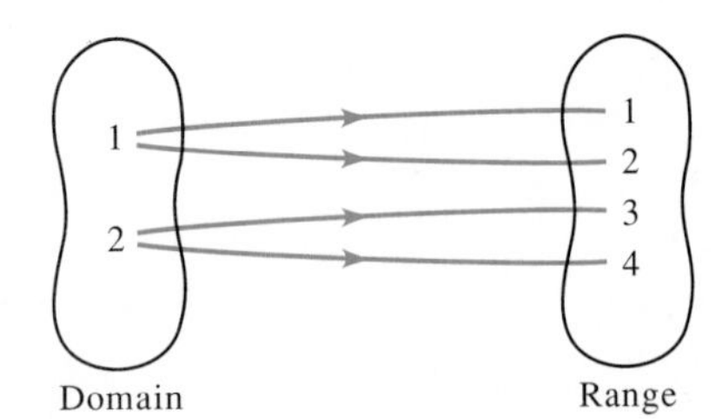

4.

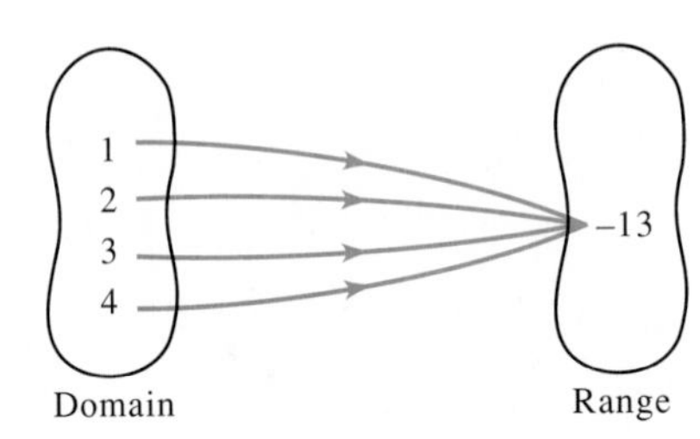

In Problems 5 and 6, illustrate each function with a set of ordered pairs, and find the range of the function. See Example 2.

5. The rule for f is square the number and add 2. The domain is $\{-1, 0, 1\}$.

6. The rule for g is assign 0 if the number is even and 1 if it is odd. The domain is $\{1, 2, 3, 4\}$.

In Problems 7 through 9, assume the domain of each function is the set of real numbers. See Example 3.

7. If $f(x) = x^2 - x$, find $f(-1)$, $f(0)$, $f(1)$, and $f(7)$.

8. If $g(x) = \dfrac{x+1}{x^2+1}$, find $g(0)$, $g(1)$, $g(-1)$, and

9. If $h(x) = x^2 - 3x + 1$, find $h(0)$, $h(3)$, $h(-1)$, $h(-3)$. $h(0) = 1$;

In Problems 10 through 13, write the natural domain of each rule. See Example 4.

10. $f(x) = x^3 + x^2 + x + 1$

11. $g(x) = \dfrac{x-1}{(x-2)(x-3)}$

12. $h(x) = \dfrac{1-x^2}{1+x^2}$

13. $F(x) = \sqrt{x-5}$

In Problems 14 and 15, graph the given function. See Example 5.

14. $f(x) = 4x - 2$

15. $g(x) = -\frac{2}{3}x + 1$

In Problems 16 and 17, decide if the graphs could be graphs of functions. See Example 6.

16.

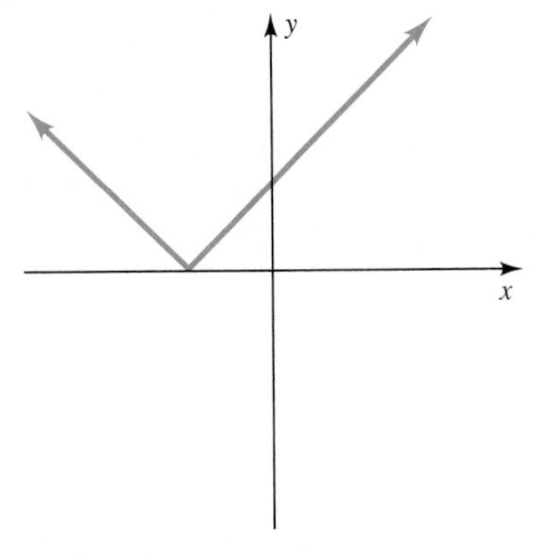

17.

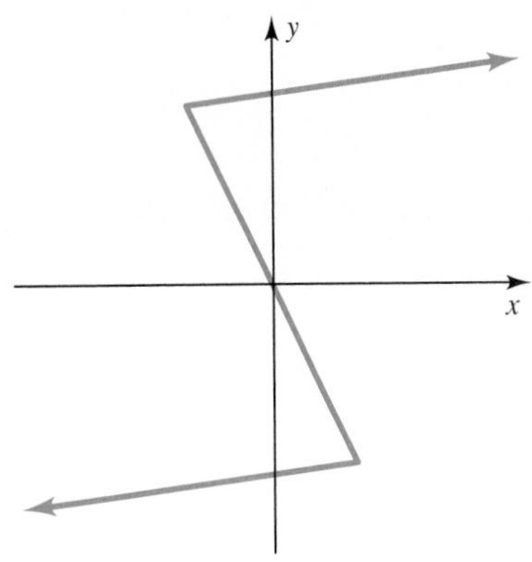

In Problems 18 through 21, determine the domain and range of each function by examining its graph. See Example 7.

18.

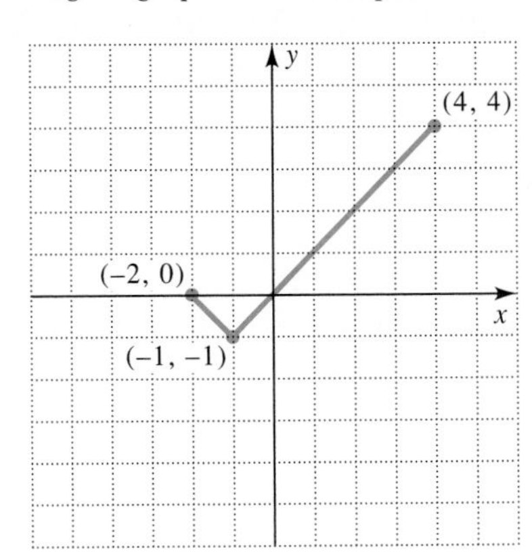

19.

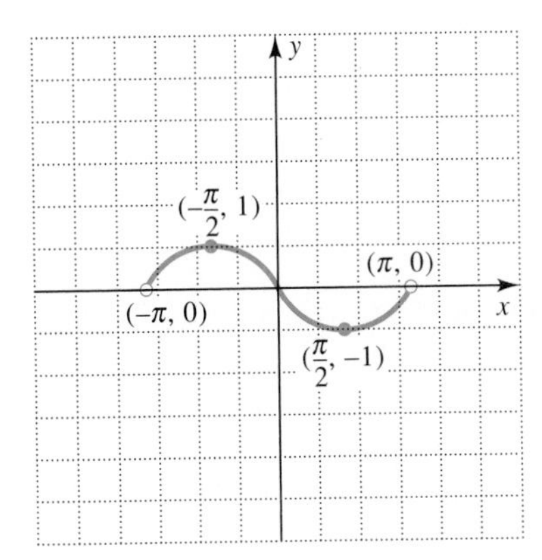

20.

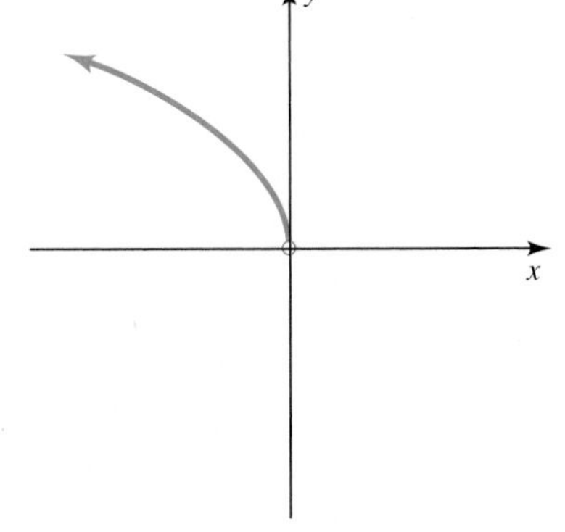

21.

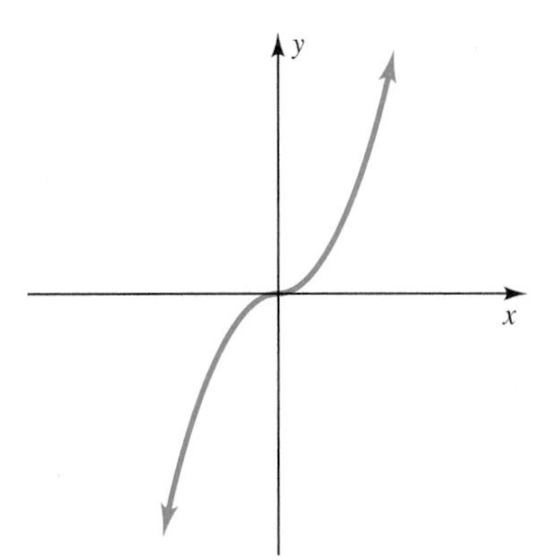

Practice Exercises

In Problems 22 through 25, illustrate each function with a set of ordered pairs, and find the range of the function.

22. The rule for f is, add twice the number to -5. The domain is $\{-2, -1, 0\}$.

23. The rule for g is divide 12 by the number. The domain is $\{1, 2, 3, 4, 5\}$.

24. The rule for F is divide the number by one more than the number. The domain is $\{0, 1, 4\}$.

25. The rule for G is square the number and divide by 2. The domain is $\{-1, 0, 1\}$.

In Problems 26 through 29, assume the domain of each function is the set of real numbers.

26. If $H(x) = x^3 + 2x^2 + 3x + 4$, find $H(-2)$, $H(-1)$, $H(0)$, $H(1)$, and $H(2)$.

27. If $f(x) = x^2 - 2x$, find $f(-2)$, $f(-1)$, $f(0)$, $f(1)$, and $f(2)$.

28. If $g(x) = \dfrac{x}{x^2 + x + 1}$, find $g(-5)$, $g(-1)$, $g(0)$, $g(1)$, and $g(5)$.

29. If $h(x) = \sqrt{2x^2 + x + 1}$, find $h(-3)$, $h(0)$, and $h(1)$.

In Problems 30 through 34, write the natural domain of each rule.

30. $f(x) = 5x^3 - 3x^2 + 2x - 1$

31. $g(x) = \dfrac{x - 1}{x^2 + 23}$

32. $h(x) = \dfrac{1 + x^2}{1 - x^2}$

33. $F(t) = \dfrac{5 - t}{4 + t}$

34. $G(s) = \sqrt{5 - 2s}$

In Problems 35 through 40, graph the given function.

35. $f(x) = -2x + 1$

36. $g(x) = 2x - 1$

37. $h(x) = \dfrac{3}{2}x + 2$

38. $F(x) = -\dfrac{2}{3}x - 2$

39. $G(x) = 4 - \dfrac{1}{3}x$

40. $H(x) = -\dfrac{1}{5}x$

In Problems 41 through 46, determine if the graphs could be graphs of functions.

41.

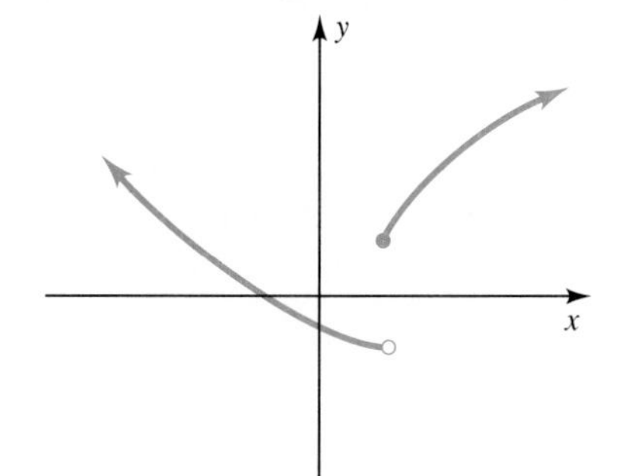

42.

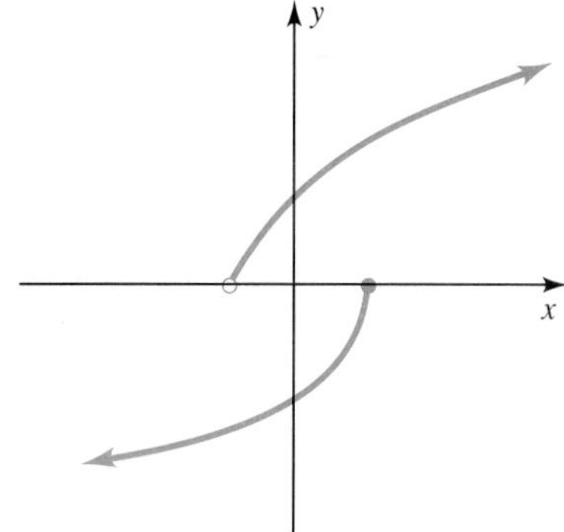

43.

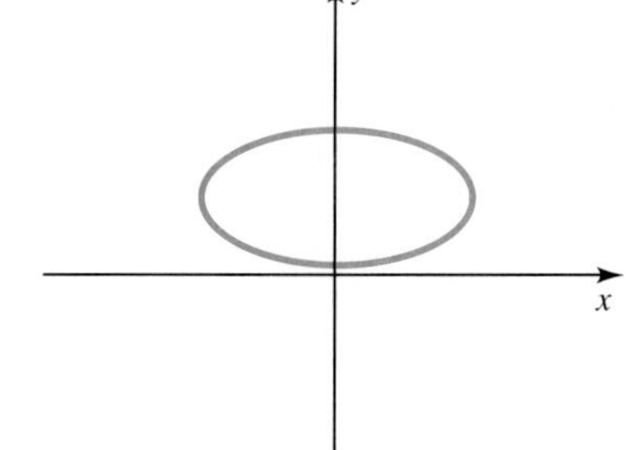

44.

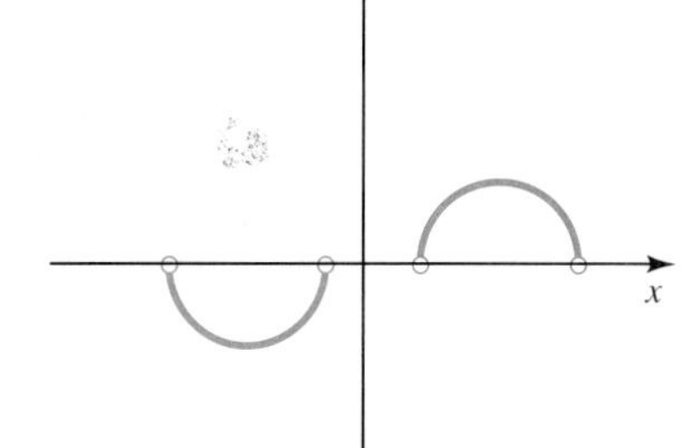

45.

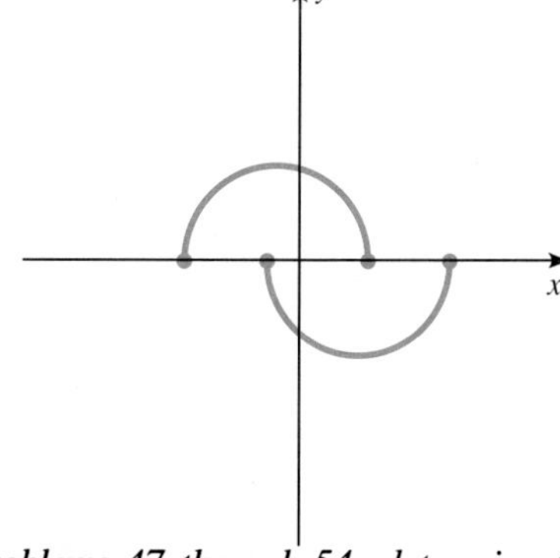

46.

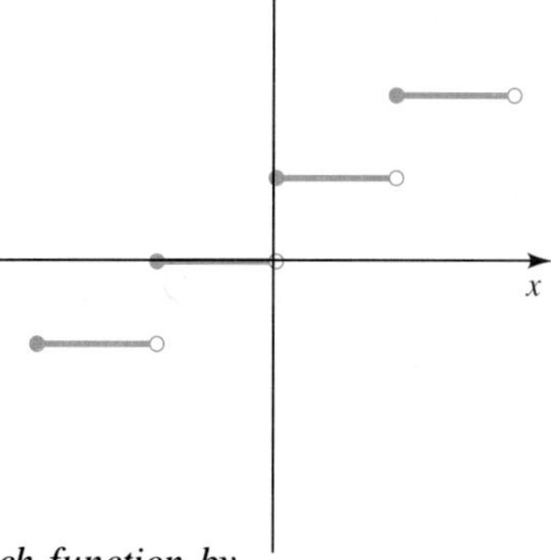

In Problems 47 through 54, determine the domain and range of each function by examining its graph.

47.

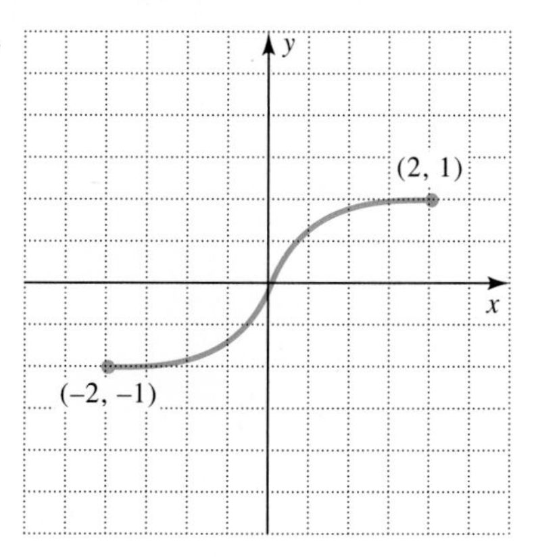

48.

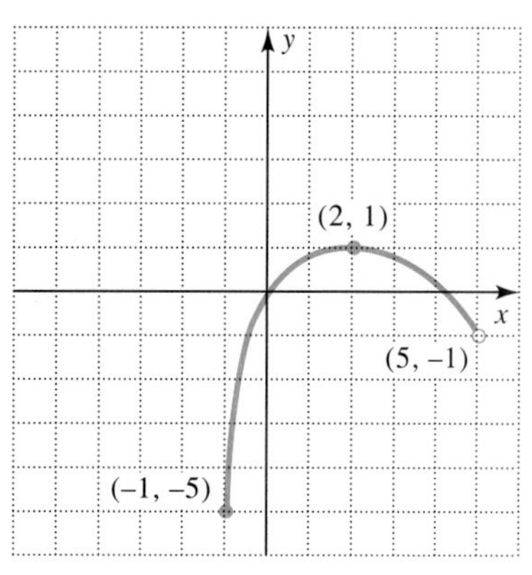

49.

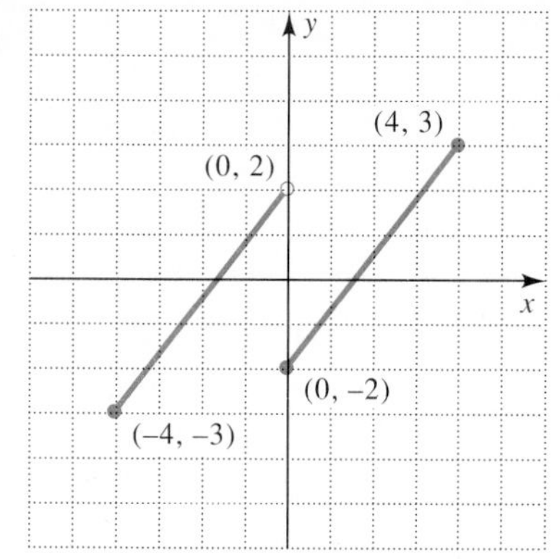

50.

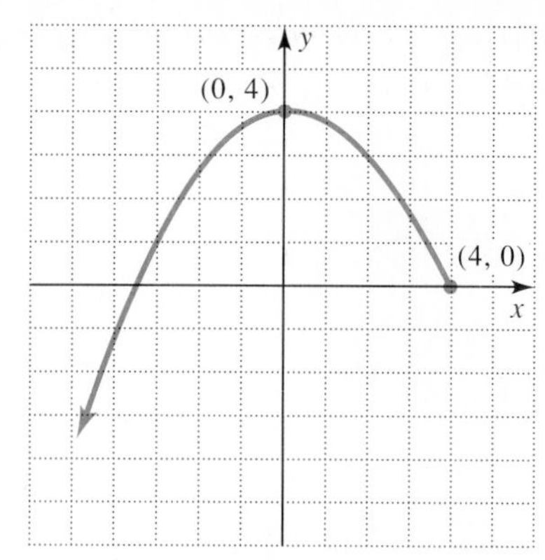

51.

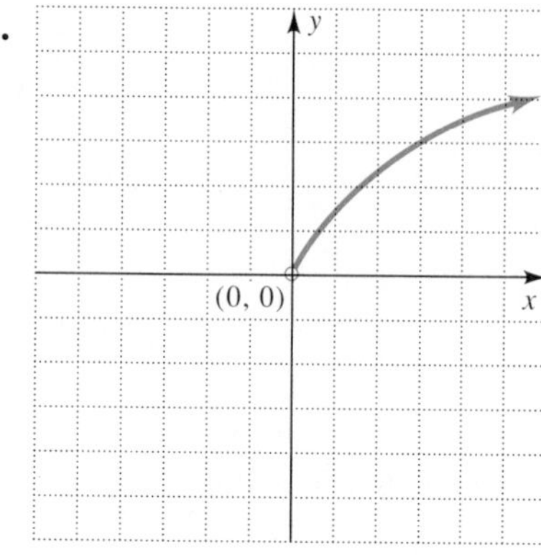

52.

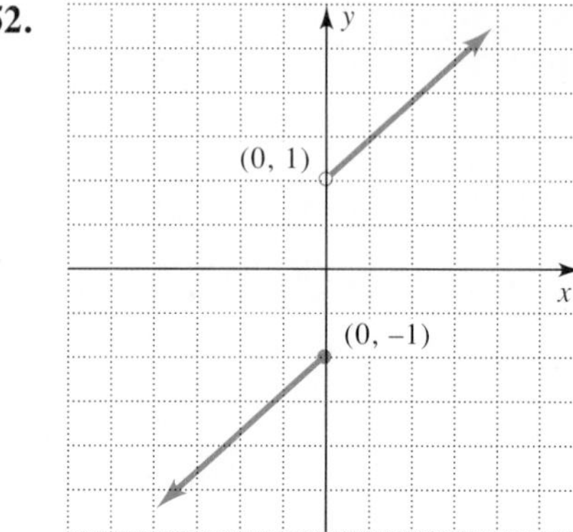

53.

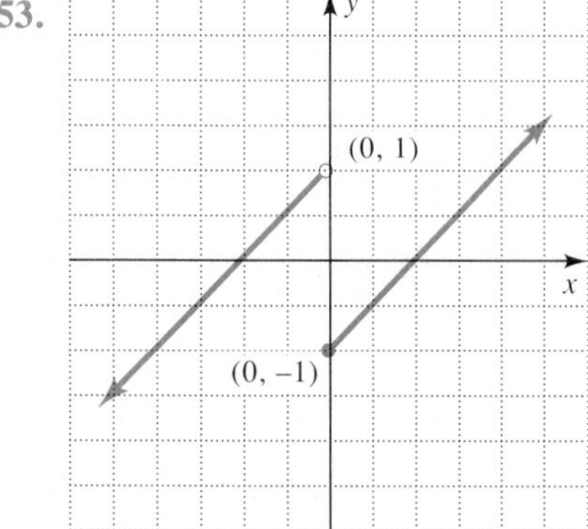

54.

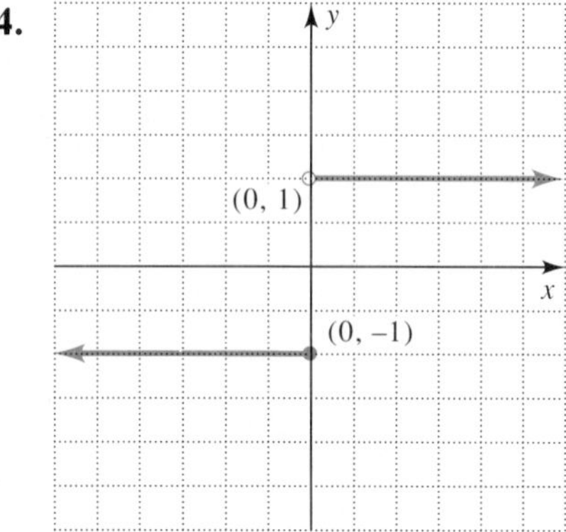

Challenge Problems

55. The following graph shows the relationship between outside temperature in degrees Fahrenheit (T) and time in hours (t) for a 24-hr period last winter.

(a) Is this graph the graph of a function?

(b) Estimate the domain and range for this particular period.

(c) What was the temperature at 2 A.M? 4 P.M.?

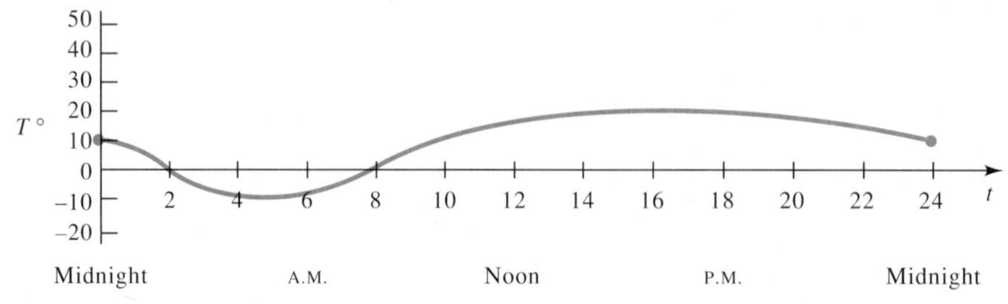

56. If $f(x) = x^2 + 2x - 3$, find each value.

(a) $f(a)$

(b) $f(a + h)$

(c) $f(x + h) - f(x)$

(d) $\frac{f(x + h) - f(x)}{h}$

IN YOUR OWN WORDS...

57. What is a function?

58. What is the domain of a function?

59. What is the range of a function?

CHAPTER SUMMARY

GLOSSARY

Rectangular coordinate system: Two number lines joined at their origins, perpendicular to each other.

x-axis: The horizontal number line of a rectangular coordinate system.

y-axis: The vertical number line of a rectangular coordinate system.

Ordered pair: Two numbers, written in the form (a, b), where the order in which they are written is important.

x-intercept: The x-coordinate of the point on the x-axis at which a graph crosses the axis.

y-intercept: The y-coordinate of the point on the y-axis at which a graph crosses the axis.

Boundary line: A line separating two regions of a linear inequality.

OPTIONAL SECTION

Function: A rule that assigns to each number in one set (the domain) exactly one number in another set (the range).

LINEAR EQUATIONS IN TWO VARIABLES

1. The graph of a linear equation in two variables is a line.
2. Three common forms of the linear equation in two variables.

(a) Standard form

$$Ax + By = C \qquad (A \text{ and } B \text{ not both } 0)$$

(b) Slope-intercept form

$$y = mx + b \qquad (m \text{ is the slope and } b \text{ is the } y\text{-intercept})$$

(c) Point-slope form

$$y - y_1 = m(x - x_1) \qquad (m \text{ is the slope and } (x_1, y_1) \text{ is a point on the line})$$

SLOPE

1. The slope of a line is given by the slope formula,

$$m = \frac{y_2 - y_1}{x_2 - x_1} \qquad \text{Where } (x_1, y_1) \text{ and } (x_2, y_2) \text{ are two points on the line.}$$

2. Parallel lines have the same slope.
3. Perpendicular lines have slopes whose product is -1.

HORIZONTAL AND VERTICAL LINES

1. Horizontal lines have a slope of 0 and equations of the form

$$y = q$$

where q is the y-coordinate of any point on the line.

2. Vertical lines have no slope and equations of the form

$$x = p$$

where p is the x-coordinate of any point on the line.

LINEAR INEQUALITIES IN TWO VARIABLES

To graph the solution set of a linear inequality, follow these steps.

1. Graph the boundary line.
2. Test the regions formed by the boundary line in the original inequality.
3. Shade the regions that tested true.

VERTICAL LINE TEST (Optional Section)

If any vertical line intersects a graph in more than one point, the graph is not the graph of a function.

CHECKUPS

1. Graph the points $(0, 2)$, $(-1, 0)$, $(-1, 2)$, $(1, -2)$ and $(-1, -2)$. — Section 6.1; Example 1
2. Draw the graph of $x + 2y = 2$. — Section 6.1; Example 5
3. Draw the graph of $x = 3$. — Section 6.1; Example 8
4. Find the slope of the line containing the points $(-2, 4)$ and $(3, -1)$. — Section 6.2; Example 3
5. Find the slope of the graph of $x + 2y = 4$. — Section 6.2; Example 7
6. Find an equation of the line containing the points $(-1, 2)$ and $(3, -4)$. — Section 6.3; Example 6
7. Find an equation of the vertical line through the point $(3, -2)$. — Section 6.3; Example 9
8. Graph the solution set for $x + 2y \geq 2$. — Section 6.4; Example 2
9. If $g(x) = 2x - 7$ with domain all real numbers, find each value.
   (a) $g(2)$ (b) $g(0)$ (c) $g(-3)$ (d) $g(\pi)$ — Section 6.5; Example 3 (Optional)
10. Give the natural domain of the rule $h(x) = 7 + \sqrt{x}$. — Section 6.5; Example 4c (Optional)

REVIEW PROBLEMS

Give the quadrant in which each point lies.

1. $(2, 5)$ 2. $(-1, 3)$ 3. $(-2, -4)$ 4. $(1, -3)$

Determine if the point is in the solution set of the equation or inequality.

5. $x + y = 0$; $(-1, 1)$
6. $x + y \leq 0$; $(2, 1)$
7. $y = -2x$; $(1, -2)$
8. $x - y \geq 1$; $(2, -1)$
9. $2x - 3y = 5$; $(1, -1)$
10. $4x + 5y = -7$; $(1, -5)$

Find the intercepts of each graph.

11. $x + y = -2$
12. $2x + y = 6$
13. $3x - y = 3$
14. $2x + y = 0$
15. $y = -4$
16. $x = 6$

Sketch the graph of each equation.

17. $x + y = 4$
18. $y = 2x - 1$
19. $2x - y = 2$
20. $y = -\frac{2}{3}x + 3$

21. $x = 5$
22. $y = \frac{1}{2}x$
23. $x - 2y = 0$
24. $y = 7$

Find the slope of the line containing the two points.

25. $(-2, 7)$; $(4, 5)$
26. $(0, -3)$; $(5, 0)$
27. $(3, -1)$; $(3, 4)$

28. $(2, -1)$; $(-5, 2)$
29. $(-4, 1)$; $(2, 1)$
30. $(0, 0)$; $(5, -4)$

Write each equation in slope-intercept form. Give the slope- and y-intercept for each graph.

31. $2x + y = 4$
32. $x - y = 3$

33. $2x + 3y = 6$
34. $3x - 4y = 8$

35. $2x - 3y = 0$
36. $x - 4y = -4$

Write an equation of the line that satisfies the given conditions.

37. Through the point $(2, 3)$ with slope 4.
38. Through the points $(-1, 3)$ and $(0, -2)$.
39. Through the points $(1, 1)$ and $(0, 3)$.
40. Vertical line through the point $(-2, 3)$.
41. Horizontal line through the point $(4, -3)$.
42. y-intercept of 3 with slope -2.
43. Through the point $(-1, 4)$ with no slope.
44. Through the point $(4, -2)$ with slope 0.

Graph the solution set for each inequality on the xy-plane.

45. $x + y \geq 4$
46. $y < -2$
47. $2x - y > 2$

48. $x \geq 4$

49. $3x - 4y \leq 12$

50. $3x + 2y < -12$

In Problems 51 through 53, find $f(2)$ and $f(-1)$ *(optional).*

51. $f(x) = x^3 - 2x^2 + 3x - 4$

52. $f(x) = \sqrt{x + 2}$

53. $f(x) = \dfrac{x + 1}{x - 1}$

...LET'S NOT FORGET...

In Problems 54 through 61, perform the operations indicated.

54. -2^2

55. $(-3)^2$

56. $x - (5x + 3)$

57. $\dfrac{a^2 - b^2}{b - a}$

58. $\dfrac{0}{7}$

59. $\dfrac{x^2}{x^2 - 4} \cdot (x - 2)$

60. $(x + 2y)^2$

61. $(a + 4)(a - 4)$

In Problems 62 through 64, simplify each expression. Watch the role of parentheses!

62. $(x + 2)x - 2$

63. $(x + 2)(x - 2)$

64. $\dfrac{1}{x + 1} - \dfrac{1 - x}{x + 1}$

In Problems 65 through 68, determine whether x^2 *is a* factor *or a* term.

65. $x^2 + 2$

66. $3(x^2 + 2)$

67. $\dfrac{1}{x^2(x + 2)}$

68. $x^2(y + 3)$

In Problems 69 through 71, multiply each expression by $\dfrac{1}{x + 1}$.

69. $x + 1$

70. $2 + 5(x + 1)$

71. $(x + 1)(x^2 - 1)$

In Problems 72 through 77, identify which expressions are factored. Factor those which are not factored, if possible.

72. $4y + y^2$

73. $x^2 - y^2$

74. $(x - 2)^3$

75. $2(ab + c)$

76. $x^2 + 2xy + y^2$

77. $x^3 - 8$

In Problems 78 through 82, identify whether each problem is an expression, equation, *or* inequality. *Solve the equations and inequalities and perform the operations indicated with the expressions.*

78. $\dfrac{1}{x + 4} - 4$

79. $\dfrac{1}{x + 4} - 4 = \dfrac{x}{x + 4}$

80. $-\frac{1}{3}x \leq 5$

81. $\dfrac{\frac{2}{x} + x}{\frac{1}{x} + 3}$

82. $2(x - 3) - (1 - x)$

CHAPTER 6 TEST

In Problems 1 through 7, choose the letter of the correct answer.

1. In which quadrant does the point $(-3, -4)$ lie?
A. I C. III
B. II D. IV

2. Which ordered pair is in the solution set of $x - y = 3$?
A. $(0, 3)$ C. $(-3, 0)$
B. $(1, -2)$ D. $(1, 3)$

3. The slope of the line parallel to the line with equation $2x - 5y = 10$ is (?).
A. $\frac{2}{5}$ C. 2
B. $-\frac{2}{5}$ D. -2

4. Which of the following is an equation of the line through the point $(1, -5)$ with slope -2?
A. $2x + y = -3$ C. $2x + y = -6$
B. $2x + y = 7$ D. $2x + y = 4$

5. The slope of the graph of $x = 6$ is (?).
A. 1 C. 6
B. 0 D. Not defined

6. The y-intercept of the graph of $2x + 3y = 9$ is (?).
A. 9 C. $\frac{9}{2}$
B. 3 D. $-\frac{2}{3}$

7. The slope of the line containing the points $(2, -3)$ and $(-1, 5)$ is (?)
A. $-\frac{8}{3}$ C. $\frac{8}{3}$
B. $-\frac{3}{8}$ D. $\frac{3}{8}$

In Problems 8 through 11, sketch the graph of each equation.

8. $2x + 3y = 6$

9. $x - 2y = 0$

10. $x = -\frac{3}{2}$

11. $y = -\frac{2}{3}x + 2$

In Problems 12 and 13, graph the solution set.

12. $x - 3y \leq 6$.

13. $2x + y > 0$

14. Write an equation of the line through the points $(1, -2)$ and $(3, 4)$.

15. Write an equation of the horizontal line through the point $(2, -3)$.

OPTIONAL SECTION

16. If $f(x) = 3 - 2x + 3x^2 - x^3$, find $f(-2)$.

17. Which of the following graphs are graphs of functions?

(a)

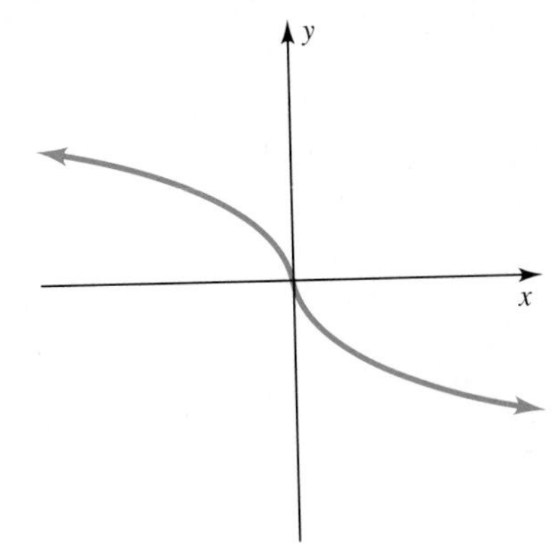

(b)

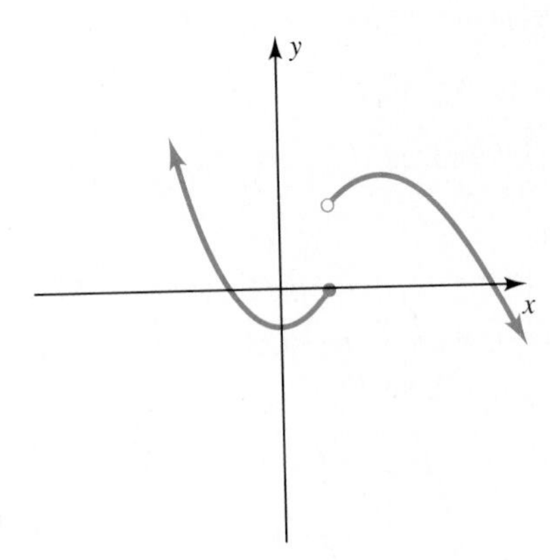

(c)

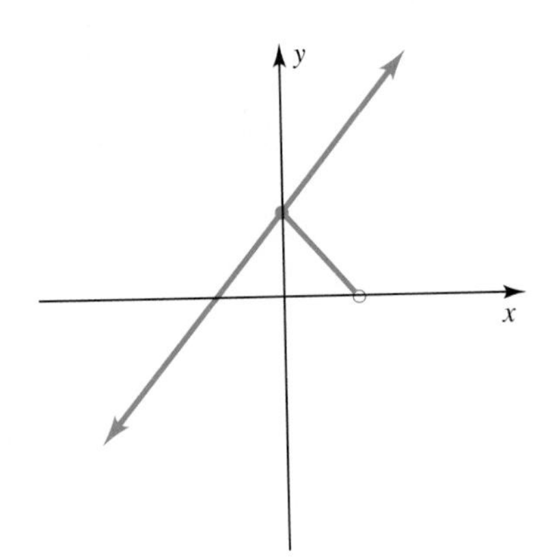

(d)

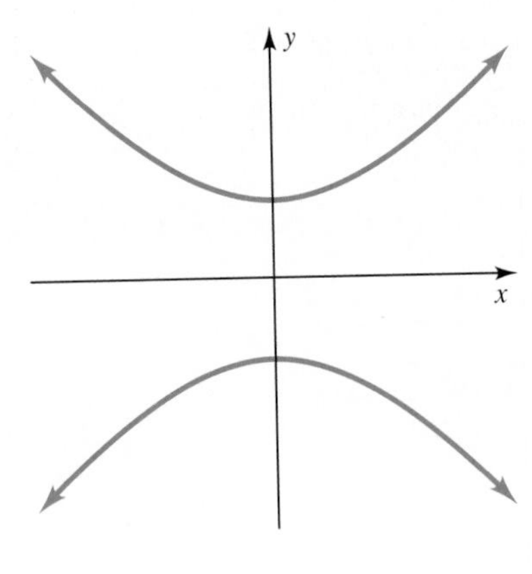

(e)

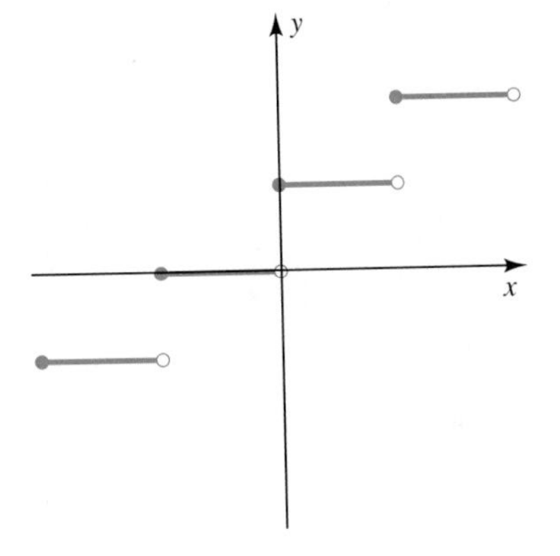

(f)

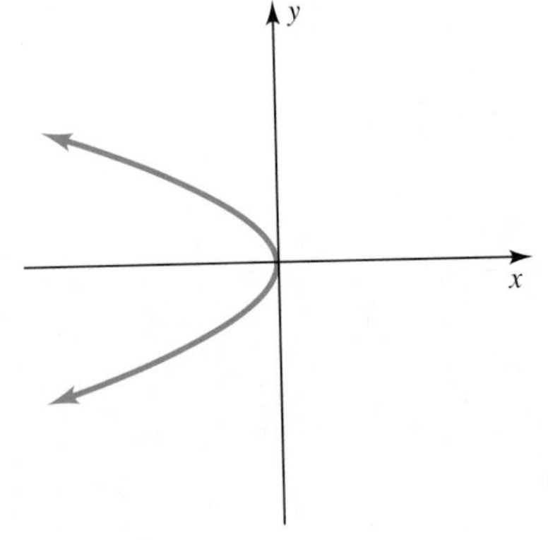

See Problem Set 7.4, Exercise 19.

Systems of Linear Equations and Inequalities

CONNECTIONS

Suppose an engineer wishes to cut a 14-foot truss into two sections, ending up with one piece 4 feet longer than the other. The engineer could solve this relatively simple problem by using trial-and-error, or by using his mathematical skills. However, engineers routinely deal with much more complicated problems where trial-and-error will not work, but the mathematical approach will. Such problems often lead to equations in two variables. Engineers are not the only people who use systems of linear equations, for these ideas can be applied to many complex mathematical concepts.

In this chapter we will learn three methods for solving systems of equations and examine applications of systems. We will also introduce systems of two inequalities in two variables.

7.1 GRAPHICAL METHOD

Suppose we are looking for two numbers whose sum is 6. If we let x and y be the two numbers, then we are looking for solutions to the equation

$$x + y = 6$$

There are an infinite number of such solutions. We learned in the last chapter that a picture of the solution set of this equation is the line shown next.

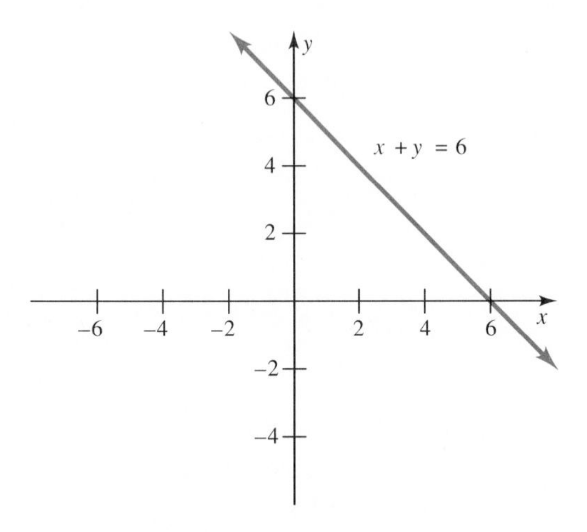

Now suppose that we also know that the difference in these two numbers is 2. That is,

$$x - y = 2$$

All solutions to this equation would lie on the following line.

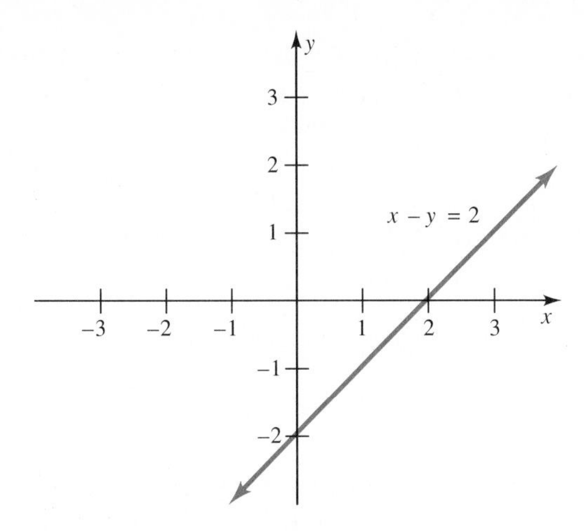

We are looking for ordered pairs that make *both* $x + y = 6$ and $x - y = 2$ true statements. If we draw the graph of each equation on the same coordinate system, we have the following graph.

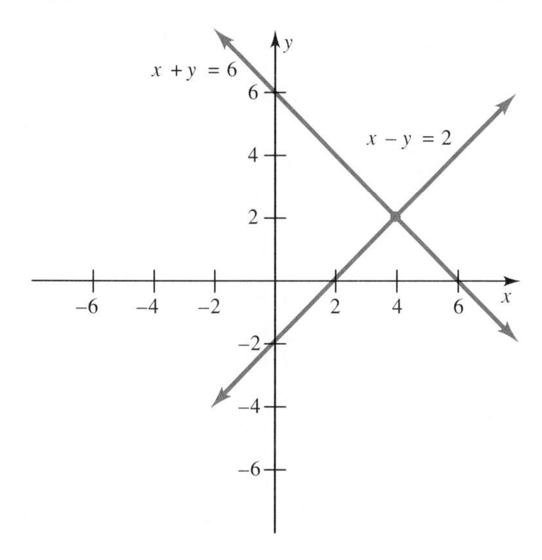

The graphs intersect in a single point. This ordered pair is in the solution set of *both* equations. Thus, the coordinates of this point are the numbers for which we are looking. From the graph we guess this ordered pair to be (4, 2).

To see that (4, 2) is correct, we replace x with 4 and y with 2 in both equations.

$x + y = 6$	$4 + 2 = 6$	True statement
$x - y = 2$	$4 - 2 = 2$	True statement

We see that (4, 2) is in the solution set of both equations. Two numbers whose sum is 6 and difference is 2 are 4 and 2.

Situations such as this arise very often in the real world. To indicate that x represents the same number in both equations and that y represents the same number in both equations, we write the system in the following manner:

$$\begin{cases} x + y = 6 \\ x - y = 2 \end{cases}$$

We call this a **system of two equations in two variables.** The set of all ordered pairs that make *both* equations true statements is called the **solution set.** To **solve** a system means to find the solution set.

Determining the solution set from a graph is called the **graphical method** of solving a system of equations. It requires judgement and estimation and is thus not a very satisfactory technique to use. We will learn two other techniques in Sections 7.2 and 7.3.

EXAMPLE 1. Using the graphical method, find the solution of the system of equations:

$$\begin{cases} x + y = 2 \\ 2x - y = 1 \end{cases}$$

Solution:

We draw the graph of each equation on the same coordinate system.

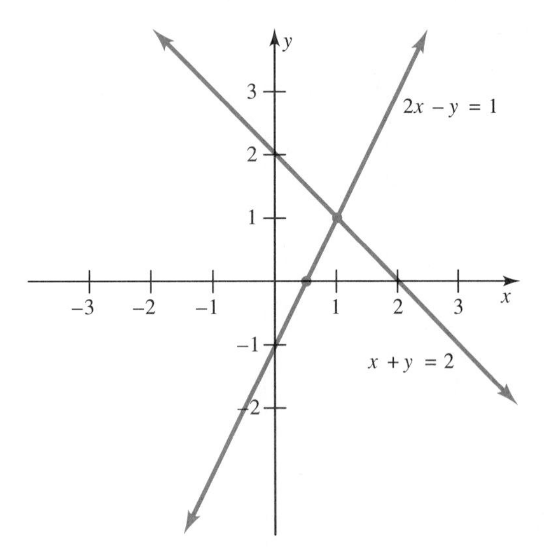

From the graph we guess the solution to be (1, 1). If we replace x with 1 and y with 1 in both equations,

$$x + y = 2 \qquad 2x - y = 1$$

$$1 + 1 = 2 \qquad 2(1) - 1 = 1$$

we see that both are true statements. Thus (1, 1) is the solution. □

The graphing method is not very satisfactory for finding the solution set, but it is very helpful in describing the solution set of a system of equations.

EXAMPLE 2. How many solutions are in the solution set for the system $\begin{cases} x - y = 1 \\ x + y = 2 \end{cases}$?

Solution:

We draw the graph of each equation.

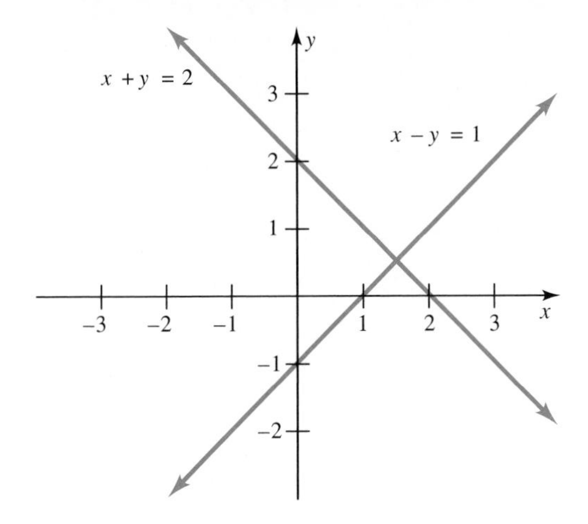

From the graph we see that the lines intersect in one point. So, there is exactly one solution to the system. Such a system is called an **independent** system. □

EXAMPLE 3. How many solutions are in the solution set of the system

$$\begin{cases} x - y = 1 \\ 2x - 2y = 4 \end{cases}?$$

Solution:

We graph each equation.

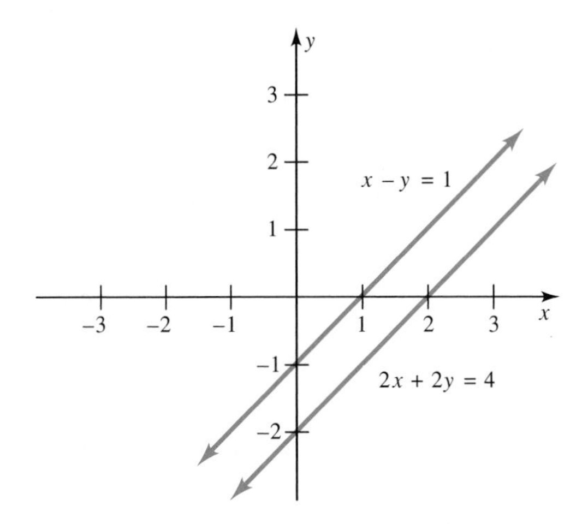

From the graph we see that the lines appear to be parallel. In fact, since each line has a slope of 1, they *are* parallel. Thus, the lines do not intersect. This means that there is *no* ordered pair that is in the solution set of both equations. So, there is no solution to the system. We call such a system **inconsistent.** □

EXAMPLE 4. How many solutions are in the solution set for the system

$$\begin{cases} x + y = 2 \\ 2x + 2y = 4 \end{cases}?$$

Solution:

We graph each line.

(continued)

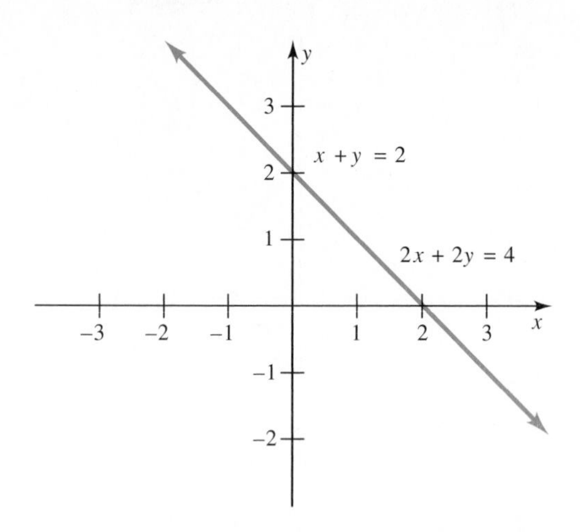

The graph of both equations is the same line. This means that every point on the line is in the solution set for *both* equations. There are an infinite number of points on this line, so there are an infinite number of solutions to the system. We call such a system **dependent.** ☐

As Examples 2, 3, and 4 suggest, there are three situations that can occur when solving systems of two linear equations in two variables.

Independent

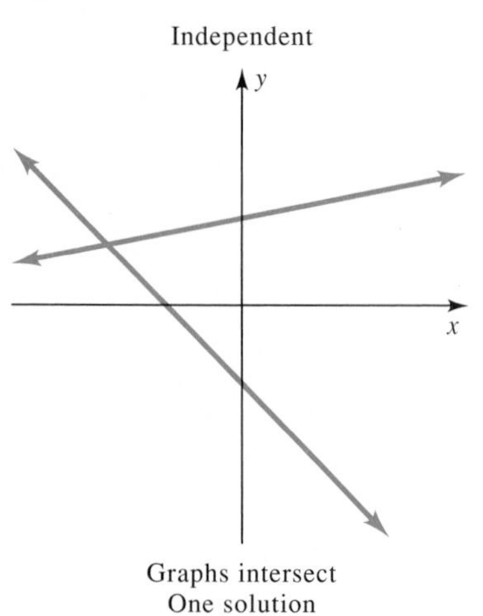

Graphs intersect
One solution

Inconsistent

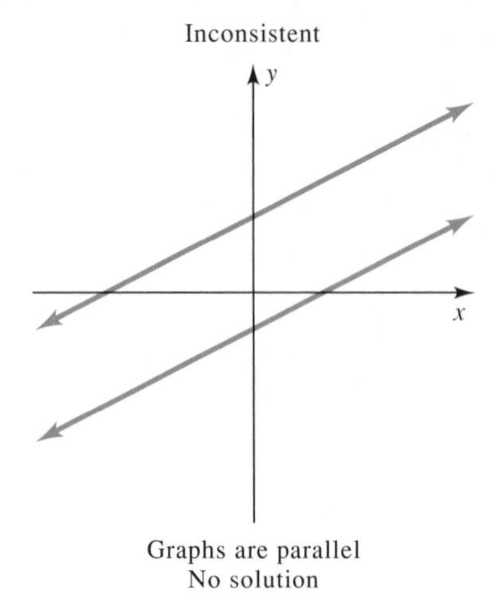

Graphs are parallel
No solution

Dependent

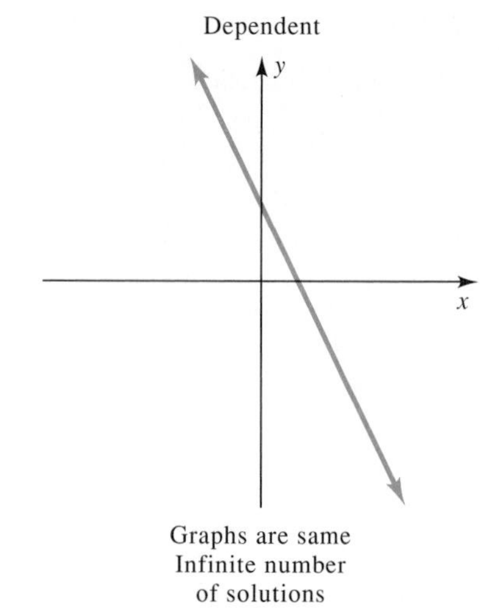

Graphs are same
Infinite number
of solutions

PROBLEM SET 7.1

Warm-ups

Find the solution set for each system by the graphical method. See Example 1.

1. $\begin{cases} x - y = 1 \\ x + y = 3 \end{cases}$

2. $\begin{cases} 2x + y = -1 \\ x + y = 1 \end{cases}$

3. $\begin{cases} 2x - y = 4 \\ x + y = -1 \end{cases}$

4. $\begin{cases} x - y = 1 \\ x + y = 5 \end{cases}$

5. $\begin{cases} 2x + y = -3 \\ x + y = -2 \end{cases}$

6. $\begin{cases} 2x - y = 1 \\ x + y = 2 \end{cases}$

7. $\begin{cases} x = 3 \\ x + y = 5 \end{cases}$

8. $\begin{cases} x - y = 3 \\ y = -2 \end{cases}$

9. $\begin{cases} x + 2y = 1 \\ x = -1 \end{cases}$

Determine the number of solutions in the solution set for each system. Classify each system as independent, inconsistent, or dependent. See Examples 2, 3, and 4.

10. $\begin{cases} x - y = 2 \\ x + y = 2 \end{cases}$

11. $\begin{cases} 2x + 2y = 4 \\ x + y = 0 \end{cases}$

12. $\begin{cases} 2x - y = -1 \\ x - y = 1 \end{cases}$

13. $\begin{cases} x - y = -2 \\ x + y = 6 \end{cases}$

14. $\begin{cases} x + y = 1 \\ 2x + 2y = 2 \end{cases}$

15. $\begin{cases} 2x - y = 1 \\ x + y = -1 \end{cases}$

16. $\begin{cases} x - y = 1 \\ 2x - 2y = 3 \end{cases}$

17. $\begin{cases} 2x + 4y = 2 \\ x + 2y = 1 \end{cases}$

18. $\begin{cases} 2x - y = 3 \\ x - y = 2 \end{cases}$

Practice Exercises

Find the solution set for each system by the graphical method.

19. $\begin{cases} x - y = 0 \\ 3x + y = 4 \end{cases}$

20. $\begin{cases} 2x + y = 3 \\ x + y = 0 \end{cases}$

21. $\begin{cases} 2x - y = 0 \\ x + y = 0 \end{cases}$

22. $\begin{cases} x - y = 3 \\ x + y = 1 \end{cases}$

23. $\begin{cases} 2x + y = 4 \\ x + y = 2 \end{cases}$

24. $\begin{cases} 2x - y = -2 \\ x + y = -4 \end{cases}$

25. $\begin{cases} x - y = -1 \\ x + y = 3 \end{cases}$

26. $\begin{cases} 2x + y = -1 \\ x + y = -1 \end{cases}$

27. $\begin{cases} 2x + y = 0 \\ x - y = 0 \end{cases}$

28. $\begin{cases} x - y = -1 \\ x + y = -3 \end{cases}$

29. $\begin{cases} 2x + y = 1 \\ x - y = 2 \end{cases}$

30. $\begin{cases} 2x - y = 2 \\ x + y = -2 \end{cases}$

31. $\begin{cases} x = -2 \\ x - y = -5 \end{cases}$

32. $\begin{cases} x - y = -1 \\ y = 2 \end{cases}$

33. $\begin{cases} x - 3y = 3 \\ x = 0 \end{cases}$

Determine the number of solutions in the solution set for each system. Classify each system as independent, inconsistent, or dependent.

34. $\begin{cases} x - y = 0 \\ x + y = 0 \end{cases}$

35. $\begin{cases} 2x - y = 6 \\ x - y = 3 \end{cases}$

36. $\begin{cases} 2x - 2y = 2 \\ x - y = 1 \end{cases}$

37. $\begin{cases} x - y = 4 \\ x + y = 2 \end{cases}$

38. $\begin{cases} x - y = 1 \\ 4x - 4y = 4 \end{cases}$

39. $\begin{cases} 2x - y = -3 \\ x + y = -6 \end{cases}$

40. $\begin{cases} x - y = 1 \\ 3x - 3y = 1 \end{cases}$

41. $\begin{cases} 2x - 12y = 12 \\ x - 6y = 6 \end{cases}$

42. $\begin{cases} 2x - y = 2 \\ x - y = 2 \end{cases}$

Challenge Problems

Find the solution set for each system graphed below.

43.

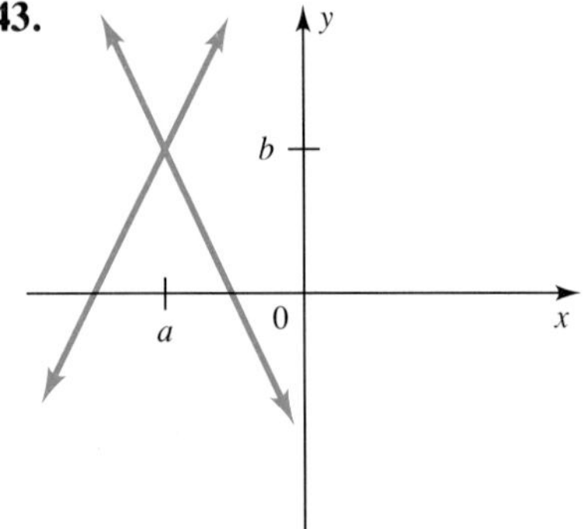

44.

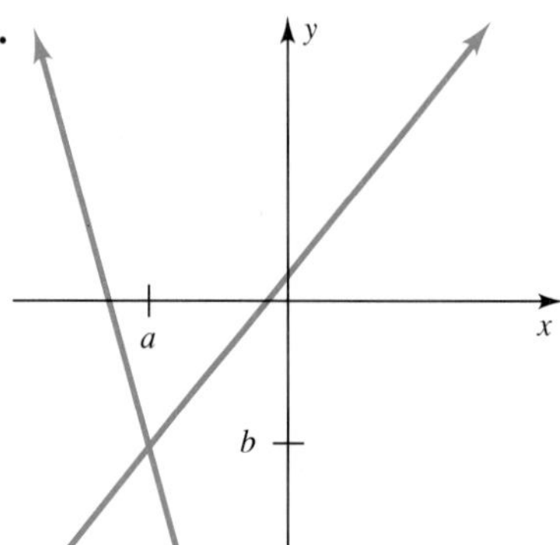

45.

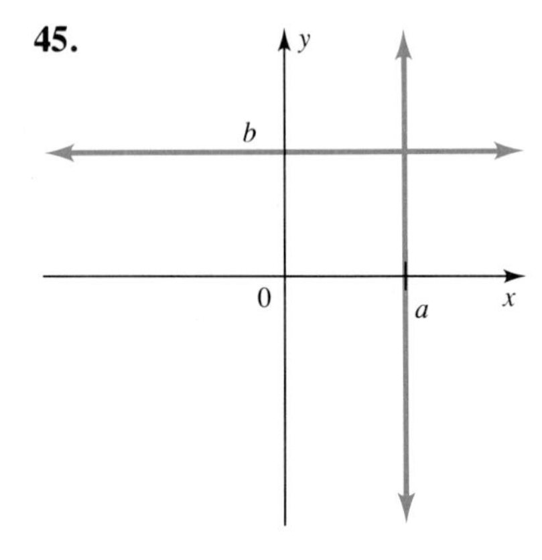

IN YOUR OWN WORDS...

46. What is a system of two equations in two variables?

47. Describe the graphical method for solving a system of equations.

7.2 SUBSTITUTION

Let's consider how we can solve the system

$$\begin{cases} x + y = 1 \\ y = x + 3 \end{cases}$$

by a method other than the graphical method. Notice that the second equation is solved for y. If y is $x + 3$ in the second equation, it must be $x + 3$ in the first equation, since y represents the same number in both equations. So, we replace y by $x + 3$ in the first equation.

$$x + (x + 3) = 1$$

Now we have an equation in one variable, which we solve.

$$2x + 3 = 1$$
$$2x = -2$$
$$x = -1$$

This gives us the x-value for the solution. To find the y-value, we replace x with -1 in the second equation.

$$y = -1 + 3$$
$$y = 2$$

So, the solution set is $\{(-1, 2)\}$.

This method is called **substitution.**

Method of Substitution

1. Solve one of the equations for one of the variables.
2. Substitute this expression for the same variable in the other equation.
3. Solve the resulting equation.
4. Substitute the result in the equation formed in Step 1 to find the other variable.
5. Check both values in both equations, if required.
6. Write the solution set.

EXAMPLE 1. Solve the system by substitution.

$$\begin{cases} x + y = 1 \\ y = x - 1 \end{cases}$$

Solution:

Step 1 Solve one equation for one variable.

The second equation is already solved for y.

Step 2 Substitute into the other equation.

We replace y in *the first equation* with $x - 1$.

$$x + (x - 1) = 1$$

Step 3 Solve the resulting equation.

$$x + (x - 1) = 1$$
$$2x - 1 = 1$$
$$2x = 2$$
$$x = 1$$

We have found the x-value of the solution.

(continued)

Step 4 Substitute the result into the equation of Step 1.

To find the y-value, we replace x in the second equation with 1.

$$y = 1 - 1$$
$$y = 0$$

Step 5 Check the result in both equations.

We are checking the ordered pair (1, 0). The left side of the first equation has the value

$$x + y = 1 + 0 = 1$$

which agrees with the right side; in the second equation we have

$$\text{LS: } y = 0$$
$$\text{RS: } x - 1 = 1 - 1 = 0$$

which also checks.

Step 6 Write the solution set.

$$\{(1, 0)\}$$

□

EXAMPLE 2. Solve the system by substitution.

$$\begin{cases} x + y = 5 \\ x - y = -1 \end{cases}$$

Solution:

Step 1 Solve one equation for one variable.

Let's solve the first equation for x.

$$x = 5 - y$$

Step 2 Substitute into the other equation.

We substitute $5 - y$ for x in the *second* equation.

$$(5 - y) - y = -1$$

Step 3 Solve the resulting equation.

$$\begin{aligned} (5 - y) - y &= -1 \\ 5 - 2y &= -1 \\ -2y &= -6 \\ y &= 3 \end{aligned}$$

Step 4 Substitute the result into the equation of Step 1.

We substitute 3 for y in Step 1.

$$x = 5 - 3$$
$$x = 2$$

Step 5 Check in both equations.

We are checking the ordered pair (2, 3). In the first equation the left side has the value

$$x + y = 2 + 3 = 5$$

which agrees with the right side. The left side of the second,

$$x - y = 2 - 3 = -1$$

also agrees with the right. The solution checks.

Step 6 Write the solution set.

$$\{(2, 3)\}$$

□

EXAMPLE 3. Solve the system by substitution.

$$\begin{cases} 2x + y = 4 \\ 3x + 2y = 7 \end{cases}$$

Solution:

Step 1 Solve one equation for one variable.

We can avoid fractions by solving the first equation for y.

$$y = 4 - 2x$$

Step 2 Substitute into the other equation.

Substitute $4 - 2x$ for y in the second equation.

$$3x + 2(4 - 2x) = 7$$

Step 3 Solve the resulting equation.

$$3x + 2(4 - 2x) = 7$$

$$3x + 8 - 4x = 7$$

$$-x + 8 = 7$$

$$-x = -1$$

$$x = 1$$

Step 4 Substitute the result into the equation of Step 1.

We substitute 1 for x in Step 1.

$$y = 4 - 2(1)$$

$$y = 4 - 2$$

$$y = 2$$

Step 5 Check in both equations.

In the first equation,

LS: $2x + y = 2(1) + 2 = 4$

RS: 4

In the second equation,

LS: $3x + 2y = 3(1) + 2(2) = 7$

RS: 7

Step 6 Write the solution set.

$$\{(1, 2)\}$$

□

EXAMPLE 4. Solve the system by substitution.

$$\begin{cases} 2x - y = 5 \\ 3x + 2y = 11 \end{cases}$$

Solution:

Step 1 Solve one equation for one variable.

We solve the first equation for y.

$$2x - y = 5$$
$$-y = 5 - 2x$$
$$y = -5 + 2x \quad \text{or} \quad y = 2x - 5$$

Step 2 Substitute into the other equation and solve.

Step 3 Substitute $2x - 5$ for y in the second equation.

$$3x + 2(2x - 5) = 11$$
$$3x + 4x - 10 = 11$$
$$7x - 10 = 11$$
$$7x = 21$$
$$x = 3$$

Step 4 Substitute into the equation of Step 1.

Substitute 3 for x in $y = 2x - 5$.

$$y = 2(3) - 5$$
$$y = 1$$

Step 5 Check.

(3, 1) checks in both equations.

Step 6 Write the solution set.

$$\{(3, 1)\}$$

□

EXAMPLE 5. Solve the system by substitution.

$$\begin{cases} 3x - 2y = 4 \\ 2x + 4y = 8 \end{cases}$$

Solution:

In deciding for which variable to solve, we notice that to solve the first equation for x or y would introduce fractions. However, if we solve the second equation for x, we can avoid fractions.

$$2x + 4y = 8$$
$$2x = 8 - 4y$$
$$x = 4 - 2y$$

Substitute $4 - 2y$ for x in the first equation and solve.

$$3(4 - 2y) - 2y = 4$$
$$12 - 6y - 2y = 4$$
$$12 - 8y = 4$$
$$-8y = -8$$
$$y = 1$$

We replace y with 1 in $x = 4 - 2y$.

$$x = 4 - 2(1)$$
$$x = 2$$

As (2, 1) checks in both equations, we write the solution set.

$$\{(2, 1)\}$$ □

EXAMPLE 6. Solve the system by substitution.

$$\begin{cases} x - 2y = 0 \\ 2x - 4y = 3 \end{cases}$$

Solution:

We solve the first equation for x.

$$x = 2y$$

We substitute $2y$ for x in the second equation.

$$2(2y) - 4y = 3$$
$$4y - 4y = 3$$
$$0 = 3$$

Notice that both variables have dropped out of the equation. If we draw the graphs of the original two equations, we see what the problem is.

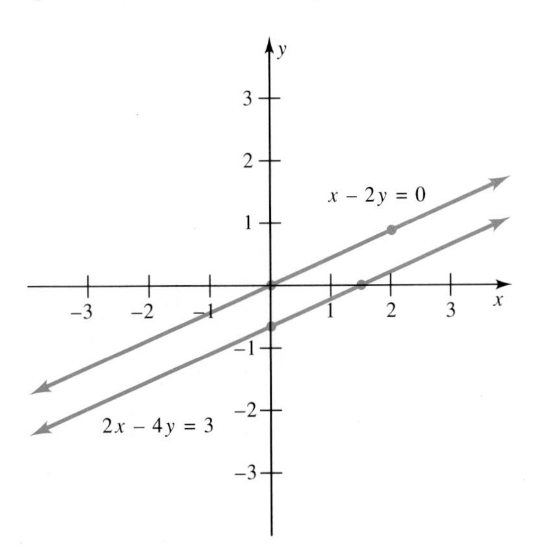

The lines appear to be parallel. In fact, since they have the same slope, they are parallel. This means that the system has no solution. Thus the solution set is the empty set, which we write as

$$\varnothing$$

Notice that when the variable disappeared, we were left with the equation $0 = 3$. This is a false statement, which tells us the system is *inconsistent* and that the solution set is the empty set. □

EXAMPLE 7. Solve the system $\begin{cases} x - 2y = 1 \\ 3x - 6y = 3 \end{cases}$ by substitution.

Solution:

We solve the first equation for x.

$$x = 1 + 2y$$

We substitute $1 + 2y$ for x in the second equation.

$$\begin{aligned} 3(1 + 2y) - 6y &= 3 \\ 3 + 6y - 6y &= 3 \\ 3 &= 3 \end{aligned}$$

Again the variables have all dropped out of the equation. Notice, however, that we are left with a *true* statement. If we draw the graph of each equation, we will see that the graph of both equations is the same line. This means that the system is *dependent* and that the solution set has an infinite number of elements. At this stage we will not write the solution set for dependent systems but will simply mark them *dependent.* □

PROBLEM SET 7.2

Warm-ups

Solve each system by substitution. If the system is dependent, write dependent. In Problems 1 through 6, see Example 1.

1. $\begin{cases} x + y = 2 \\ y = x \end{cases}$
2. $\begin{cases} x - y = 4 \\ y = 3x \end{cases}$
3. $\begin{cases} x + 2y = 0 \\ y = 4x \end{cases}$
4. $\begin{cases} x - 2y = -5 \\ y = 5 - 2x \end{cases}$
5. $\begin{cases} 2x + y = 3 \\ x = 4 - y \end{cases}$
6. $\begin{cases} 3x + 2y = -2 \\ x = 4 - 3y \end{cases}$

In Problems 7 through 12, see Examples 2, 3, and 4.

7. $\begin{cases} x - 2y = 5 \\ x = 3 \end{cases}$
8. $\begin{cases} x + y = 1 \\ x - y = 3 \end{cases}$
9. $\begin{cases} x + y = 0 \\ 2x + y = 1 \end{cases}$
10. $\begin{cases} 2x - 2y = 14 \\ x - 2y = 10 \end{cases}$
11. $\begin{cases} 5x - 3y = 25 \\ 4x - y = 13 \end{cases}$
12. $\begin{cases} 5x - y = -4 \\ 3x + 2y = -5 \end{cases}$

In Problems 13 through 15, see Example 5.

13. $\begin{cases} 2x + 3y = -4 \\ 3x + 6y = -9 \end{cases}$
14. $\begin{cases} 3x - 6y = 0 \\ 6x - 16y = -4 \end{cases}$
15. $\begin{cases} 3x + 5y = -23 \\ 2x - 4y = -8 \end{cases}$

In Problems 16 through 18, see Examples 6 and 7.

16. $\begin{cases} x + y = 5 \\ x + y = 3 \end{cases}$
17. $\begin{cases} x + y = 2 \\ 2x + 2y = 4 \end{cases}$
18. $\begin{cases} 3x - 6y = -2 \\ 6x - 12y = 4 \end{cases}$

Practice Exercises

Solve each system by substitution. If the system is dependent, write dependent.

19. $\begin{cases} x + y = 4 \\ y = 3x \end{cases}$

20. $\begin{cases} x - y = 6 \\ x = 7y \end{cases}$

21. $\begin{cases} 4x + y = 0 \\ y = -4x \end{cases}$

22. $\begin{cases} x - 3y = -17 \\ y = -1 - 3x \end{cases}$

23. $\begin{cases} 3x + y = 4 \\ x = 6 - y \end{cases}$

24. $\begin{cases} 3x - 2y = 8 \\ x = 3y + 12 \end{cases}$

25. $\begin{cases} x + y = 7 \\ x - y = 3 \end{cases}$

26. $\begin{cases} x + y = 0 \\ 2x - y = -9 \end{cases}$

27. $\begin{cases} x - 2y = 7 \\ x - 2y = 5 \end{cases}$

28. $\begin{cases} 3x - 3y = 2 \\ x - y = 1 \end{cases}$

29. $\begin{cases} x - 2y = 3 \\ 2x - 4y = 6 \end{cases}$

30. $\begin{cases} x - y = -1 \\ 3x - 5y = -4 \end{cases}$

31. $\begin{cases} 2x - 3y = 19 \\ x + 2y = 13 \end{cases}$

32. $\begin{cases} x + 2y = -6 \\ 3x - 4y = 2 \end{cases}$

33. $\begin{cases} 4x - y = 17 \\ 2x - 3y = 1 \end{cases}$

34. $\begin{cases} 2x + 4y = 20 \\ 3x + 5y = 24 \end{cases}$

35. $\begin{cases} 3x - 2y = 2 \\ 5x - 10y = 50 \end{cases}$

36. $\begin{cases} 4x + 2y = -18 \\ 3x - 7y = -22 \end{cases}$

37. $\begin{cases} 5x + 6y = -2 \\ 3x - 3y = 12 \end{cases}$

38. $\begin{cases} 2x - 5y = 7 \\ 4x - 10y = 14 \end{cases}$

39. $\begin{cases} 6x + 8y = 6 \\ 2x + 4y = 2 \end{cases}$

40. $\begin{cases} x + 2y = 3 \\ x = 3 \end{cases}$

41. $\begin{cases} 3x - 2y = 0 \\ y = 3 \end{cases}$

42. $\begin{cases} 3x - 4y = 9 \\ x = -1 \end{cases}$

Challenge Problems

43. Solve the system by substitution.
$\begin{cases} x + y = 4 \\ y = k \end{cases}$

44. Solve the system by substitution.
$\begin{cases} x - y = a \\ x + y = b \end{cases}$

45. Solve the system by substitution.
$\begin{cases} x^2 + y^2 = 10 \\ y = 3x \end{cases}$

IN YOUR OWN WORDS...

46. Describe the substitution method for solving two equations with two variables.

47. How do we detect inconsistent and dependent systems?

7.3 ELIMINATION

The next method for solving systems of equations is called the method of **elimination.** Because adding the same number to both sides of an equation does not change its solution set, we can add two equations together. This procedure is used in the following examples.

EXAMPLE 1. Solve $\begin{cases} x + y = 1 \\ x - y = 3 \end{cases}$ by elimination.

Solution:

We add the two equations together:

$$\begin{cases} x + y = 1 \\ x - y = 3 \end{cases}$$
$$\overline{2x \qquad = 4}$$

Then we solve the resulting equation:

$$x = 2$$

(continued)

We have found the x-value of the solution. To find the y-value of the solution, we replace x with 2 in either of the original equations and solve for y. Suppose we use the first equation.

$$x + y = 1$$
$$2 + y = 1$$
$$y = -1$$

The solution set is $\{(2, -1)\}$. □

EXAMPLE 2. Solve $\begin{cases} 2x + 3y = 1 \\ 3x + y = -2 \end{cases}$ by elimination.

Solution:

If we add the equations as they are, we get another equation containing x and y. This will not help us solve the system. However, if we multiply the second equation by -3, the solution set for the system does not change. Notice what happens.

$$\begin{cases} 2x + 3y = 1 & \text{First equation unchanged.} \\ -9x - 3y = 6 & \text{Multiplied the second equation by } -3. \end{cases}$$

The solution set for this system is the same as the solution set for the original system. But, when we add the equations the variable y will be eliminated.

$$\begin{cases} 2x + 3y = 1 \\ -9x - 3y = 6 \end{cases}$$
$$-7x \quad = 7$$
$$x = -1$$

We replace x with -1 in either of the original equations and solve for y. Using the second, we have

$$3x + y = -2$$
$$3(-1) + y = -2$$
$$-3 + y = -2$$
$$y = 1$$

The ordered pair $(-1, 1)$ checks in both of the original equations. Thus the solution set is

$$\{(-1, 1)\}$$ □

Method of Elimination

1. Write the equations in the form $Ax + By = C$.
2. If necessary, multiply both sides of each equation by a suitable number so one of the variables will be eliminated by addition of the equations.
3. Add the equations together and solve the resulting equation.
4. Substitute the value found in Step 2 into one of the original equations and solve this equation.
5. Check, if required.
6. Write the solution set.

EXAMPLE 3. Solve $\begin{cases} 2x + 5y = 6 \\ 3x = 2y - 10 \end{cases}$ by elimination.

Solution:

Step 1 Write each equation in the form $Ax + By = C$.

$$\begin{cases} 2x + 5y = 6 \\ 3x = 2y - 10 \end{cases} \qquad \begin{cases} 2x + 5y = 6 \\ 3x - 2y = -10 \end{cases}$$

Step 2 Multiply each equation by a suitable number so addition will eliminate a variable.

If we multiply the first equation by -3 and the second equation by 2, the system becomes

$$\begin{cases} 2x + 5y = 6 \\ 3x - 2y = -10 \end{cases} \xrightarrow[\text{Multiply by } 2]{\text{Multiply by } -3} \begin{cases} -6x - 15y = -18 \\ 6x - 4y = -20 \end{cases}$$

Step 3 Add the equations together and solve.

$$\begin{cases} -6x - 15y = -18 \\ 6x - 4y = -20 \end{cases}$$

$$-19y = -38$$

$$y = 2$$

Step 4 Substitute this value into one of the original equations and solve.

We replace y with 2 in the first equation of the original system and solve for x.

$$2x + 5(2) = 6$$
$$2x + 10 = 6$$
$$2x = -4$$
$$x = -2$$

Step 5 Check.

Since we found x with the first of the original equations, we need to check only the second.

LS: $3x = 3(-2) = -6$

RS: $2y - 10 = 2(2) - 10 = 4 - 10 = -6$

The ordered pair $(-2, 2)$ checks.

Step 6 Write the solution set.

$$\{(-2, 2)\}$$

□

EXAMPLE 4. Solve $\begin{cases} x - 3y = 0 \\ 2x - 6y = 5 \end{cases}$ by elimination.

Solution:

Step 1 Write each equation in the form $Ax + By = C$.

Both equations are already in the proper form. *(continued)*

Step 2 Multiply each equation by a suitable number so addition will eliminate a variable.

We multiply the first equation by -2.

$$\begin{cases} x - 3y = 0 \\ 2x - 6y = 5 \end{cases} \xrightarrow{\text{Multiply by } -2} \begin{cases} -2x + 6y = 0 \\ 2x - 6y = 5 \end{cases}$$

Step 3 Add the equations together and solve.

$$\begin{cases} -2x + 6y = 0 \\ 2x - 6y = 5 \end{cases}$$
$$0 = 5$$

As before, a false statement indicates that the system is *inconsistent* and the solution set is the empty set.

Step 6 Write the solution set.

$$\varnothing$$

□

EXAMPLE 5. Solve $\begin{cases} 9x - 6y = 18 \\ 3x - 2y = 6 \end{cases}$ by elimination.

Solution:

The equations are in the proper form.

Step 2 Multiply each equation by a suitable number so addition will eliminate a variable.

We multiply the second equation by -3.

$$\begin{cases} 9x - 6y = 18 \\ 3x - 2y = 6 \end{cases} \xrightarrow[\text{Multiply by } -3]{} \begin{cases} 9x - 6y = 18 \\ -9x + 6y = -18 \end{cases}$$

Step 3 Add the equations together and solve.

$$\begin{cases} 9x - 6y = 18 \\ -9x + 6y = -18 \end{cases}$$
$$0 = 0$$

This is a true statement, so the system is *dependent.*

□

PROBLEM SET 7.3

Warm-ups

Solve the following systems by elimination. If the system is dependent, write dependent.

For Problems 1 through 3, see Example 1.

1. $\begin{cases} x + y = 4 \\ x - y = 4 \end{cases}$

2. $\begin{cases} 2x - y = 3 \\ x + y = 6 \end{cases}$

3. $\begin{cases} x - y = 0 \\ 4x + y = 10 \end{cases}$

For Problems 4 through 9, see Example 2.

4. $\begin{cases} x + y = 4 \\ x + 2y = 5 \end{cases}$

5. $\begin{cases} x - y = 3 \\ x - 2y = 4 \end{cases}$

6. $\begin{cases} 3x + y = -2 \\ 2x + y = -1 \end{cases}$

7. $\begin{cases} 2x - y = 8 \\ 3x + 2y = 5 \end{cases}$

8. $\begin{cases} 3x - y = 1 \\ x + 2y = 12 \end{cases}$

9. $\begin{cases} 4x + 3y = 9 \\ 2x - y = 7 \end{cases}$

For Problems 10 through 12, see Examples 4 and 5.

10. $\begin{cases} 2x + 3y = 1 \\ 2x + 3y = 0 \end{cases}$

11. $\begin{cases} 3x - 2y = 4 \\ 9x - 6y = 12 \end{cases}$

12. $\begin{cases} 6x - 4y = -24 \\ 3x - 2y = 10 \end{cases}$

Practice Exercises

Solve the following systems by elimination. If the system is dependent, write dependent.

13. $\begin{cases} x + y = 3 \\ x - y = 3 \end{cases}$

14. $\begin{cases} 3x - y = 11 \\ x + y = 5 \end{cases}$

15. $\begin{cases} x - y = 0 \\ 3x + y = 12 \end{cases}$

16. $\begin{cases} x + y = 1 \\ x + 2y = 2 \end{cases}$

17. $\begin{cases} x - y = 8 \\ x + 2y = -7 \end{cases}$

18. $\begin{cases} 3x - y = 2 \\ 2x - y = -4 \end{cases}$

19. $\begin{cases} 2x - 3y = 4 \\ 2x - 3y = 2 \end{cases}$

20. $\begin{cases} x - 3y = -3 \\ 2x - y = 4 \end{cases}$

21. $\begin{cases} 9x + 6y = 4 \\ 3x + 2y = 1 \end{cases}$

22. $\begin{cases} 2x - y = 8 \\ 4x + 2y = 0 \end{cases}$

23. $\begin{cases} 3x - y = 22 \\ x - 3y = 2 \end{cases}$

24. $\begin{cases} -6x - 3y = 9 \\ 2x + y = 4 \end{cases}$

25. $\begin{cases} 2x + 3y = 2 \\ 5x + 4y = -9 \end{cases}$

26. $\begin{cases} 5x - 3y = 21 \\ 6x - 5y = 35 \end{cases}$

27. $\begin{cases} 7x + 2y = 7 \\ 3x + 4y = -19 \end{cases}$

28. $\begin{cases} 2x - 5y = 8 \\ 4x - 10y = -10 \end{cases}$

29. $\begin{cases} 4x - 3y = 11 \\ 3x + 2y = -13 \end{cases}$

30. $\begin{cases} 6x - 9y = 12 \\ 2x - 3y = 4 \end{cases}$

Challenge Problems

Solve the following systems by elimination.

31. $\begin{cases} x - y = a \\ x + y = a \end{cases}$

32. $\begin{cases} x + y = b \\ x - y = a \end{cases}$

33. $\begin{cases} x + y = a \\ x - 2y = b \end{cases}$

IN YOUR OWN WORDS...

34. Why is the procedure outlined in this section called the method of *elimination?*

35. Describe the method of elimination.

7.4 APPLICATIONS

Often word problems are easier to solve if we use more than one variable. In this section we examine solving word problems by using two variables. We follow a procedure similar to that given in Section 3.4. When we use two variables, the mathematical model should be a system of two equations.

A Procedure for Solving Word Problems with Two Variables

1. **Read** the problem and determine two quantities that are to be found.
2. **Assign** two variables, such as x and y, to represent the quantities to be found.
3. **Draw** a figure or picture if possible.
4. **Reread** the problem and form a system of two equations.
5. **Solve** the system of equations by substitution or elimination.
6. **Translate** the solution of the system to answer the question in the word problem and check this answer in the original problem. It should make sense and answer the question.
7. **Answer** the question asked in the word problem.

We will look at some examples that illustrate how this procedure works.

EXAMPLE 1. If the smaller of two numbers is subtracted from the larger, the result is 8. If the larger number is added to twice the smaller number, the result is 53. Find the two numbers.

Solution:

Step 1 Read the problem and determine two quantities to be found.

We are to find two numbers.

Step 2 Assign a variable to each number to be found.

Let L be the larger number and S be the smaller.

Step 4 Form a system of two equations.

Since L is the larger number, the smaller subtracted from the larger is $L - S$. One equation is

$$L - S = 8$$

Twice the smaller number is $2S$. The larger number added to twice the smaller is $L + 2S$. So, a second equation is

$$L + 2S = 53$$

Thus we have the system

$$\begin{cases} L - S = 8 \\ L + 2S = 53 \end{cases}$$

Step 5 Solve the system.

Elimination is a good method to use if both equations are linear.

We multiply the first equation by 2 and add the two equations.

Thus,

$$\begin{cases} L - S = 8 \\ L + 2S = 53 \end{cases} \xrightarrow{\text{Multiply by 2}} \begin{cases} 2L - 2S = 16 \\ L + 2S = 53 \end{cases}$$

$$\begin{cases} 2L - 2S = 16 \\ L + 2S = 53 \end{cases} \quad \text{Add the equations.}$$

$$3L = 69 \quad \text{Solve.}$$

$$L = 23$$

We replace L with 23 in the first equation and get

$$L - S = 8$$

$$23 - S = 8$$

$$-S = -15$$

$$S = 15$$

Step 6 Translate into an answer and check.

We can check these two numbers in the original word problem: $23 - 15$ is 8, and $2(15) + 23$ is 53.

Step 7 Answer the question.

The numbers are 23 and 15. □

The next example deals with the perimeter of a rectangle. Recall the formula: $P = 2L + 2W$.

EXAMPLE 2. The perimeter of a rectangle is 128 ft. If the length is three times the width, find the dimensions of the rectangle.

Solution:

Step 1 Read the problem and determine two quantities to be found.

We are to find the length and width of a rectangle.

Step 2 Assign a variable to each number to be found.

Let L be the length and W be the width in feet.

Step 3 Draw a picture.

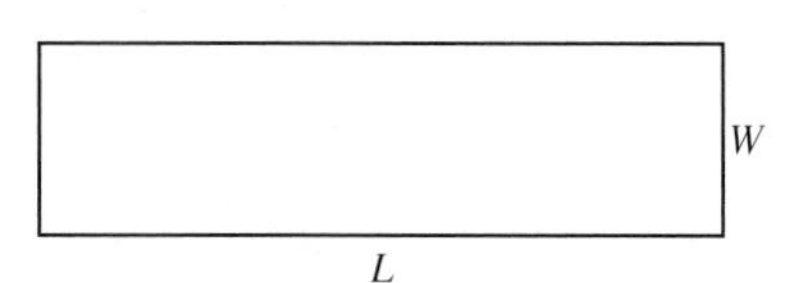

Step 4 Form a system of two equations.

Using the formula for perimeter, we form the equation

$$2L + 2W = 128$$

(continued)

We also know that the length is three times the width, so

$$L = 3W$$

This gives us the system

$$\begin{cases} 2L + 2W = 128 \\ L = 3W \end{cases}$$

Step 5 Solve the system.

If one of the equations of the system is already solved for one of the variables, such as the second equation of this system, substitution is a convenient method to use.

Replace L in the first equation with $3W$.

$$2L + 2W = 128$$

$$2(3W) + 2W = 128$$

$$8W = 128$$

$$W = 16$$

Since $L = 3W$, we see that L is 48.

Step 6 Check.

We notice that 48 is three times 16, and if the rectangle has dimensions of 16 ft and 48 ft, its perimeter is 2(16) + 2(48), or 128 ft.

Step 7 Answer the question.

The rectangle is 16 ft by 48 ft. □

EXAMPLE 3. Tickets to a concert are \$5 for balcony and \$10 for orchestra. If attendance at one show was 800 and if total receipts were \$7000, how many people bought orchestra seats?

Solution:

Step 1 Read the problem and determine two quantities to be found.

We are asked for the number of *orchestra* seats sold. However, in solving the problem, we will also find the number of balcony seats sold.

Step 2 Assign a variable to each quantity to be found.

Let x be the number of balcony tickets sold.
Let y be the number of orchestra tickets sold.

Step 4 Form a system of two equations.

As a total of 800 tickets were sold, one equation is

$$x + y = 800$$

The total receipts are found by finding the value of x balcony tickets plus the value of y orchestra tickets. The value of the balcony tickets is $5x$ dollars and the value of the orchestra tickets is $10y$ dollars. So, the total receipts are $5x + 10y$ dollars. This gives the equation

$$5x + 10y = 7000$$

We have the following system.

$$\begin{cases} x + y = 800 \\ 5x + 10y = 7000 \end{cases}$$

Step 5 Solve the system.

We choose the method of elimination. If we multiply the first equation by -5 and add the two equations, we will eliminate x.

$$\begin{cases} x + y = 800 \\ 5x + 10y = 7000 \end{cases} \xrightarrow{\text{Multiply by } -5} \begin{cases} -5x - 5y = -4000 \\ 5x + 10y = 7000 \end{cases}$$

$$\begin{cases} -5x - 5y = -4000 \\ 5x + 10y = 7000 \end{cases} \quad \text{Add the equations.}$$

$$5y = 3000$$

$$y = 600 \quad \text{Solve.}$$

It is not necessary to find x to answer the question. However, to check we need its value. Replacing y with 600 in the first equation gives a value of 200 for x.

Step 6 Check.

Since $200 + 600$ is 800 and $5(200) + 10(600)$ is 7000, the solution checks.

Step 7 Answer the question.

There were 600 orchestra tickets sold. ☐

The next example uses the formula for simple interest,

$$\text{Interest} = \text{principal} \cdot \text{rate} \cdot \text{time}$$

$$I = PRT$$

EXAMPLE 4. Madame Ross invested \$150,000 for one year in two different accounts, which pay 10% and 12% simple interest. If she earned a total of \$15,600 in interest for the year, how much did she invest at each rate?

Solution:

Step 1 Determine two quantities to be found.

We are to find the amount invested at 10% and the amount invested at 12%.

Step 2 Assign variables.

Let x be the number of dollars invested at 10%.
Let y be the number of dollars invested at 12%.

Step 4 Find two equations.

One equation is $x + y = 150{,}000$.
A table will help find another equation.

(continued)

	PRINCIPAL P	RATE R	TIME T	INTEREST $I = PRT$
10% investment	x	10%	1 year	$x(0.10)(1)$
12% investment	y	12%	1 year	$y(0.12)(1)$

From the table we see that the interest earned by the 10% investment is $0.10x$, or $\frac{10}{100}x$, dollars, and the interest earned by the 12% investment is $0.12y$, or $\frac{12}{100}y$, dollars. But we know she earned \$15,600 on these two investments. That provides the second equation.

$$\frac{10}{100}x + \frac{12}{100}y = 15600$$

Our system of equations is

$$\begin{cases} x + y = 150{,}000 \\ \frac{10}{100}x + \frac{12}{100}y = 15{,}600 \end{cases}$$

Step 5 Solve the system.

First, we clear fractions in the second equation.

$$\begin{cases} x + y = 150{,}000 \\ \frac{10}{100}x + \frac{12}{100}y = 15{,}600 \end{cases} \xrightarrow{\text{Multiply by 100}} \begin{cases} x + y = 150{,}000 \\ 10x + 12y = 1{,}560{,}000 \end{cases}$$

We solve by elimination.

$$\begin{cases} x + y = 150{,}000 \\ 10x + 12y = 1{,}560{,}000 \end{cases} \xrightarrow{\text{Multiply by } -10} \begin{cases} -10x - 10y = -1{,}500{,}000 \\ 10x + 12y = 1{,}560{,}000 \end{cases}$$

$$\begin{cases} -10x - 10y = -1{,}500{,}000 \\ 10x + 12y = 1{,}560{,}000 \end{cases} \quad \text{Add the equations.}$$

$$2y = 60{,}000 \quad \text{Solve.}$$
$$y = 30{,}000$$

Putting this into the first equation gives

$$x + y = 150{,}000$$
$$x + 30{,}000 = 150{,}000$$
$$x = 120{,}000$$

Step 6 Check.

Since $120{,}000 + 30{,}000 = 150{,}000$, the total investment is \$150,000. Also, 10% of \$120,000 is \$12,000 and 12% of \$30,000 is \$3,600, so the total interest is \$15,600. The solution checks.

Step 7 Answer the question.

Madame Ross invested \$120,000 at 10% and \$30,000 at 12%. □

PROBLEM SET 7.4

Warm-ups

Use two variables to solve the following problems. In Problems 1 through 4, see Example 1.

1. The sum of two numbers is 48. One of the numbers is twice the other number. Find the two numbers.
2. If twice a number is added to another number, the result is 16. If the sum of the numbers is 10, find the numbers.
3. There were 6000 fans at the basketball game. If the home team fans outnumber the visiting team fans by 540, how many fans were for the home team?
4. Beverly made 78 red and blue ribbons. If she made twice as many blue as red, how many of each did she make?

In Problems 5 through 7, see Example 2.

5. The parking lot of Freeman's Hardware Store is a rectangle with a perimeter of 72 m. If the length is 6 m more than the width, find the dimensions of the parking lot.
6. The perimeter of a triangle is 46 ft. The shortest side is 8 ft. The longest side is 1 ft less than twice the middle side. Find the lengths of the longest and middle sides.
7. The perimeter of a rectangle is 200 yd. If the length is three times the width, find the dimensions of the rectangle.

In Problems 8 through 11, see Example 3.

8. Jane has \$625 in \$5 and \$10 bills. If she has a total of 105 bills, how many of each kind does she have?
9. Tickets to the senior play were \$6 on Friday and \$8 on Saturday. The attendance for both performances was 2850. If the total receipts were \$20,300, how many attended the senior play each night?
10. Sam bought \$6.52 worth of 22¢ and 15¢ stamps. If he bought 4 more 15¢ stamps than 22¢ stamps, how many of each kind did he buy?
11. Two cans of Wilson tennis balls and three cans of Spalding tennis balls cost a total of \$10.55, whereas one can of Wilson and two cans of Spalding balls cost \$6.37. How much is a can of Spalding balls?

In Problems 12 through 14, see Example 4.

12. Annie Brown invested a total of \$50,000, some at 8% and the rest at 11%. If she earned a total of \$4900 in 1 yr, how much did she invest at each rate?
13. Glen Cunningham borrowed a total of \$15,000 for 1 yr. On one part he pays 10% interest and on the rest he pays 14% interest. If the total interest he paid in the year was \$1740, how much did he borrow at each rate?
14. Sarah borrowed \$6000, some at 12% interest and the rest at 15% interest, for 1 yr. If she paid a total of \$795 in interest, how much did she borrow at each rate?

Practice Exercises

Use two variables to solve the following problems.

15. The sum of two numbers is 6 and one number is 18 more than the other number. Find the two numbers.
16. There are 48 elm and oak trees on the north campus. The number of oaks subtracted from three times the number of elms is 12. Find the number of elms and the number of oaks.
17. A crystal salt and pepper set sells for \$35. If the salt shaker sells for \$5 more than the pepper shaker, how much does each cost alone?
18. The sum of two numbers is 9. If the larger is subtracted from twice the smaller, the result is 3. Find the numbers.
19. There are a total of 2000 riders in the Baltimore bike race. If the female riders outnumber the male riders by 400, how many female riders are in the race?

20. One number is 7 more than another number. If the smaller number is subtracted from twice the larger number, the result is 25. Find the two numbers.
21. Two dozen doughnuts and 6 dozen sweet rolls cost $16.52 whereas 3 dozen doughnuts and 1 dozen sweet rolls cost $8.06. How much is a dozen doughnuts?
22. The sum of two numbers is 1. If the sum of one of the numbers and twice the other number is 0, find the numbers.
23. Maria bought 2 acorn squash and 1 spaghetti squash for $2.61. Jose bought 3 acorn and 2 spaghetti for $4.41. How much is a spaghetti squash?
24. Three times the smaller of two numbers is one more than the larger number. If they differ by 37, find the numbers.
25. Tanya served 84 Coca-Colas, some Classic and some caffeine-free. If she served 10 fewer Classic than caffeine-free, how many of each kind did she serve?
26. Tom and John sold 15 boxes of Christmas fruit cake to raise funds for their fraternity. Tom sold twice as many boxes as John. How many boxes did each sell?
27. The length of a rectangle is 6 cm longer than twice the width. The perimeter is 54 cm. Find the dimensions of the rectangle.
28. The width of a rectangle is 4 m less than the length. The perimeter is 40 m. Find the dimensions of the rectangle.
29. A box contains 120 black and white marbles. There are twice as many black as white. How many of each kind are in the box?
30. If four times a number is added to twice another number, the result is 30. The sum of the numbers is 6. Find the numbers.
31. The perimeter of a triangle is 38 ft. The longest side is 18 ft. The shortes side is 4 ft less than the middle side. Find the lengths of the middle and short sides.
32. Odette has $1500 in $50 and $20 bills. If she has a total of 45 bills, how many of each kind does she have?
33. Ricardo bought 4 lb of mixed peppermints and chocolates for $13.16. If peppermints are $1.19 a pound and chocolates are $3.99 a pound, how many pounds of each kind did he buy?
34. The sum of two numbers is 0. The sum of three times the first number and the second number is 6. Find the numbers.
35. Sharon bought 3 pens and 2 pencils for $2.95, and Linda bought 2 pens and 3 pencils for $2.45. What is the price of a pencil?
36. If 1 tuna and 2 chicken sandwiches cost $3.27, but 3 tuna and 1 chicken cost $3.86, how much is a tuna sandwich?
37. If 3 yd of wool and 2 yd of cotton cost $29.92, but 1 yd of wool and 1 yd of cotton cost $10.97, find the price per yard of wool and cotton.
38. Jim LeVert has $60,000 invested at simple interest in two accounts, which pay 11% and 14% interest. If the total annual interest is $7650, how much has Jim invested in each account?
39. Mrs. Gould borrowed three times as much money at 12% interest as she did at 16% interest. If she paid $104 in interest in one year, how much money did she borrow at each rate?
40. Joan Baxter invested a total of $100, some at 10% interest and the rest at 15% interest. If her total annual interest was $11.25, how much did she invest at each rate?

Challenge Problems

41. The length of a rectangular field is twice the width. The length is 75 m more than the width. If fencing cost $10.60 per meter, find the cost to fence the field.
42. Work Problems 6, 11, and 35 using the one-variable technique of Section 3.4. Which method is easier?

IN YOUR OWN WORDS...

43. What is the two-variable procedure for solving word problems?
44. What is the difference between the two-variable procedure and the one-variable procedure?
45. In what ways are the two-variable procedure and the one-variable procedure the same?

7.5 SYSTEMS OF LINEAR INEQUALITIES IN TWO VARIABLES

So far in this chapter we have studied systems of linear *equations* in two variables. If we consider two linear inequalities together, we have a **system of linear inequalities.** To solve such a system means to find all ordered pairs that make both of the inequalities true. Since there is usually an infinite number of solutions, we don't try to list them. Instead, we graph the solution set.

EXAMPLE 1. Graph the solution set of the system of linear inequalities:

$$\begin{cases} x - y < 1 \\ x + y \geq 1 \end{cases}$$

Solution:

We graph the solution set of the inequality $x - y < 1$. As we did in Section 6.4, we draw the boundary line, $x - y = 1$. We use a dashed line since equality is not included.

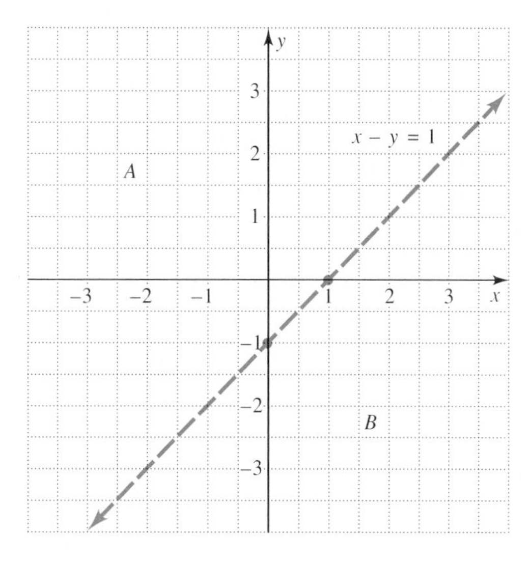

The boundary line divides the plane into two regions, A and B. We test each region.

REGION	TEST POINT IN REGION	STATEMENT $x - y < 1$	TRUTH OF STATEMENT	REGION IN SOLUTION SET?
A	(0, 0)	$0 - 0 < 1$	True	Yes
B	(2, 0)	$2 - 0 < 1$	False	No

(continued)

We shade the solution set.

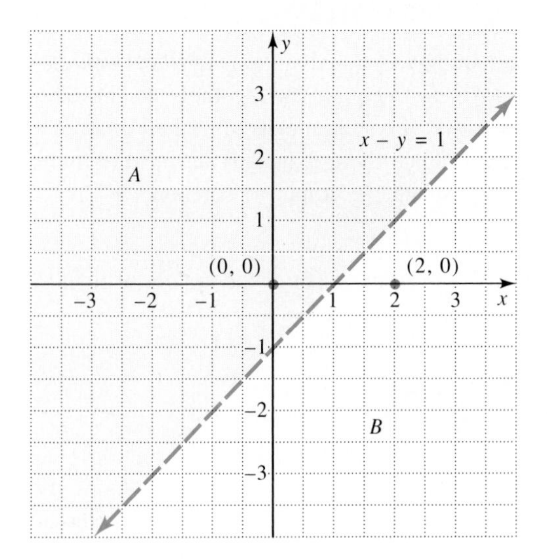

Next we draw the solution set of the second inequality $x + y \geq 1$. We draw the boundary line, $x + y = 1$, as a solid line because equality *is* included and test the two regions.

REGION	TEST POINT IN REGION	STATEMENT $x + y \geq 1$	TRUTH OF STATEMENT	REGION IN SOLUTION SET?
A	(0, 0)	$0 + 0 \geq 1$	False	No
B	(2, 0)	$2 + 0 \geq 1$	True	Yes

We shade the solution set.

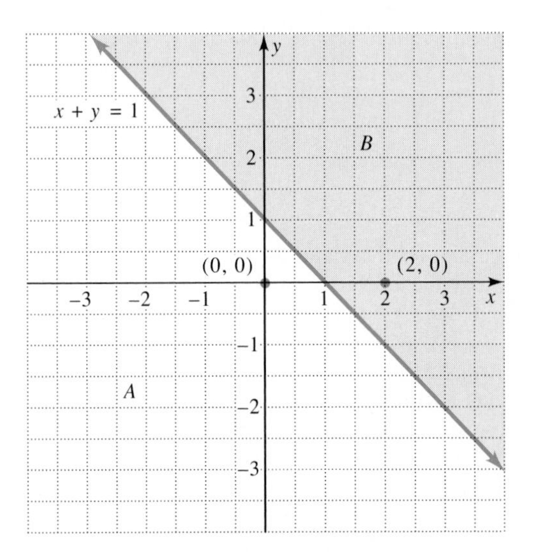

The solution set for the system consists of all ordered pairs that make *both* inequalities true statements. If we graph the solution set for both inequalities on the same coordinate system, we see both inequalities are true where the shading overlaps.

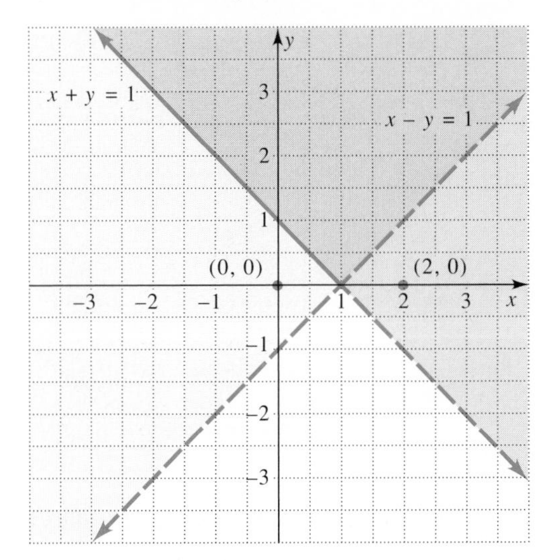

Notice the line $x + y = 1$, in the final graph of the solution set. We have changed the portion of that line *not in the solution set* from solid to dashed. □

> **A Procedure for Graphing the Solution Set of a System of Linear Inequalities**
>
> 1. Graph each inequality on the same coordinate system, shading each solution set lightly.
> 2. Darken where the lightly shaded regions overlap. Change any portions of solid lines that are not in the solution set to dashed lines.

EXAMPLE 2. Graph the solution set of the following system of inequalities.

$$\begin{cases} 2x + y < 2 \\ x \geq 1 \end{cases}$$

Solution:

We graph the solution set for $2x + y < 2$. The boundary line, $2x + y = 2$, is dashed because equality is not included. We test each region.

REGION	TEST POINT IN REGION	STATEMENT $2x + y < 2$	TRUTH OF STATEMENT	REGION IN SOLUTION SET?
A	(0, 0)	$0 + 0 < 2$	True	Yes
B	(1, 1)	$2 + 1 < 2$	False	No

(continued)

We lightly shade the solution set.

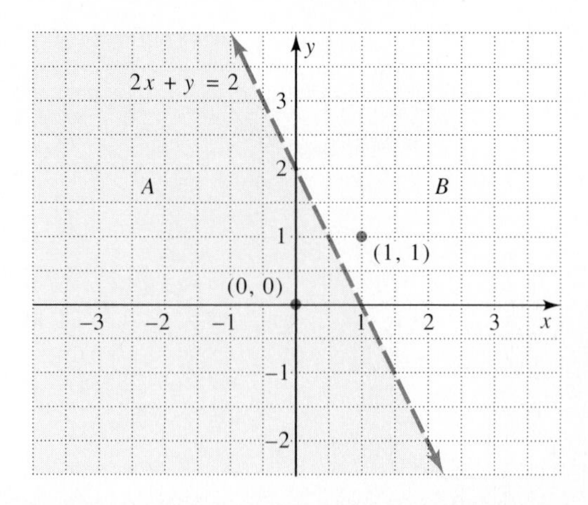

Next we graph the solution set for $x \geq 1$. We draw the boundary $x = 1$ as a solid line because equality is included and test each region.

REGION	TEST POINT IN REGION	STATEMENT $x \geq 1$	TRUTH OF STATEMENT	REGION IN SOLUTION SET?
A	(0, 0)	$0 \geq 1$	False	No
B	(2, 0)	$2 \geq 1$	True	Yes

We lightly shade the solution set.

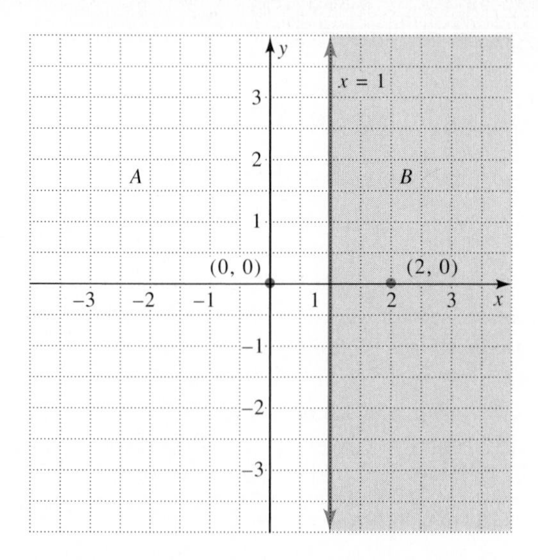

If we graph the two solutions on the same coordinate system and darken the area where the lightly shaded regions overlapped, we have the following graph of the solution set of the system.

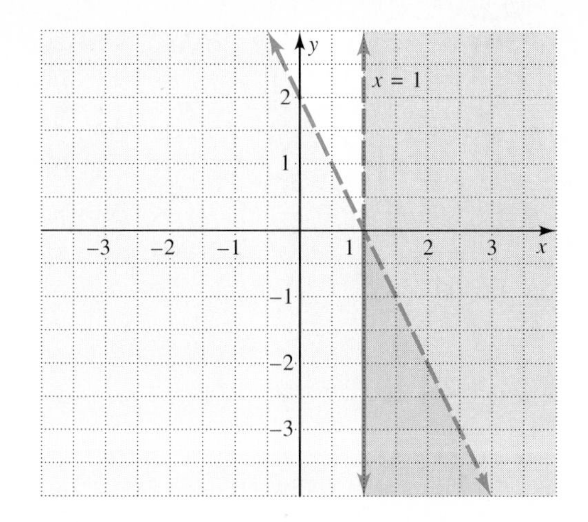

Notice that we changed the portion of the line $x = 1$ that is *above* the x-axis from solid to dashed. □

PROBLEM SET 7.5

Warm-ups

Graph the solution set of each system of inequalities. See Examples 1 and 2.

1. $\begin{cases} x + y \geq 3 \\ x - y < 2 \end{cases}$

2. $\begin{cases} x + y \leq 2 \\ x + 2y \geq 2 \end{cases}$

3. $\begin{cases} 2x + y < 2 \\ x - 3y > 3 \end{cases}$

4. $\begin{cases} x - 3y \leq 3 \\ 2x - y \leq 2 \end{cases}$

5. $\begin{cases} x - 3y < 3 \\ 4x + y < 2 \end{cases}$

6. $\begin{cases} 3x - y \geq 3 \\ x - 2y \geq 2 \end{cases}$

7. $\begin{cases} x - y < 0 \\ x + y > 0 \end{cases}$

8. $\begin{cases} 4x - y \geq 2 \\ x + 5y < 2 \end{cases}$

9. $\begin{cases} x - y < 3 \\ x + 2y \geq 0 \end{cases}$

10. $\begin{cases} x + y < 2 \\ x < 2 \end{cases}$

11. $\begin{cases} y \geq 2 \\ 2x - y \leq 2 \end{cases}$

12. $\begin{cases} x \leq 2 \\ x + y > 0 \end{cases}$

Practice Exercises

Graph the solution set of each system of inequalities.

13. $\begin{cases} x - y \geq 1 \\ x + y < 1 \end{cases}$

14. $\begin{cases} x - y \leq 2 \\ x - 2y \geq 1 \end{cases}$

15. $\begin{cases} 2x - y < 2 \\ x + 3y > 3 \end{cases}$

16. $\begin{cases} 4x - 3y > 6 \\ 2x + 3y \leq -6 \end{cases}$

17. $\begin{cases} 2x - 3y \geq 6 \\ 3x + 4y < 4 \end{cases}$

18. $\begin{cases} 4x - 7y < 7 \\ 7x + 4y \geq 4 \end{cases}$

19. $\begin{cases} 2x + 3y \leq 6 \\ 2x + y \leq 2 \end{cases}$

20. $\begin{cases} x + 3y < 3 \\ 4x - y < 2 \end{cases}$

21. $\begin{cases} 3x - y \geq 3 \\ x - 2y \geq 2 \end{cases}$

22. $\begin{cases} x - y \geq 0 \\ x + y \geq 0 \end{cases}$

23. $\begin{cases} 4x + y \geq 4 \\ x - 5y < 5 \end{cases}$

24. $\begin{cases} x + y \leq 3 \\ x - 3y \geq 0 \end{cases}$

25. $\begin{cases} x + 2y > 2 \\ 2x - 3y < 6 \end{cases}$

26. $\begin{cases} 3x + 2y \leq 12 \\ x + 3y \geq -6 \end{cases}$

27. $\begin{cases} 2x - 5y < 10 \\ x + 5y < -5 \end{cases}$

28. $\begin{cases} 2x - 3y > 6 \\ 3x + 5y < 15 \end{cases}$

29. $\begin{cases} 4x + 3y \geq 12 \\ 2x - 7y \leq 14 \end{cases}$

30. $\begin{cases} 3x + 4y < 12 \\ 2x - 5y \geq 10 \end{cases}$

31. $\begin{cases} 4x + 3y > 12 \\ 5x - 3y \leq 15 \end{cases}$

32. $\begin{cases} 2x + 3y \geq -6 \\ 3x - 4y < 12 \end{cases}$

33. $\begin{cases} 4x + 7y < 28 \\ 7x - 4y \geq 28 \end{cases}$

34. $\begin{cases} x - y < 3 \\ x \leq -2 \end{cases}$

35. $\begin{cases} y > 3 \\ 2x + y \leq 4 \end{cases}$

36. $\begin{cases} x \leq 1 \\ x + y \leq 0 \end{cases}$

Challenge Problems

Graph the solution set.

37. $\begin{cases} y > 0 \\ x < 0 \end{cases}$

38. $\begin{cases} y < 0 \\ x \geq 0 \end{cases}$

39. $\begin{cases} x \geq 0 \\ y \leq 0 \end{cases}$

IN YOUR OWN WORDS...

40. Describe how to graph a system of two linear inequalities in two variables.

CHAPTER SUMMARY

GLOSSARY

System of two equations in two variables: Two equations linked so that a solution of the system is a solution of *both* equations.

Independent system of linear equations: A system of linear equations with exactly one solution.

Dependent system: A system of equations with an infinite number of solutions.

Inconsistent system: A system of equations with no solutions.

METHOD OF GRAPHING FOR SOLVING SYSTEMS OF EQUATIONS

Graph both equations on the same coordinate system and estimate their point of intersection, if any.

SUBSTITUTION METHOD FOR SOLVING SYSTEMS OF EQUATIONS

1. Solve one of the equations for one of the variables.
2. Substitute this expression for the same variable in the other equation.
3. Solve the resulting equation.
4. Substitute the solution of Step 3 into the equation formed in Step 1 to find the value of the other variable.
5. Write the solution set.

ELIMINATION METHOD FOR SOLVING SYSTEMS OF EQUATIONS

1. Multiply both sides of each equation by a suitable number so that adding the equations eliminates a variable.
2. Add the equations and solve the resulting equation.
3. Substitute this value into one of the original equations to find the value of the other variable.
4. Write the solution set.

GRAPHING SYSTEMS OF LINEAR INEQUALITIES

1. Graph each inequality on the same coordinate system. Lightly shade the solution set of each.
2. Darken where the lightly shaded portions overlap. Change any portions of solid lines that are not in the solution set to dashed lines.

CHECKUPS

1. How many solutions are in the solution set for the following system?

$$\begin{cases} x + y = 2 \\ 2x + 2y = 4 \end{cases}$$

Section 7.1; Example 4

2. Solve the system by substitution.

$$\begin{cases} x + y = 5 \\ x - y = -1 \end{cases}$$

Section 7.2; Example 2

3. Solve $\begin{cases} 2x + 3y = 1 \\ 3x + y = -2 \end{cases}$ by elimination.

Section 7.3; Example 2

4. If the smaller of two numbers is subtracted from the larger, the result is 8. If the larger number is added to twice the smaller number, the result is 53. Find the two numbers.

Section 7.4; Example 1

5. Graph the solution set of the system of inequalities.

$$\begin{cases} x - y < 1 \\ x + y \geq 1 \end{cases}$$

Section 7.5; Example 1

REVIEW PROBLEMS

Use the graphical method to classify each system as independent, inconsistent, or dependent.

1. $\begin{cases} x - 2y = 4 \\ 2x - y = 2 \end{cases}$
2. $\begin{cases} 3x - y = 3 \\ 3x + y = 3 \end{cases}$
3. $\begin{cases} x - y = 0 \\ 2x - 2y = 0 \end{cases}$
4. $\begin{cases} 4x + 2y = 4 \\ 2x + y = 0 \end{cases}$
5. $\begin{cases} x + y = 3 \\ y = 2x \end{cases}$
6. $\begin{cases} x - y = 0 \\ 2x + 2y = 4 \end{cases}$

Use either substitution or elimination to solve each system. If the system is dependent, write dependent.

7. $\begin{cases} x + y = 3 \\ y = x \end{cases}$
8. $\begin{cases} x + 2y = 4 \\ x - 2y = 2 \end{cases}$
9. $\begin{cases} x - 3y = 5 \\ x + y = -3 \end{cases}$

10. $\begin{cases} y = x + 1 \\ 2x - y = 1 \end{cases}$

11. $\begin{cases} 2x + 3y = 1 \\ x + y = 0 \end{cases}$

12. $\begin{cases} 2x - 4y = 3 \\ x = 2y \end{cases}$

13. $\begin{cases} 2x + 5y = 1 \\ 5x + 2y = -8 \end{cases}$

14. $\begin{cases} 3x - 2y = 4 \\ 6x - 4y = 8 \end{cases}$

15. $\begin{cases} 2x + 3y = 5 \\ y = 2x - 1 \end{cases}$

16. $\begin{cases} 3x - 4y = 18 \\ 2x + 3y = -5 \end{cases}$

17. $\begin{cases} 2x + 7y = 0 \\ 5x - 2y = 0 \end{cases}$

18. $\begin{cases} 3x - 5y = 6 \\ 2x - 3y = 4 \end{cases}$

Graph the solution set for each system of inequalities.

19. $\begin{cases} x - 2y < 0 \\ x + y \geq 3 \end{cases}$

20. $\begin{cases} 2x + y \geq 2 \\ x + 2y \geq -2 \end{cases}$

21. $\begin{cases} x + 3y < 0 \\ 2x - y > 0 \end{cases}$

22. $\begin{cases} 2x - y \leq 4 \\ x + y > 5 \end{cases}$

23. $\begin{cases} 2x + 3y > 6 \\ x \geq -1 \end{cases}$

24. $\begin{cases} x + 2y \geq 2 \\ y > 3 \end{cases}$

25. Joan and Sue have 36 books. If Joan has three times as many as Sue, how many books does each have?

26. One number is 8 less than another number. If the sum of the numbers is 12, find the numbers.

27. The width of a rectangle is 5 m less than the length. If the perimeter is 38 m, find the dimensions of the rectangle.

28. Terry invested \$800 for one year, some at 11% simple interest and the rest at 14% simple interest. If he earned a total of \$103 in interest, how much did he invest at 11%?

...LET'S NOT FORGET...

In Problems 29 through 36, perform the operations indicated.

29. -2^4

30. $(-2)^4$

31. $6x - (x + 3y)$

32. $\dfrac{a^2 - ab}{b - a}$

33. $(3x - 2)^2$

34. $(2x - 5)(2x + 5)$

35. $(3a + b)(2a - b)$

36. $\dfrac{x^2 + y^2}{x^2y^3} \cdot \dfrac{xy^2}{x^4 + x^2y^2}$

In Problems 37 through 39, simplify each expression. Watch the role of parentheses.

37. $2x + y(2x - y)$

38. $(2x + y)(x - y)$

39. $\dfrac{1}{x + 1} - \dfrac{1}{x - 1}$

In Problems 40 through 43, determine if b is a factor *or a* term.

40. $b + 2$

41. $b(x + 2)$

42. $\dfrac{1}{b + 2}$

43. $x(y + b)$

In Problems 44 through 46, multiply each expression by $\frac{x+2}{x-2}$.

44. $\frac{x-2}{x+2}$

45. $1+\frac{x-2}{x+2}$

46. $\frac{x-2}{x+2}\left(1+\frac{x-2}{x+2}\right)$

In Problems 47 through 52, identify which expressions are factored. Factor those that are not factored, if possible.

47. xy^2

48. $x^2 - 4z^2$

49. $(a+5)^3$

50. $a^2b - b^2c$

51. $a^2 + 25$

52. $\frac{x}{2} + x^2$

In Problems 53 through 57, identify whether each problem is an expression, equation, *or* inequality. *Solve the equations and inequalities, and perform the operation indicated with the expressions.*

53. $\frac{2}{x+1} - 3$

54. $\frac{2}{x+1} - 3 = \frac{-2x}{x+1}$

55. $-\frac{2}{3}x > 1$

56. $\frac{\frac{2}{xy^2}}{\frac{1}{xy}}$

57. $\frac{x^2+4x+4}{x^2} \div \frac{x^2+x-2}{x^2-x}$

CHAPTER 7 TEST

In Problems 1 through 5, choose the correct answer.

1. The system $\begin{cases} x - y = 0 \\ 2x - 2y = 0 \end{cases}$ is (?).

A. Independent B. Inconsistent
C. Dependent D. None of the above

2. The x-coordinate of the solution to the system of equations $\begin{cases} 2x + 3y = 16 \\ 3x - y = 2 \end{cases}$ is (?).

A. $\frac{18}{11}$ B. $\frac{18}{5}$
C. 1 D. 2

3. The solution set for the system $\begin{cases} 2x - 4y = 8 \\ x - 2y = 8 \end{cases}$ is (?).

A. $\{(16, 4)\}$ B. $\{(0, -2)\}$
C. $\{\ \}$ D. None of the above

4. The y-coordinate of the solution to the system of equations $\begin{cases} y = 2 - x \\ 2x - 3y = 9 \end{cases}$ is (?).

A. 3 B. -1
C. -13 D. 5

5. The solution set for the system $\begin{cases} x + y = 0 \\ x - 2y = 0 \end{cases}$ is (?).

A. $\{(0, 0)\}$ B. $\{(-3, 3)\}$
C. $\varnothing$ D. None of the above

In Problems 6 through 9, solve the system of equations.

6. $\begin{cases} x = 3y \\ 3x - y = 8 \end{cases}$

7. $\begin{cases} x + 2y = 4 \\ 2x + 4y = 0 \end{cases}$

8. $\begin{cases} 3x + 2y = 0 \\ 2x - 3y = 13 \end{cases}$

9. $\begin{cases} y = x + 1 \\ 2x - y = -2 \end{cases}$

In Problems 10 through 13, graph the solution set for each system of inequalities.

10. $\begin{cases} 2x - y \le 0 \\ x + 3y \ge 0 \end{cases}$

11. $\begin{cases} x + y \le 3 \\ x - y > 1 \end{cases}$

12. $\begin{cases} x - 2y > 1 \\ x + y < 2 \end{cases}$

13. $\begin{cases} x \ge -2 \\ x + y < 1 \end{cases}$

14. The length of a rectangle is 7 ft more than the width. If the perimeter is 46 ft, find the dimensions of the rectangle.

15. If a head of lettuce and 2 avocados cost \$2.27 and 2 heads of lettuce and 3 avocados cost \$3.85, how much does 1 avocado cost?

CHAPTER 8

See Problem Set 8.2, Exercise 66.

Roots and Radicals

CONNECTIONS

The first numbers we encounter as we become educated in mathematics are the *natural numbers,* or counting numbers: 1, 2, 3, ... Soon we need zero and the negatives of the natural numbers to indicate such things as temperatures and bank balances. The set of these numbers together with the natural numbers is called the *integers.* Next, we add the fractions, or *rational numbers,* in order to measure real things such as weight and length. Rational numbers are sufficient to solve simple equations such as the linear equations studied in Chapter 3, and to solve many day-to-day problems.

Unfortunately, the rational numbers solve only the simplest equations. In the next chapter we will study equations involving numbers such as $\sqrt{2}$, $\sqrt[3]{5}$, $2 - \sqrt{3}$, which are irrational. In preparation, this chapter continues the study of square roots, introduced in Section 0.5, and extends this idea to the nth root of a number. We examine the arithmetic of such numbers and when they can be simplified. The last section (optional) covers fractional exponents and their relationship to radicals.

8.1 DEFINITIONS

We know that 2 and -2 are opposites and that if we square 2 or -2 the result is 4. Using an exponent of 2 indicates squaring a number. So,

$$2^2 = 4$$

$$(-2)^2 = 4$$

Thus 4 is the result of squaring 2 or -2. Integers such as 4, 9, and 16 are called perfect squares because they can be expressed as the square of an integer.

The reverse of squaring is called finding the square root. We use the symbol $\sqrt{4}$ to represent the *positive* number whose square is 4. That is,

$$\sqrt{4} = 2$$

We call $\sqrt{4}$ the **principal square root** of 4.

Using this notation, we can write $-\sqrt{4}$ to indicate the opposite of the principal square root of 4. It is the *other* number whose square is 4. Thus

$$-\sqrt{4} = -2$$

EXAMPLE 1. Find the roots indicated.

(a) $\sqrt{9}$ (b) $-\sqrt{9}$ (c) $\sqrt{\frac{4}{9}}$

Solutions:

(a) $\sqrt{9} = 3$ because 3^2 is 9 and 3 is *positive.*

(b) $-\sqrt{9} = -3$ because $\sqrt{9} = 3$ and the opposite of 3 is -3.

(c) $\sqrt{\frac{4}{9}} = \frac{2}{3}$ because $\left(\frac{2}{3}\right)^2 = \frac{4}{9}$ and $\frac{2}{3}$ is *positive.* □

Is there a number whose square is 7? Yes, there are two such numbers, and they are opposites. We write $\sqrt{7}$ to indicate the *positive* number whose square is 7 and $-\sqrt{7}$ to indicate the negative number whose square is 7. That is,

$$(\sqrt{7})^2 = 7 \quad \text{and} \quad (-\sqrt{7})^2 = 7$$

Because 7 is not a perfect square, it is difficult to write $\sqrt{7}$ in a simpler form. However, $\sqrt{7}$ can be written as a nonrepeating, nonterminating decimal. We can *approximate* its value with a calculator or a square root table. The square root of seven is approximately 2.6, rounded to the nearest tenth. We write

$$\sqrt{7} \approx 2.6$$

Notice the *approximately equal* symbol, $\approx$.

EXAMPLE 2. Approximate each square root to two decimal places.

(a) $\sqrt{11}$ (b) $-\sqrt{5}$ (c) $\sqrt{3}$

Solutions:

(a) $\sqrt{11} \approx 3.32$

(b) $-\sqrt{5} \approx -2.24$

(c) $\sqrt{3} \approx 1.73$ □

What is $\sqrt{-4}$? Is there a real number whose square is -4? Since $(-2)^2 = 4$, there is no such real number. *Square roots of negative real numbers do not exist in our system of real numbers.*

Square Root of a Positive Number

If x is a positive real number, $\sqrt{x}$ is the *positive* number whose square is x. That is,

$$(\sqrt{x})^2 = x$$

Square Root of Zero

$$\sqrt{0} = 0$$

If we cube 2, the result is 8. An exponent of 3 indicates cubing. Thus

$$2^3 = 8$$

There is not another real number whose cube is 8, since

$$(-2)^3 = -8$$

The reverse of cubing is called finding a *cube root.* We indicate the cube root of 8 by $\sqrt[3]{8}$. So,

$$\sqrt[3]{8} = 2$$

EXAMPLE 3. Find $\sqrt[3]{-27}$.

Solution:

$$\sqrt[3]{-27} = -3 \text{ because } (-3)^3 = -27. \quad \square$$

Notice that the cube root of a negative number does exist and is a negative number. In fact, if x is *any real number,* $\sqrt[3]{x}$ is the number whose cube is x. That is,

$$(\sqrt[3]{x})^3 = x$$

EXAMPLE 4. Find the following cube roots.

(a) $\sqrt[3]{27}$ (b) $\sqrt[3]{-8}$ (c) $-\sqrt[3]{-8}$ (d) $\sqrt[3]{-\frac{8}{27}}$

Solutions:

(a) $\sqrt[3]{27} = 3$ because $3^3 = 27$.
(b) $\sqrt[3]{-8} = -2$ because $(-2)^3 = -8$.
(c) $-\sqrt[3]{-8} = 2$ because $-\sqrt[3]{-8} = -(-2)$.
(d) $\sqrt[3]{-\frac{8}{27}} = -\frac{2}{3}$ because $\left(-\frac{2}{3}\right)^3 = -\frac{8}{27}$. $\square$

In addition to square roots and cube roots, we have fourth roots, fifth roots and, in general, n^{th} roots.

n^{th} Root of a Real Number

Suppose n is a natural number and x is a real number.

1. If *n is even* and x is not negative, $\sqrt[n]{x}$ is the *nonnegative* number whose n^{th} power is x.
2. If n is *odd,* $\sqrt[n]{x}$ is the number whose n^{th} power is x.

That is,

$$(\sqrt[n]{x})^n = x$$

Be Careful!

Remember, if n is even, x must be *nonnegative,* and $\sqrt[n]{x}$ is also *nonnegative.*

The expression $\sqrt[n]{x}$ is called a **radical.** We call the symbol, $\sqrt[n]{}$, a **radical symbol.** Also, n is called the **index** and x is called the **radicand.** Since square roots occur so often, we omit the index of 2 and use $\sqrt{}$ instead of $\sqrt[2]{}$.

EXAMPLE 5. Find the following roots, if they exist as real numbers.

(a) $\sqrt{36}$ (b) $\sqrt[3]{-64}$ (c) $\sqrt{-16}$ (d) $\sqrt[4]{16}$ (e) $\sqrt[5]{-32}$

Solutions:

(a) $\sqrt{36} = 6$ because $6^2 = 36$ and 6 is *positive.*
(b) $\sqrt[3]{-64} = -4$ because $(-4)^3 = -64$.

(c) $\sqrt{-16}$ is not a real number because -16 is *negative*.
(d) $\sqrt[4]{16} = 2$ because $2^4 = 16$ and 2 is *positive*.
(e) $\sqrt[5]{-32} = -2$ because $(-2)^5 = -32$. □

Suppose we want to find $\sqrt{x^2}$. Is there a *nonnegative* number whose square is x^2? The square of x is x^2. However, we do not know if x represents a nonnegative number. To avoid this problem, we shall assume in this chapter that *all variables represent nonnegative real numbers.* So, we write

$$\sqrt{x^2} = x$$

because $(x)^2 = x^2$ and we are assuming x is *nonnegative*.

EXAMPLE 6. Find the following roots. Assume that all variables represent *nonnegative* real numbers.

(a) $\sqrt{x^4}$ (b) $\sqrt{x^{10}}$ (c) $\sqrt[3]{x^3}$ (d) $\sqrt[5]{x^{10}}$ (e) $\sqrt{25x^2}$

Solutions:

(a) $\sqrt{x^4} = x^2$ because $(x^2)^2 = x^4$ and x^2 is nonnegative.
(b) $\sqrt{x^{10}} = x^5$ because $(x^5)^2 = x^{10}$ and x^5 is nonnegative.
(c) $\sqrt[3]{x^3} = x$ because $(x)^3 = x^3$.
(d) $\sqrt[5]{x^{10}} = x^2$ because $(x^2)^5 = x^{10}$.
(e) $\sqrt{25x^2} = 5x$ because $(5x)^2 = 25x^2$ and $5x$ is nonnegative. □

In working with radicals, it is very helpful to recognize some powers of 2, 3, 4, and 5. Some of them are listed next.

n	2^n	3^n	4^n	5^n
1	2	3	4	5
2	4	9	16	25
3	8	27	64	125
4	16	81	256	625
5	32	243	1,024	3,125
6	64	729	4,096	15,625
7	128	2,187	16,384	78,125
8	256	6,561	65,536	390,625
9	512	19,683	262,144	1,953,125

PROBLEM SET 8.1

Warm-ups

In Problems 1 through 5, find the indicated roots if they exist. See Example 1.

1. $\sqrt{25}$ **2.** $\sqrt{-81}$ 3. $-\sqrt{36}$

4. $\sqrt{\dfrac{4}{25}}$ **5.** $-\sqrt{\dfrac{4}{9}}$

In Problems 6 through 10, approximate the indicated roots to two decimal places if they exist. See Example 2.

6. $\sqrt{2}$ **7.** $\sqrt{-2}$ **8.** $-\sqrt{2}$

9. $\sqrt{8}$ **10.** $-\sqrt{13}$

In Problems 11 through 15, find the indicated roots if they exist. See Examples 3 and 4.

11. $\sqrt[3]{64}$ **12.** $\sqrt[3]{-64}$ **13.** $-\sqrt[3]{64}$

14. $\sqrt[3]{\frac{1}{64}}$ **15.** $-\sqrt[3]{\frac{1}{64}}$

In Problems 16 through 25, find the indicated roots if they exist. See Example 5.

16. $\sqrt[4]{16}$ 17. $\sqrt[4]{-16}$ 18. $-\sqrt[4]{16}$

19. $\sqrt[4]{\frac{1}{16}}$ **20.** $-\sqrt[4]{\frac{1}{16}}$ 21. $\sqrt[5]{32}$

22. $\sqrt[5]{-32}$ **23.** $-\sqrt[5]{32}$ **24.** $\sqrt[5]{\frac{1}{32}}$

25. $-\sqrt[5]{\frac{1}{32}}$

In Problems 26 through 35, find the indicated roots. Assume all variables represent nonnegative real numbers. See Example 6.

26. $\sqrt{x^6}$ **27.** $\sqrt{y^2}$ **28.** $\sqrt{t^8}$

29. $\sqrt{z^{12}}$ **30.** $\sqrt{x^{16}}$ **31.** $\sqrt{4x^8}$

32. $\sqrt{25y^4}$ 33. $\sqrt{9t^6}$ **34.** $\sqrt{100z^{18}}$

35. $\sqrt[3]{8z^3}$

Practice Exercises

Find the indicated roots if they exist.

36. $\sqrt{16}$ 37. $\sqrt{-25}$ **38.** $-\sqrt{49}$

39. $-\sqrt{16}$ **40.** $\sqrt{\frac{9}{25}}$ **41.** $-\sqrt{\frac{4}{81}}$

42. $\sqrt{100}$ **43.** $\sqrt{-400}$ **44.** $\sqrt{49}$

45. $\sqrt{-100}$ **46.** $-\sqrt{1}$ **47.** $\sqrt{\frac{49}{1}}$

48. $\sqrt[3]{-1}$ **49.** $\sqrt[3]{1}$ **50.** $-\sqrt[3]{-1}$

51. $\sqrt[3]{-\frac{1}{8}}$ **52.** $\sqrt{0}$ 53. $\sqrt{\frac{64}{49}}$

54. $-\sqrt{\frac{36}{225}}$ **55.** $\sqrt[3]{1000}$ **56.** $\sqrt[3]{-125}$

57. $-\sqrt[3]{8}$ **58.** $\sqrt[3]{\frac{1}{27}}$ **59.** $-\sqrt[3]{\frac{8}{27}}$

60. $-\sqrt[3]{-\frac{8}{27}}$ **61.** $-\sqrt[3]{1}$ **62.** $-\sqrt[3]{-27}$

63. $\sqrt[3]{-\frac{1}{125}}$ **64.** $-\sqrt[3]{\frac{27}{8}}$ **65.** $\sqrt[4]{81}$

66. $\sqrt[4]{-625}$ 67. $-\sqrt[4]{625}$ **68.** $\sqrt[4]{\frac{1}{81}}$

69. $-\sqrt[4]{\frac{1}{81}}$

70. $\sqrt[5]{243}$

71. $\sqrt[5]{-243}$

72. $-\sqrt[5]{243}$

73. $\sqrt[5]{\frac{1}{243}}$

74. $-\sqrt[5]{\frac{1}{243}}$

75. $-\sqrt[3]{-27}$

76. $\sqrt[4]{256}$

77. $\sqrt[5]{0}$ 0

78. $\sqrt{\frac{1}{16}}$

79. $\sqrt[4]{-1}$

Find each root. Assume that all variables represent nonnegative real numbers.

80. $\sqrt{x^4}$

81. $\sqrt{y^{10}}$

82. $\sqrt{t^{18}}$

83. $\sqrt{z^{20}}$

84. $\sqrt{x^{36}}$

85. $\sqrt{16x^6}$

86. $\sqrt{36y^{16}}$

87. $\sqrt{49t^2}$

88. $\sqrt{64x^8y^4}$

89. $\sqrt{25y^4z^2}$

90. $\sqrt{81s^8t^6}$

91. $\sqrt{121x^4y^2z^{18}}$

92. $\sqrt{144x^6y^4z^{12}}$

93. $\sqrt{169a^2b^2c^2}$

94. $\sqrt{121z^{14}}$

95. $\sqrt{16x^4y^2}$

96. $\sqrt{81y^6z^8}$

97. $\sqrt{144s^4t^2}$

98. $\sqrt{100x^2y^2z^2}$

99. $\sqrt{x^8y^4z^{10}}$

Approximate the indicated roots to two decimal places if they exist.

100. $\sqrt{3}$

101. $\sqrt{-3}$

102. $-\sqrt{3}$

103. $\sqrt{10}$

Challenge Problems

Estimate each root to the nearest integer.

104. $\sqrt{24}$

105. $\sqrt{15}$

106. $\sqrt{10}$

107. $\sqrt{12}$

108. $\sqrt[3]{7}$

109. $\sqrt[3]{-9}$

110. $\sqrt[3]{30}$

111. $\sqrt[3]{-26}$

IN YOUR OWN WORDS...

112. What do we mean by the square root of a number?

113. What do we mean by the seventh root of a number?

8.2 SIMPLIFYING RADICALS

We know that $\sqrt{64} = 8$ and that $8 = 2 \cdot 4$. Notice that 64 is $4 \cdot 16$,

$$\sqrt{4} = 2$$

and

$$\sqrt{16} = 4$$

So we have

$$\sqrt{64} = \sqrt{4 \cdot 16} = \sqrt{4}\sqrt{16}$$

This leads to an important property of radicals.

> **Multiplication Property of Radicals**
>
> If $\sqrt[n]{p}$ and $\sqrt[n]{q}$ exist as real numbers,
>
> $$\sqrt[n]{pq} = \sqrt[n]{p} \cdot \sqrt[n]{q}$$

The nth root of a product is the product of the nth roots.

We would not use this property to find $\sqrt{64}$ because 64 is a perfect square. However, it can be used to simplify radicals such as $\sqrt{75}$. Since 75 is not a perfect square, $\sqrt{75}$ is not an integer. However, 75 has a *factor* that is a perfect square.

$$\begin{aligned}\sqrt{75} &= \sqrt{25 \cdot 3} \\ &= \sqrt{25} \cdot \sqrt{3} && \text{Multiplication Property} \\ &= 5\sqrt{3}\end{aligned}$$

So, we see that we can write $\sqrt{75}$ as $5\sqrt{3}$. We say that $5\sqrt{3}$ is the *simplified form* of $\sqrt{75}$. The Multiplication Property of Radicals allows us to rewrite radical expressions with any perfect square factors of the radicand removed. We call a square root **simplified** when its written with all perfect square factors removed from the radicand. The following examples should clarify this.

EXAMPLE 1. Simplify each square root.

(a) $\sqrt{24}$ (b) $\sqrt{18}$ (c) $\sqrt{48}$

Solutions:

(a) The largest perfect square factor of 24 is 4.

$$\begin{aligned}\sqrt{24} &= \sqrt{4 \cdot 6} \\ &= \sqrt{4} \cdot \sqrt{6} && \text{Multiplication Property} \\ &= 2\sqrt{6} && \sqrt{4} = 2\end{aligned}$$

Notice that we could have factored 24 as $3 \cdot 8$. However, neither 3 nor 8 is a perfect square, so no simplification could take place.

(b) The largest perfect square factor of 18 is 9.

$$\begin{aligned}\sqrt{18} &= \sqrt{9 \cdot 2} \\ &= \sqrt{9} \cdot \sqrt{2} && \text{Multiplication Property} \\ &= 3\sqrt{2} && \sqrt{9} = 3\end{aligned}$$

(c) The largest perfect square factor of 48 is 16.

$$\begin{aligned}\sqrt{48} &= \sqrt{16 \cdot 3} \\ &= \sqrt{16} \cdot \sqrt{3} && \text{Multiplication Property} \\ &= 4\sqrt{3} && \sqrt{16} = 4\end{aligned}$$

Notice that we could have factored 48 as $4 \cdot 12$ and removed the perfect square factor 4.

$$\begin{aligned}\sqrt{48} &= \sqrt{4 \cdot 12} \\ &= \sqrt{4} \cdot \sqrt{12} \\ &= 2\sqrt{12}\end{aligned}$$

However, this last expression is *not simplified*. There is a perfect square factor of 12, which is 4. We could continue on and remove it:

$$\begin{aligned} 2\sqrt{12} &= 2\sqrt{4 \cdot 3} \\ &= 2\sqrt{4} \cdot \sqrt{3} \\ &= 2 \cdot 2\sqrt{3} \\ &= 4\sqrt{3} \end{aligned}$$

which is the same result as we got when we removed the *largest perfect square factor,* 16. Notice how much easier it is to remove the largest perfect square factor first. □

Cube roots can also be simplified using the Multiplication Property of Radicals. Instead of looking for perfect squares as factors of the radicand, we look for perfect cube factors.

EXAMPLE 2. Simplify each cube root.

(a) $\sqrt[3]{24}$ (b) $\sqrt[3]{-54}$

Solutions:

(a) The largest *perfect cube* factor of 24 is 8.

$$\begin{aligned} \sqrt[3]{24} &= \sqrt[3]{8 \cdot 3} & \\ &= \sqrt[3]{8} \cdot \sqrt[3]{3} & \text{Multiplication Property} \\ &= 2\sqrt[3]{3} & \sqrt[3]{8} = 2 \end{aligned}$$

(b) Because the index, 3, is *odd,* the negative radicand is permitted. However, as 27 and -27 are both perfect cube factors of -54, we have a choice to make.

$$\sqrt[3]{-54} = \sqrt[3]{(27)(-2)} \quad \text{or} \quad \sqrt[3]{-54} = \sqrt[3]{(-27)(2)}$$

In this situation, we choose to remove the negative perfect cube factor.

$$\begin{aligned} \sqrt[3]{-54} &= \sqrt[3]{-27 \cdot 2} & \\ &= \sqrt[3]{-27} \cdot \sqrt[3]{2} & \text{Multiplication Property} \\ &= -3\sqrt[3]{2} & \sqrt[3]{-27} = -3 \end{aligned}$$

□

EXAMPLE 3. Simplify each radical. Assume that all variables represent nonnegative real numbers.

(a) $\sqrt{8x^2}$ (b) $\sqrt[3]{16y^6}$ (c) $\sqrt[3]{-16x^3}$

Solutions:

(a)
$$\begin{aligned} \sqrt{8x^2} &= \sqrt{4 \cdot 2 \cdot x^2} & \\ &= \sqrt{4 \cdot x^2 \cdot 2} & \text{Commutative Property} \\ &= \sqrt{4} \cdot \sqrt{x^2} \cdot \sqrt{2} & \text{Multiplication Property} \\ &= 2x\sqrt{2} & \end{aligned}$$

(continued)

(b) $\sqrt[3]{16y^6} = \sqrt[3]{8 \cdot 2y^6}$

$= \sqrt[3]{8 \cdot y^6 \cdot 2}$ Commutative Property

$= \sqrt[3]{8} \cdot \sqrt[3]{y^6} \cdot \sqrt[3]{2}$ Multiplication Property

$= 2y^2\sqrt[3]{2}$

(c) $\sqrt[3]{-16x^3} = \sqrt[3]{-8 \cdot 2x^3}$

$= \sqrt[3]{-8 \cdot x^3 \cdot 2}$

$= \sqrt[3]{-8} \cdot \sqrt[3]{x^3} \cdot \sqrt[3]{2}$

$= -2x\sqrt[3]{2}$

Again, to simplify a root of *odd* index and negative radicand, we remove a negative factor from the radicand. □

Let's examine the quotient $\sqrt{\frac{36}{25}}$. Because $\left(\frac{6}{5}\right)^2 = \frac{36}{25}$, $\sqrt{\frac{36}{25}} = \frac{6}{5}$. However, $\frac{6}{5}$ also equals $\frac{\sqrt{36}}{\sqrt{25}}$. So, we can write $\sqrt{\frac{36}{25}} = \frac{\sqrt{36}}{\sqrt{25}}$. This is the idea behind the Division Property of Radicals.

Division Property of Radicals

If $\sqrt[n]{p}$ and $\sqrt[n]{q}$ exist as real numbers and $q \neq 0$, then

$$\sqrt[n]{\frac{p}{q}} = \frac{\sqrt[n]{p}}{\sqrt[n]{q}}$$

The nth root of a quotient is the quotient of nth roots.

EXAMPLE 4. Simplify each radical.

(a) $\sqrt{\frac{7}{4}}$ (b) $\sqrt{\frac{8}{25}}$ (c) $\sqrt[3]{\frac{5}{8}}$ (d) $\sqrt[3]{\frac{16}{27}}$

Solutions:

(a) $\sqrt{\frac{7}{4}} = \frac{\sqrt{7}}{\sqrt{4}}$ Division Property

$= \frac{\sqrt{7}}{2}$ $\sqrt{4} = 2$

(b) $\sqrt{\frac{8}{25}} = \frac{\sqrt{8}}{\sqrt{25}}$ Division Property

$= \frac{\sqrt{4 \cdot 2}}{5}$ $\sqrt{25} = 5$

$= \frac{\sqrt{4}\sqrt{2}}{5}$ Multiplication Property

$= \frac{2\sqrt{2}}{5}$

$= \frac{2}{5}\sqrt{2}$

(c) $\sqrt[3]{\frac{5}{8}} = \frac{\sqrt[3]{5}}{\sqrt[3]{8}}$ Division Property

$= \frac{\sqrt[3]{5}}{2}$ $\sqrt[3]{8} = 2$

(d) $\sqrt[3]{\frac{16}{27}} = \frac{\sqrt[3]{16}}{\sqrt[3]{27}}$ Division Property

$= \frac{\sqrt[3]{8 \cdot 2}}{3}$ $\sqrt[3]{27} = 3$

$= \frac{\sqrt[3]{8}\sqrt[3]{2}}{3}$ Multiplication Property

$= \frac{2\sqrt[3]{2}}{3}$

$= \frac{2}{3}\sqrt[3]{2}$ □

Procedure for Simplifying a Square Root or a Cube Root

1. If the radicand contains a fraction, use the Quotient Property to write the radical as the quotient of two radicals.
2. Remove all perfect square factors from any square roots. Remove perfect cube factors from cube roots.
3. If cube roots contain a negative radicand, be sure to remove a negative factor.

EXAMPLE 5. Simplify each expression.

(a) $\frac{6\sqrt{5}}{15}$ (b) $\frac{\sqrt{24}}{12}$

Solutions:

(a) This expression contains a radical, but the radical cannot be simplified further. However, 3 is a common factor of both the numerator and denominator of this fraction, so the fraction can be simplified.

$$\frac{6\sqrt{5}}{15} = \frac{2\sqrt{5}}{5}$$

This expression cannot be simplified further. The 5 in the denominator and the $\sqrt{5}$ in the numerator *are not* common factors!

(b) Although 12 *is* a factor of 24, it is *not* a factor of $\sqrt{24}$. However, $\sqrt{24}$ can be simplified.

$$\frac{\sqrt{24}}{12} = \frac{\sqrt{4 \cdot 6}}{12}$$

$= \frac{\sqrt{4} \cdot \sqrt{6}}{12}$ Multiplication Property

(continued)

$$= \frac{2\sqrt{6}}{12}$$

$$= \frac{\sqrt{6}}{6} \qquad \text{Common factor of 2}$$

As before, 6 and $\sqrt{6}$ *do not* have any common factors. This fraction *cannot be* reduced further. □

PROBLEM SET 8.2

Warm-ups

Simplify each radical.
In Problems 1 through 5, see Example 1.

1. $\sqrt{8}$
2. $\sqrt{12}$
3. $\sqrt{50}$
4. $\sqrt{200}$
5. $\sqrt{40}$

In Problems 6 through 9, see Example 2.

6. $\sqrt[3]{24}$
7. $\sqrt[3]{54}$
8. $\sqrt[3]{-32}$
9. $\sqrt[3]{40}$

In Problems 10 through 13, see Example 3. Assume all variables represent nonnegative real numbers.

10. $\sqrt{8x^4}$
11. $\sqrt[3]{16y^3}$
12. $\sqrt{20x^6}$
13. $\sqrt[3]{-24x^6}$

In Problems 14 through 17, see Example 4.

14. $\sqrt{\frac{3}{16}}$
15. $\sqrt{\frac{6}{49}}$
16. $\sqrt{\frac{12}{25}}$
17. $\sqrt{\frac{27}{4}}$

Practice Exercises

Simplify each radical. Assume all variables represent nonnegative real numbers.

18. $\sqrt{88}$
19. $\sqrt{125}$
20. $\sqrt{500}$
21. $\sqrt{600}$
22. $\sqrt{44}$
23. $\sqrt{108}$
24. $\sqrt{52}$
25. $\sqrt{68}$
26. $\sqrt{360}$
27. $\sqrt{99}$
28. $\sqrt{20}$
29. $\sqrt{112}$
30. $\sqrt{180}$
31. $\sqrt{300}$
32. $\sqrt{162}$
33. $\sqrt{32}$
34. $\sqrt{72}$
35. $\sqrt{54}$
36. $\sqrt{250}$
37. $\sqrt{96}$
38. $\sqrt[3]{-48}$
39. $\sqrt[3]{250}$
40. $\sqrt[3]{-56}$
41. $\sqrt[3]{-16}$
42. $\sqrt[3]{\frac{4}{125}}$
43. $\sqrt[3]{\frac{5}{8}}$
44. $\sqrt[3]{\frac{24}{125}}$
45. $\sqrt[3]{\frac{32}{27}}$
46. $\sqrt[3]{72}$
47. $\sqrt[3]{108}$
48. $\sqrt[3]{-96}$
49. $\sqrt[3]{56}$
50. $\sqrt[3]{-135}$
51. $\sqrt[3]{500}$
52. $\sqrt[3]{-168}$
53. $\sqrt[3]{-144}$
54. $\sqrt[3]{-250}$
55. $\sqrt[3]{160}$
56. $\sqrt[3]{-88}$
57. $\sqrt[3]{-104}$
58. $\sqrt{27x^2}$
59. $\sqrt[3]{54y^6}$
60. $\sqrt{40x^4}$
61. $\sqrt[3]{-32x^3}$
62. $\sqrt{24x^4y^2}$
63. $\sqrt[3]{72x^6y^3}$
64. $\sqrt{20x^4y^6}$
65. $\sqrt{-8t^{12}}$

66. The baseball "diamond" at Candlestick Park is not a diamond at all but a square with the bases at the corners. The distance between the bases along the base path is 90 ft. The length of the hypotenuse of a right triangle is given by the formula, $c = \sqrt{a^2 + b^2}$ where c is the length of the hypotenuse and a and b are the lengths of the other two sides. How far is it from home plate to second base?

67. The navigator of a ship can find the distance to the horizon in miles with the formula, $d = \sqrt{1.5h}$ where h is the height of the ship's bridge in feet. How far away is the horizon if the height of the ship's bridge is 150 ft?

68. Weather forecasters estimate the duration of some storms with the formula, $t = \sqrt{\frac{d^3}{216}}$ where t is the duration of the storm in hours and d is the diameter of the storm in miles. Find the approximate duration of storm of this type whose diameter is 8 miles.

Challenge Problems

Simplify each radical.

69. $\sqrt[4]{32}$

70. $\sqrt[5]{64}$

71. $\sqrt[6]{128}$

72. $\sqrt[7]{256}$

IN YOUR OWN WORDS...

73. What do we mean by a *simplified* radical expression?

74. Describe the steps in simplifying $\frac{\sqrt{75}}{15}$.

8.3 ADDITION AND SUBTRACTION OF RADICALS

Since radicals represent real numbers, we can add and subtract them. We know that

$$2x + 3x = 5x$$

because we can add *like* terms, but

$$2x + 3y$$

cannot be simplified because $2x$ and $3y$ are *not* like terms.

Addition of radicals is performed in the same way. For example,

$$2\sqrt{11} + 3\sqrt{11} = 5\sqrt{11}$$

because $2\sqrt{11}$ and $3\sqrt{11}$ are *like* radicals. However,

$$2\sqrt{5} + 3\sqrt{11}$$

cannot be simplified because $2\sqrt{5}$ and $3\sqrt{11}$ are *not* like radicals.

Likewise, $3\sqrt[3]{2} + 4\sqrt[3]{5}$ cannot be simplified because $3\sqrt[3]{2}$ and $4\sqrt[3]{5}$ are *not* like radicals. **Like radicals** are radicals with the same index and the same radicand.

To add or subtract radicals, combine like terms.

EXAMPLE 1. Perform the operations indicated.

(a) $\sqrt{3} + \sqrt{3}$ (b) $3\sqrt{2} - \sqrt{2}$ (c) $8\sqrt[3]{7} + \sqrt[3]{7} - 4\sqrt[3]{7}$

Solutions:

(a) $\sqrt{3} + \sqrt{3} = 2\sqrt{3}$ (b) $3\sqrt{2} - \sqrt{2} = 2\sqrt{2}$

(c) $8\sqrt[3]{7} + \sqrt[3]{7} - 4\sqrt[3]{7} = 5\sqrt[3]{7}$ □

Radicals that cannot be combined because they have different radicands sometimes can be simplified to a form in which they can be combined.

EXAMPLE 2. Perform the operations indicated and simplify.

(a) $\sqrt{8} + \sqrt{18}$ (b) $\sqrt[3]{16} - \sqrt[3]{2}$ (c) $\sqrt{12} - \sqrt{48} + \sqrt{75}$

Solutions:

(a) Because the radicands are different, these are not like radicals and cannot be combined as they are. However, they both can be simplified.

$$\sqrt{8} + \sqrt{18} = \sqrt{4 \cdot 2} + \sqrt{9 \cdot 2}$$
$$= 2\sqrt{2} + 3\sqrt{2}$$

Now we *do* have like terms.

$$= 5\sqrt{2}$$

(b) $\sqrt[3]{16} - \sqrt[3]{2} = \sqrt[3]{8 \cdot 2} - \sqrt[3]{2}$
$= 2\sqrt[3]{2} - \sqrt[3]{2}$
$= \sqrt[3]{2}$

(c) $\sqrt{12} - \sqrt{48} + \sqrt{75} = \sqrt{4 \cdot 3} - \sqrt{16 \cdot 3} + \sqrt{25 \cdot 3}$
$= 2\sqrt{3} - 4\sqrt{3} + 5\sqrt{3}$
$= 3\sqrt{3}$ □

EXAMPLE 3. Perform the operations indicated and simplify. Assume all variables represent nonnegative real numbers.

(a) $\sqrt{12x} + \sqrt{27x}$ (b) $\sqrt[3]{16y} - \sqrt[3]{54y}$ (c) $\sqrt{4x} - \sqrt{9x}$
(d) $\sqrt[3]{-8y} - \sqrt[3]{27y}$ (e) $\sqrt{8x^2} + \sqrt{18x^2}$

Solutions:

(a) $\sqrt{12x} + \sqrt{27x} = \sqrt{4 \cdot 3x} + \sqrt{9 \cdot 3x}$
$= 2\sqrt{3x} + 3\sqrt{3x}$
$= 5\sqrt{3x}$

(b) $\sqrt[3]{16y} - \sqrt[3]{54y} = \sqrt[3]{8 \cdot 2y} - \sqrt[3]{27 \cdot 2y}$
$= 2\sqrt[3]{2y} - 3\sqrt[3]{2y}$
$= -\sqrt[3]{2y}$

(c) $\sqrt{4x} - \sqrt{9x} = 2\sqrt{x} - 3\sqrt{x}$
$= -\sqrt{x}$

(d) $\sqrt[3]{-8y} - \sqrt[3]{27y} = -2\sqrt[3]{y} - 3\sqrt[3]{y}$
$= -5\sqrt[3]{y}$

(e) $\sqrt{8x^2} + \sqrt{18x^2} = \sqrt{4x^2 \cdot 2} + \sqrt{9x^2 \cdot 2}$
$= 2x\sqrt{2} + 3x\sqrt{2}$
$= 5x\sqrt{2}$ □

PROBLEM SET 8.3

Warm-ups

Perform the operations indicated and simplify. Assume that all variables represent nonnegative real numbers. For Problems 1 through 6, see Example 1.

1. $3\sqrt{5} + 6\sqrt{5}$
2. $7\sqrt{x} - 2\sqrt{x}$
3. $4\sqrt[3]{6} + 7\sqrt[3]{6}$
4. $\sqrt[3]{3} - 3\sqrt[3]{3}$
5. $2\sqrt{3} + 4\sqrt{3} - 7\sqrt{3}$
6. $5\sqrt[3]{7} - 6\sqrt[3]{7} - 3\sqrt[3]{7}$

For Problems 7 through 12, see Example 2.

7. $\sqrt{8} + \sqrt{2}$
8. $\sqrt{24} - \sqrt{54}$
9. $\sqrt{48} - 2\sqrt{12}$
10. $2\sqrt{50} + \sqrt{8} + \sqrt{18}$
11. $3\sqrt{20} - 2\sqrt{45} + \sqrt{80}$
12. $\sqrt[3]{16} - \sqrt[3]{54} + \sqrt[3]{-128}$

For Problems 13 through 18, see Example 3.

13. $\sqrt{8x} + \sqrt{2x}$
14. $\sqrt{48x^2} - 3\sqrt{12x^2}$
15. $\sqrt{4x} - 2\sqrt{16x}$
16. $5\sqrt{20x^2} - 3\sqrt{80x^2}$
17. $\sqrt[3]{-54x^3} - 3\sqrt[3]{-16x^3}$
18. $4\sqrt[3]{250x^3} - \sqrt[3]{-16x^3}$

Practice Exercises

Perform the operations indicated and simplify. Assume that all variables represent nonnegative real numbers.

19. $2\sqrt{3} + \sqrt{3}$
20. $5\sqrt{6} - 8\sqrt{6}$
21. $3\sqrt[3]{11} + 5\sqrt[3]{11}$
22. $\sqrt[3]{y} - 3\sqrt[3]{y}$
23. $6\sqrt{5} + \sqrt{5} - 2\sqrt{5}$
24. $8\sqrt[3]{4} - 3\sqrt[3]{4} - 2\sqrt[3]{4}$
25. $\sqrt{18} + \sqrt{2}$
26. $\sqrt{12} + \sqrt{27}$
27. $\sqrt{75} - 2\sqrt{12}$
28. $3\sqrt{72} + \sqrt{8} + \sqrt{32}$
29. $3\sqrt{45} - 3\sqrt{20} + \sqrt{5}$
30. $\sqrt{200} - 2\sqrt{8} + 3\sqrt{50}$
31. $2\sqrt{18} - \sqrt{98} - \sqrt{8}$
32. $\sqrt[3]{54} - 3\sqrt[3]{2}$
33. $\sqrt{25x} + 3\sqrt{36x}$
34. $3\sqrt{72x} - \sqrt{8x}$
35. $\sqrt[3]{16x} - 2\sqrt[3]{54x}$
36. $\sqrt[3]{-8x} + 2\sqrt[3]{27x}$
37. $\sqrt[3]{-16} + 5\sqrt[3]{-54}$
38. $4\sqrt[3]{2} - 3\sqrt[3]{-16}$
39. $\sqrt[3]{-16} - \sqrt[3]{54} - \sqrt[3]{-128}$
40. $\sqrt[3]{250} - 4\sqrt[3]{16} - 3\sqrt[3]{54}$
41. $\sqrt{20x} + \sqrt{45x}$
42. $\sqrt{24x^2} - 3\sqrt{54x^2}$
43. $\sqrt{25x} - 2\sqrt{36x}$
44. $5\sqrt{12x^2} - 3\sqrt{27x^2}$
45. $\sqrt{64x} + 3\sqrt{49x}$
46. $3\sqrt{98x} - 2\sqrt{8x}$
47. $\sqrt[3]{24x} - 2\sqrt[3]{81x}$
48. $\sqrt[3]{-27x} + 2\sqrt[3]{8x}$
49. $\sqrt[3]{-81x^3} - 3\sqrt[3]{-24x^3}$
50. $4\sqrt[3]{250x^3} + \sqrt[3]{16x^3}$

IN YOUR OWN WORDS...

51. When can expressions containing radicals be combined?

8.4 MULTIPLICATION AND DIVISION OF RADICALS

The Multiplication Property of Radicals that we studied in Section 8.2 tells us how to multiply radicals with the *same* index.

> **Multiplication Property of Radicals**
>
> If $\sqrt[n]{p}$ and $\sqrt[n]{q}$ exist as real numbers,
>
> $$\sqrt[n]{p} \cdot \sqrt[n]{q} = \sqrt[n]{pq}$$

A product of nth roots is the nth root of the product of the radicands. If the index is the same, we just multiply the radicands.

EXAMPLE 1. Perform the multiplication indicated and simplify.

(a) $\sqrt{2} \cdot \sqrt{3}$ (b) $\sqrt{3} \cdot \sqrt{6}$ (c) $\sqrt[3]{4} \cdot \sqrt[3]{2}$

(d) $\sqrt[3]{-9} \cdot \sqrt[3]{6}$ (e) $2\sqrt{3} \cdot 5\sqrt{7}$ (f) $(5\sqrt{6})^2$

Solutions:

(a) $\sqrt{2} \cdot \sqrt{3} = \sqrt{6}$

$\sqrt{6}$ cannot be simplified further.

(b) $\sqrt{3} \cdot \sqrt{6} = \sqrt{18}$

$\sqrt{18}$ can be simplified further.

$\sqrt{18} = \sqrt{9 \cdot 2}$

$= 3\sqrt{2}$ $\quad \sqrt{9} = 3$

(c) $\sqrt[3]{4} \cdot \sqrt[3]{2} = \sqrt[3]{8}$

$= 2$

(d) $\sqrt[3]{-9} \cdot \sqrt[3]{6} = \sqrt[3]{-54}$

$= \sqrt[3]{-27 \cdot 2}$

$= -3\sqrt[3]{2}$ $\quad \sqrt[3]{-27} = -3$

(e) $2\sqrt{3} \cdot 5\sqrt{7} = 2 \cdot 5 \cdot \sqrt{3} \cdot \sqrt{7}$ Commutative Property

$= 10\sqrt{21}$

(f) $(5\sqrt{6})^2 = 5^2(\sqrt{6})^2$ Property of Exponents

$= 25 \cdot 6$

$= 150$ □

The Distributive Property and the special products that we learned when we studied polynomials tell us how to perform other multiplications.

EXAMPLE 2. Perform the multiplications indicated and simplify.

(a) $\sqrt{3}(\sqrt{5} + \sqrt{3})$

(b) $(2\sqrt{7} + \sqrt{3})(2\sqrt{7} - \sqrt{3})$

(c) $(2\sqrt{3} + \sqrt{5})^2$

Solutions:

(a) Use the Distributive Property.

$$\sqrt{3}(\sqrt{5}+\sqrt{3}) = \sqrt{3}\cdot\sqrt{5}+\sqrt{3}\cdot\sqrt{3}$$
$$= \sqrt{15}+3$$

(b) Use the special product, *difference of two squares:*

$$(p+q)(p-q) = p^2 - q^2$$
$$(2\sqrt{7}+\sqrt{3})(2\sqrt{7}-\sqrt{3}) = (2\sqrt{7})^2 - (\sqrt{3})^2$$
$$= 2^2\cdot(\sqrt{7})^2 - (\sqrt{3})^2$$
$$= 4\cdot 7 - 3$$
$$= 28 - 3$$
$$= 25$$

(c) Use the special product, *square of a binomial:*

$$(p+q)^2 = p^2 + 2pq + q^2$$
$$(2\sqrt{3}+\sqrt{5})^2 = (2\sqrt{3})^2 + 2\cdot 2\sqrt{3}\cdot\sqrt{5} + (\sqrt{5})^2$$
$$= 4\cdot 3 + 4\sqrt{15} + 5$$
$$= 12 + 4\sqrt{15} + 5$$
$$= 17 + 4\sqrt{15}$$

□

To divide radicals with the same index, we use the Division Property of Radicals.

Division Property of Radicals

If $\sqrt[n]{p}$ and $\sqrt[n]{q}$ exist as real numbers and $q \neq 0$,

$$\frac{\sqrt[n]{p}}{\sqrt[n]{q}} = \sqrt[n]{\frac{p}{q}}$$

A quotient of nth roots is the nth root of the quotient of the radicands. If the index is the same, we just divide the radicands.

EXAMPLE 3. Perform the divisions indicated and simplify.

(a) $\dfrac{\sqrt{4}}{\sqrt{2}}$ (b) $\dfrac{\sqrt{16}}{\sqrt{2}}$ (c) $\dfrac{\sqrt[3]{8}}{\sqrt[3]{2}}$ (d) $\dfrac{\sqrt[3]{-54}}{\sqrt[3]{2}}$

Solutions:

(a) $\dfrac{\sqrt{4}}{\sqrt{2}} = \sqrt{\dfrac{4}{2}}$ Division Property

$= \sqrt{2}$

(b) $\dfrac{\sqrt{16}}{\sqrt{2}} = \sqrt{8}$ Division Property

$= \sqrt{4\cdot 2}$

$= 2\sqrt{2}$

(continued)

(c) $\dfrac{\sqrt[3]{8}}{\sqrt[3]{2}} = \sqrt[3]{4}$ Division Property

(d) $\dfrac{\sqrt[3]{-54}}{\sqrt[3]{2}} = \sqrt[3]{-27}$ Division Property

$= -3$ □

If we divide $\sqrt{5}$ by $\sqrt{3}$ we get $\dfrac{\sqrt{5}}{\sqrt{3}}$, which cannot be simplified using the procedure of Section 8.2. However, we do not consider an expression simplified if it contains a radical in the denominator. We can use the Fundamental Principle of Fractions to write the expression without a fraction in the radicand and without a radical in the denominator.

Notice what happens if we use the Fundamental Principle of Fractions and multiply numerator and denominator by $\sqrt{3}$.

$$\frac{\sqrt{5}}{\sqrt{3}} = \frac{\sqrt{5}\cdot\sqrt{3}}{\sqrt{3}\cdot\sqrt{3}}$$
$$= \frac{\sqrt{15}}{3}$$
$$= \frac{1}{3}\sqrt{15}$$

This procedure is called **rationalizing the denominator.** Notice that the radicand is not a fraction and the denominator does not contain a radical.

If the denominator contains a square root as a factor, the denominator can be rationalized by multiplying both numerator and denominator by the square root factor.

EXAMPLE 4. Rationalize the denominator and simplify. Assume that all variables represent nonnegative numbers.

(a) $\dfrac{\sqrt{3}}{\sqrt{5}}$ (b) $\dfrac{\sqrt{2t}}{\sqrt{2}}$ (c) $\dfrac{6}{5\sqrt{2}}$

Solutions:

(a) $\dfrac{\sqrt{3}}{\sqrt{5}} = \dfrac{\sqrt{3}\cdot\sqrt{5}}{\sqrt{5}\cdot\sqrt{5}}$

$= \dfrac{\sqrt{15}}{5}$

(b) $\dfrac{\sqrt{2t}}{\sqrt{2}} = \dfrac{\sqrt{2t}\cdot\sqrt{2}}{\sqrt{2}\cdot\sqrt{2}}$

$= \dfrac{\sqrt{4t}}{2}$

$= \dfrac{2\sqrt{t}}{2}$

$= \sqrt{t}$

(c) $\frac{6}{5\sqrt{2}} = \frac{6\sqrt{2}}{5\sqrt{2}\cdot\sqrt{2}}$

$= \frac{6\sqrt{2}}{5\cdot 2}$

$= \frac{3\sqrt{2}}{5}$, or $\frac{3}{5}\sqrt{2}$

Notice in (c) that it was not necessary to multiply by $5\sqrt{2}$. □

If the denominator contains a square root *term,* then we use the special product, *difference of two squares,*

$$(p + q)(p - q) = p^2 - q^2$$

to rationalize the denominator.

EXAMPLE 5. Rationalize the denominator and simplify. Assume that all variables represent nonnegative real numbers.

(a) $\frac{3}{\sqrt{2} - 5}$ (b) $\frac{\sqrt{2} - \sqrt{3}}{\sqrt{2} + \sqrt{3}}$ (c) $\frac{\sqrt{y}}{2\sqrt{y} - 1}$

Solutions:

(a) We multiply numerator and denominator by $\sqrt{2} + 5$ to set up the special product $(p - q)(p + q)$ in the denominator.

$$\frac{3}{\sqrt{2} - 5} = \frac{3(\sqrt{2} + 5)}{(\sqrt{2} - 5)(\sqrt{2} + 5)}$$

$$= \frac{3\sqrt{2} + 15}{(\sqrt{2})^2 - 5^2}$$

$$= \frac{3\sqrt{2} + 15}{2 - 25}$$

$$= \frac{3\sqrt{2} + 15}{-23}$$

(b) Here we multiply the numerator and the denominator by $\sqrt{2} - \sqrt{3}$ to set up the difference of two squares.

$$\frac{\sqrt{2} - \sqrt{3}}{\sqrt{2} + \sqrt{3}} = \frac{(\sqrt{2} - \sqrt{3})(\sqrt{2} - \sqrt{3})}{(\sqrt{2} + \sqrt{3})(\sqrt{2} - \sqrt{3})}$$

$$= \frac{(\sqrt{2})^2 - 2\sqrt{2}\sqrt{3} + (\sqrt{3})^2}{(\sqrt{2})^2 - (\sqrt{3})^2}$$

$$= \frac{2 - 2\sqrt{6} + 3}{2 - 3}$$

$$= \frac{5 - 2\sqrt{6}}{-1}$$

$$= -5 + 2\sqrt{6}$$

(continued)

(c) We multiply the numerator and the denominator by $2\sqrt{y} + 1$.

$$\begin{aligned}\frac{\sqrt{y}}{2\sqrt{y} - 1} &= \frac{\sqrt{y}(2\sqrt{y} + 1)}{(2\sqrt{y} - 1)(2\sqrt{y} + 1)} \\ &= \frac{\sqrt{y} \cdot 2\sqrt{y} + \sqrt{y}}{(2\sqrt{y})^2 - 1^2} \\ &= \frac{2y + \sqrt{y}}{4y - 1}\end{aligned}$$

□

An application of radicals arises in the study of right triangles. Two special triangles are very common. The first is an isosceles right triangle. The legs are each of the same length and the angles are 45°, 45°, and 90°. In such a triangle, the length of the hypotenuse is $\sqrt{2}$ times the length of a leg. Another special right triangle occurs when the length of the hypotenuse is exactly twice the length of the shorter leg of the triangle. The angles in such a triangle are 30°, 60°, and 90°.

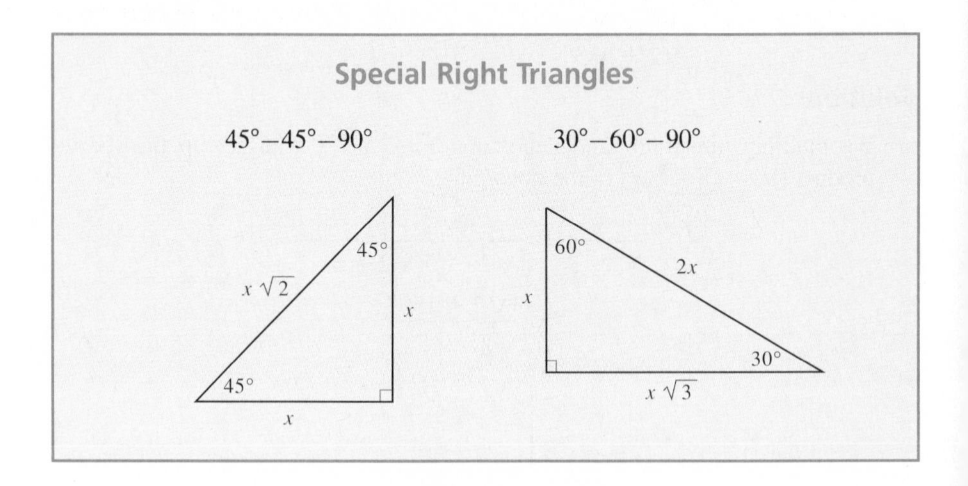

EXAMPLE 6. Given that the hypotenuse of a 45°–45°–90° triangle is 4 units long, what is the length of the legs?

Solution:

Let the length of the legs be x units.

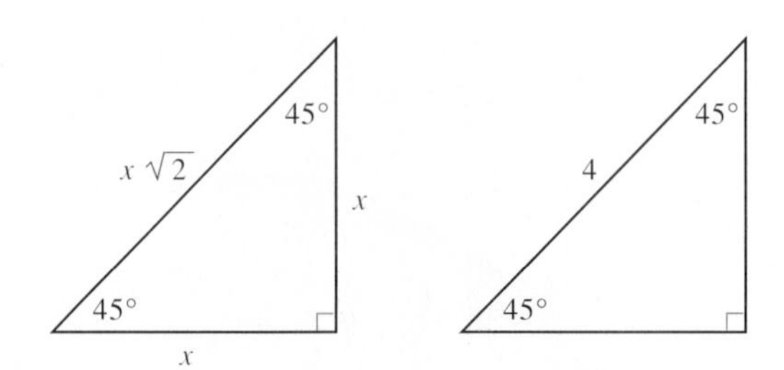

From the picture, we see that since the hypotenuse is 4,

$$x\sqrt{2} = 4$$

$$x = \frac{4}{\sqrt{2}}$$

$$x = \frac{4 \cdot \sqrt{2}}{\sqrt{2} \cdot \sqrt{2}}$$

$$= \frac{4\sqrt{2}}{2}$$

$$= 2\sqrt{2}$$

The legs are $2\sqrt{2}$ units long. □

EXAMPLE 7. If the shortest side of a 30°–60°–90° triangle is $4\sqrt{2}$ ft long, what are the lengths of the other two sides?

Solution:

Let x be the length of the shorter leg.

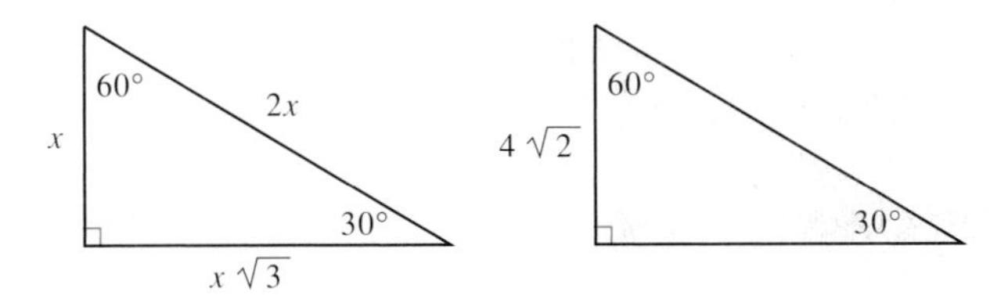

From our special 30°–60°–90° relationships, we see that

$$x = 4\sqrt{2}$$

$$2x = 8\sqrt{2}$$

$$x\sqrt{3} = 4\sqrt{2} \cdot \sqrt{3}$$

$$= 4\sqrt{6}$$

The lengths of the other sides are $8\sqrt{2}$ ft and $4\sqrt{6}$ ft. □

PROBLEM SET 8.4

Warm-ups

Perform the operations indicated and simplify. Assume that all variables represent nonnegative real numbers.
For Problems 1 through 9, see Example 1.

1. $\sqrt{5} \cdot \sqrt{7}$
2. $\sqrt{3} \cdot \sqrt{11}$
3. $\sqrt[3]{3} \cdot \sqrt[3]{5}$
4. $\sqrt[3]{-2} \cdot \sqrt[3]{4}$
5. $2\sqrt{3} \cdot 4\sqrt{5}$
6. $3\sqrt{x} \cdot 2\sqrt{x}$
7. $4\sqrt{5} \cdot 3\sqrt{5}$
8. $(\sqrt{5})^2$
9. $(8\sqrt{x})^2$

For Problems 10 through 19, see Example 2.

10. $\sqrt{3}(\sqrt{2}+\sqrt{5})$
11. $\sqrt{7}(\sqrt{6}-1)$
12. $\sqrt{5}(\sqrt{5}+\sqrt{7})$
13. $\sqrt{t}(\sqrt{t}-\sqrt{y})$
14. $(2+\sqrt{3})(5-\sqrt{3})$
15. $(\sqrt{5}+2\sqrt{2})(3\sqrt{2}-\sqrt{5})$
16. $(\sqrt{3}-1)(\sqrt{3}+1)$
17. $(\sqrt{5}-\sqrt{3})(\sqrt{5}+\sqrt{3})$
18. $(\sqrt{3}+2)^2$
19. $(2\sqrt{6}+5\sqrt{2})^2$

For Problems 20 through 27, see Example 3.

20. $\dfrac{\sqrt{6}}{\sqrt{3}}$
21. $\dfrac{\sqrt{10}}{\sqrt{2}}$
22. $\dfrac{\sqrt[3]{21}}{\sqrt[3]{7}}$
23. $\dfrac{\sqrt[3]{12}}{\sqrt[3]{2}}$
24. $\dfrac{\sqrt{32}}{\sqrt{2}}$
25. $\dfrac{\sqrt{45}}{\sqrt{5}}$
26. $\dfrac{\sqrt[3]{16}}{\sqrt[3]{2}}$
27. $\dfrac{\sqrt[3]{81}}{\sqrt[3]{-3}}$

Rationalize the denominators in Problems 28 through 38.
For Problems 28 through 33, see Example 4.

28. $\dfrac{1}{\sqrt{3}}$
29. $\dfrac{6}{\sqrt{6}}$
30. $\dfrac{\sqrt{2}+1}{\sqrt{2}}$
31. $\dfrac{\sqrt{3}}{4\sqrt{2}}$
32. $\dfrac{\sqrt{3}-\sqrt{2}}{2\sqrt{3}}$
33. $\dfrac{\sqrt{x}+\sqrt{y}}{\sqrt{x}}$

For Problems 34 through 38, see Example 5.

34. $\dfrac{1}{\sqrt{2}+3}$
35. $\dfrac{3}{\sqrt{2}-\sqrt{3}}$
36. $\dfrac{\sqrt{5}}{\sqrt{5}-2\sqrt{3}}$
37. $\dfrac{\sqrt{3}+\sqrt{6}}{\sqrt{3}-\sqrt{6}}$
38. $\dfrac{3\sqrt{x}+2\sqrt{y}}{3\sqrt{x}-2\sqrt{y}}$

Practice Exercises

Perform the operations indicated and simplify. Assume that all variables represent nonnegative real numbers.

39. $\sqrt{3}\cdot\sqrt{5}$
40. $\sqrt{2}\cdot\sqrt{13}$
41. $\sqrt[3]{7}\cdot\sqrt[3]{4}$
42. $\sqrt[3]{9}\cdot\sqrt[3]{7}$
43. $\sqrt{3}\cdot\sqrt{6}$
44. $\sqrt{5}\cdot\sqrt{8}$
45. $\sqrt[3]{3}\cdot\sqrt[3]{9}$
46. $\sqrt[3]{-4}\cdot\sqrt[3]{4}$
47. $2\sqrt{3}\cdot4\sqrt{2}$
48. $3\sqrt{3}\cdot2\sqrt{2}$
49. $3\sqrt[3]{6}\cdot2\sqrt[3]{3}$
50. $5\sqrt[3]{6}\cdot2\sqrt[3]{2}$
51. $3\sqrt{3}\cdot4\sqrt{8}$
52. $4\sqrt{8}\cdot3\sqrt{2}$
53. $2\sqrt{y}\cdot5\sqrt{y}$
54. $4\sqrt{2}\cdot3\sqrt{2}$
55. $(\sqrt{7})^2$
56. $(\sqrt{x})^2$
57. $(3\sqrt{2})^2$
58. $(5\sqrt{2})^2$
59. $(6\sqrt{z})^2$
60. $\sqrt{2}(\sqrt{3}+\sqrt{7})$
61. $\sqrt{5}(\sqrt{3}-1)$
62. $\sqrt{3}(\sqrt{3}+\sqrt{7})$
63. $\sqrt{x}(\sqrt{x}-\sqrt{y})$
64. $(\sqrt{3}+2)(\sqrt{2}-3)$
65. $(\sqrt{2}-\sqrt{5})(\sqrt{2}+3\sqrt{5})$
66. $(3\sqrt{3}-\sqrt{5})(\sqrt{3}+\sqrt{5})$
67. $(\sqrt{x}-\sqrt{y})(\sqrt{x}-3\sqrt{y})$
68. $(3\sqrt{6}-\sqrt{3})(\sqrt{2}+2)$
69. $(\sqrt{2}-1)(\sqrt{2}+1)$
70. $(4\sqrt{7}-\sqrt{5})(2\sqrt{7}+\sqrt{5})$

71. $(\sqrt{6}-\sqrt{3})(\sqrt{6}+\sqrt{3})$

72. $(\sqrt{x}+2\sqrt{y})(\sqrt{x}-2\sqrt{y})$

73. $(\sqrt{2}+3)^2$

74. $(\sqrt{5}-\sqrt{3})^2$

75. $(2\sqrt{2}+\sqrt{6})^2$

76. $(3\sqrt{3}-4\sqrt{5})^2$

77. $(2\sqrt{10}+3\sqrt{2})^2$

78. $\dfrac{\sqrt{15}}{\sqrt{3}}$

79. $\dfrac{\sqrt{8}}{\sqrt{2}}$

80. $\dfrac{\sqrt[3]{21}}{\sqrt[3]{3}}$

81. $\dfrac{\sqrt[3]{6}}{\sqrt[3]{2}}$

82. $\dfrac{\sqrt{80}}{\sqrt{5}}$

83. $\dfrac{\sqrt{75}}{\sqrt{3}}$

84. $\dfrac{\sqrt[3]{54}}{\sqrt[3]{-2}}$

85. $\dfrac{\sqrt[3]{-32}}{\sqrt[3]{4}}$

86. $\dfrac{\sqrt{24}}{\sqrt{2}}$

87. $\dfrac{\sqrt{54}}{\sqrt{3}}$

88. $\dfrac{\sqrt[3]{96}}{\sqrt[3]{6}}$

89. $\dfrac{\sqrt[3]{108}}{\sqrt[3]{-2}}$

Rationalize the denominators in Problems 90 through 106.

90. $\dfrac{1}{\sqrt{7}}$

91. $\dfrac{5}{\sqrt{5}}$

92. $\dfrac{\sqrt{3}+1}{\sqrt{3}}$

93. $\dfrac{5}{2\sqrt{3}}$

94. $\dfrac{\sqrt{3}}{3\sqrt{5}}$

95. $\dfrac{\sqrt{5}-\sqrt{3}}{2\sqrt{3}}$

96. $\dfrac{w}{5\sqrt{w}}$

97. $\dfrac{\sqrt{x}-\sqrt{y}}{\sqrt{y}}$

98. $\dfrac{1}{\sqrt{3}+2}$

99. $\dfrac{3}{\sqrt{2}+\sqrt{3}}$

100. $\dfrac{\sqrt{2}}{\sqrt{2}-3\sqrt{3}}$

101. $\dfrac{\sqrt{2}+\sqrt{5}}{\sqrt{2}-\sqrt{5}}$

102. $\dfrac{\sqrt{7}-2\sqrt{3}}{2\sqrt{7}+\sqrt{3}}$

103. $\dfrac{\sqrt{5}-\sqrt{6}}{\sqrt{5}+\sqrt{6}}$

104. $\dfrac{\sqrt{x}-\sqrt{y}}{\sqrt{x}+\sqrt{y}}$

105. $\dfrac{\sqrt{s}+2\sqrt{t}}{3\sqrt{s}-\sqrt{t}}$

106. $\dfrac{3\sqrt{x}-2\sqrt{y}}{3\sqrt{x}+2\sqrt{y}}$

In Problems 107 through 114, find the lengths of the sides.

107.

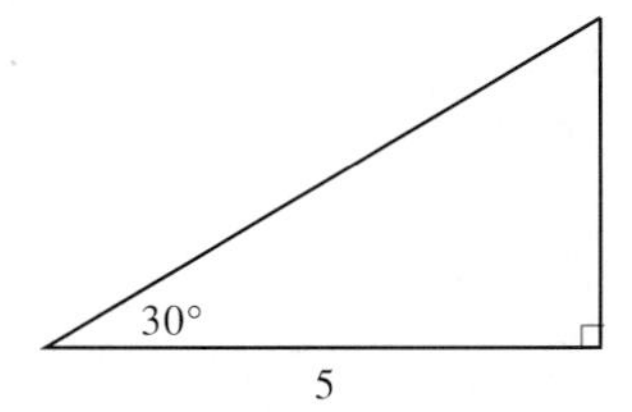

108.

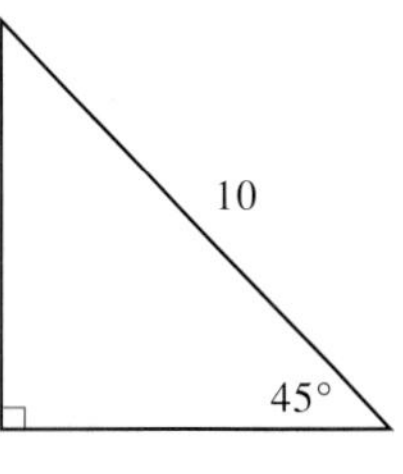

109.

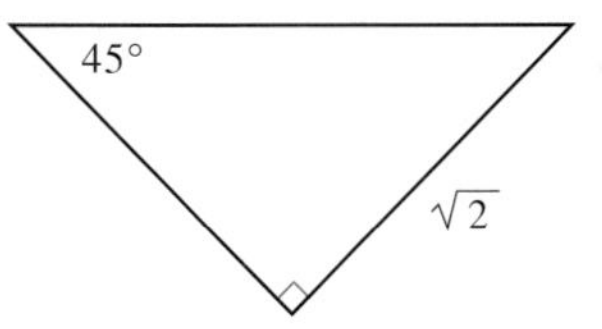

110.

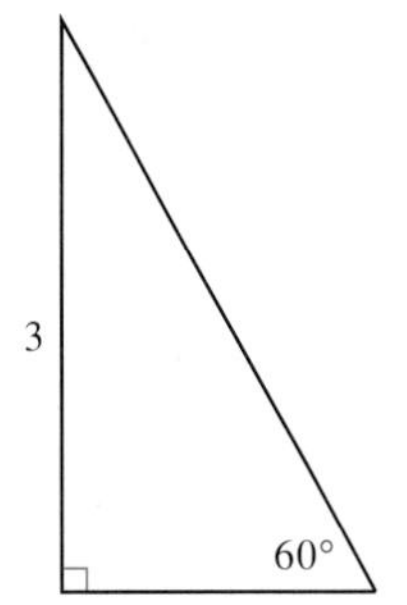

111.

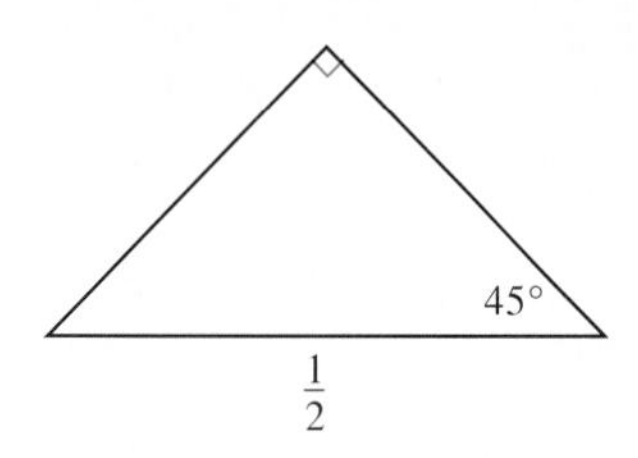

112.

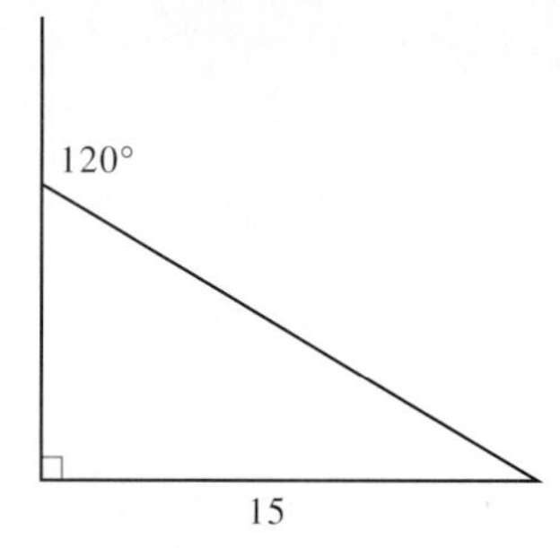

113.

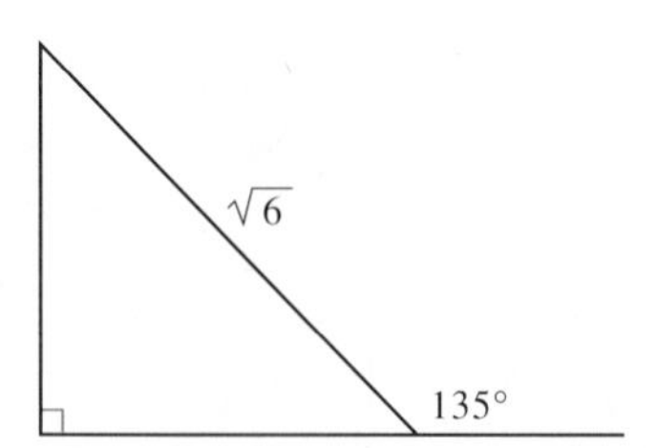

114.

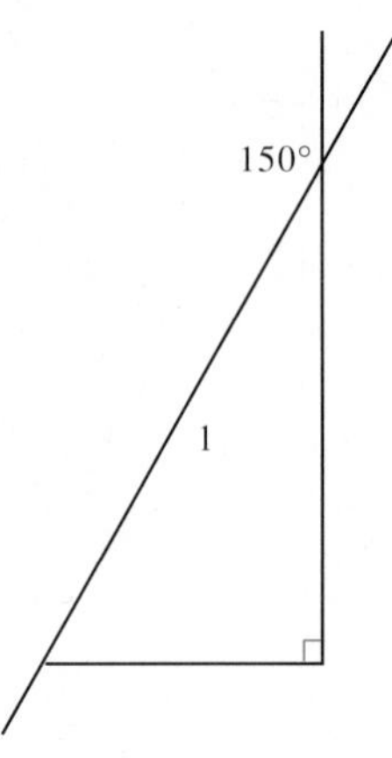

IN YOUR OWN WORDS...

115. State the Multiplication Property of Radicals.

116. State the Division Property of Radicals.

8.5 RATIONAL EXPONENTS (Optional)

In Sections 2.1 and 2.2 we studied integer exponents. In this section we examine exponents that are fractions.

The major reason we return to exponents is to capitalize on their properties. So, in defining fractional exponents, we ensure that the five properties of exponents that we developed in Chapter 2 remain valid. In review, they are listed below.

Five Properties of Exponents

If m and n are integers and x and y are real numbers,

1. $x^m x^n = x^{m+n}$	Product with same base
2. $(x^m)^n = x^{mn}$	Power of a power
3. $(xy)^n = x^n y^n$	Product to a power
4. $\left(\frac{x}{y}\right)^n = \frac{x^n}{y^n}; y \neq 0$	Quotient to a power
5. $\frac{x^m}{x^n} = x^{m-n}; x \neq 0$	Quotient with same base

Now suppose we define an exponent of $\frac{1}{2}$ in such a way that the properties of exponents are still valid. Then Property 2 will be true when m is $\frac{1}{2}$ and n is 2. Therefore, it must be true that

$$(x^{1/2})^2 = x^1$$
$$= x$$

So, $x^{1/2}$ is a number that when squared is x. Because x is a real number squared, it must be positive. We know such a number. It is $\sqrt{x}$. Thus we define $x^{1/2}$ by

$$x^{1/2} = \sqrt{x}$$

By a similar argument,

$$(x^{1/3})^3 = x^1$$
$$= x$$

Thus, $x^{1/3}$ is a number that when cubed is x. That number is $\sqrt[3]{x}$. So we define

$$x^{1/3} = \sqrt[3]{x}$$

Notice that x can be negative in this case.

These two examples suggest the following definition.

Exponent of the Form $\frac{1}{n}$

For a natural number n,

$$x^{1/n} = \sqrt[n]{x} \qquad x \geq 0 \text{ if } n \text{ is even.}$$

EXAMPLE 1. Simplify each.

(a) $4^{1/2}$ (b) $8^{1/3}$ (c) $(-27)^{1/3}$ (d) $-9^{1/2}$ (e) $-64^{1/3}$

Solutions:

(a) $4^{1/2} = \sqrt{4}$
$= 2$

(b) $8^{1/3} = \sqrt[3]{8}$
$= 2$

(c) $(-27)^{1/3} = \sqrt[3]{-27}$
$= -3$

Notice that the base is -27.

(d) $-9^{1/2} = -\sqrt{9}$
$= -3$

Notice that the base is 9 and *not* -9.

(e) $-64^{1/3} = -\sqrt[3]{64}$
$= -4$

Again note that the base is 64 and *not* -64. ☐ *Be Careful!*

We can use our definition and the laws of exponents to simplify expressions such as $8^{2/3}$. Notice how this is done in the next example.

EXAMPLE 2. Simplify $8^{2/3}$.

Solution:

The second property of exponents allows us to rewrite this expression so that we can employ the definition of the power $1/n$.

$$\begin{aligned} 8^{2/3} &= (8^{1/3})^2 && \text{Power of a power} \\ &= (\sqrt[3]{8})^2 && \text{Definition of exponent } \tfrac{1}{3} \\ &= (2)^2 \\ &= 4 \end{aligned}$$

□

The technique employed in Example 2 leads to the following rule.

> **Rule for Calculating $x^{m/n}$**
>
> For m and n natural numbers,
>
> $$x^{m/n} = (\sqrt[n]{x})^m$$

EXAMPLE 3. Simplify each of the following using the rule for calculating $x^{m/n}$.

(a) $4^{3/2}$ (b) $-16^{3/2}$ (c) $(-27)^{2/3}$ (d) $-27^{2/3}$

Solutions:

(a)
$$\begin{aligned} 4^{3/2} &= (\sqrt{4})^3 \\ &= (2)^3 \\ &= 8 \end{aligned}$$

(b) The base of the exponent is 16, *not* -16.

$$\begin{aligned} -16^{3/2} &= -(\sqrt{16})^3 \\ &= -4^3 \\ &= -64 \end{aligned}$$

(c) Notice that the base here is -27.

$$\begin{aligned} (-27)^{2/3} &= (\sqrt[3]{-27})^2 \\ &= (-3)^2 \\ &= 9 \end{aligned}$$

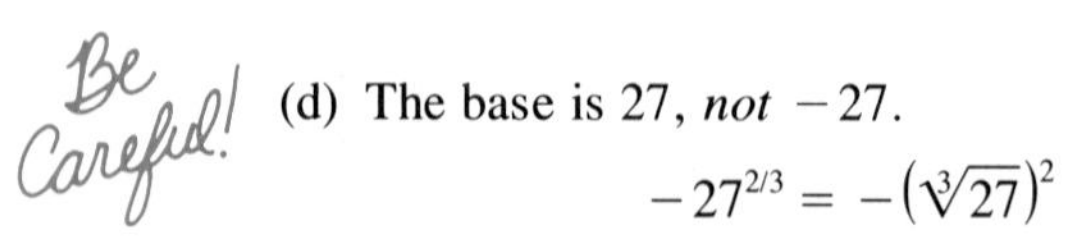

(d) The base is 27, *not* -27.

$$\begin{aligned} -27^{2/3} &= -(\sqrt[3]{27})^2 \\ &= -3^2 \\ &= -9 && \text{The base is 3, } \textit{not } -3. \end{aligned}$$

□

A Summary of Exponent Definitions

Given n is a natural number,

1. $x^n = \underbrace{x \cdot x \cdot x \cdots x}_{n \text{ factors of } x}$
2. $x^0 = 1; \quad x \neq 0$
3. $x^{-n} = \frac{1}{x^n}; \quad x \neq 0$
4. $x^{1/n} = \sqrt[n]{x}; \quad x \geq 0$ if n is even.

PROBLEM SET 8.5

Warm-ups

In Problems 1 through 12, simplify each expression.
In Problems 1 through 6, see Example 1.

1. $16^{1/2}$
2. $36^{1/2}$
3. $125^{1/3}$
4. $-36^{1/2}$
5. $-8^{1/3}$
6. $(-8)^{1/3}$

In Problems 7 through 12, see Example 3.

7. $27^{2/3}$
8. $16^{3/2}$
9. $9^{3/2}$
10. $(-1000)^{2/3}$
11. $-8^{2/3}$
12. $-4^{3/2}$

Practice Exercises

Simplify each.

13. $49^{1/2}$
14. $100^{1/2}$
15. $81^{1/2}$
16. $121^{1/2}$
17. $216^{1/3}$
18. $1000^{1/3}$
19. $343^{1/3}$
20. $(-1000)^{1/3}$
21. $(-125)^{1/3}$
22. $-81^{1/2}$
23. $-100^{1/2}$
24. $-121^{1/2}$
25. $-343^{1/3}$
26. $-1000^{1/3}$
27. $-216^{1/3}$
28. $64^{2/3}$
29. $25^{3/2}$
30. $36^{3/2}$
31. $(-8)^{2/3}$
32. $-125^{2/3}$
33. $-9^{3/2}$
34. $(-216)^{2/3}$
35. $-81^{3/2}$
36. $-216^{2/3}$

Challenge Problems

A negative fraction, such as $-\frac{2}{3}$*, can always be written with a* positive *denominator and a* negative *numerator:* $-\frac{2}{3} = \frac{-2}{3}$*. Use this fact and the rule to calculate* $x^{m/n}$ *to evaluate each of the following.*

37. $8^{-2/3}$
38. $16^{-3/2}$
39. $-64^{-2/3}$
40. $(-64)^{-2/3}$
41. $125^{-1/3}$

IN YOUR OWN WORDS...

42. Describe how to evaluate a number raised to a fractional power.

CHAPTER SUMMARY

GLOSSARY

Principal square root of a nonnegative number, q: The positive number whose square is q; written $\sqrt{q}$.

Like radicals: Radicals with the same index and radicand.

THE nth ROOT OF A REAL NUMBER

If n is a natural number and x is a real number

1. If n is even and x is not negative, $\sqrt[n]{x}$ is the nonnegative number whose n^{th} power is x.
2. If n is odd, $\sqrt[n]{x}$ is the number whose n^{th} power is x.

MULTIPLICATION PROPERTY OF RADICALS

If $\sqrt[n]{p}$ and $\sqrt[n]{q}$ exist as real numbers, $\quad \sqrt[n]{pq} = \sqrt[n]{p}\sqrt[n]{q}$.

DIVISION PROPERTY OF RADICALS

If $\sqrt[n]{p}$ and $\sqrt[n]{q}$ exist as real numbers and $q \neq 0$,

$$\sqrt[n]{\frac{p}{q}} = \frac{\sqrt[n]{p}}{\sqrt[n]{q}}.$$

FRACTIONAL EXPONENTS (OPTIONAL)

$x^{1/n} = \sqrt[n]{x}; \qquad x \geq 0$ if n is even.

$x^{m/n} = \left(\sqrt[n]{x}\right)^m; \qquad x \geq 0$ if n is even.

CHECKUPS

1. Find $\sqrt{\frac{4}{9}}$. — Section 8.1, Example 1c
2. Find $\sqrt{x^{10}}$. — Section 8.1, Example 6b
3. Simplify $\sqrt{48}$. — Section 8.2, Example 1c
4. Simplify $\sqrt[3]{\frac{5}{8}}$. — Section 8.2, Example 4c
5. Perform the operation indicated. — Section 8.3, Example 3b

$$\sqrt[3]{16y} - \sqrt[3]{54y}$$

6. Perform the operation indicated. — Section 8.4, Example 2b

$$\left(2\sqrt{7} + \sqrt{3}\right)\left(2\sqrt{7} - \sqrt{3}\right)$$

7. Rationalize the denominator. — Section 8.4, Example 5a

$$\frac{3}{\sqrt{2} - 5}$$

8. Simplify $(-27)^{2/3}$. — Section 8.5, Example 3c (Optional)

REVIEW PROBLEMS

Perform the operations indicated and simplify.

1. $\sqrt{18} + 5\sqrt{2}$

2. $\sqrt{7} \cdot \sqrt{8}$

3. $\sqrt{2}(\sqrt{2} + \sqrt{3})$

4. $(\sqrt{2} + 5)(\sqrt{2} - 3)$

5. $(\sqrt{7} + 3)^2$

6. $(\sqrt{7} - \sqrt{11})(\sqrt{7} + \sqrt{11})$

7. $\sqrt{49x^2} + \sqrt{25x^2} - \sqrt{16x^2}$

8. $(\sqrt{7})^2$

9. $(2\sqrt{3})(6\sqrt{3})$

10. $2\sqrt{5}(3\sqrt{2} - \sqrt{5})$

11. $(\sqrt{6} - \sqrt{2})(\sqrt{6} + \sqrt{3})$

12. $\dfrac{\sqrt{18}}{\sqrt{2}}$

13. $2\sqrt[3]{5} \cdot \sqrt[3]{25}$

14. $\sqrt{48} - 2\sqrt{12} + \sqrt{27}$

15. $\dfrac{\sqrt[3]{40}}{\sqrt[3]{5}}$

16. $\sqrt[3]{27x^3} - \sqrt[3]{-8x^3}$

17. $(2\sqrt{6} - \sqrt{2})^2$

18. $(\sqrt{10} - \sqrt{2})(2\sqrt{10} + 3\sqrt{2})$

19. $\sqrt[3]{54} - \sqrt[3]{-16} + \sqrt[3]{-128}$

20. $4\sqrt{\dfrac{3}{25}} - \sqrt{\dfrac{12}{25}}$

21. $\dfrac{\sqrt{60}}{\sqrt{5}}$

22. $\sqrt[3]{54x^6y^3} - \sqrt[3]{16x^6y^3}$

23. $\dfrac{\sqrt[3]{-48}}{\sqrt[3]{-2}}$

24. $(2\sqrt{5} + 3)^2$

25. $2\sqrt{2}(\sqrt{3} - 2\sqrt{2})$

26. $(3\sqrt{11})^2$

27. $(2\sqrt{3} + 3\sqrt{2})(\sqrt{3} - \sqrt{2})$

28. $\sqrt{\dfrac{25}{4}} - \sqrt{\dfrac{16}{9}}$

29. $3\sqrt{6} - 2\sqrt{6} + \sqrt{6}$

30. $\sqrt{x}(2\sqrt{x} + \sqrt{y})$

Rationalize the denominator.

31. $\dfrac{4}{\sqrt{2}}$

32. $\dfrac{3}{2\sqrt{5}}$

33. $\dfrac{2}{\sqrt{x}}$

34. $\dfrac{4}{3\sqrt{2}}$

35. $\dfrac{1}{3\sqrt{5}}$

36. $\dfrac{\sqrt{3}}{\sqrt{3} + \sqrt{2}}$

37. $\dfrac{2\sqrt{2}}{\sqrt{2} - 1}$

38. $\dfrac{\sqrt{5} + \sqrt{3}}{\sqrt{5} - \sqrt{3}}$

39. $\dfrac{\sqrt{3} + 5\sqrt{2}}{2\sqrt{3} - \sqrt{2}}$

40. $\dfrac{x + 2\sqrt{y}}{x - 2\sqrt{y}}$

Simplify each (optional).

41. $-16^{1/2}$

42. $\left(\dfrac{8}{125}\right)^{1/3}$

43. $-\left(\dfrac{1}{4}\right)^{3/2}$

44. $\left(-\dfrac{1}{8}\right)^{2/3}$

45. $25^{3/2}$

...LET'S NOT FORGET...

In Problems 46 through 53, perform the operations indicated.

46. -3^2

47. $(-3)^2$

48. $\dfrac{-2}{-2^0}$

49. $\dfrac{(a+b)(a-b)}{b^2-a^2}$

50. $(\sqrt{5}-\sqrt{3})^2$

51. $(\sqrt{7}+\sqrt{5})(\sqrt{7}-\sqrt{5})$

52. $\dfrac{a^2-a}{b^3}\cdot\dfrac{b^3}{a^4}$

53. $(2\sqrt{6}-\sqrt{2})(\sqrt{6}+\sqrt{2})$

In Problems 54 through 56, perform the operations indicated. Watch the role of parentheses.

54. $\dfrac{x-1}{x+1}-1$

55. $\dfrac{x-1}{x+1}-\dfrac{x-2}{x+1}$

56. $\dfrac{x-1}{x+1}-\dfrac{1}{x-1}$

In Problems 57 through 60, determine if $\sqrt{2}$ is a factor *or a* term.

57. $\sqrt{2}+2$

58. $\sqrt{2}(\sqrt{3})$

59. $\dfrac{1}{\sqrt{2}+2}$

60. $(y-\sqrt{2})(y+\sqrt{2})$

In Problems 61 through 63, multiply each expression by $2\sqrt{3}$.

61. $2\sqrt{3}$

62. $2+\sqrt{3}$

63. $\dfrac{1}{2\sqrt{3}}(2+\sqrt{3})$

In Problems 64 through 67, identify which expressions are factored. Factor those which are not factored, if possible.

64. $a^2\sqrt{3}-4\sqrt{3}$

65. x^2y^2

66. $(x-y)^3$

67. $2(b-c)$

68. x^3+y^3

In Problems 69 through 72, identify whether each problem is an expression, equation, *or* inequality. *Solve the equations and inequalities and perform the operation indicated with the expressions.*

69. $\dfrac{2}{x-1}-\dfrac{1}{x+3}$

70. $\dfrac{2}{x-1}-\dfrac{1}{x+3}=\dfrac{2x}{(x-1)(x+3)}$

71. $2x-3(x+3)\leq 0$

72. $\dfrac{\dfrac{2}{x-1}}{\dfrac{1}{x+3}}$

CHAPTER 8 TEST

In Problems 1 through 6, choose the correct answer.

1. $(2\sqrt{3})^2=(?)$

   A. 6 B. 12
   C. 18 D. 36

2. $\sqrt{-4}=(?)$

   A. 2 B. -2
   C. -4 D. Is not a real number

3. $(\sqrt{5}+\sqrt{3})^2=(?)$

   A. 8 B. 34
   C. $8+2\sqrt{15}$ D. $8+\sqrt{15}$

4. $\sqrt[3]{-\dfrac{8}{27}}=(?)$

   A. $\dfrac{3}{2}$ B. $-\dfrac{3}{2}$
   C. $\dfrac{2}{3}$ D. $\dfrac{-2}{3}$

5. $\dfrac{\sqrt{54}}{\sqrt{3}} = (?)$

A. $2\sqrt{3}$ B. $3\sqrt{2}$

C. $3\sqrt{6}$ D. $2\sqrt{6}$

6. $3\sqrt{48} - \sqrt{12} = (?)$

A. $10\sqrt{3}$ B. 18

C. $\sqrt{3}$ D. $44\sqrt{3}$

In Problems 7 through 13, perform the operations indicated and simplify.

7. $(3\sqrt{7})(4\sqrt{5})$

8. $\sqrt{6}(2\sqrt{2} - \sqrt{3})$

9. $(\sqrt{3} - 2\sqrt{7})(\sqrt{3} + 2\sqrt{7})$

10. $(2\sqrt{7} + \sqrt{3})^2$

11. $(\sqrt{5} + 2\sqrt{6})(\sqrt{3} - \sqrt{5})$

12. $\dfrac{\sqrt[3]{-81}}{\sqrt[3]{3}}$

13. $\sqrt{32x^2} + \sqrt{8x^2} - 2\sqrt{18x^2}$

In Problems 14 and 15, rationalize the denominator.

14. $\dfrac{3}{2\sqrt{3}}$

15. $\dfrac{\sqrt{3} - 2}{\sqrt{2} + \sqrt{3}}$

Optional section

Simplify each of the following.

16. $-49^{1/2}$

17. $125^{2/3}$

CHAPTER 9

See Problem Set 9.1, Exercise 39.

Quadratic Equations

CONNECTIONS

The usefulness of mathematics lies in our ability to match a mathematical concept to a physical, social, or economic problem. In Chapter 3, we saw how linear equations became mathematical models that could be used to solve applications. Hindu engineers in the ninth century were confronted with the problem of changing the size of the altars in temples. This led them to seek solutions to equations such as $x^2 - x - 6 = 0$. After electricity was discovered, scientists produced mathematical models such as $x^2 + 1 = 0$. These are examples of an important class of equations called quadratic equations.

There are several methods that can be used to solve such equations. Factoring is one method that we studied in Chapter 4. However, not all quadratic equations can be solved by factoring. In this chapter, we will see how to use the square root property and completing the square. We will derive the quadratic formula which can be used to solve any quadratic equation.

9.1 SQUARE ROOT PROPERTY

In this section we study solving second-degree equations. We saw in Section 4.5 that some second-degree equations can be solved by factoring. We now focus on the Square Root Property, which will allow us to solve second-degree equations that cannot be solved by factoring. We call second-degree equations **quadratic equations.**

Quadratic Equation

A **quadratic equation** in one variable is an equation that can be written in the form:

$$ax^2 + bx + c = 0$$

where a, b, and c are real numbers with $a \neq 0$. (If $a = 0$, the equation is not quadratic.)

We refer to a as the *leading coefficient, b,* as the *coefficient of x* and c as the constant term.

Let's begin by looking at how $x^2 = 4$ was solved in Section 4.5.

$$x^2 = 4$$

$$x^2 - 4 = 0$$

$$(x + 2)(x - 2) = 0$$

$$\{\pm 2\} \quad \text{or} \quad \{-2, 2\}$$

Notice that the solution set is $\{\pm\sqrt{4}\}$.

How could we solve $x^2 = 5$? It looks much like $x^2 = 4$. We begin by writing the equation so that one side is 0.

$$x^2 - 5 = 0$$

We were able to factor $x^2 - 4$ as the difference of two squares. Since 5 is not a perfect square, we must extend our ideas about factoring. There is an irrational number that, when squared, is 5. Recall that

$$(\sqrt{5})^2 = 5$$

Now we can write the equation as

$$x^2 - (\sqrt{5})^2 = 0$$

This is the difference of two squares, which factors as

$$(x + \sqrt{5})(x - \sqrt{5}) = 0$$

Thus, the solution set is

$$\{\pm\sqrt{5}\}$$

If the factoring steps are omitted, we can see a similarity in the two equations and in the two solution sets.

$$x^2 = 4 \qquad x^2 = 5$$

$$\{\pm\sqrt{4}\} \qquad \{\pm\sqrt{5}\}$$

Each equation is of the form $x^2 = A$, where A is a positive number. The solution set for each is $\{\pm\sqrt{A}\}$. These equations lead to the following result, which is very useful. It allows us to leave out the factoring steps.

Square Root Property

If A is a nonnegative real number, the equation

$$\mathbf{X}^2 = A$$

has the same solution set as the pair of equations

$$\mathbf{X} = \pm\sqrt{A}$$

If A is a negative number such as -3, then $\sqrt{A}$ is $\sqrt{-3}$. Since $\sqrt{-3}$ is not a real number, it is necessary to require that A be nonnegative. If A is 0, then $\sqrt{A}$ is $\sqrt{0}$, which has value 0.

EXAMPLE 1. Solve the following equations using the Square Root Property.

(a) $x^2 - 9 = 0$ (b) $x^2 = 6$

Solutions:

(a) We must first write the equation in the form $\mathbf{X}^2 = A$ before we can use the Square Root Property.

$$x^2 - 9 = 0$$

$$x^2 = 9$$

$$x = \pm\sqrt{9} \quad \text{Square Root Property}$$

$$x = \pm 3$$

$$\{\pm 3\}$$

Be Careful!

(continued)

(b) $x^2 = 6$

$$x = \pm\sqrt{6} \qquad \text{Square Root Property}$$

$$\{\pm\sqrt{6}\}$$

□

Be careful not to omit the ± in using the Square Root Property. To omit it would mean losing a solution. The solutions can be checked in the original equation.

Be Careful!

EXAMPLE 2. Solve $x^2 - 20 = 7$ by the Square Root Property.

Solution:

The equation must be in the form $\mathbf{X}^2 = A$.

$$x^2 - 20 = 7$$

$$x^2 = 27 \qquad \text{Add 20 to both sides.}$$

$$x = \pm\sqrt{27} \qquad \text{Square Root Property}$$

$$x = \pm 3\sqrt{3} \qquad \sqrt{27} = \sqrt{9(3)} = 3\sqrt{3}$$

$$\{\pm 3\sqrt{3}\}$$

□

EXAMPLE 3. Solve $3x^2 = 3$ by the Square Root Property.

Solution:

The equation must be in the form $\mathbf{X}^2 = A$. We must divide by 3.

$$3x^2 = 3$$

$$x^2 = 1 \qquad \text{Divide by 3.}$$

$$x = \pm\sqrt{1} \qquad \text{Square Root Property}$$

$$x = \pm 1$$

$$\{\pm 1\}$$

□

The **X** in the Square Root Property need not be just a single variable; it can be an expression, such as $x - 3$.

EXAMPLE 4. Solve $(x - 3)^2 = 4$ by the Square Root Property.

Solution:

We treat $(x - 3)$ like the **X** in the Square Root Property. So

$$(x - 3)^2 = 4$$

$$x - 3 = \pm\sqrt{4} \qquad \text{Square Root Property}$$

$$x - 3 = \pm 2$$

$$x = 3 \pm 2 \qquad \text{Add 3 to both sides.}$$

Solve the two equations.

$$x = 3 + 2 \quad \text{or} \quad x = 3 - 2$$

$$x = 5 \quad \text{or} \quad x = 1$$

$$\{1, 5\}$$

□

EXAMPLE 5. Solve $(x - \frac{1}{2})^2 = \frac{1}{25}$ by using the Square Root Property.

Solution:

$$\left(x - \frac{1}{2}\right)^2 = \frac{1}{25}$$

$$x - \frac{1}{2} = \pm\sqrt{\frac{1}{25}} \qquad \text{Square Root Property}$$

$$x - \frac{1}{2} = \pm\frac{1}{5}$$

$$x = \frac{1}{2} \pm \frac{1}{5}$$

$$x = \frac{5}{10} \pm \frac{2}{10} \qquad \text{Common denominator}$$

$$x = \frac{5}{10} + \frac{2}{10} \quad \text{or} \quad x = \frac{5}{10} - \frac{2}{10}$$

$$x = \frac{7}{10} \quad \text{or} \quad x = \frac{3}{10}$$

$$\left\{\frac{3}{10}, \frac{7}{10}\right\}$$

□

EXAMPLE 6. The hypotenuse of an isosceles right triangle is 12 m long. Find the length of the legs.

Solution:

Let x be the length of a leg. (in meters)

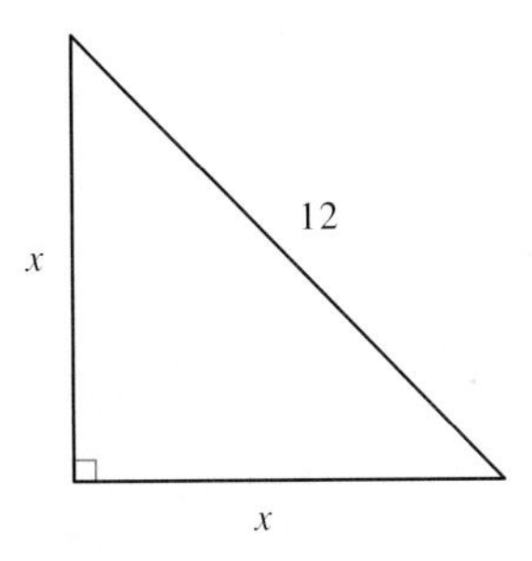

Using the Pythagorean Theorem, we form an equation.

$$x^2 + x^2 = 12^2$$

$$2x^2 = 144$$

$$x^2 = 72 \qquad \text{Divide by 2.}$$

$$x = \pm\sqrt{72} \qquad \text{Square Root Property}$$

$$x = \pm\sqrt{36 \cdot 2}$$

$$= \pm\sqrt{36}\sqrt{2}$$

$$= \pm 6\sqrt{2}$$

(continued)

Since a length cannot be negative, we use $6\sqrt{2}$. The legs are $6\sqrt{2}$ m long. Check by noting

$$(6\sqrt{2})^2 + (6\sqrt{2})^2 = 72 + 72 = 144$$

and that $12^2 = 144$. □

PROBLEM SET 9.1

Warm-ups

Use the Square Root Property to solve each quadratic equation.
In Problems 1 through 10, see Example 1.

1. $x^2 = 25$ **2.** $z^2 - 81 = 0$ 3. $x^2 = 100$

4. $x^2 = 0$ **5.** $w^2 - 120 = 1$ **6.** $x^2 - 169 = 0$

7. $y^2 = 225$ **8.** $x^2 = 17$ 9. $x^2 - 3 = 0$

10. $y^2 - 12 = 2$

In Problems 11 through 16, see Example 2.

11. $z^2 = 24$ **12.** $x^2 = 48$ 13. $x^2 - 32 = 0$

14. $y^2 - 25 = 100$ **15.** $w^2 - 54 = 0$ **16.** $x^2 - 72 = 0$

In Problems 17 through 22, see Example 3.

17. $5y^2 = 5$ **18.** $2x^2 = 12$ 19. $4x^2 - 9 = 0$

20. $\frac{1}{2}x^2 = 4$ 21. $\frac{1}{7}x^2 = 7$ **22.** $25w^2 = 36$

In Problems 23 through 30, see Example 4.

23. $(x + 3)^2 = 4$ **24.** $(w - 7)^2 = 1$ **25.** $(z + 3)^2 = 16$

26. $(x + 1)^2 = 25$ 27. $(x + 2)^2 = 5$ **28.** $(y - 9)^2 = 7$

29. $(x + 5)^2 = 12$ **30.** $(z - 4)^2 = 27$

In Problems 31 through 36, see Example 5.

31. $\left(t - \frac{1}{2}\right)^2 = \frac{1}{9}$ **32.** $\left(w - \frac{2}{3}\right)^2 = \frac{1}{4}$ **33.** $\left(x + \frac{2}{7}\right)^2 = \frac{4}{9}$

34. $\left(y + \frac{3}{5}\right)^2 = \frac{9}{16}$ 35. $\left(x - \frac{1}{3}\right)^2 = \frac{3}{4}$ **36.** $\left(y + \frac{2}{5}\right)^2 = \frac{7}{9}$

In Problems 37 through 41, see Example 6.

37. The legs of a right triangle are 1 ft in length. Find the length of the hypotenuse.

38. Find the length of a diagonal of a square with side of 10 m.

39. Glenn leans a 20-ft ladder against his house. If the base of the ladder is 4 ft from the house, how many feet above the ground is the top of the ladder?

40. The square of 3 less than a number is 16. Find all such numbers.

41. The diagonal of a square place mat has length $7\sqrt{2}$ in. Find the length of a side of the place mat.

Practice Exercises

Use the Square Root Property to solve each quadratic equation.

42. $x^2 = 36$ 43. $\frac{1}{2}y^2 = 8$ **44.** $x^2 - 4 = 0$

45. $z^2 - 280 = 9$

46. $x^2 = 144$

47. $x^2 = 64$

48. $2w^2 - 512 = 0$

49. $x^2 - 324 = 0$

50. $\frac{1}{4}y^2 = 100$

51. $x^2 = 484$

52. $x^2 = 21$

53. $4x^2 = 81$

54. $2x^2 - 4 = 0$

55. $y^2 - 23 = 0$

56. $y^2 - 250 = 0$

57. $w^2 - 63 = 0$

58. $(x + 5)^2 = 9$

59. $(w - 8)^2 = 16$

60. $(z + 1)^2 = 36$

61. $(x + 2)^2 = 64$

62. $(x + 3)^2 = 7$

63. $(y - 5)^2 = 3$

64. $(w - 2)^2 = 17$

65. $(x + 7)^2 = 11$

66. $(x + 2)^2 = 18$

67. $(z - 5)^2 = 24$

68. $(z + 4)^2 = 27$

69. $(t - 3)^2 = 125$

70. $\left(t - \frac{1}{2}\right)^2 = \frac{1}{16}$

71. $\left(w - \frac{2}{5}\right)^2 = \frac{1}{9}$

72. $\left(x + \frac{2}{9}\right)^2 = \frac{9}{16}$

73. $\left(y + \frac{3}{4}\right)^2 = \frac{81}{25}$

74. $\left(x - \frac{1}{5}\right)^2 = \frac{2}{9}$

75. $\left(y + \frac{2}{3}\right)^2 = \frac{5}{4}$

76. The legs of a right triangle are 1 ft and 3 ft in length. Find the length of the hypotenuse.

77. The length of the hypotenuse of an isosceles right triangle is 6 m. Find the length of the legs.

78. The length of the height of a right triangle is three times the length of the base. If the area of the triangle is 24 cm^2, find the lengths of the base and the height.

79. The square of two more than a number is 25. Find all such numbers.

80. The diagonal of a square picture is $4\sqrt{2}$ ft. Find the length of the side of the picture.

81. Find the radius of a circular table cloth if its area is 2025π in^2.

82. Glenn has a 30-ft ladder leaning against his house. If the top of the ladder is 20 ft above the ground, how far away from the house is the foot of the ladder?

83. Warren Mason purchased a radio antenna tower 15 m tall. He wishes to stabilize the tower in a vertical position with three guy wires. Each guy wire is to extend from the top of the tower to the ground 6 m from the base of the tower. How much wire should Warren buy?

84. A rocket accelerates from rest for 300 ft with a constant acceleration of $\frac{50}{3}$ ft per second per second. How much time has elapsed? (A formula for distance under these conditions is $d = \frac{1}{2}at^2$, where a is acceleration and t is time.)

85. The length of the diagonal of a rectangular placemat is 18 in. If the length of one side is 12 in., what are its dimensions?

86. The cost (in dollars) of producing x radios is given by the formula $c = (x + 3)^2$. If Radio Shack budgets \$1600 for producing radios, how many radios can they produce?

87. Find the radius of a circular rug of area 128π ft^2.

88. The length of the height of a right triangle is five times the length of the base. If the area of the triangle is 250 m^2, find the lengths of the base and the height.

89. Find the radius of a coffee can if its volume is 9π in^3 and its height is 4 in.

90. The distance a walnut falls from a tree to the ground is given by the formula $s = 16t^2$, where s is the distance measured in feet and t is the number of seconds that the walnut falls. How many seconds will it take a walnut to fall 81 ft to the ground?

Challenge Problems

Solve by the Square Root Property.

91. $x^2 - c = 0;\quad c > 0$

92. $(x - 2)^2 = a^2;\quad a > 0$

93. $(x - a)^2 = 4$

IN YOUR OWN WORDS...

94. Explain when the Square Root Property can be used to solve a quadratic equation.

9.2 COMPLETING THE SQUARE

In Section 9.1 the Square Root Property allowed us to solve quadratic equations in the form $X^2 = A$. In this section we examine how to write any quadratic equation in this form. Recall the following from the previous section.

> **Square Root Property**
>
> If A is a nonnegative real number, the equation
>
> $$\mathbf{X}^2 = A$$
>
> has the same solution set as the pair of equations,
>
> $$\mathbf{X} = \pm\sqrt{A}$$

EXAMPLE 1. Solve $x^2 + 2x + 1 = 5$ by the Square Root Property.

Solution:

We must write the equation in the form $\mathbf{X}^2 = A$ before we can apply the Square Root Property. Notice that the left side can be factored as $(x + 1)^2$. In other words, it is a perfect trinomial square.

$$x^2 + 2x + 1 = 5$$

$$(x + 1)^2 = 5 \qquad \text{Factor.}$$

$$x + 1 = \pm\sqrt{5} \qquad \text{Square Root Property}$$

$$x = -1 \pm \sqrt{5} \qquad \text{Subtract 1 from both sides.}$$

$$\{-1 \pm \sqrt{5}\} \quad \text{or} \quad \{-1 + \sqrt{5}, -1 - \sqrt{5}\}$$

□

The Square Root Property can be used to solve a quadratic equation if the equation can be written in the form $\mathbf{X}^2 = A$, where A is nonnegative. Example 1 suggests that we need a perfect trinomial square as $\mathbf{X}^2$.

If we make $x^2 + 6x$ into a perfect trinomial square, we will have a binomial squared. Recall how to square a binomial:

$$(x + p)^2 = x^2 + 2px + p^2$$

We need to add a number to $x^2 + 6x$ to make the expression a perfect trinomial square. That is,

$$(x + \underline{\quad\quad})^2 = x^2 + 6x + \underline{\quad\quad}$$

We must find the value of p. To do this, we need to unravel the middle term, $6x$. In squaring $x + p$, the middle term is $2px$. Since $6x$ is $2 \cdot 3x$, p must be 3. Thus the number for which we are looking is 3^2. So,

$$(x + 3)^2 = x^2 + 6x + 9$$

How can we look at $x^2 + 6x$ and compute the number that must be added to make a perfect trinomial square? We take the coefficient of x, which is 6. To undo the squaring, we divide 6 by 2 and then we square this result.

What number must be added to $x^2 - 2x$ to make it into a perfect trinomial square?

$$x^2 - 2x + \_\_\_\_ = (x + \_\_\_\_)^2$$

The coefficient of x is -2. We divide -2 by 2 and square the result.

$$\left(\frac{-2}{2}\right)^2 = (-1)^2 = 1$$

The number to add is 1.

$$x^2 - 2x + 1 = (x - 1)^2$$

The process of making an expression into a perfect trinomial square is called *completing the square*. We must learn how to do this before we can use the Square Root Property to solve more quadratic equations.

EXAMPLE 2. What number must be added to $x^2 + 4x$ to make it a perfect trinomial square?

Solution:

The coefficient of x is 4. We divide 4 by 2 and square the result.

$$\left(\frac{4}{2}\right)^2 = (2)^2 = 4$$

The number that we should add is 4. $x^2 + 4x + 4$ is a perfect trinomial square. This can be checked by noting that $(x + 2)^2 = x^2 + 4x + 4$. □

EXAMPLE 3. What number must be added to $y^2 - 7y$ to make it a perfect trinomial square?

Solution:

The coefficient of y is -7. We divide -7 by 2 and square the result.

$$\left(\frac{-7}{2}\right)^2 = \frac{49}{4}$$

The number to add is $\frac{49}{4}$. Thus $y^2 - 7y + \frac{49}{4}$ is a perfect square.

Check by noting that $y^2 - 7y + \frac{49}{4} = \left(y - \frac{7}{2}\right)^2$. □

EXAMPLE 4. What number must be added to $x^2 + \frac{2}{3}x$ to make a perfect trinomial square?

Solution:

The coefficient of x is $\frac{2}{3}$. We must divide $\frac{2}{3}$ by 2.

$$\frac{2}{3} \div 2 = \frac{2}{3} \cdot \frac{1}{2}$$

$$= \frac{1}{3}$$

(continued)

Now we must square $\frac{1}{3}$.

$$\left(\frac{1}{3}\right)^2 = \frac{1}{9}$$

We must add $\frac{1}{9}$.

$x^2 + \frac{2}{3}x + \frac{1}{9}$ is a perfect square.

This can be checked by noting that

$$\left(x + \frac{1}{3}\right)^2 = x^2 + \frac{2}{3}x + \frac{1}{9}$$ □

To complete the square, we must compute the number to add to make a perfect square. This involves dividing the coefficient of x by 2. As Example 4 indicates, dividing by 2 is the same as multiplying by $\frac{1}{2}$. Sometimes it is more convenient to multiply the coefficient of x by $\frac{1}{2}$. If the coefficient of x is an integer, then dividing by 2 is the best strategy. However, if the coefficient of x is a fraction, then multiplying by $\frac{1}{2}$ is easier.

EXAMPLE 5. Solve $x^2 + 6x - 7 = 0$ by completing the square.

Solution:

Adding 7 to both sides, the equation becomes,

$$x^2 + 6x = 7$$

Now complete the square on the left side. First calculate the number to add. We divide the coefficient of x by 2 and square the result.

$$\left(\frac{6}{2}\right)^2 = 3^2 = 9$$

We must add 9 to *both* sides of the equation.

$$x^2 + 6x = 7$$

$$x^2 + 6x + 9 = 7 + 9 \qquad \text{Add 9 to both sides.}$$

$$(x + 3)^2 = 16 \qquad \text{Factor.}$$

$$x + 3 = \pm\sqrt{16} \qquad \text{Square Root Property}$$

$$x + 3 = \pm 4$$

$$x = -3 \pm 4 \qquad \text{Subtract 3 from both sides.}$$

Solve the two equations.

$$x = -3 + 4 \quad \text{or} \quad x = -3 - 4$$

$$x = 1 \quad \text{or} \quad x = -7$$

$$\{-7, 1\}$$ □

The coefficient of the squared term must be 1 when completing the square. If the coefficient of the squared term in the quadratic equation is not 1, make it 1 and then complete the square.

EXAMPLE 6. Solve $3x^2 + 6x = 3$ by completing the square.

Solution:

First we divide both sides of the equation by 3.

$$x^2 + 2x = 1$$

Now we complete the square by dividing 2 by 2 and squaring the result.

$$2 \div 2 \text{ is } 1 \quad \text{and} \quad 1^2 \text{ is } 1.$$

We add 1 to both sides:

$$x^2 + 2x + 1 = 1 + 1$$
$$(x + 1)^2 = 2$$

Now we apply the Square Root Property.

$$x + 1 = \pm\sqrt{2}$$
$$x = -1 \pm \sqrt{2} \qquad \text{Subtract 1.}$$
$$\{-1 \pm \sqrt{2}\}$$

□

Solving a Quadratic Equation by Completing the Square

1. Write the equation in the form $ax^2 + bx + c = 0$.
2. Divide both sides by the coefficient of x^2 if it is not 1.
3. Subtract the constant term from both sides.
4. Divide the coefficient of x by 2, square this result, then add this number to both sides.
5. Factor the left side and simplify the right side.
6. Apply the Square Root Property.
7. Check if required.
8. Write the solution set.

It may not always be necessary to carry out Steps 1, 2, or 3.

EXAMPLE 7. Solve $3x^2 - 2 = 10x$ by completing the square.

Solution:

Step 1 Write the equation in the form $ax^2 + bx + c = 0$.

$$3x^2 - 10x - 2 = 0 \qquad \text{Make the right side 0.}$$

Step 2 Divide both sides by the coefficient of x^2.

$$x^2 - \frac{10}{3}x - \frac{2}{3} = 0 \qquad \text{Divide by 3.}$$

(continued)

Step 3 Place the constant term on the right side.

$$x^2 - \frac{10}{3}x = \frac{2}{3} \qquad \text{Add } \frac{2}{3} \text{ to both sides.}$$

Step 4 Divide the coefficient of x by 2, square this result, and add this number to both sides.

We calculate the number to add:

$$\left(-\frac{10}{3}\cdot\frac{1}{2}\right)^2 = \left(-\frac{5}{3}\right)^2 = \frac{25}{9}.$$

$$x^2 - \frac{10}{3}x + \frac{25}{9} = \frac{2}{3} + \frac{25}{9} \qquad \text{Add } \frac{25}{9} \text{ to both sides.}$$

Step 5 Factor the left side and simplify the right side.

$$\left(x - \frac{5}{3}\right)^2 = \frac{6}{9} + \frac{25}{9} \qquad \text{Factor the left side.}$$

$$\left(x - \frac{5}{3}\right)^2 = \frac{31}{9} \qquad \text{Simplify the right side.}$$

Step 6 Apply the Square Root Property.

$$x - \frac{5}{3} = \pm\sqrt{\frac{31}{9}} \qquad \text{Square Root Property}$$

$$x - \frac{5}{3} = \pm\frac{\sqrt{31}}{3}$$

$$x = \frac{5}{3} \pm \frac{\sqrt{31}}{3} \qquad \text{Add } \frac{5}{3} \text{ to both sides.}$$

Step 7 Check in the original equation.

Step 8 Write the solution set.

$$\left\{\frac{5}{3} \pm \frac{\sqrt{31}}{3}\right\}$$

This may also be written as $\left\{\frac{5 \pm \sqrt{31}}{3}\right\}$, $\left\{\frac{5}{3} + \frac{\sqrt{31}}{3}, \frac{5}{3} - \frac{\sqrt{31}}{3}\right\}$, or $\left\{\frac{5 + \sqrt{31}}{3}, \frac{5 - \sqrt{31}}{3}\right\}$. □

PROBLEM SET 9.2

Warm-ups

In Problems 1 through 18, calculate the term that must be added to each binomial to make each one a perfect trinomial square.
In Problems 1 through 6, see Example 2.

1. $x^2 - 4x$

2. $x^2 + 8x$

3. $x^2 - 6x$

4. $y^2 - 14y$

5. $w^2 + 12w$

6. $t^2 - 22t$

In Problems 7 through 12, see Example 3.

7. $x^2 + 5x$
8. $z^2 + 9z$
9. $w^2 - w$
10. $y^2 - 3y$
11. $x^2 + 13x$
12. $x^2 - 15x$

In Problems 13 through 18, see Example 4.

13. $t^2 + \frac{1}{2}t$
14. $y^2 - \frac{1}{3}y$
15. $x^2 - \frac{2}{5}x$
16. $z^2 + \frac{2}{7}z$
17. $x^2 + \frac{4}{3}x$
18. $x^2 - \frac{6}{5}x$

In Problems 19 through 42, solve the quadratic equations by completing the square.
In Problems 19 through 28, see Example 5.

19. $x^2 + 4x - 5 = 0$
20. $x^2 - 2x - 8 = 0$
21. $y^2 + 6y = 7$
22. $z^2 - 2z = 3$
23. $w^2 = 2w - 1$
24. $x^2 = 4x - 3$
25. $y^2 + 4y + 4 = 0$
26. $x^2 - 6x + 9 = 0$
27. $x^2 - 5x = 14$
28. $x^2 = 7x - 10$

In Problems 29 through 34, see Example 6.

29. $x^2 + 6x - 2 = 0$
30. $x^2 + 4x = 1$
31. $x^2 - 8x = -11$
32. $y^2 - 10y = -22$
33. $2x^2 - 8x = -4$
34. $3x^2 + 6x - 3 = 0$

In Problems 35 through 42, see Example 7.

35. $z^2 + z - 2 = 0$
36. $y^2 - 3y = 4$
37. $2x^2 + 4x - 16 = 0$
38. $3x^2 - 3x - 36 = 0$
39. $2x^2 - 5x = 3$
40. $3z^2 = 7z - 4$
41. $2x^2 - 10x + 1 = 0$
42. $4x^2 = -8x + 1$

Practice Exercises

Solve each quadratic equation by completing the square.

43. $x^2 + 4x - 12 = 0$
44. $x^2 - 2x - 35 = 0$
45. $y^2 + 6y = 16$
46. $z^2 - 2z = 24$
47. $w^2 = 4w - 4$
48. $x^2 = 6x - 5$
49. $x^2 - 8x + 12 = 0$
50. $x^2 + 12x = -32$
51. $z^2 + z - 12 = 0$
52. $y^2 - 3y = 10$
53. $x^2 - 5x = -6$
54. $x^2 = 7x + 18$
55. $y^2 + 10y + 25 = 0$
56. $x^2 - 8x + 16 = 0$
57. $4y^2 + 8y + 4 = 0$
58. $5z^2 - 10z - 15 = 0$
59. $2x^2 = 3x + 2$
60. $4y^2 + y = 3$

61. $2x^2 + 10x - 28 = 0$

62. $3x^2 + 15x + 18 = 0$

63. $4y^2 + 8y - 12 = 0$

64. $5z^2 - 20z - 20 = 0$

65. $2x^2 + 3x = 1$

66. $3z^2 = 7z - 3$

67. $2x^2 = 5x + 1$

68. $4y^2 - y = 2$

69. $2z^2 - 3z + 1 = 0$

70. $3y^2 + 2y - 2 = 0$

71. $3x^2 - 2x - 1 = 0$

72. $5x^2 + 4x - 2 = 0$

Challenge Problems

Solve each by completing the square.

73. $x^2 + 2x = c; \quad c \geq 0$

74. $x^2 + 4bx = 4$

75. $x^2 + bx + c = 0$

76. $ax^2 + bx + c = 0; \quad a > 0$

IN YOUR OWN WORDS...

77. Explain how to solve $2x^2 - 6x + 8 = 0$ by completing the square.

9.3 THE QUADRATIC FORMULA

Completing the square can be used to solve any quadratic equation. However, the method is often messy and time consuming. In this section we solve the general quadratic equation, $ax^2 + bx + c = 0$, by completing the square. This gives us a formula that we can use to solve all quadratic equations and avoid having to complete the square.

Standard Form

A quadratic equation written in the form

$$ax^2 + bx + c = 0$$

where a, b, and c are real numbers with $a > 0$, is said to be in **standard form.**

For example,

$$3x^2 + 5x + 8 = 0$$

is a quadratic equation in standard form with $a = 3$, $b = 5$, and $c = 8$. Likewise,

$$2x^2 + (-6)x + (-7) = 0$$

is a quadratic equation in standard form. However, we write it in the form,

$$2x^2 - 6x - 7 = 0$$

Notice that $a = 2$, $b = -6$, and $c = -7$. The equation

$$x^2 - 2x = 5$$

is also a quadratic equation. To write it in standard form, we subtract 5 from both sides.

$$x^2 - 2x = 5$$

$$x^2 - 2x - 5 = 0 \qquad \text{Subtract 5 from both sides.}$$

Thus, $a = 1$, $b = -2$, and $c = -5$.

As we study quadratic equations, we will see that it is often useful to write them in standard form.

EXAMPLE 1. Write each quadratic equation in standard form and give the values of a, b, and c.

(a) $2x^2 + 3x = -7$ (b) $t^2 = 1$

(c) $2x = -5x^2 + 6$ (d) $\frac{1}{2}x^2 - 2x = 0$

(e) $\sqrt{2}x^2 - \sqrt{7} + x = 0$ (f) $x(x + 3) = 5$

Solutions:

(a) $2x^2 + 3x = -7$

We write the equation in standard form,

$$2x^2 + 3x + 7 = 0$$

and we see that $a = 2$, $b = 3$, and $c = 7$.

(b) $t^2 = 1$

In standard form this is

$$t^2 - 1 = 0$$

Notice that having no t-term means that we could write

$$t^2 + 0t - 1 = 0$$

So $a = 1$, $b = 0$, and $c = -1$.

(c) $2x = -5x^2 + 6$

We can subtract $2x$ from both sides and write

$$0 = -5x^2 - 2x + 6$$

Standard form requires that a is positive. So, we multiply both sides by -1.

$$0 = 5x^2 + 2x - 6$$

Thus $a = 5$, $b = 2$, and $c = -6$.

We could add $5x^2$ to both sides of the original equation and then subtract 6 from both sides to get

$$5x^2 + 2x - 6 = 0$$

giving $a = 5$, $b = 2$, and $c = -6$.

(continued)

(d) $\frac{1}{2}x^2 - 2x = 0$

This quadratic equation is in standard form with $a = \frac{1}{2}$, $b = -2$, and $c = 0$.

(e) $\sqrt{2}x^2 - \sqrt{7} + x = 0$

We rearrange the terms of the quadratic equation to put it in standard form.

$$\sqrt{2}x^2 + x - \sqrt{7} = 0.$$

We can see that $a = \sqrt{2}$, $b = 1$, $c = -\sqrt{7}$.

(f) $x(x + 3) = 5$

To see that this equation is a quadratic equation, we must use the distributive property.

$$x(x + 3) = 5$$

$$x^2 + 3x = 5$$

We write it in standard form,

$$x^2 + 3x - 5 = 0$$

and see that $a = 1$, $b = 3$, and $c = -5$. □

Let's solve $ax^2 + bx + c = 0$ $(a > 0)$ by completing the square. We follow the steps outlined in Section 9.1.

$ax^2 + bx + c = 0$	Standard form.
$x^2 + \frac{b}{a}x + \frac{c}{a} = 0$	Divide by a.
$x^2 + \frac{b}{a}x = \frac{c}{-a}$	Subtract $\frac{c}{a}$.

To complete the square, multiply $\frac{b}{a}$ by $\frac{1}{2}$, which gives $\frac{b}{2a}$. The number that we must add to both sides of the equation is $\left(\frac{b}{2a}\right)^2$, or $\frac{b^2}{4a^2}$.

$x^2 + \frac{b}{a}x + \frac{b^2}{4a^2} = -\frac{c}{a} + \frac{b^2}{4a^2}$	Add $\frac{b^2}{4a^2}$.
$\left(x + \frac{b}{2a}\right)^2 = \frac{b^2 - 4ac}{4a^2}$	Factor the left side; simplify the right side.
$x + \frac{b}{2a} = \pm\sqrt{\frac{b^2 - 4ac}{4a^2}}$	Square Root Property
$x + \frac{b}{2a} = \pm\frac{\sqrt{b^2 - 4ac}}{2a}$	Simplify.
$x = -\frac{b}{2a} \pm \frac{\sqrt{b^2 - 4ac}}{2a}$	Subtract $\frac{b}{2a}$.
$x = \frac{-b \pm \sqrt{b^2 - 4ac}}{2a}$	

This equation is called the **Quadratic Formula.** It should be memorized. Although we derived the formula for $a > 0$, it works just as well is $a < 0$.

The Quadratic Formula

The quadratic equation, $ax^2 + bx + c = 0$, with $a \neq 0$, has the same solution set as

$$x = \frac{-b \pm \sqrt{b^2 - 4ac}}{2a}$$

The Quadratic Formula can be used to solve any quadratic equation.

Using the Quadratic Formula to Solve Quadratic Equations

1. Write the equation in standard form.
2. Identify the values of a, b, and c.
3. Substitute these values into the formula and simplify.
4. Check if required.
5. Write the solution set.

EXAMPLE 2. Solve $x^2 + 4x - 5 = 0$ by the Quadratic Formula.

Solution:

We note that the equation is in standard form. So $a = 1$, $b = 4$, and $c = -5$. Substitute these values into the Quadratic Formula.

$$\begin{aligned} x &= \frac{-b \pm \sqrt{b^2 - 4ac}}{2a} \\ &= \frac{-4 \pm \sqrt{4^2 - 4(1)(-5)}}{2(1)} \\ &= \frac{-4 \pm \sqrt{16 + 20}}{2} \\ &= \frac{-4 \pm \sqrt{36}}{2} \\ &= \frac{-4 \pm 6}{2} \end{aligned}$$

$$x = \frac{-4 + 6}{2} \quad \text{or} \quad x = \frac{-4 - 6}{2}$$

$$x = 1 \quad \text{or} \quad x = -5$$

$$\{-5, 1\}.$$

□

EXAMPLE 3. Solve $2x^2 - 3x = 3$ by the quadratic formula.

Solution:

Writing the equation in standard form, $2x^2 - 3x - 3 = 0$, we see that $a = 2$, $b = -3$, and $c = -3$. Substituting these into the quadratic formula gives

$$x = \frac{-b \pm \sqrt{b^2 - 4ac}}{2a}$$

$$= \frac{-(-3) \pm \sqrt{(-3)^2 - 4(2)(-3)}}{2(2)}$$

Be Careful!

Notice that b has a value of -3, so $-b$ is $-(-3)$.

$$x = \frac{3 \pm \sqrt{9 + 24}}{4}$$

$$= \frac{3 \pm \sqrt{33}}{4}$$

$$\left\{\frac{3 \pm \sqrt{33}}{4}\right\}$$

We may also write the solution set as $\left\{\frac{3}{4} \pm \frac{\sqrt{33}}{4}\right\}$, $\left\{\frac{3 + \sqrt{33}}{4}, \frac{3 - \sqrt{33}}{4}\right\}$, or $\left\{\frac{3}{4} + \frac{\sqrt{33}}{4}, \frac{3}{4} - \frac{\sqrt{33}}{4}\right\}$. □

The quadratic formula often leads to expressions such as

$$\frac{-6 \pm \sqrt{24}}{8}$$

Many mistakes are made in simplifying these expressions. Notice how this can be simplified.

EXAMPLE 4. Simplify $\frac{-6 \pm \sqrt{24}}{8}$

Solution:

Notice first that $\sqrt{24}$ can be simplified.

$$\sqrt{24} = \sqrt{4 \cdot 6} = \sqrt{4}\sqrt{6} = 2\sqrt{6}$$

Therefore,

$$\frac{-6 \pm \sqrt{24}}{8} = \frac{-6 \pm 2\sqrt{6}}{8}$$

Be Careful!

There are two ways to reduce these expressions. One way is to factor the numerator and divide out all common factors.

$$\frac{-6 \pm 2\sqrt{6}}{8} = \frac{2(-3 \pm \sqrt{6})}{2 \cdot 4}$$

$$= \frac{-3 \pm \sqrt{6}}{4}$$

Another way is to divide each term in the numerator by the denominator.

$$\frac{-6 \pm 2\sqrt{6}}{8} = \frac{-6}{8} \pm \frac{2\sqrt{6}}{8}$$

$$= \frac{-3}{4} \pm \frac{\sqrt{6}}{4}$$

Notice that both answers are the same, although they are written in different forms. That is,

$$\frac{-3 \pm \sqrt{6}}{4} = \frac{-3}{4} \pm \frac{\sqrt{6}}{4}$$

□

EXAMPLE 5. Solve $x^2 = 4x - 2$ by the quadratic formula.

Solution:

We write the equation in standard form, $x^2 - 4x + 2 = 0$. From this we see that $a = 1$, $b = -4$, and $c = 2$. Be careful in substituting the value of b into the formula. Since b is -4, $-b$ is $-(-4)$.

$$x = \frac{-b \pm \sqrt{b^2 - 4ac}}{2a}$$

$$= \frac{-(-4) \pm \sqrt{(-4)^2 - 4(1)(2)}}{2(1)}$$

$$= \frac{4 \pm \sqrt{16 - 8}}{2}$$

$$= \frac{4 \pm \sqrt{8}}{2}$$

$$= \frac{4 \pm 2\sqrt{2}}{2}$$

$$= \frac{4}{2} \pm \frac{2\sqrt{2}}{2}$$

$$= 2 \pm \sqrt{2}$$

Be Careful!

$$\{2 \pm \sqrt{2}\}$$

□

Each time that we use the Quadratic Formula, we must find the value of the expression $\sqrt{b^2 - 4ac}$. In Examples 2, 3, and 5, the number under the square root was a positive number. If we solve $x^2 - x + 1 = 0$ by the Quadratic Formula, $\sqrt{b^2 - 4ac}$ is $\sqrt{1^2 - 4(1)(1)}$, which is $\sqrt{-3}$. Since $\sqrt{-3}$ is not a real number, there are no real solutions to the equation $x^2 - x + 1 = 0$. This idea is examined further in Section 9.4.

PROBLEM SET 9.3

Warm-ups

In Problems 1 through 14, write each quadratic equation in standard form and give the values of a, b, and c for each equation. See Example 1.

1. $4x^2 - 5x + 7 = 0$
2. $x^2 = 6x$
3. $x^2 - 8x = 7$
4. $t^2 - t + 1 = 0$

5. $x^2 = 0$

6. $x^2 = 16$

7. $y = 3y^2 + 5$

8. $\frac{1}{2}x^2 = x - \frac{2}{3}$

9. $-2x^2 = 4x - 7$

10. $2w = w^2 - 3$

11. $x(x + 1) = 7$

12. $5 - 2x(x + 2) = 2x$

13. $(y + 4)(y - 5) = 0$

14. $z^2 + z(z - 4) = 0$

In Problems 15 through 34, solve each of the quadratic equations by using the Quadratic Formula.
In Problems 15 through 26, see Example 2.

15. $x^2 + 3x + 2 = 0$

16. $x^2 + 2x - 3 = 0$

17. $x^2 - 3x = 4$

18. $x^2 - 5x = 6$

19. $2x^2 + 7x + 3 = 0$

20. $2x^2 - 5x + 2 = 0$

21. $6x^2 + x - 2 = 0$

22. $9x^2 + 3x - 2 = 0$

23. $6x^2 + 7x - 3 = 0$

24. $2x^2 = 3x + 5$

25. $2x - 1 = x^2$

26. $x^2 + 6x + 9 = 0$

In Problems 27 through 34, see Example 3.

27. $x^2 - 3x - 1 = 0$

28. $x^2 + 3x - 2 = 0$

29. $2x^2 = 7x - 4$

30. $(3x + 4)(x - 1) = -3$

31. $x^2 + 5x = -2$

32. $3x - 1 = x^2$

33. $x(x - 3) = 3$

34. $x(2x + 3) = 1$

In Problems 35 through 38, reduce each expression. See Example 4.

35. $\frac{4 \pm \sqrt{8}}{2}$

36. $\frac{-6 \pm \sqrt{12}}{4}$

37. $\frac{-8 \pm \sqrt{32}}{4}$

38. $\frac{-1 \pm \sqrt{8}}{2}$

In Problems 39 through 48, solve each quadratic equation by using the Quadratic Formula. See Example 5.

39. $x^2 + 2x - 4 = 0$

40. $x^2 - 3x - 9 = 0$

41. $2x^2 - 4x = -1$

42. $3x^2 = 2x + 1$

43. $2x^2 + 6x = -3$

44. $5x^2 - 4x = 4$

45. $3x^2 + 6x + 1 = 0$

46. $3x^2 - 8x + 4 = 0$

47. $2x^2 - 2x - 3 = 0$

48. $5x^2 + 2x - 1 = 0$

Practice Exercises

In Problems 49 through 74, solve each quadratic equation by using the Quadratic Formula.

49. $x^2 + 5x + 4 = 0$

50. $x^2 + 6x - 7 = 0$

51. $x(x - 2) = 8$

52. $x(x + 5) = 6$

53. $3x^2 + 2x - 5 = 0$

54. $14x^2 - 3x - 2 = 0$

55. $6x^2 + 11x + 4 = 0$

56. $8x^2 - 2x - 15 = 0$

57. $x^2 - 5x - 1 = 0$

58. $x^2 + 5x - 2 = 0$

59. $2x^2 = 9x - 4$

60. $3x^2 = 2 - x$

61. $x^2 + 5x = -3$

62. $5x - 5 = x^2$

63. $4x - 4 = x^2$

64. $x^2 + 8x + 16 = 0$

65. $x^2 + 2x - 1 = 0$

66. $x^2 - 4x - 2 = 0$

67. $2x^2 - 6x = -1$

68. $3x^2 = 4x + 2$

69. $2x^2 + 8x = -5$

70. $5x^2 - 2x = 2$

71. $3x^2 + 4x + 1 = 0$

72. $7x^2 - 12x - 4 = 0$

73. $2x^2 - 2x - 1 = 0$

74. $5x^2 + 6x - 2 = 0$

Challenge Problems

Write each quadratic equation in standard form and give the values of a, b, and c in each.

75. $x^2 + x + k = 0$

76. $gx^2 - 7x = -5$

77. $\pi x^2 - x = k$

78. $x^2 = \pi^2$

Solve by using the Quadratic Formula.

79. $x^2 + kx + 1 = 0$

80. $x^2 + 3x + k = 0$

81. $x^2 + px + q = 0$

IN YOUR OWN WORDS...

82. Explain what the Quadratic Formula is and why it is useful.

9.4 QUADRATIC EQUATIONS WITH COMPLEX SOLUTIONS

In Section 8.1 we saw that expressions such as $\sqrt{-4}$ have no meaning as real numbers because there is no real number that equals -4 when squared. The quadratic equation $x^2 + 4 = 0$ has no solution in the set of real numbers. Any solution to this equation must be a number whose square is -4. Since there is no real number with this property, we will have to *enlarge* our set of numbers to include solutions to such equations. We call this new, larger set of numbers the **complex numbers.**

We construct the complex numbers by introducing a single new number to the set of real numbers. The new number is the number that when squared is -1. We call it i. That is, i is a number such that

$$i^2 = -1.$$

The Set of Complex Numbers

The set of complex numbers is the set of all numbers that can be written as

$$a + bi$$

where a and b are real numbers and i is a number with the property

$$i^2 = -1$$

Some examples of complex numbers are:

$$5 + 3i \qquad \frac{-6}{7} + 13i \qquad 1 - 2i$$

$$1 + \sqrt{22}i \qquad 6i \qquad -17$$

Note that the set of real numbers is a subset of the set of complex numbers. (The real number -17 can be written as $-17 + 0i$; thus it belongs to the set of complex numbers.)

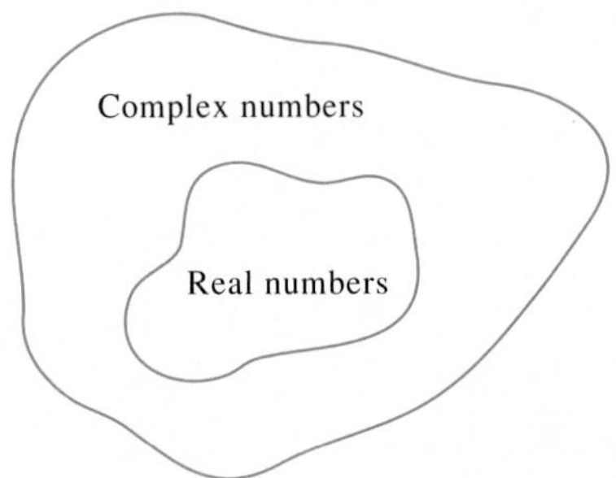

Next, let's look at $\sqrt{-4}$. This should be the number that when squared is -4. Of course, it cannot be a real number, but is it a complex number? To answer this question notice that

$$(2i)^2 = (2)^2 \cdot i^2 = 4 \cdot (-1) = -4$$

Thus a sensible definition is $\sqrt{-4} = \sqrt{4}i$.

Negative Radicand

If k is any positive real number,

$$\sqrt{-k} = \sqrt{k}i$$

Please note that the i is *not* under the radical. Sometimes $\sqrt{k}i$ is written as $i\sqrt{k}$.

Be Careful!

EXAMPLE 1. Simplify the following complex numbers.

(a) $\sqrt{-4}$ (b) $\sqrt{-7}$ (c) $\sqrt{-12}$

Solutions:

(a) $\sqrt{-4} = \sqrt{4}i = 2i$

(b) $\sqrt{-7} = \sqrt{7}i$

(c) $\sqrt{-12} = \sqrt{12}i$

$= \sqrt{4 \cdot 3}i$

$= 2\sqrt{3}i$ □

When dealing with symbols like $\sqrt{-5}$, it is *important* to rewrite the radical as $\sqrt{5}i$ *before* doing any arithmetic.

Arithmetic with complex numbers is done like arithmetic with polynomials, remembering to replace i^2 with -1.

EXAMPLE 2. Perform the operations indicated.

(a) $(3 + 5i) + (2 - 3i)$
(b) $(3 + 5i) - (2 - 3i)$
(c) $(3 + 5i)(2 - 3i)$

Solutions:

(a) $(3 + 5i) + (2 - 3i) = (3 + 2) + (5i - 3i)$

$= 5 + 2i$

(b) $(3 + 5i) - (2 - 3i) = (3 + 5i) + (-2 + 3i)$

$= (3 - 2) + (5i + 3i)$

$= 1 + 8i$

(c) $(3 + 5i)(2 - 3i) = 6 - 9i + 10i - 15i^2$

$= 6 + i - 15(-1)$

$= 6 + i + 15$

$= 21 + i$ □

Special products work with complex numbers as well.

EXAMPLE 3. Perform the operations indicated.

(a) $(2 + 3i)(2 - 3i)$

(b) $(2 + 3i)^2$

Solutions:

(a) The complex numbers $a + bi$ and $a - bi$ are called **complex conjugates.** The product of conjugates is an important special product.

$$\begin{aligned}(2 + 3i)(2 - 3i) &= 2^2 - (3i)^2 \\ &= 4 - 9i^2 = 4 - 9(-1) \\ &= 4 + 9 = 13\end{aligned}$$

(b)
$$\begin{aligned}(2 + 3i)^2 &= 4 + 12i + 9i^2 \\ &= 4 + 12i + 9(-1) \qquad (i^2 = -1) \\ &= 4 + 12i - 9 \\ &= -5 + 12i\end{aligned}$$

□

To do a division problem in complex numbers, we write it as a fraction and multiply numerator and denominator by the conjugate of the denominator.

EXAMPLE 4. Perform the operation indicated.

$(2 + 3i) \div (1 + i)$

Solution:

Rewrite the problem as a fraction.

$$(2 + 3i) \div (1 + i) = \frac{2 + 3i}{1 + i}$$

Next rationalize the denominator by multiplying the numerator and denominator by the *conjugate* of the *denominator.*

$$\begin{aligned}\frac{2 + 3i}{1 + i} &= \frac{(2 + 3i)(1 - i)}{(1 + i)(1 - i)} \\ &= \frac{2 - 2i + 3i - 3i^2}{1^2 - i^2} \\ &= \frac{2 - 2i + 3i + 3}{1 - (-1)} \\ &= \frac{5 + i}{2}, \quad \text{or} \quad \frac{5}{2} + \frac{1}{2}i\end{aligned}$$

□

Now we can solve more quadratic equations.

EXAMPLE 5. Solve the equations by the Square Root Property.

(a) $x^2 + 9 = 0$ (b) $(t - 1)^2 = -8$

Solution:

(a) Write the equation in the form $x^2 = A$.

$$x^2 + 9 = 0$$

$$x^2 = -9$$
$$x = \pm\sqrt{-9} \qquad \text{Square Root Property}$$
$$x = \pm\sqrt{9}i$$
$$x = \pm 3i$$
$$\{\pm 3i\}$$

(b) $(t-1)^2 = -8$

$$t - 1 = \pm\sqrt{-8} \qquad \text{Square Root Property}$$
$$t - 1 = \pm\sqrt{8}i$$
$$t - 1 = \pm 2\sqrt{2}i$$
$$t = 1 \pm 2\sqrt{2}i \qquad \text{Add 1 to both sides.}$$
$$\{1 \pm 2\sqrt{2}i\}$$

□

EXAMPLE 6. Solve the equations using the Quadratic Formula.

(a) $x^2 - 2x = -5$ (b) $3x^2 + 4x + 2 = 0$

Solutions:

(a) Write the equation in standard form first.

$$x^2 - 2x = -5$$
$$x^2 - 2x + 5 = 0 \qquad \text{Standard form}$$
$$x = \frac{-(-2) \pm \sqrt{(-2)^2 - 4(1)(5)}}{2(1)}$$
$$= \frac{2 \pm \sqrt{4 - 20}}{2}$$
$$= \frac{2 \pm \sqrt{-16}}{2}$$
$$= \frac{2 \pm \sqrt{16}i}{2}$$
$$= \frac{2 \pm 4i}{2}$$
$$= 1 \pm 2i$$
$$\{1 \pm 2i\}, \quad \text{or} \quad (1 + 2i, 1 - 2i\}$$

(b) $3x^2 + 4x + 2 = 0$

$$x = \frac{-4 \pm \sqrt{4^2 - 4(3)(2)}}{2(3)}$$
$$= \frac{-4 \pm \sqrt{16 - 24}}{6}$$
$$= \frac{-4 \pm \sqrt{-8}}{6}$$

(continued)

$$= \frac{-4 \pm \sqrt{8}i}{6}$$

$$= \frac{-4 \pm 2\sqrt{2}i}{6}$$

$$= \frac{-4}{6} \pm \frac{2\sqrt{2}}{6}i$$

$$= -\frac{2}{3} \pm \frac{\sqrt{2}}{3}i$$

$$\left\{-\frac{2}{3} \pm \frac{\sqrt{2}}{3}i\right\}$$

□

PROBLEM SET 9.4

Warm-ups

In Problems 1 through 6, simplify each complex number. See Example 1.

1. $\sqrt{-9}$ **2.** $\sqrt{-49}$ **3.** $\sqrt{-5}$

4. $\sqrt{-3}$ **5.** $\sqrt{-8}$ **6.** $\sqrt{-24}$

In Problems 7 through 26, perform the operations indicated.
In Problems 7 through 16, see Example 2.

7. $(1 + i) + (2 + 3i)$ **8.** $(3 + 5i) - (2 + 3i)$ **9.** $(5 - 2i) + (3 - 5i)$

10. $(-4 - i) + (1 + 5i)$ 11. $(2 + 3i) - (6 + 5i)$ **12.** $(5 - 6i) - (3 + 6i)$

13. $(1 + i)(2 + 3i)$ **14.** $(1 - i)(5 + 2i)$ 15. $(1 + 2i)(3 - 2i)$

16. $(2 + 3i)(3 - 4i)$

In Problems 17 through 20, see Example 3.

17. $(4 - 3i)^2$ **18.** $(3 + i)^2$ **19.** $(2 - i)(2 + i)$

20. $(5 + 2i)(5 - 2i)$

In Problems 21 through 26, see Example 4.

21. $\frac{1}{1 + i}$ **22.** $\frac{1}{2 - i}$ **23.** $\frac{2}{1 - 3i}$

24. $\frac{1 + 2i}{1 + 3i}$ 25. $\frac{2 - i}{3 + 2i}$ **26.** $\frac{2 - i}{2 + i}$

In Problems 27 through 32, find the solution set by the Square Root Property.
See Example 5.

27. $x^2 + 1 = 0$ **28.** $w^2 = -16$

29. $2x^2 + 14 = 0$ **30.** $\frac{1}{2}y^2 = -3$

31. $(x - 2)^2 = -12$ **32.** $(y + 3)^3 = -27$

In Problems 33 through 40, find the solution set by using the Quadratic Formula. See Example 6.

33. $t^2 + t + 1 = 0$

34. $x^2 + 3x = -3$

35. $5x^2 + 4x = -1$

36. $9x^2 = 3x - 1$

37. $x^2 + 2x + 5 = 0$

38. $4x^2 = 2x - 3$

39. $x^2 + x + 2 = 0$

40. $3x^2 + 2x = -1$

Practice Exercises

In Problems 41 through 60, perform the operations indicated.

41. $(1 - \sqrt{-1}) + (2 - \sqrt{-9})$

42. $(3 - \sqrt{-25}) - (2 - \sqrt{-9})$

43. $(5 + 2i) + (3 + 5i)$

44. $(-4 + i) - (1 - 5i)$

45. $(3 + \sqrt{-9}) - (5 + \sqrt{-25})$

46. $(6 - \sqrt{-49}) - (4 + \sqrt{-25})$

47. $(1 - i)(3 + 2i)$

48. $(1 + i)(2 + 4i)$

49. $(1 - 3i)(2 - 3i)$

50. $(3 + 2i)(2 - 5i)$

51. $(1 - i)^2$

52. $(2 + 3i)^2$

53. $(3 - i)(3 + i)$

54. $(4 + 3i)(4 - 3i)$

55. $\dfrac{1}{1 - i}$

56. $\dfrac{1}{3 + i}$

57. $\dfrac{3}{1 - 2i}$

58. $\dfrac{2 + i}{1 - 3i}$

59. $\dfrac{3 - i}{2 + 3i}$

60. $\dfrac{5 - i}{5 + i}$

In Problems 61 through 66, find the solution set by the Square Root Property.

61. $x^2 + 36 = 0$

62. $w^2 = -49$

63. $2x^2 + 10 = 0$

64. $\frac{1}{3}y^2 = -7$

65. $(x - 3)^2 = -8$

66. $(y + 1)^2 = -12$

In Problems 67 through 74, find the solution set by using the Quadratic Formula.

67. $t^2 - t + 3 = 0$

68. $x^2 - 3x = -3$

69. $5x^2 - 4x = -1$

70. $x^2 = 4x - 8$

71. $x^2 - 2x + 5 = 0$

72. $4x^2 = 5x - 3$

73. $2x^2 + 4x + 5 = 0$

74. $3x^2 + 6x = -4$

Challenge Problems

75. Factor $x^2 + 9$. (*Hint:* $-9 = (3i)^2$)

76. Show that $(a + bi)(a - bi) = a^2 + b^2$.

IN YOUR OWN WORDS...

77. Why was it necessary to introduce the number i?

9.5 SUMMARY OF SOLVING QUADRATIC EQUATIONS

We have learned how to solve a quadratic equation by factoring, using the Square Root Property, completing the square, and using the Quadratic Formula. In this section we examine how to choose the most appropriate method to use in solving a quadratic equation.

Any quadratic equation can be solved by using the Quadratic Formula or by completing the square. However, factoring is much easier if it can be used. The Square Root Property saves some steps if it can be used directly.

Choosing a Method to Use in Solving a Quadratic Equation

1. Determine if the equation is of the form $\mathbf{X}^2 = A$ or can be written as $\mathbf{X}^2 = A$ without completing the square. If so, use the Square Root Property.
2. Write the equation in standard form.
3. Try to factor the nonzero side. If it can be factored easily, use factoring.
4. Use the Quadratic Formula if factoring cannot be used.
5. Use completing the square when directed to do so.

EXAMPLE 1. Solve each quadratic equation.

(a) $3x^2 = 54$ (b) $(x - 5)^2 + 8 = 0$

Solutions:

(a) This equation can be written in the form $\mathbf{X}^2 = A$. The Square Root Property is appropriate.

$x^2 = 18$ Divide both sides by 3.

$x = \pm\sqrt{18}$ Square Root Property

$x = \pm\sqrt{9 \cdot 2}$

$x = \pm\sqrt{9}\sqrt{2}$

$x = \pm 3\sqrt{2}$

$\{\pm 3\sqrt{2}\}$

(b) This equation can be written in the form $\mathbf{X}^2 = A$.

$(x - 5)^2 = -8$

$x - 5 = \pm\sqrt{-8}$ Square Root Property

$x - 5 = \pm\sqrt{8}i$

$x - 5 = \pm 2\sqrt{2}i$

$x = 5 \pm 2\sqrt{2}i$

$\{5 \pm 2\sqrt{2}i\}$ □

EXAMPLE 2. Solve each equation.

(a) $2x^2 - 13x + 15 = 0$ (b) $2(y^2 - y) = 24$ (c) $z^2 + z = 0$

Solutions:

(a) This equation is not in the form $\mathbf{X}^2 = A$ as in Example 1. We try factoring. Notice that it is in standard form.

$$(2x - 3)(x - 5) = 0 \qquad \text{Factor.}$$

$$\left\{\frac{3}{2}, 5\right\}$$

(b) This equation is not in the form $\mathbf{X}^2 = A$. We try factoring. Notice that 2 is a factor of each side of the equation. We can divide both sides by 2 before we begin.

$$\frac{2(y^2 - y)}{2} = \frac{24}{2}$$

$$y^2 - y = 12$$

Be careful to write the equation in standard form.

$$y^2 - y - 12 = 0$$

$$(y - 4)(y + 3) = 0$$

$$\{-3, 4\}$$

(c) $z^2 + z = 0$

$z(z + 1) = 0$ Factor.

$\{-1, 0\}$

□

EXAMPLE 3. Solve $(t + 1)(t + 2) = 5$.

Solution:

Be careful to write the equation in standard form first.

$$(t + 1)(t + 2) = 5$$

$$t^2 + 3t + 2 = 5$$

$$t^2 + 3t - 3 = 0$$

Try to factor. This cannot be factored. Use the Quadratic Formula.

$$t = \frac{-3 \pm \sqrt{3^2 - 4(1)(-3)}}{2(1)}$$

$$= \frac{-3 \pm \sqrt{9 + 12}}{2}$$

$$= \frac{-3 \pm \sqrt{21}}{2}$$

$$\left\{\frac{-3 \pm \sqrt{21}}{2}\right\}$$

□

PROBLEM SET 9.5

Warm-ups

In Problems 1 through 20, solve the equations.
In Problems 1 through 8, see Example 1.

1. $2x^2 = 10$
2. $3x^2 = -18$
3. $(x - 7)^2 = -5$
4. $(y + 1)^2 = 4$
5. $(x - 8)^2 + 12 = 0$
6. $(z - 2)(z + 2) = 4$
7. $\frac{1}{2}x^2 - 4 = 0$
8. $4t^2 - 25 = 0$

In Problems 9 through 13, see Example 2.

9. $x^2 + 5x = 0$
10. $t^2 = t$
11. $x^2 - 3x = 10$
12. $2z^2 = 5z + 3$
13. $(x + 4)(2x + 1) = 0$

In Problems 14 through 20, see Example 3.

14. $x^2 + x - 1 = 0$
15. $x^2 - x + 1 = 0$
16. $x(x - 2) = 5$
17. $x^2 + 6 = 2(x - 3)$
18. $2x^2 + 1 = -x$
19. $3x^2 - 2x + 1 = 0$
20. $4(x^2 + 4) = 12x$

Practice Exercises

In Problems 21 through 40, solve the equations.

21. $3x^2 = 15$
22. $5x^2 = -45$
23. $(x + 7)^2 = -3$
24. $(y - 8)^2 = 9$
25. $2x^2 + 12x = 0$
26. $t^2 = 4t$
27. $x^2 + 3x = 28$
28. $2z^2 = 3z + 5$
29. $x^2 - x - 1 = 0$
30. $x^2 + x + 1 = 0$
31. $9t^2 - 49 = 0$
32. $\frac{1}{3}x^2 - 9 = 0$
33. $(x - 4)(2x - 1) = 0$
34. $(z - 3)(z + 3) = 9$
35. $x(x + 2) = 5$
36. $x^2 + 4 = 2(x - 4)$
37. $(x - 1)^2 + 24 = 0$
38. $3x^2 + 2 = -x$
39. $4x^2 - 2x + 3 = 0$
40. $5(x^2 + 4) = 15x$

41. A rectangular swimming pool with dimensions 7 m by 14 m is to have a walk of uniform width around it. Find the width of the walk if the combined area of the pool and walk is 260 m^2.
42. The area of the following figure is x^2 square units. Find x.

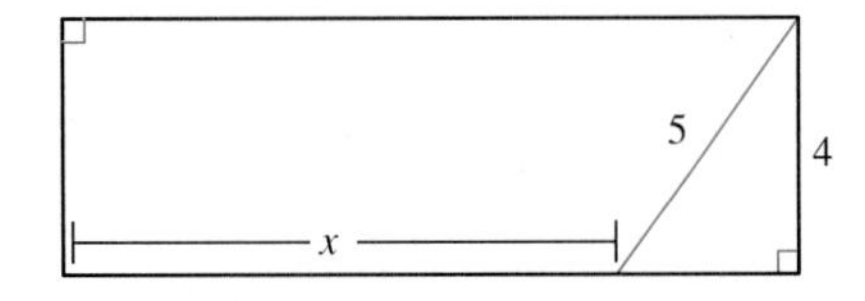

43. A rectangular garden with dimensions of 8 yd by 12 yd is to have a walk of uniform width around it. Find the width of the walk if the area of the walk is equal to the area of the garden.
44. The area of the following figure is 230 square units. Find x.

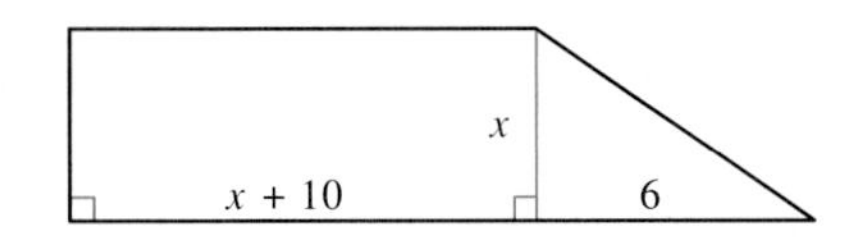

Challenge Problems

45. Solve $s = \frac{1}{2}gt^2$ for t.

46. Solve $t^2 + v_0t + s_0 = 0$ for t

IN YOUR OWN WORDS...

47. Explain how to choose a method to use in solving a quadratic equation.

9.6 EQUATIONS CONTAINING SQUARE ROOTS

In this section we solve equations containing square roots. Adding the same number to both sides of an equation or multiplying both sides of an equation by the same nonzero number will not lead us to the solution set. Somehow we must remove the square root from the equation. Recall from Section 8.1 that squaring a square root removes the radical. That is, if x is a nonnegative real number.

$$(\sqrt{x})^2 = x$$

This property of square roots along with the following result will allow us to solve equations with square roots.

Squaring Property

If $A = B$ has solutions, these solutions are found in the solutions of $A^2 = B^2$.

Let's solve the equation $\sqrt{x} = 2$. To remove the square root, we square both sides of the equation.

$$(\sqrt{x})^2 = 2^2$$
$$x = 4$$

The Squaring Property says to find the solution set of the equation $\sqrt{x} = 2$, we must find the solution set of the squared equation. The solution of the squared equation is 4. This means that 4 is the only possible solution of $\sqrt{x} = 2$. We can determine if it is a solution only by checking 4 in the original equation, $\sqrt{x} = 2$.

To check 4 in the equation $\sqrt{x} = 2$, we evaluate each side when x is 4.

LS: $\sqrt{x} = \sqrt{4} = 2$

RS: 2

Since the value we get on both sides is 2, we say that 4 checks. Thus 4 is in the solution set of $\sqrt{x} = 2$. There are no other possible solutions. The solution set is {4}.

A Procedure for Solving Equations Containing Square Roots

1. Isolate a square root.
2. Square both sides.
3. Solve the resulting equation for all possible solutions.
4. Check each possible solution in the original equation.
5. Write the solution set.

EXAMPLE 1. Solve $\sqrt{x+1} = 5$.

Solution:

Step 1 The square root is already isolated.

$$\sqrt{x+1} = 5$$

Step 2 Square both sides.

$$(\sqrt{x+1})^2 = 5^2$$

Step 3 Solve the resulting equation.

$$x + 1 = 25 \qquad \text{Remember } (\sqrt{x+1})^2 = x + 1.$$

$$x = 24$$

Step 4 Check 24 in the original equation.

Check 24 in $\sqrt{x+1} = 5$.

LS: $\sqrt{x+1} = \sqrt{24+1} = \sqrt{25} = 5$

RS: 5

24 checks and thus is in the solution set.

Step 5 Write the solution set.

$$\{24\}$$

□

The next example shows that checking is necessary in solving equations with square roots.

EXAMPLE 2. Solve $\sqrt{x-3} + 3 = 0$.

Solution:

$$\sqrt{x-3} = -3 \qquad \text{Isolate a square root.}$$

$$(\sqrt{x-3})^2 = (-3)^2 \qquad \text{Square both sides.}$$

$$x - 3 = 9$$

$$x = 12$$

Check 12 in $\sqrt{x-3} + 3 = 0$.

LS: $\sqrt{x-3} + 3 = \sqrt{12-3} + 3 = \sqrt{9} + 3 = 3 + 3 = 6$

RS: 0

The value on the left side is 6 and the value on the right side is 0, so 12 does *not* check. This means that 12 is *not* in the solution set. However, since 12 is the only possible solution, the solution set must be the empty set.

$$\varnothing$$

□

We could have noticed that the solution set would be the empty set because the square root of a real number is never negative.

EXAMPLE 3. Solve $\sqrt{2x} - \sqrt{x+2} = 0$.

Solution:

$$\sqrt{2x} = \sqrt{x+2} \qquad \text{Isolate a square root.}$$

$$(\sqrt{2x})^2 = (\sqrt{x+2})^2 \qquad \text{Square both sides.}$$

$$2x = x + 2$$

$$x = 2$$

Check 2 in $\sqrt{2x} - \sqrt{x+2} = 0$.

$$\text{LS:} \quad \sqrt{2x} - \sqrt{x+2} = \sqrt{2(2)} - \sqrt{2+2}$$

$$= \sqrt{4} - \sqrt{4} = 2 - 2 = 0$$

$$\text{RS:} \quad 0$$

Thus 2 checks.

$$\{2\}$$

□

The equation that results from squaring an equation with square roots may also be a quadratic equation.

EXAMPLE 4. Solve $\sqrt{x+3} + 3 = x$.

Solution:

$$\sqrt{x+3} = x - 3 \qquad \text{Isolate the radical.}$$

$$(\sqrt{x+3})^2 = (x-3)^2 \qquad \text{Square both sides.}$$

The next step requires us to square a binomial. Remember that $(x-3)^2 = x^2 - 6x + 9$. Many mistakes are made in finding the middle term. Thus we obtain,

Be Careful!

$$x + 3 = x^2 - 6x + 9$$

This is a quadratic equation. We put it in standard form first and then solve by factoring.

$$0 = x^2 - 7x + 6$$

$$0 = (x-6)(x-1)$$

$$x = 6 \quad \text{or} \quad x = 1$$

There are two possible solutions to our original equation. We *must* check *both* of them.

Check 6 in $\sqrt{x+3} + 3 = x$.

$$\text{LS:} \quad \sqrt{x+3} + 3 = \sqrt{6+3} + 3 = \sqrt{9} + 3$$

$$= 3 + 3 = 6$$

$$\text{RS:} \quad x = 6$$

Thus 6 checks and is in the solution set.

(continued)

Check 1 in $\sqrt{x + 3} + 3 = x$.

$$\text{LS:} \quad \sqrt{x+3} + 3 = \sqrt{1+3} + 3 = \sqrt{4} + 3$$
$$= 2 + 3 = 5$$
$$\text{RS:} \quad 1$$

Thus 1 does *not* check and is not in the solution set.

$$\{6\}$$

□

PROBLEM SET 9.6

Warm-ups

Solve each equation.
In Problems 1 through 6, see Example 1.

1. $\sqrt{x + 2} = 4$ **2.** $\sqrt{x - 1} = 5$ 3. $\sqrt{5x + 1} - 1 = 3$

4. $\sqrt{3x + 1} = 1$ **5.** $\sqrt{x - 4} = 0$ **6.** $\sqrt{2x - 3} = 3$

In Problems 7 through 10, see Example 2.

7. $\sqrt{1 - 2x} = -5$ **8.** $\sqrt{x + 3} = -4$ **9.** $\sqrt{4 - x} + 5 = 0$

10. $\sqrt{2x + 5} - 1 = -6$

In Problems 11 through 16, see Example 3.

11. $\sqrt{3x + 1} = \sqrt{5 - x}$ **12.** $\sqrt{x + 9} = \sqrt{3x - 5}$

13. $\sqrt{5 - x} - \sqrt{2x + 17} = 0$ **14.** $\sqrt{2x + 9} + \sqrt{9 - x} = 0$

15. $\sqrt{x^2 - x - 3} = \sqrt{x^2 - 7}$ **16.** $\sqrt{x^2 + x + 4} - \sqrt{x^2 - 3x + 12} = 0$

In Problems 17 through 28, see Example 4.

17. $\sqrt{x + 3} = x + 1$ **18.** $\sqrt{x + 9} = x - 3$

19. $\sqrt{2 - x} = x + 4$ **20.** $\sqrt{2 - x} = x$

21. $\sqrt{4x + 9} = x + 1$ **22.** $\sqrt{2x + 20} = x + 10$

23. $\sqrt{4x + 5} = x$ **24.** $\sqrt{1 - 2x} = x + 1$

25. $\sqrt{2x + 6} - 3 = x$ **26.** $x = \sqrt{x + 4} - 2$

27. $3 = \sqrt{2x^2 + 2} - x$ **28.** $\sqrt{x + 5} + 7 = x$

Practice Exercises

Solve each equation.

29. $\sqrt{x + 7} = 3$ **30.** $\sqrt{x - 1} = 2$

31. $\sqrt{x + 1} = -2$ **32.** $\sqrt{5x + 4} = 2$

33. $\sqrt{x - 5} = 0$ **34.** $\sqrt{3x + 1} = 4$

35. $\sqrt{1 - 2x} = 3$ **36.** $\sqrt{2x + 8} - 1 = 1$

37. $\sqrt{10 - x} + 3 = 0$ **38.** $\sqrt{8x + 1} - 2 = 1$

39. $\sqrt{x - 3} = \sqrt{11 - x}$ **40.** $\sqrt{x + 8} = \sqrt{3x + 16}$

41. $\sqrt{7 - x} - \sqrt{2x + 10} = 0$ **42.** $\sqrt{2x + 4} + \sqrt{4 - x} = 0$

43. $\sqrt{x^2 - x + 2} = \sqrt{x^2 + 3}$ **44.** $\sqrt{x^2 - x - 5} - \sqrt{x^2 - 3x + 7} = 0$

45. $\sqrt{x^2 + 1} = 1 - x$

46. $\sqrt{x^2 + 2x + 1} = x + 3$

47. $\sqrt{x^2 - 3x + 2} + 2 = x$

48. $\sqrt{x^2 - 5} - 5 = x$

49. $\sqrt{x^2 - 1} = 1 + x$

50. $\sqrt{x^2 + 2x + 4} = x + 2$

51. $\sqrt{x^2 - x - 4} + 9 = x$

52. $\sqrt{x^2 + 7} - 7 = x$

53. $\sqrt{x - 1} = x - 3$

54. $\sqrt{x + 8} = x + 2$

55. $\sqrt{3 - x} = x + 3$

56. $\sqrt{3 - 2x} = x$

57. $\sqrt{3x + 4} = x - 2$

58. $\sqrt{2x + 8} = x + 4$

59. $\sqrt{6x - 9} = x$

60. $\sqrt{1 - 3x} = x - 1$

61. $\sqrt{2x + 2} - 1 = x$

62. $x = \sqrt{x + 6} + 6$

63. $2 = \sqrt{3x^2 - 2} - x$

64. $\sqrt{x + 10} + 2 = x$

Challenge Problems

Solve each equation (a and b are positive constants).

65. $\sqrt{ax} = b$

66. $\sqrt{x^2 + b} = x + a$

IN YOUR OWN WORDS...

67. Explain how to solve an equation that contains a square root.

68. Why is checking not an optional part of the procedure when squaring both sides of an equation?

CHAPTER SUMMARY

GLOSSARY

Quadratic equation: A second-degree equation.

Complex number: A number that can be written in the form $a + bi$, where a and b are real numbers and i is a number such that $i^2 = -1$.

SQUARE ROOT PROPERTY

The equation $\mathbf{X}^2 = A$ has the same solution set as the pair of equations $\mathbf{X} = \pm\sqrt{A}$.

SOLVING QUADRATIC EQUATIONS BY COMPLETING THE SQUARE

1. Write the equation in the form $ax^2 + bx + c = 0$.
2. Divide both sides by the coefficient of x^2 if it is not 1.
3. Subtract the constant term from both sides.
4. Divide the coefficient of x by 2, square this result and then add this number to both sides.
5. Factor the left side and simplify the right side.
6. Apply the Square Root Property.
7. Check if required.
8. Write the solution set.

QUADRATIC FORMULA

To solve the quadratic equation $ax^2 + bx + c = 0$ $(a \neq 0)$, we use the Quadratic Formula:

$$x = \frac{-b \pm \sqrt{b^2 - 4ac}}{2a}$$

It should be used if the quadratic equation cannot be factored easily.

SOLVING EQUATIONS CONTAINING SQUARE ROOTS

1. Isolate a square root.
2. Square both sides.
3. Solve the resulting equation for all possible solutions.
4. Check each possible solution in the original equation.
5. Write the solution set.

CHECKUPS

1. Solve $x^2 - 20 = 7$.	Section 9.1; Example 2
2. Solve $3x^2 + 6x = 3$ by completing the square.	Section 9.2; Example 6
3. Solve $x^2 = 4x - 2$.	Section 9.3; Example 5
4. The hypotenuse of an isosceles right triangle is 12 m. Find the lengths of the legs.	Section 9.1; Example 6
5. Simplify $\sqrt{-4}$.	Section 9.4; Example 1a
6. Perform the operation indicated: $(3 + 5i)(2 - 3i)$.	Section 9.4; Example 2c
7. Solve $(t - 1)^2 = -8$ using the Square Root Property.	Section 9.4; Example 5b
8. Solve $x^2 - 2x = -5$.	Section 9.4; Example 6a
9. Solve $2(y^2 - y) = 24$ using any method.	Section 9.5; Example 2b
10. Solve $\sqrt{x + 3} + 3 = x$.	Section 9.6; Example 4

REVIEW PROBLEMS

Find the solution set.

1. $(x - 7)(x - 3) = 0$
2. $x(x + 4) = 0$
3. $x(x + 6) = 0$
4. $(t + 6)(t - 5) = 0$
5. $(3z + 4)(z + 7) = 0$
6. $(4x - 7)(3x + 5) = 0$
7. $x(5x - 9) = 0$
8. $(z - 5)(z + 5) = 0$
9. $(t - \sqrt{5})(t + \sqrt{5}) = 0$
10. $(z + \sqrt{15})(z - \sqrt{15}) = 0$

Find the solution set by factoring.

11. $x^2 + 7x + 10 = 0$
12. $t^2 - 7t + 6 = 0$
13. $z^2 + 2z = 24$
14. $x^2 - 3x - 28 = 0$
15. $x^2 = 2x$
16. $2x^2 + x = 10$
17. $3x^2 + 2x - 1 = 0$
18. $10z^2 = z + 3$
19. $w^2 = 169$
20. $4x^2 - 13x + 9 = 0$

Find the solution set by the Square Root Property.

21. $x^2 = -49$
22. $(x + 9)^2 = -4$
23. $(t - 2)^2 = 7$
24. $(x + 2)^2 = 32$

Find the solution set by completing the square.

25. $x^2 + 6x + 8 = 0$
26. $x^2 - 2x - 15 = 0$
27. $2x^2 + 4x - 16 = 0$
28. $3x^2 - 6x = 12$
29. $2x^2 + 6x = 6$
30. $3x^2 + 18x + 9 = 0$

Find the solution set by the Quadratic Formula.

31. $x^2 + x - 3 = 0$
32. $x^2 - 3x + 1 = 0$
33. $x^2 + 5x + 4 = 0$
34. $t^2 - 7t + 5 = 0$
35. $2x^2 - x + 1 = 0$
36. $3x^2 + 6x + 4 = 0$
37. $4x^2 = 6x - 2$
38. $2z^2 + 8z = -3$
39. $5x^2 = 7x - 1$
40. $4t + 2 = 3t^2$

Find the solution set by any method.

41. $(x - 3)(x + 4) = 0$
42. $z^2 - z = 0$
43. $(x - 5)^2 = 4$
44. $x^2 + 7x = 8$
45. $z^2 = 3z + 2$
46. $(x - \sqrt{3})(x + \sqrt{3}) = 0$
47. $(x + 8)^2 = -8$
48. $(x + 2)(x - 3) = -4$
49. $t^2 + t = 3$
50. $x^2 = 7x$
51. $x(x - 3) = 0$
52. $2x^2 + 9x - 5 = 0$
53. $3x^2 + 4x = 2$
54. $4z^2 = z - 1$

55. $(3x - 2)^2 = 0$

56. $(x - 4)^2 = 1$

57. $3x^2 + 6x + 1 = 0$

58. $x^2 - 12x + 36 = 0$

59. $(5x - 4)(3x + 7) = 0$

60. $(2x - 1)(3x + 2) = 10$

61. $6x^2 = 11x + 7$

62. $w^2 = 100$

63. $7x^2 = 8x - 1$

64. $10x^2 - 9x + 2 = 0$

65. $3x^2 - 5x = 0$

66. $(x + 7)(x - 3) = 0$

67. $2x^2 - 3x = 35$

68. $x^2 + 10x + 25 = 0$

69. $3x^2 + 7x + 3 = 0$

70. $3x^2 - 5x = 25$

71. $x^2 - 121 = 0$

72. $(x - 5)^2 = 7$

73. $(x + 1)^2 = 8$

74. $(x + 11)^2 = 24$

75. $x^2 = 144$

76. $x^2 + x = 0$

Find the solution set.

77. $\sqrt{x + 4} = 5$

78. $\sqrt{x + 12} = x$

79. $\sqrt{x^2 + 7} = x + 1$

80. $\sqrt{7x + 15} - x = 3$

81. $\sqrt{x} + 3 = 0$

82. $\sqrt{5x + 11} = x + 1$

83. $\sqrt{2x + 8} = x + 4$

84. $\sqrt{x^2 - 2x + 1} = \sqrt{2x - 2}$

85. $\sqrt{x + 2} = \sqrt{x - 7}$

86. $\sqrt{x^2} = 5$

Solve each word problem.

87. Find two consecutive negative integers whose product is 306.

88. One leg in a right triangle is 3 units less than the other leg. If the length of the hypotenuse is $\sqrt{65}$, find the length of the legs.

89. The length of a rectangle is 5 units more than the width of the rectangle. If the area of the rectangle is 204 square units, find the dimensions of the rectangle.

90. One leg of a right triangle is 2 meters less than the other leg. If the area of the triangle is 24 square meters, find the lengths of both legs.

...LET'S NOT FORGET...

In Problems 91 through 95, perform the operations indicated.

91. -4^2

92. $(\sqrt{2} + \sqrt{3})^2$

93. $\dfrac{-2x^2}{y} \div \dfrac{x^3}{-y^3}$

94. $\dfrac{4(a - b)}{b - a}$

95. $\dfrac{0}{-7}$

In Problems 96 through 98, simplify each expression. Watch the role of the parentheses.

96. $(\sqrt{2} + \sqrt{5})(\sqrt{2} - \sqrt{5})$

97. $\sqrt{2} + \sqrt{5}(\sqrt{2} - \sqrt{5})$

98. $\dfrac{x - 4}{x + 3} - \dfrac{x - 2}{x + 3}$

In Problems 99 through 102, determine if $\sqrt{3}$ *is a* factor *or a* term.

99. $\sqrt{3} + 2$

100. $\sqrt{3}(x + 2)$

101. $\dfrac{1}{\sqrt{3} + 2}$

102. $x(y + \sqrt{3})$

In Problems 103 through 105, multiply each expression by $\dfrac{x}{7}$.

103. $-\dfrac{14}{x^2}$

104. $\dfrac{21}{x}(x - 7)$

105. $x^2 + 7x$

In Problems 106 through 110, label each problem as an expression, equation, *or* inequality. *Solve the equations and inequalities and perform the indicated operation with the expressions.*

106. $\dfrac{2}{x + 1} - \dfrac{1}{x + 3}$

107. $\dfrac{2}{x + 1} - \dfrac{1}{x + 3} = \dfrac{2x}{(x + 1)(x + 3)}$

108. $(x + 1)(x + 3) \leq 0$

109. $\dfrac{\dfrac{2}{x + 1}}{\dfrac{1}{x + 3}}$

110. $\sqrt{x + 2} = x$

In Problems 111 through 114, identify which expressions are factored. Factor those that are not factored, if possible.

111. $y\sqrt{3} + x\sqrt{3}$

112. $x^3 + y^3$

113. $(x + y)^3$

114. $a(b + c)$

CHAPTER 9 TEST

In Problems 1 through 5, choose the letter of the correct answer.

1. The solution set for $x^2 - 2x - 8 = 0$ is (?).

A. $\{4, 8\}$ B. $\{-2, 4\}$
C. $\{-4, 2\}$ D. $\{2, 8\}$

2. The solution set for $2x^2 - 3x - 3 = 0$ is (?).

A. $\left\{\dfrac{3}{2}, -1\right\}$ B. $\left\{\dfrac{-3 \pm \sqrt{33}}{4}\right\}$
C. $\left\{\dfrac{3 \pm \sqrt{33}}{4}\right\}$ D. $\left\{\dfrac{1}{2} \pm \dfrac{\sqrt{33}}{4}\right\}$

3. $\dfrac{-3 \pm \sqrt{12}}{12} = (?)$.

A. $-\dfrac{1}{4} \pm \dfrac{\sqrt{3}}{6}$ B. $-\dfrac{1}{4} \pm \dfrac{\sqrt{3}}{2}$

C. $-\dfrac{1}{4} \pm \dfrac{\sqrt{3}}{3}$ D. $\dfrac{-1 \pm \sqrt{3}}{2}$

4. The solution set for $\sqrt{x + 1} = \sqrt{x^2 - x - 2}$ is (?).

A. $\{\pm 3\}$ B. $\{3\}$

C. $\{-1, 3\}$ D. $\{\ \}$

5. The solution set for $(x + 2)^2 = 9$ is (?).

A. $\{\pm 3\}$ B. $\{-5, 1\}$

C. $\{-11, 7\}$ D. $\{1\}$

In Problems 6 through 12 find the solution set.

6. $x^2 - 4x - 12 = 9$

7. $(x - 5)^2 = -7$

8. $\sqrt{2x^2 - 2x} = x + 3$

9. $2x^2 - 3x + 2 = 0$

10. $x^2 = 100$

11. $2x^2 - 2x - 1 = 0$

12. $x^2 - 6x = 0$

13. Find the solution set for $3x^2 - 6x - 3 = 0$ *by completing the square.*

In Problems 14 and 15, perform the operations indicated.

14. $(5 - 3i) - (2 + 7i)$

15. $\dfrac{2 - i}{3 + i}$

16. The hypotenuse of a right triangle is 13 m. If one leg is 7 m more than the other leg, find the length of the legs.

Answers to Selected Problems

CHAPTER 0

PROBLEM SET 0.1

Warm-ups

1. $3 + 4$ **3.** $3(9)$ **5.** $21/8$ **7.** $\frac{1}{2}(17)$ **9.** $1 + 5$ **11.** $2(5)$ **13.** 2^3 **15.** $2(13) - 1$ **17.** 9
19. 125 **21.** 12 **23.** 2 **25.** 38 **27.** 4 **29.** 11 **31.** 1 **33.** 9 **35.** 1

Practice Exercises

37. 12 **39.** 56 **41.** 7 **43.** 144 **45.** 3 **47.** 30 **49.** 163 **51.** 16 **53.** 1 **55.** 14
57. 14 **59.** 16 **61.** 5 **63.** 1 **65.** 7 **67.** 5 **69.** 36 **71.** 8 **73.** 1 **75.** 24
77. They have 14 books. **79.** Ellen has 14 quarters. **81.** They have 16 records. **83.** Choi has 36 nickels.

PROBLEM SET 0.2

Warm-ups

1. $2^3 \cdot 3$ **3.** 5^2 **5.** 2^3 **7.** $5 \cdot 7$ **9.** $3^2 \cdot 5$

Practice Exercises

11. $2^2 \cdot 3$ **13.** $2 \cdot 5^2$ **15.** 3^4 **17.** $2 \cdot 5 \cdot 7$ **19.** $2 \cdot 3^2 \cdot 5$ **21.** $2^3 \cdot 5^2$ **23.** $2^3 \cdot 7$ **25.** $2 \cdot 3 \cdot 5^2$
27. $2^5 \cdot 3$ **29.** $2^2 \cdot 3^3$ **31.** $3^2 \cdot 7^2$ **33.** $3^4 \cdot 7^2$ **35.** $2^2 \cdot 3^5$ **37.** 13^2 **39.** $2^6 \cdot 3 \cdot 5$
41. He bought 18 pictures. **43.** Branden has 6 coins. **45.** Irene bought 14 pencils. **47.** The board is 15 m.

PROBLEM SET 0.3

Warm-ups

1. $\frac{2}{3}$ **3.** $\frac{5}{10}$ **5.** $\frac{24}{18}$ **7.** $\frac{1}{4}$ **9.** $\frac{4}{7}$ **11.** $3\frac{1}{16}$ **13.** $\frac{9}{4}$ **15.** $1\frac{1}{33}$ **17.** 21 **19.** 24 **21.** 108
23. $\frac{3}{5}$ **25.** $\frac{5}{12}$ **27.** $\frac{7}{24}$ **29.** $\frac{29}{36}$ **31.** $1\frac{6}{7}$ **33.** $\frac{10}{7}$ **35.** $\frac{1}{12}$ **37.** $\frac{2}{5}$ **39.** $\frac{53}{10}$ **41.** 27
43. The sum is $6\frac{7}{9}$. **45.** Yes, she has more than 15 lbs.

Practice Exercises

47. $\frac{4}{15}$ **49.** $\frac{9}{5}$ **51.** $\frac{1}{8}$ **53.** $\frac{1}{100}$ **55.** $\frac{5}{14}$ **57.** $\frac{5}{18}$ **59.** $\frac{32}{45}$ **61.** 4 **63.** 4 **65.** $2\frac{1}{4}$
67. $\frac{7}{4}$ **69.** 8 **71.** $\frac{1}{6}$ **73.** $\frac{1}{4}$ **75.** $\frac{15}{16}$ **77.** $\frac{3}{25}$ **79.** $\frac{1}{4}$ **81.** $\frac{13}{36}$ **83.** $\frac{7}{144}$
85. He has $12\frac{1}{4}$ lb of fertilizer. **87.** She talked $21\frac{1}{2}$ minutes.

PROBLEM SET 0.4

Warm-ups

1. 0.6 **3.** $1.1\overline{6}$ **5.** $2.\overline{6}$ **7.** 111.45 **9.** 2 **11.** 22.565 **13.** 91% **15.** 9% **17.** 14%
19. $66\frac{2}{3}\%$ **21.** 37.5% **23.** $\frac{11}{100}$; 0.11 **25.** $\frac{2}{25}$; 0.08 **27.** $\frac{9}{40}$; 0.225 **29.** $\frac{1}{30}$; $0.0\overline{3}$ **31.** 40
33. 196 **35.** 5227.2 **37.** $4250 **39.** 148.4

Practice Exercises

41. 2.25 **43.** $1.8\overline{3}$ **45.** $\frac{39}{50}$, 0.78 **47.** $\frac{1}{10}$, 0.1 **49.** $\frac{1}{50}$, 0.02 **51.** $\frac{11}{100}$, 0.11 **53.** $\frac{23}{200}$, 0.115
55. $\frac{1}{15}$, $0.0\overline{6}$ **57.** 97% **59.** 3% **61.** 22% **63.** $22.\overline{2}\%$ **65.** 62.5% **67.** 3400 **69.** 21
71. 4.7 **73.** 642 **75.** 8.84 **77.** 46.9945 **79.** 10.1 **81.** 193.7 **83.** 20.5 **85.** 19.32
87. 3 **89.** 4.41 **91.** 0.01 **93.** 209.1204 **95.** 9.3875 **97.** 46.033 **99.** He has $51.55.
101. No, the sum of her purchases is $9.12. **103.** The ring is worth $5,900. **105.** It will be $52.50.
107. He has $141.20. **109.** It is $9.16 per pound. **111.** No, she bought 14.97 lb.
113. The sale price was $47.20. **115.** He should study $\frac{1}{2}$ hour longer each day.

PROBLEM SET 0.5

Warm-ups

1. 5 **3.** 10 **5.** 7 **7.** 13 **9.** 5 **11.** 14 **13.** 15 **15.** 16

Practice Exercises

17. 8 **19.** 30 **21.** 100 **23.** 60 **25.** $\sqrt{17} \approx 4.123$ **27.** $\sqrt{48} \approx 6.928$ **29.** 1
31. It is 13 units. **33.** It is 204.

Challenge Problems

35. $\sqrt{200} \approx 14.14$; $\sqrt{2} \approx 1.41$

PROBLEM SET 0.6

Warm-ups

1. $\angle a = 45°$ $\angle b = 135°$ $\angle c = 135°$ $\angle d = 45°$ $\angle e = 135°$ $\angle f = 45°$ $\angle g = 45°$ $\angle h = 135°$ $\angle i = 90°$ $\angle j = 90°$ $\angle k = 90°$ $\angle l = 90°$ $\angle m = 135°$ $\angle n = 45°$ $\angle o = 45°$ $\angle p = 135°$ $\angle q = 45°$ $\angle r = 135°$ $\angle s = 135°$ $\angle t = 45°$ **3.** 120° **5.** 30°

7. (a) Two angles of one triangle are equal to two angles of the other triangle. (b) Two angles of one triangle are equal to two angles of the other triangle.

Practice Exercises

9. False **11.** False **13.** False **15.** False **17.** False **19.** False **21.** 125° **23.** 20° **25.** 64° **27.** $13^2 = 169$; $5^2 + 12^2 = 169$ **29.** Corresponding sides are proportional. (Similar, but not congruent) **31.** Side-Angle-Side (Congruent) **33.** False **35.** True **37.** True

Challenge Problems

39. Three less than the number of sides

PROBLEM SET 0.7

Warm-ups

1. The perimeter is $11\frac{1}{2}$ ft. The area is $5\frac{5}{8}$ sq ft. **3.** The perimeter is 32 cm. The area is $32\frac{15}{16}$ sq cm. **5.** The perimeter is $12 + \pi$ in. The area is $10 + \frac{\pi}{2}$ sq in. **7.** The volume is 8π m^3.

Practice Exercises

9. The perimeter is $13\frac{1}{2}$ cm. The area is $7\frac{7}{8}$ sq cm. **11.** Allen needs 310 sq ft of paint. **13.** Yes, he needs only $232\frac{1}{2}$ m. **15.** She needs 372 sq ft of wall paper. **17.** The area of the poster is 324.3 sq in. **19.** The perimeter is 10 cm. **21.** The area is 0.16π sq in. **23.** The volume is 288π in^3.

REVIEW PROBLEMS CHAPTER 0

1. $\frac{4}{21}$ **3.** 15 **5.** $\frac{9}{2}$ **7.** 0 **9.** $\frac{19}{3}$ **11.** $2(5) + 3$ **13.** $\frac{1}{2} \cdot 7 + 2 \cdot 4$ **15.** $15 - (4 + 6)$ **17.** She bought 17.22 lb of meat. **19.** It is 150. **21.** $\frac{3}{20}$; 0.15 **23.** $\frac{23}{200}$; 0.115 **25.** $\frac{5}{6}$; $0.8\overline{3}$ **27.** 18% **29.** $83\frac{1}{3}$% **31.** 2^4 **33.** $2^2 \cdot 5^2$ **35.** $2^4 \cdot 5^2$ **37.** 70 **39.** 80 **41.** ≈ 6.325 **43.** 21 **45.** The circumference is $13\frac{1}{2}\pi$ m and the area is $45\frac{9}{16}\pi$ sq m.

CHAPTER 0 TEST

1. A **2.** D **3.** A **4.** B **5.** B **6.** 11 **7.** 10.81 **8.** $\frac{11}{27}$ **9.** 13 **10.** 3 **11.** 13.5 **12.** The area is $3\frac{3}{10}$ sq in. **13.** 37.5% **14.** $3\frac{1}{4}$ **15.** She bought $13\frac{5}{8}$ yd.

CHAPTER 1

PROBLEM SET 1.1

Warm-ups

1. ∅, {1}, {4}, {1, 4} **3.** 1, 2, 4, 6, 8 **5.** 4 **7.** rational **9.** irrational **11.** T **13.** F **15.** F

17. 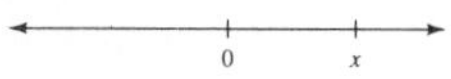

19. 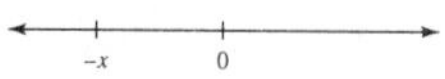

21. 3.5 **23.** −11/13 **25.** 6 **27.** π **29.** 12 **31.** 16 **33.** −4 **35.** −16

Practice Exercises

37. 1/2, 2, 9/4, 5 **39.** ∅, {9/4}, {5}, {9/4, 5} **41.** T **43.** T **45.** T **47.** F **49.** T

51. 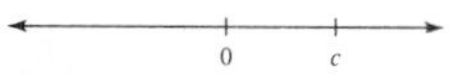

53. 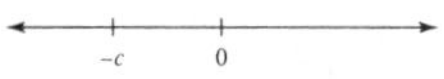

55. T **57.** T **59.** T **61.** 2 **63.** 8 **65.** 5 **67.** 3.14 **69.** −8 **71.** −37 **73.** 48, 49, 50
75. 37, 39, 41

Challenge Problems

77. 1/2 **79.** 1, 2, 3, 4, 5, 6, 7, 8, 9 **81.** 5, 10, 15, . . . **83.** $\{x \mid x$ is an even natural number$\}$
85. $\{x \mid x$ is an integer between -4 and $4\}$

PROBLEM SET 1.2

Warm-ups

1. 8 **3.** 4 **5.** −13 **7.** −1 **9.** 4 **11.** 0 **13.** −1 **15.** $\frac{5}{24}$ **17.** $-\frac{5}{3}$ **19.** 2.2 **21.** 14

23. 1 **25.** −4 **27.** −4.9 **29.** −4 **31.** $-\frac{7}{8}$ **33.** 10 **35.** −19.8 **37(a)** 17 **(b)** 23

39. 2 **41.** 4° **43.** no

Practice Exercises

45. 9 **47.** 4 **49.** 5 **51.** −5 **53.** 0 **55.** 16 **57.** 4 **59.** 3 **61.** 0 **63.** 50 **65.** −3

67. 0 **69.** $-\frac{2}{5}$ **71.** $-\frac{1}{15}$ **73.** $-\frac{7}{4}$ **75.** 0.6 **77.** 10.5 **79.** 0 **81.** −4 **83.** −19 **85.** $\frac{2}{3}$

87. −7 **89.** −18 **91.** −10.1 **93.** 1 **95.** −14 **97.** −12 **99.** −2 **101(a)** 11 **(b)** 23
103. −22 **105.** 15 **107.** 6° **109.** $60

PROBLEM SET 1.3

Warm-ups

1. 2 **3.** −10 **5.** 26 **7.** 3 **9.** −1 **11.** 0 **13.** −25 **15.** 21 **17.** $\frac{17}{8}$ **19.** $-\frac{5}{3}$

21. $-\frac{1}{2}$ **23.** −7.1 **25.** 7 **27.** −4 **29.** $\frac{7}{8}$ **31.** 15 **33.** $\frac{2}{15}$ **35.** −7 **37.** 1 **39.** 12

41. −17 **43.** −15°

Practice Exercises

45. 3 **47.** −5 **49.** 4 **51.** 0 **53.** −4 **55.** 0 **57.** 12 **59.** 7 **61.** −5 **63.** 27 **65.** 0

67. 50 **69.** 14 **71.** -14 **73.** $\frac{1}{8}$ **75.** $-\frac{8}{5}$ **77.** $-\frac{1}{2}$ **79.** -5.9 **81.** 13 **83.** 6 **85.** -23
87. 20 **89.** 58 **91.** 9 **93.** -4 **95.** $-18°$

PROBLEM SET 1.4

Warm-ups

1. 6 **3.** $-\frac{1}{4}$ **5.** -11 **7.** 25 **9.** 20 **11.** 0 **13.** 5 **15.** undefined **17.** -5 **19.** 0
21. -2 **23.** -15 **25.** -5 **27.** undefined **29.** -2 **31.** 12 **33.** 1 **35.** -12 **37.** -4
39. 6 **41.** 97 liters

Practice Exercises

43. 0 **45.** -24 **47.** 49 **49.** 36 **51.** 32 **53.** -24 **55.** d **57.** 100 **59.** -35 **61.** $\frac{25}{36}$
63. -2 **65.** -4 **67.** -9 **69.** 0 **71.** 1 **73.** 8 **75.** 0 **77.** -9 **79.** -1
81. undefined **83.** 72 **85.** -16 **87.** -5 **89.** 25 **91.** -12 **93.** 11.2 lbs

Challenge Problems

95. $14 \div (5 \cdot 11)$

PROBLEM SET 1.5

Warm-ups

1. -3 **3.** 2 **5.** 3 **7.** $-\frac{29}{3}$ **9.** -2 **11.** -1 **13.** 0 **15.** -1 **17.** 38 **19.** 8

21. term **23.** term **25.** factor **27.** 18 **29.** 4 **31.** -3 **33.** 12 **35.** $\frac{1}{4}$ **37.** -27

Practice Exercises

39. -4 **41.** 7 **43.** 14 **45.** 6 **47.** -13 **49.** 1 **51.** -2 **53.** -6 **55.** 1 **57.** 20
59. 1 **61.** term **63.** term **65.** term **67.** 23, 19 **69.** 20, 20 **71.** $-4, -4$ **73.** $48, -96$

Challenge Problems

75. not equal **77.** equal

PROBLEM SET 1.6

Warm-ups

1. commutative for multiplication **3.** associative for addition **5.** distributive **7.** multiplicative inverse
9. commutative for addition **11.** identity for multiplication **13.** additive inverse **15.** multiplicative inverse
17. x **19.** y **21.** 874 **23.** -4 **25.** $-x, \frac{1}{x}$ **27.** $-\pi, \frac{1}{\pi}$ **29.** $-1, 1$ **31.** $3x - 12$
33. $-4x + 8$ **35.** $xy - xz$

Practice Exercises

37. associative for addition **39.** associative for multiplication **41.** identity for multiplication
43. multiplicative inverse **45.** identity for addition **47.** identity for addition **49.** multiplicative inverse
51. $\frac{x}{3}$ **53.** 148 **55.** 12 **57.** $\frac{3}{4}z$ **59.** $-\frac{5}{3}, \frac{3}{5}$ **61.** 0, no multiplicative inverse **63.** $8 + 2x$
65. $-12 - 2y$ **67.** $xy + 3x$ **69.** $xy - 5x$ **71.** $3t - tx$ **73.** $3n + 2nx$ **75.** $-9c + cx$

Challenge Problems

77. no **79.** $x^2 + 3x$ **81.** $6x^3$

PROBLEM SET 1.7

Warm-ups

1. $41 + 6y^2$ **3.** $8xy$ **5.** $10/x^2$ **7.** $2x - 2y$ **9.** $\dfrac{x+4}{x-4}$ **11.** $x - 4$ in. are left. **13.** $x - 17$ green marbles
15. It has $x - 70$ quarters. **17.** Don is 51. **19.** $(x + 6)$ yrs old **21.** $7x^2$ **23.** $-3abc$
25. $-3xy + 3$ **27.** $13x^2y - 6xy^2$ **29.** $-x^2z + 39z^2$ **31.** $2\sqrt{x}$ **33.** $29x - 35$ **35.** $-x - 7$
37. $24b - 9$ **39.** $10\sqrt{x-1}$ **41.** The sum is 219. **43.** The sum is $3x + 6$. **45.** The sum is $3x - 6$.

Practice Exercises

47. $6x + 23$ **49.** $\frac{1}{3}(2t)$ **51.** $-4x^3$ **53.** $10z$ **55.** $10x$ **57.** $-15x^2$ **59.** $24 + 11v$ **61.** $7x^2$
63. $2x^3 + 3x^2$ **65.** $-5x - 3$ **67.** $6x + 3$ **69.** $3x + 2$ **71.** $11\sqrt{2x}$ **73.** $10x - 25$
75. $17y^2 - 20$ **77.** $15x - 4$ **79.** $5x - 18$ **81.** $8 + 14a$ **83.** $-4 + 4y$ **85.** $10 - 12x$
87. \$5,700 **89.** $\angle B$ is 74°and $\angle C$ is 69° **91.** 37 miles **93.** 60 quarters **95.** $102 - x$ quarters
97. Its measure is 124°. **99.** Som is 25 yrs old. **101.** $x + 13$ yrs old
103. They are $x + 2$, $x + 4$ and $x + 6$. **105.** The sum is 84.

Challenge Problems

107. $12\sqrt{x^2+3} - 5\sqrt{x^2-5}$ **109.** $-3\sqrt{x+4} - 2\sqrt{2y}$

PROBLEM SET 1.8

Warm-ups

1. no **3.** no **5.** no **7.** yes **9.** $\{5\}$ **11.** $\{5\}$ **13.** $\{7\}$ **15.** $\{-5\}$ **17.** $\{0\}$ **19.** $\{6\}$
21. $\{2\}$ **23.** $\{26\}$ **25.** $\left\{\frac{27}{2}\right\}$ **27.** $\{-5\}$ **29.** $\left\{-\frac{1}{2}\right\}$ **31.** $\left\{-\frac{5}{3}\right\}$ **33.** y is -18 **35.** x is -5
37. y is 6 **39.** t is 75

Practice Exercises

41. yes **43.** no **45.** no **47.** no **49.** $\{-1\}$ **51.** $\{15\}$ **53.** $\{9\}$ **55.** $\{6\}$ **57.** $\{20\}$ **59.** $\left\{\frac{5}{2}\right\}$
61. $\left\{\frac{7}{4}\right\}$ **63.** y is -2 **65.** x is -7 **67.** 13 **69.** t is -35

Challenge Problems

71. $p + 2$ **73.** $\frac{p}{5}$ **75.** $\frac{6}{p}, p \neq 0$

REVIEW PROBLEMS CHAPTER 1

1. -2 **3.** -9.2 **5.** -3 **7.** 17 **9.** -8 **11.** -1 **13.** 0 **15.** -5 **17.** -18 **19.** 29
21. -7 **23.** 30 **25.** They gained 9 yds. **27.** $|8 + 11|$ **29.** $\{2 \cdot 7\}^3$ **31.** $\{10\}$ **33.** $\{15\}$
35. $\{-8\}$ **37.** T **39.** T **41.** F **43.** F **45.** T

Let's Not Forget

46. 1 **47.** 9 **48.** -1 **49.** 0 **50.** -5 **51.** 0 **52.** 6 **53.** -80 **54.** -53 **55.** -17
56. 7 **57.** $\pi - 2$ **58.** $2\pi - 1$ **59.** expression **60.** equation, $\{-6\}$ **61.** equation, $\{5\}$
62. expression **63.** expression **64.** equation, $\{11\}$

CHAPTER 1 TEST

1. D **2.** D **3.** A **4.** B **5.** D **6.** T **7.** T **8.** T **9.** F **10.** F **11.** 9 **12.** 2
13. -7 **14.** 37 **15.** -1 **16.** 12 **17.** 6 **18.** 41 **19.** 4 **20.** -7
21. a. 5, b. 4, c. 3, d. 4, e. 2, f. 1 **22.** a. $-5, 0, 102$ b. $-5, \frac{-3}{5}, 0, \frac{17}{3}, 12.7, 102$ c. $\sqrt{3}$ **23.** $\{25\}$
24. $\{-4\}$ **25.** $\{6\}$

CHAPTER 2

PROBLEM SET 2.1

Warm-ups

1. 4; 16 **3.** -4; 16 **5.** 4; -64 **7.** x^4 **9.** Can't be simplified. **11.** y^{25} **13.** z^6
15. Can't be simplified. **17.** y^6 **19.** x^7y^7 **21.** $121k^2$ **23.** $-27t^3$ **25.** $\frac{x^8}{y^8}$ **27.** $\frac{x^3}{125}$
29. $\frac{-125}{p^3}$ **31.** x^8 **33.** k^{21} **35.** 1 **37.** xz^{14} **39.** $\frac{125}{8t^3}$ **41.** $\frac{1}{49a^6}$ **43.** $4x^2y^2$
45. Can't be simplified. **47.** $4x^4y^6$ **49.** $4x^4y^6$ **51.** $64p^9q^3$ **53.** $\frac{36x^{14}t^2}{s^{12}}$ **55.** 1 **57.** 1
59. -12

Practice Exercises

61. 7; -49 **63.** -7; 49 **65.** -7; 1 **67.** $-k^3$ **69.** 1 **71.** y^{36} **73.** p^9
75. Can't be simplified. **77.** 2^6 **79.** s^4t^4 **81.** $169x^2$ **83.** $\frac{r^6}{64}$ **85.** x^2 **87.** q **89.** 1
91. ab^{36} **93.** $\frac{64r^3}{27}$ **95.** $256x^{12}y^{16}$ **97.** $-8x^6y^9$ **99.** $\frac{1}{7a}$ **101.** $\frac{-216x^{21}t^3}{s^{18}}$

Challenge Problems

103. (a) 1; -1; 1; -1; 1 (b) $(-1)^n = 1$ if n is even and $(-1)^n = -1$ if n is odd. **105.** 1

PROBLEM SET 2.2

Warm-ups

1. $\frac{1}{16}$ **3.** $\frac{1}{2}$ **5.** $\frac{1}{y^5}$ **7.** $\frac{1}{16}$ **9.** $-\frac{1}{4}$ **11.** $-\frac{1}{125}$ **13.** $-\frac{1}{25}$ **15.** $\frac{2}{x^3}$ **17.** $\frac{-2}{x^3}$ **19.** 2
21. $\frac{1}{16}$ **23.** x^4 **25.** $\frac{1}{x^2}$ **27.** $\frac{1}{64}$ **29.** 64 **31.** $\frac{1}{x^{10}}$ **33.** $\frac{1}{x^3y^3}$ **35.** $\frac{1}{9x^2}$ **37.** $\frac{1}{16t^2}$ **39.** $\frac{1}{100}$
41. 125 **43.** $\frac{1}{r^7}$ **45.** $\frac{1}{k^2}$ **47.** $\frac{32}{b^2}$ **49.** 1.2345×10^{-6} **51.** 7.7722×10^{13} **53.** -2.1367×10^{-8}
55. 0.0000543 **57.** 681 **59.** -0.0432101 **61.** 200 **63.** 0.00000055

Practice Exercises

65. $\frac{1}{27}$ **67.** $\frac{1}{x^{11}}$ **69.** $\frac{1}{T}$ **71.** $-\frac{1}{27}$ **73.** $-\frac{1}{4}$ **75.** $-\frac{1}{125}$ **77.** $-\frac{1}{64}$ **79.** $\frac{4}{x^2}$ **81.** $\frac{6}{x^7}$
83. $\frac{-8}{x^2}$ **85.** $\frac{b}{t^3}$ **87.** $\frac{k^2}{x^2}$ **89.** $\frac{3}{a^3t^2}$ **91.** 3 **93.** 4 **95.** $\frac{1}{81}$ **97.** 100 **99.** $\frac{1}{z}$ **101.** $\frac{1}{v^5}$
103. $\frac{1}{3^6}$ **105.** 49 **107.** $\frac{1}{5^{15}}$ **109.** $\frac{1}{x^6}$ **111.** x^{36} **113.** $\frac{1}{32x^5}$ **115.** $\frac{1}{64r^6}$ **117.** $-\frac{1}{b^4w^4}$
119. $\frac{1}{x^6}$ **121.** $\frac{1}{w^2}$ **123.** 3 **125.** 1.86×10^5 **127.** 1.429×10^{-3} **129.** 2.03×10^4
131. 1.0×10^{-9} **133.** 0.00123 **135.** 0.091 **137.** 34,500,000,000

Challenge Problems

139. 4 **141.** $\frac{11}{3}$ **143.** $\frac{27}{y^3}$ **145.** $\frac{15}{4}$ **147.** -1

PROBLEM SET 2.3

Warm-ups

1. $x + 6$; 1; binomial **3.** $-3x^3 + 4x^2 + \frac{1}{2}x - 7$; 3 **5.** $x^5 - 25$; 5; binomial **7.** $y^7 - 5y^5 + 6y - 1$; 7
9. 13 **11.** 6 **13.** 13 **15.** 1 **17.** 7 **19.** 29 **21.** $10x$ **23.** $4x$ **25.** $3w^3$ **27.** $3a$
29. $-21x^2$ **31.** $8x + 3$ **33.** $7r^3 - 5$ **35.** $2x^2 - 4x + 6$ **37.** $y^2 - 5y + 1$ **39.** $2x^2 + 3y^2$
41. $8x - 3$ **43.** $-10x^3 - 1$ **45.** 0 **47.** $-10x + 2$ **49.** $3.0z - 6.3$

Practice Exercises

51. $-x + 5$; 1; binomial **53.** $x^9 - 2x^8$; 9; binomial **55.** $-y^7 + 5y^5 - 7y - 1$; 7 **57.** 16 **59.** 9
61. -48 **63.** -2 **65.** 21 **67.** 105 **69.** $4x$ **71.** $-x$ **73.** $-4w^3$ **75.** $2c$ **77.** $-6z$
79. $5x$ **81.** $5r^3 - 6$ **83.** $2x + 31$ **85.** $5y^2 - 9y - 5$ **87.** $6x^2 + 4y^2$ **89.** $x + 2$

Challenge Problems

91. 0 if n is odd; 2 if n is even

PROBLEM SET 2.4

Warm-ups

1. $7x + 4$ **3.** $4x^2 - 7$ **5.** $8z^2 - 3$ **7.** $12x - 4$ **9.** $-t^2 - 7$ **11.** $3y^2 + 6y - 12$
13. $5x^2 - 3x - 3$ **15.** $4x^5 + 4x^4 + x^3 - x^2 + 8x + 2$ **17.** $-3x - 1$ **19.** $-6x + 9$ or $9 - 6x$
21. $-x^2 + 8x + 7$ **23.** $-7x^2 - 5$ **25.** $x - 4$ **27.** $x + 3$ **29.** $-2x^2 - 13$ **31.** $4x^4 - 3$
33. $x^3 - 2x^2 + x$ **35.** $x^2 + 2x - 1$ **37.** $4x^3 + 6x^2 + 7x - 10$ **39.** $2x + 6$ **41.** $-2x^2 - 12$
43. $-6x + 11$ **45.** $-10x - 2$ **47.** $5x^2 + 3x - 2$ **49.** $3x^2 - 4x + 9$ **51.** $6 - 2x$
53. The perimeter is $8s + 12$ ft. **55.** The circumference is $2\pi r - 4\pi$ inches. **57.** They have $7x - 2$ dollars.

Practice Exercises

59. $24x + 4$ **61.** $6y^2 + 4y - 17$ **63.** $-2x^2 - 11x + 4$ **65.** $15x^3 + 2x^2 + 4x - 8$
67. $8x^5 - 3x^4 + 13x^3 - x^2 + 7x + 12$ **69.** $6x^3 - 9x^2 + 15x - 13$ **71.** $24x^5 + 14x^4 - 4x^3 - 10x$
73. $x^2 - 15x + 10$ **75.** $z^2 - 2z - 7$ **77.** $-2x^2 - 3x + 17$ **79.** $2v^4 - 2v^3 - 4v^2 + 6v - 12$
81. $5x^2 - 4$ **83.** $6x^2 - 4x + 8$ **85.** $8t^3 - 3t^2 + 12$ **87.** $7x^2 + 5x - 5$
89. The length of the ribbon is $r^2 + 4r + 2$ inches. **91.** The distance between them is $900x + 400$ km.
93. The sum of their ages is $3y + 7$ years.

Challenge Problems

95. $7x^n$

PROBLEM SET 2.5

Warm-ups

1. $10x^5$ **3.** $-24x^7$ **5.** z^6 **7.** $-12s^4$ **9.** $8z^3$ **11.** $8x + 12$ **13.** $9x^8 - 3x^5 - 6x^4$
15. $-15x^4 + 35x^7$ **17.** $x^2 - x - 20$ **19.** $y^2 - 8y + 12$ **21.** $5x^2 + 13x + 8$ **23.** $5z^2 + 31z - 72$
25. $6x^2 + 29x + 9$ **27.** $x^3 - x^2 - 2x - 12$ **29.** $6t^3 - 7t^2 - 10t + 6$ **31.** $x^4 + 7x^3 - 4x^2 - 21x + 3$
33. $x^3 - 9x^2 + 27x - 27$ **35.** $2y^3 + 3y^2 - 18y - 27$ **37.** $2y^4 - y^3 - 3y^2$ **39.** The area is $x^2 + 3x$ sq ft.
41. The area is $3h^2 - 2h$ sq yd. **43.** The product is $x^2 + 2x$. **45.** The product is $y^2 - y$.

Practice Exercises

47. $18r^6$ **49.** $-24w^5$ **51.** $21x^{10}$ **53.** $-2x^2$ **55.** $4x - 20$ **57.** $6y - 18y^2 + 2y^3$
59. $-21t^3 + 24t^2 - 6$ **61.** $x^2 + 7x + 10$ **63.** $t^2 - 9$ **65.** $3t^2 + 29t + 18$ **67.** $2x^2 - x - 10$
69. $12x^2 - x - 6$ **71.** $6r^2 + 11r - 35$ **73.** $18 - 2x^2$ **75.** $x^2 - 16$ **77.** $y^2 + 10y + 25$
79. $5t^2 - 38t + 21$ **81.** $15x^2 - 19x - 10$ **83.** $6s^2 - s - 35$ **85.** $6 + 13x - 8x^2$ **87.** $x^3 - 1$
89. $x^3 + 64$ **91.** $6x^3 - 7x^2 - 11x + 12$ **93.** $s^3 - 3s^2 + 3s - 1$ **95.** $3t^3 - 75t$
97. The area is $2x^2 - x$ sq cm. **99.** The product is $z^2 + 6z + 8$.

Challenge Problems

101. $\frac{1}{4}x^2 - 1$ **103.** $\frac{2}{9}z^2 + z - 2$ **105.** $p^3 + q^3$

PROBLEM SET 2.6

Warm-ups

1. $x^2 + 2x + 1$ **3.** $x^2 - 2x + 1$ **5.** $16s^2 + 24s + 9$ **7.** $4x^2 - 4x + 1$ **9.** $81 + 18y + y^2$ **11.** $x^2 - 1$
13. $4x^2 - 9$ **15.** $36x^2 - 81$ **17.** $9x^2 - 49$ **19.** $81 - 25t^2$ **21.** $-2x - 1$ **23.** $-12x^3 + 40x^2 + 30x$
25. $2x^2 + 3x + 9$ **27.** $2y^2 - y - 16$ **29.** The area is $9y^2 + 12y + 4$ sq ft. **31.** The area is $x^2 - 9$ sq ft.
33. It is $q^2 - 2q + 1$.

Practice Exercises

35. $x^2 + 6x + 9$ **37.** $y^2 - 25$ **39.** $16t^2 + 40t + 25$ **41.** $4t^2 - 49$ **43.** $4x^2 - 12x + 9$
45. $4z^2 + 4z + 1$ **47.** $121q^2 - 25$ **49.** $9w^2 - 24w + 16$ **51.** $25 - 144k^2$ **53.** $81g^2 + 36g + 4$
55. $-6x^3 + 39x^2 + 17x$ **57.** $-4z^2 + 4z - 1$ **59.** $-8w - 23$ **61.** $-3t - 18$ **63.** $-t^3 - t^2 + 7t$
65. The area is $\pi r^2 + 14\pi r + 49\pi$ sq in. **67.** The area is $x^2 - 4$ sq m.

Challenge Problems

69. $\frac{4}{9} - \frac{1}{4}x^2$

PROBLEM SET 2.7

Warm-ups

1. $a^2 - a + 1$ **3.** $1 - \frac{1}{x} + \frac{1}{x^2}$ **5.** $-x + 1$ **7.** $3x^3 - x$ **9.** $-3x + 1 - \frac{1}{x^2}$ **11.** $x + \frac{3x^2}{2}$
13. $1 + \frac{2x}{3}$ **15.** $\frac{1}{3} - \frac{x^2}{2}$ **17.** $x + 4$ **19.** $x + 3 + \frac{1}{x + 1}$ **21.** $t - 2 + \frac{2}{t + 3}$ **23.** $3x + 2$
25. $s + 2 + \frac{1}{s - 1}$ **27.** $3z - 2 + \frac{-4}{z - 3}$ **29.** $x^2 + 2x - 7 + \frac{4}{x + 2}$ **31.** $x^2 + x - 1 + \frac{1}{x + 2}$
33. $x^2 + 2x - 2 + \frac{4}{x + 2}$ **35.** $2y^2 - y + 1 + \frac{-3}{2y + 3}$ **37.** $3v^2 - 2v - 1 + \frac{-5}{2v - 1}$
39. $z^2 - 2z + 2 + \frac{-8}{4 + 2z}$

Practice Exercises

41. $\frac{3y^4}{2} - 1$ **43.** $x^3 + 4x^4$ **45.** $t^2 - 6$ **47.** $\frac{1}{4} - \frac{x^3}{3}$ **49.** $x + 3$ **51.** $x - 2$ **53.** $y + 4 + \frac{-1}{y + 3}$
55. $x + 2$ **57.** $3x - 2 + \frac{-12}{x - 3}$ **59.** $3x + 2 + \frac{-3}{x + 2}$ **61.** $x^2 + x + 1$ **63.** $t^2 - t + 2$
65. $x^2 + x + 2 + \frac{-5}{x - 2}$ **67.** $x^2 + x - 1$ **69.** $2x^2 + 3x + 5 + \frac{4}{2x - 3}$ **71.** $4x^2 + 3x + 2 + \frac{6}{4x - 3}$

Challenge Problems

73. $2x^3 - 2x^2 - 3x + 5 + \frac{-4}{2x - 1}$ **75.** $x^2 + 3x - 4$ **77.** $x^2 + 2x + 6 + \frac{-1}{x^2 + 1}$

REVIEW PROBLEMS CHAPTER 2

1. x^7 **3.** $\frac{1}{4z^8}$ **5.** $-\frac{1}{49}$ **7.** t^{10} **9.** 1 **11**(a) 0.000000163 (b) 8,503,000 **13.** -25
15. $5x^5 - 2x^4 - 3x^3 - 12x + 5$ **17.** $-5x^2 + 25x^7$ **19.** $x^3 - 2x^2 + 2x - 1$ **21.** $3x - 5$
23. $x^3 - 6x^2 + 12x - 8$ **25.** $7x - 3$ **27.** $-3x^4y^4$ **29.** $x^2 + 14x + 49$ **31.** $4t^2 - 12t + 9$
33. The length is $3x + 5$ ft. **35.** The area is $y^2 - 9$ sq in.

Let's Not Forget

36. -16 **37.** $9x - 12$ **38.** $90 - x$ **39.** $3x^2 - 3x + 5$ **40.** $2x^3$ **41.** -16 **42.** $-\frac{1}{16}$ **43.** $\frac{1}{16}$
44. 16 **45.** -1 **46.** 0 **47.** undefined **48.** $\frac{-2}{x}$ **49.** $x - 1$ **50.** $x^2 - 9$ **51.** $-2x - 9$
52. $x^2 - 3x + 3$ **53.** factor **54.** term **55.** factor **56.** x **57.** $x - 1$ **58.** $x - 3$
59. Equation, $\{-9\}$ **60.** Expression, $2x + 6$ **61.** Expression, $8x$ **62.** Equation, $\{5\}$ **63.** Expression, -5

CHAPTER 2 TEST

1. A **2.** D **3.** C **4.** B **5.** B **6**(a) 2.33×10^{-3} (b) 5.763×10^2 **7**(a) 9,400 (b) 0.00004321
8. $6 - 4x$ **9.** $2x^2 - x - 3$ **10.** $2x^3 - 3x^2 - 4x$ **11.** $9x^2 - 12x + 4$ **12.** $x^3 + 6x^2 + 12x + 8$
13. $6x^2 - 8x + 9 + \frac{-11}{x + 1}$ **14.** $4x^3 - x^2 + 2x + 11$ **15.** 4 **16.** $4 - y^2$ **17.** $2x^7 - 3x^6 + 3x^3$
18. $x^2 - x - 16$ **19.** The area is $2y^2 + 3y - 2$ sq inches. **20.** The length is $3s + 22$ yd.

CHAPTER 3

PROBLEM SET 3.1

Warm-ups

1. $\{-9\}$ **3.** $\{0\}$ **5.** $\{-1/2\}$ **7.** $\{12\}$ **9.** $\{15\}$ **11.** $\{4\}$ **13.** $\{1/2\}$ **15.** $\{2\}$ **17.** $\{0\}$ **19.** $\{-3\}$
21. $\{5\}$ **23.** $\{-1\}$ **25.** $\{1\}$ **27.** $\{1\}$ **29.** $\{-3\}$ **31.** $\{1\}$ **33.** $\{-1\}$ **35.** $\{3\}$ **37.** $\{7\}$ **39.** $\{3\}$
41. $\{5\}$ **43.** 5 **45.** 9 **47.** -1 **49.** -1 **51.** 16 in. **53.** 11 nickels and 16 dimes **55.** 16 cm

Practice Exercises

57. $\{-1\}$ **59.** $\{7\}$ **61.** $\{-3\}$ **63.** $\{5\}$ **65.** $\{-4\}$ **67.** $\{5/2\}$ **69.** $\{-14\}$ **71.** $\{11\}$ **73.** $\{-3\}$
75. $\{1\}$ **77.** $\{3\}$ **79.** $\{-16/7\}$ **81.** $\{2/3\}$ **83.** $\{-7\}$ **85.** $\{-1\}$ **87.** -3 **89.** 8 **91.** 1
93. 7 cm **95.** 8 pairs

Challenge Problems

97. $\{2q\}$

PROBLEM SET 3.2

Warm-ups

1. {11/4} **3.** {0} **5.** {3/4} **7.** {− 1/6} **9.** {− 18} **11.** {11} **13.** {− 15} **15.** {7/3} **17.** {18}
19. {4} **21.** {2} **23.** {− 1} **25.** {1/3} **27.** {4} **29.** {− 5} **31.** {1} **33.** {3} **35.** {− 15}
37. ∅ **39.** He worked 27 hrs. **41.** She has 20 dimes. **43.** They are − 1, 0, 1, 2. **45.** She deposited $570.
47. He rode 8 hrs. **49.** It is 500. **51.** It contains 20 gal.

Practice Exercises

53. {2} **55.** {2} **57.** {3} **59.** {− 1} **61.** {3} **63.** {− 2} **65.** {− 1} **67.** {− 1/9} **69.** {− 7/3}
71. {10} **73.** {2} **75.** $\left\{\frac{1}{2}\right\}$ **77.** {1} **79.** {1/4} **81.** {1} **83.** {− 0.4} **85.** {− 1} **87.** {− 1}
89. {− 9} **91.** {− 6} **93.** {3/2} **95.** {1} **97.** {− 2/3} **99.** $\{x \mid x$ is a real number$\}$ **101.** It is 480.
103. He borrowed $750. **105.** He has 8. **107.** It is 15 ft high. **109.** He bought 8 tapes.

Challenge Problems

111. $\left\{\frac{4}{3}q\right\}$

PROBLEM SET 3.3

Warm-ups

1. $w = \frac{A}{l}$ **3.** $r = \frac{C}{2\pi}$ **5.** $h = \frac{3V}{\pi r^2}$ **7.** $C = \frac{5}{9}(F - 32)$ **9.** $y = 1 - x$ **11.** $x = a + y$

13. $x = -p/3$ **15.** $y = -2k$ **17.** $y = -\frac{z}{b}$ **19.** $y = \frac{4f + 4L}{a}$ **21.** $w = -\frac{2kq}{a}$ **23.** $y = \frac{1 + 3q}{3}$

25. $x = 6p - a$ **27.** $w = \frac{2bL - ak}{a}$

Practice Exercises

29. $m = \frac{rF}{v^2}$ **31.** $F = \frac{9}{5}C + 32$ **33.** $z = 5 - 2x$ **35.** $x = k + 3z$ **37.** $x = -\frac{q}{2}$ **39.** $x = -\frac{7}{p}$

41. $x = -\frac{a}{k}$ **43.** $y = h - 5x$ **45.** $y = \frac{k - ax}{4}$ **47.** $y = \frac{ax - z}{b}$ **49.** $x = 6 + k$ **51.** $y = -3w$

53. $z = pT + pq$ **55.** $y = \frac{-2f - 2L}{q}$ **57.** $w = \frac{k^2t}{a}$ **59.** $y = \frac{1 + 2p}{2}$ **61.** $x = \frac{kp - 2}{p}$

63. $x = \frac{C + AB}{3A}$ **65.** $s = -A$ **67.** $x = \frac{3r - 2d}{2}$ **69.** $x = -2r$ **71.** $w = \frac{5dL - ck}{c}$

Challenge Problems

73. $P_2 = \frac{P_1V_1T_2}{T_1V_2}$

PROBLEM SET 3.4

Warm-ups

1. It is 34. **3.** They are 6 ft and 10 ft. **5.** 9 m by 11 m **7.** 17 yds by 51 yds
9. Homer is 64 and Pete is 32. **11.** Ellen is 30 and Sarah is 34. **13.** 10 nickels and 12 dimes
15. 24 nickels and 30 pennies. **17.** She should add 3 liters. **19.** 630 barrels of $16 crude
21. It will take $2\frac{1}{2}$ hrs. **23.** It will take 2 hrs.

Practice Exercises

25. It is 49. **27.** 24 m by 27 m **29.** Kitty is 16 and Charlie is 32. **31.** 11 nickels and 22 dimes **33.** They are 23 and 49. **35.** It is 27 mm. **37.** Kimberly is 6 and Tom is 24. **39.** Jim is 2 and Sondra is 5.

41. 10 lbs of \$3.50 pecans **43.** It will take $3\frac{1}{3}$ hrs. **45.** She should add 6 liters. **47.** It will take 3 hrs.

49. They are 15 ft and 26 ft. **51.** 30 yds by 120 yds **53.** J.R. is 9 and Kim is 15.

55. 5 nickels and 15 pennies **57.** 25 gallons of the \$35 oil **59.** It will take $2\frac{1}{2}$ hrs. **61.** It is 43.

63. It is 19 mm. **65.** Daughter is 14 and Louise is 42. **67.** 300 nickels, 600 dimes and 1500 quarters

69. 108 lbs are required. **71.** It will take $1\frac{1}{2}$ hrs.

Challenge Problems

73. $54\frac{6}{11}$ miles

PROBLEM SET 3.5

Warm-ups

1. $-8, -2, 0, 2$ **3.** 40, 400 **5.** 1, 2, 3, 4, 5, 6 **7.** 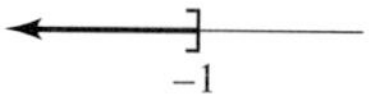 -1 **9.** 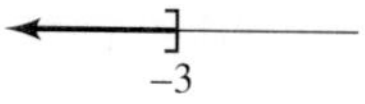 -3

11. $\{x \mid x < 18\}$ 18 **13.** $\{x \mid x \geq 20\}$ 20 **15.** $\{x \mid x \geq 6\}$ 6 **17.** $\{x \mid x \leq 3\}$ 3

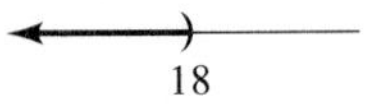

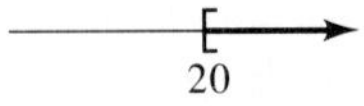

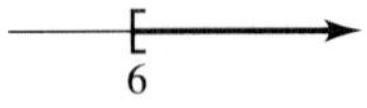

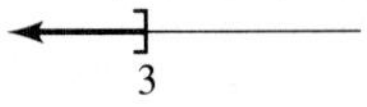

19. $\{t \mid t \leq -\frac{8}{3}\}$ $-\frac{8}{3}$ **21.** $\{x \mid x > -4\}$ -4 **23.** $\{w \mid w \geq -25\}$ 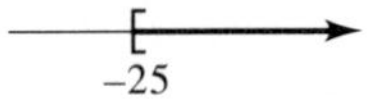 -25 **25.** $\{x \mid x < 9\}$ 9

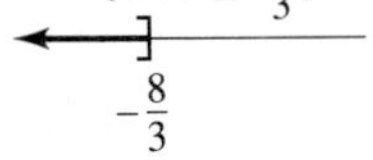

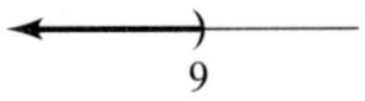

27. $\{x \mid x > -1\}$ -1 **29.** $\{t \mid t \geq -2\}$ -2 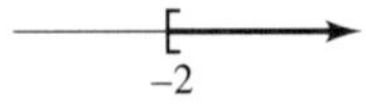 **31.** $\{k \mid k > \frac{1}{3}\}$ $\frac{1}{3}$ **33.** $\{y \mid y \geq -2\}$ -2

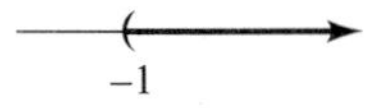

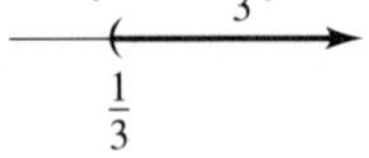

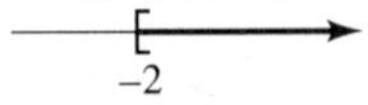

35. $\{x \mid x < -5\}$ -5 **37.** $\{w \mid w < 1\}$ 1 **39.** $\{z \mid z > -3\}$ -3 **41.** $\{x \mid x > 3\}$ 3

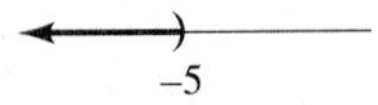

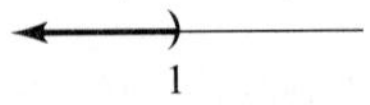

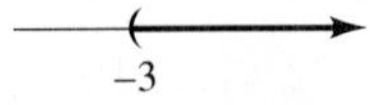

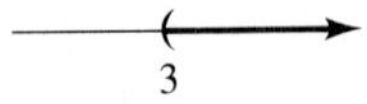

43. $\{z \mid z \geq -1\}$ -1 **45.** $\{x \mid x > \frac{9}{10}\}$ $\frac{9}{10}$ 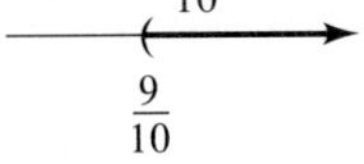 **47.** $\{x \mid x > \frac{16}{3}\}$ $\frac{16}{3}$ **49.** $\{x \mid x > 8\}$ 8

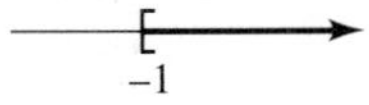

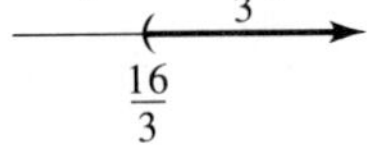

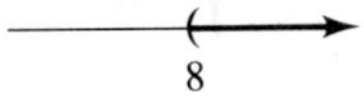

51. $\{x \mid x \geq \frac{1}{2}\}$ $\frac{1}{2}$ **53.** 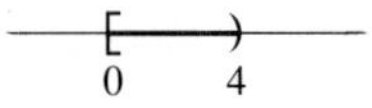 0 4 **55.** 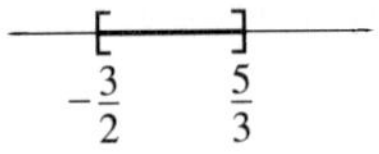 $-\frac{3}{2}$ $\frac{5}{3}$ **57.** $\{x \mid x < 30\}$ 30

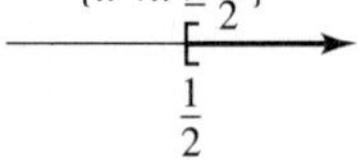

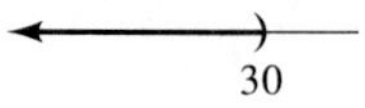

59. $\{x \mid x \geq -26\}$ -26 **61.** $\{x \mid x > 8\}$ 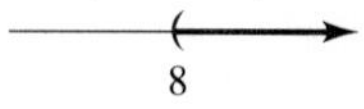 8 **63.** $\{z \mid z > -8\}$ -8 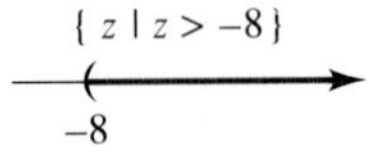 **65.** $\{x \mid x > -\frac{7}{2}\}$ $-\frac{7}{2}$

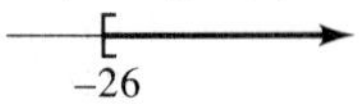

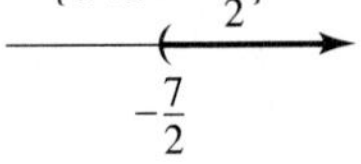

67. $\{s \mid s \geq -49\}$ -49 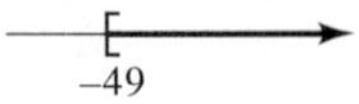 **69.** $\{x \mid x < 0\}$ 0 **71.** $\{s \mid s \leq -5\}$ -5 **73.** $\{x \mid x > 11\}$ 11

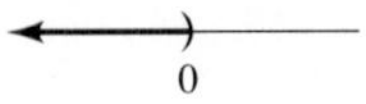

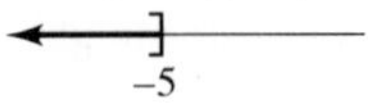

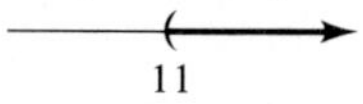

75. 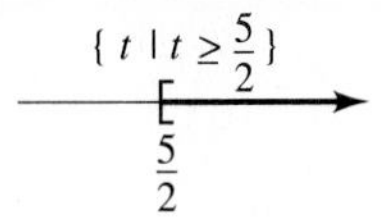

77. 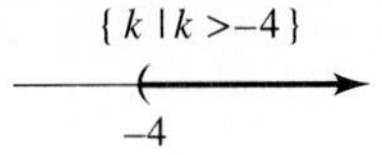

79. 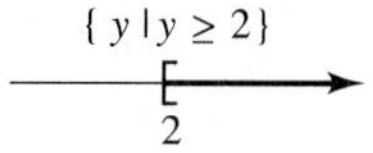

81. 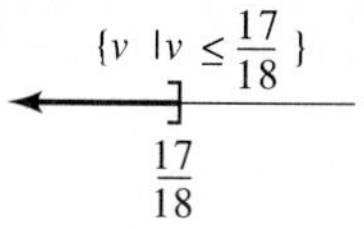

83. 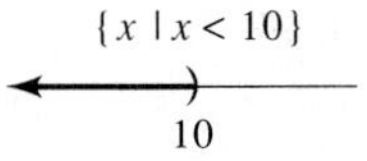

85. 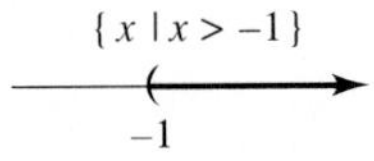

87. 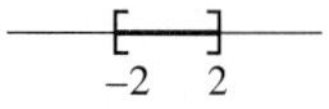

89. 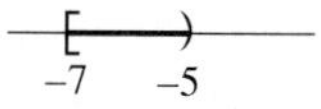

Challenge Problems

91. $\{x \mid x > p/2\}$ **93.** $\{x \mid x < 2/p\}$

PROBLEM SET 3.6

Warm-ups

1. $y < -2$ **3.** $s + (s + 1) > 0$ **5.** $-3z \leq 30$ **7.** $2x + 3 > -5$ **9.** greater than $10\frac{1}{2}$ ft
11. no longer than 12 in. **13.** at least 6.6

Practice Exercises

15. $xy \geq -3$ **17.** $\frac{1}{3}w + 5 \geq 9$ **19.** $-x < 0$ **21.** $\frac{1}{2}p \leq -4$ **23.** $\sqrt{p} > 7$ **25.** $\frac{5}{3} < x$ **27.** $|k| \geq 4$
29. at least 9.05 **31.** all numbers less than 6 **33.** at least 134 **35.** 11 or less **37.** at least 45
39. all numbers less than 5

Challenge Problems

41. He would need 118!

REVIEW PROBLEMS CHAPTER 3

1. $\{20\}$ **3.** $\{-9\}$ **5.** $\{2\}$ **7.** $\{x \mid x < 20\}$ **9.** $\{4\}$ **11.** $\{y \mid y > 5\}$ **13.** $\{1\}$ **15.** $\{x \mid x \geq 6\}$
17. $\left\{-\frac{1}{2}\right\}$ **19.** $x = t - s$ **21.** $x = \dfrac{b + c}{a}$ **23.** $r = \dfrac{K + 2at}{2}$ **25.** $x = \dfrac{3bh}{a}$ **27.** 43 and 83.
29. Carol is 60 and Russ is 64. **31.** 5 grams of 50% gold should be added.
33. all numbers less than or equal to $-\frac{6}{5}$ **35.** 15 in. by 15 in.

Let's Not Forget

36. $-6x^2$ **37.** $-6x - 2x^2$ **38.** undefined **39.** $x^2 - 18x + 81$ **40.** -36 **41.** $3z$
42. $2 + x^2 - x$ **43.** $x^2 + x - 2$ **44.** $2x + x^2 - 1$ **45.** term **46.** factor **47.** factor **48.** term
49. $-3x$ **50.** $3(x + 7)$ **51.** $3x - 2$ **52.** expression; $5x - 19$ **53.** inequality; $\{x \mid x > 5\}$
54. equation; $\{-1\}$ **55.** expression; $-2x - 7$ **56.** inequality; $\left\{x \mid x \geq -\frac{1}{2}\right\}$ **57.** equation; $\{-1\}$

CHAPTER 3 TEST

1. D **2.** C **3.** A **4.** A **5.** B **6.** B **7.** $\{1\}$ **8.** $\{x \mid x \leq 0\}$ **9.** $\{2\}$ **10.** $\left\{x \mid x < -\frac{2}{3}\right\}$

11. $x = -\dfrac{2t}{a}$ **12.** $h = \dfrac{3V}{\pi r^2}$ **13.** 17 and 33. **14.** Billy is 16 and his sister is 11. **15.** 13 nickels and 7 dimes

CHAPTER 4

PROBLEM SET 4.1

Warm-ups

1. $3(2x+1)$ **3.** $x(y+z)$ **5.** $5(x-3y)$ **7.** $3a(3b-2c)$ **9.** $6(x+1)$ **11.** 6 **13.** 9 **15.** xy
17. $3x$ **19.** $9xy$ **21.** $5x(x-3)$ **23.** $25x^2yz(z-3)$ **25.** $3rst(2rst+1)$ **27.** $13x^2(2x^4-x^2+3)$
29. $7m^2(n-3mn^2+1)$ **31.** $8rst(7r^3s+8rst-3)$ **33.** $-4(x-1)$ **35.** $-5y^2(3y^3-2y^2+4)$
37. $(s+t)(r+2)$ **39.** $(y-z)(x^2+y^2)$ **41.** $(p+r)(q-1)$ **43.** $(s^2-2)(3r^3-2w)$
45. $(s+t)(m+n)$ **47.** $(y-z)(r^2+s^2)$ **49.** $(px+qz)(r-3)$ **51.** $(z^5+x)(2-y)$
53. $(y^2-2)(3x-2w)$ **55.** $(a+b)(x^2+1)$

Practice Exercises

57. $5(x+2)$ **59.** $r(w+z)$ **61.** $3(2x-5y)$ **63.** $-3a(3b-2c)$ **65.** $2(x^3+1)$ **67.** $3xy(x-5)$
69. $15x^2yz^2(x-2)$ **71.** $-3r^3st(2t^2-1)$ **73.** prime **75.** $24r^3s^2(3r+2)$ **77.** $7x^3(2x^3-3x+4)$
79. $7ab(3a-2b+1)$ **81.** $12rst(9r^3s+8rst-7)$ **83.** $(a-b)(r+2t)$ **85.** $(x^2+1)(x-y)$
87. $(a^2-b)(2r^2-1)$ **89.** $(a-b)(s+2t)$ **91.** $(y^2+1)(y-x)$ **93.** $(q+r)(p+3)$
95. $(a^2-b)(2-z)$ **97.** $(a+2)(3+y)$ **99.** $(x+3)(x^2+1)$

Challenge Problems

101. $a(s+t)(m+n)$ **103.** $6(c+d)(b-d)$

PROBLEM SET 4.2

Warm-ups

1. $(x+3)(x-3)$ **3.** $(2x+3)(2x-3)$ **5.** $(4x+5)(4x-5)$ **7.** $(a+2b)(a-2b)$
9. $(ab+6)(ab-6)$ **11.** $(r+st)(r-st)$ **13.** $(2z-7)(2z+7)$ **15.** $(10c+3d)(10c-3d)$
17. $x(x-1)(x+1)$ **19.** $2(2x+3)(2x-3)$ **21.** $3(2b+1)(2b-1)$ **23.** $2a(a+2)(a-2)$
25. $4c(c-2)(c+2)$ **27.** $(y-x)(y^2+yx+x^2)$ **29.** $(x+5)(x^2-5x+25)$ **31.** $(z-1)(z^2+z+1)$
33. $(2x+3)(4x^2-6x+9)$ **35.** $(2p-1)(4p^2+2p+1)$ **37.** $(3t+2s)(9t^2-6st+4s^2)$
39. $2(y+x)(y^2-yx+x^2)$ **41.** $q(2p-1)(4p^2+2p+1)$ **43.** $8(z-1)(z^2+z+1)$
45. $8(2x-y)(4x^2+2xy+y^2)$ **47.** $-5(z+1)(z^2-z+1)$

Practice Exercises

49. $(y-8)(y+8)$ **51.** $(x-11)(x+11)$ **53.** $(2x+y)(2x-y)$ **55.** $(2b+9)(2b-9)$
57. $(t-2xy)(t+2xy)$ **59.** $4(x^2+1)$ **61.** $9(x+m)(x-m)$ **63.** $2(x-3)(x+3)$
65. $16(2x+1)(2x-1)$ **67.** $s(s-t)(s+t)$ **69.** $8(z^2+9)$ **71.** $2x(5y+z)(5y-z)$
73. $(c-d)(c^2+cd+d^2)$ **75.** $(x-4)(x^2+4x+16)$ **77.** $(d+2c)(d^2-2cd+4c^2)$
79. $(3k-4)(9k^2+12k+16)$ **81.** $(2x+y)(4x^2-2xy+y^2)$ **83.** $(5z+3w)(25z^2-15wz+9w^2)$
85. $3(z+2)(z^2-2z+4)$ **87.** $2(x-4)(x^2+4x+16)$ **89.** $b(2a+3)(4a^2-6a+9)$
91. $2(2p-q)(4p^2+2pq+q^2)$

Challenge Problems

93. $(x^3-2)(x^3+2)$ **95.** $(x^2+4)(x+2)(x-2)$ **97.** $(x-1)(x^2+x+1)(x^6+x^3+1)$
99. $(a+b+2)\{(a+b)^2-2(a+b)+4\}$ **101.** $(4-a-b)(4+a+b)$

PROBLEM SET 4.3

Warm-ups

1. $(x+2)(x+1)$ **3.** $(x+4)(x+1)$ **5.** $(x+6)(x+1)$ **7.** $(x+2)(x+3)$ **9.** $(x-3)(x-5)$
11. $(x-2y)(x-y)$ **13.** $(x-4)^2$ **15.** $(x-6)(x-1)$ **17.** $(a-3b)(a+b)$ **19.** $(x+5)(x-1)$
21. $(x+2)(x-3)$ **23.** $(x-5)(x+2)$ **25.** $(x+4)(x-5)$ **27.** $(3b+1)(b+1)$

29. $(3x - z)(x - 2z)$ **31.** $(3x + 1)(x + 3)$ **33.** $(5x - 1)(x - 3)$ **35.** $(3c + 1)(c - 1)$ **37.** $(3x - 2)(x - 1)$ **39.** $(2x + 3)(x + 2)$ **41.** $(3a - 2)(a - 4)$ **43.** $(7x - 2)(x + 5)$ **45.** $(4x - 7)(2x - 3)$ **47.** $2(x + 2)(x - 3)$ **49.** $2(2x + 1)(x - 4)$

Practice Exercises

51. $(x + 1)(x + 7)$ **53.** $(x + 1)(x + 9)$ **55.** $(x + 2)(x + 6)$ **57.** $(x - 1)^2$ **59.** $(x - 3)(x - 6)$ **61.** $(y + 2)(y - 1)$ **63.** prime **65.** $(x + 6)(x - 1)$ **67.** $(a + 2)(a - 4)$ **69.** $(x + 3)(x - 4)$ **71.** $(5x + 1)(x + 1)$ **73.** $(5a - b)(a - b)$ **75.** $(3x + 1)(x + 2)$ **77.** $(7b - 2)^2$ **79.** $(2a - b)(a + b)$ **81.** $(3x - 1)(x + 2)$ **83.** prime **85.** $(2c - 1)(3c - 1)$ **87.** $(3x - 2)(2x + 3)$ **89.** $(2x - 5)(3x - 4)$ **91.** $(2a + b)(8a - 9b)$ **93.** $(3x - 2)(8x - 3)$ **95.** $(6x + 5)(x - 5)$ **97.** $4a(x + 2)^2$ **99.** $a^2(a - b)(a + 2b)$

Challenge Problems

101. $(x^2 - 2)(x^2 - 3)$ **103.** $(x - 1)(x + 1)(x - 2)(x + 2)$

PROBLEM SET 4.4

Warm-ups

1. $25(x - 1)(x + 1)$ **3.** $(3b - x)(3b + x)$ **5.** $(s + 3)(s^2 - 3s + 9)$ **7.** $4(x^2 + 1)$ **9.** $(z - 4)(y + 1)$ **11.** $2(2 - b)(2 + b)$ **13.** $(4 - 5x)(4 + 5x)$ **15.** $(x - y)(ab + c)$ **17.** $16(5 - y)(25 + 5y + y^2)$ **19.** $(x + 2)(x - 1)$ **21.** $3(x - 9)(x - 1)$ **23.** $(x + 3)^2$ **25.** $(7x - 1)(3x - 2)$ **27.** $(x - 6)(x + 1)$ **29.** $(z + x)(a - 1)(a + 1)$ **31.** $(z + y)(y + a)$

Practice Exercises

33. $(x + 7)(x + 2)$ **35.** $45r^2t^4(3rt + 5)$ **37.** $(1 - 3a)(1 + 3a + 9a^2)$ **39.** $(x + y)(a + 1)$ **41.** Factored completely **43.** $2(4 - x)(4 + x)$ **45.** $(t - x)(r - 1)(r^2 + r + 1)$ **47.** $(z - xy)(z + xy)$ **49.** $(a + b)(ab + 1)$ **51.** $(x + 2)(x + 1)$ **53.** $(2x + y)(4x^2 - 2xy + y^2)$ **55.** $(x + 6)^2$ **57.** prime **59.** $(2y + z)(4y^2 - 2yz + z^2)$ **61.** $(x + 5)(x - 4)$ **63.** $(9 - x)^2$ **65.** $(a - 5)(a^2 + 5a + 25)$ **67.** $a^5b^5c^3(a^2c - abc^3 + b)$

Challenge Problems

69. $(x^2 - 3)(x^2 + 2)$ **71.** $5(x^2 + 5)(x - 2)(x + 2)$ **73.** $(x^3 - 3)(x^2 + 3)$

PROBLEM SET 4.5

Warm-ups

1. $\{-7, 9\}$ **3.** $\{-7, 1\}$ **5.** $\{0, 8\}$ **7.** $\left\{-3, \frac{5}{2}\right\}$ **9.** $\left\{-\frac{5}{6}, \frac{5}{6}\right\}$ **11.** $\left\{-4, 0, \frac{8}{7}\right\}$ **13.** $\{9\}$ **15.** $\left\{-8, -\frac{9}{2}\right\}$ **17.** $\{-1, 1\}$ **19.** $\left\{-\frac{1}{4}, \frac{5}{3}\right\}$ **21.** $\{-\sqrt{5}, \sqrt{5}\}$ **23.** $\{\pm 1\}$ **25.** $\{\pm 8\}$ **27.** $\left\{\pm\frac{2}{3}\right\}$ **29.** $\{0, 5\}$ **31.** $\{-2, -1\}$ **33.** $\{-1, 6\}$ **35.** $\{-4\}$ **37.** $\left\{\frac{3}{2}, 2\right\}$ **39.** $\{-4, 1\}$ **41.** $\{3, 5\}$ **43.** $\{-2, 0, 2\}$ **45.** $(-3, 1, 8)$

Practice Exercises

47. $\{\pm 4\}$ **49.** $\{0, \pm 9\}$ **51.** $\left\{\pm\frac{5}{4}\right\}$ **53.** $\{0, 4\}$ **55.** $\{0, 3\}$ **57.** $\{-8, -1\}$ **59.** $\{-4, -2\}$ **61.** $\{-6, 1\}$ **63.** $\{3\}$ **65.** $\left\{-4, -\frac{1}{3}\right\}$ **67.** $\left\{-\frac{5}{2}, 3\right\}$ **69.** $\left\{-\frac{2}{3}, \frac{1}{2}\right\}$ **71.** $\{3, 4\}$ **73.** $(-3,6)$

Challenge Problems

75. $\{\pm 2\}$ **77.** $\{3\}$

PROBLEM SET 4.6

Warm-ups

1. The integers are 8 and 9. **3.** The temperatures were $-9°$ and $-7°$. **5.** The integers are 3 and 4. **7.** The distance is 1300 ft. **9.** The sides are 3, 4, and 5 units. **11.** The sides are 10, 24, and 26 units. **13.** The dimensions are $\frac{1}{2}$ yd by 2 yd. **15.** The dimensions are 12 in. by 4 in. **17.** The base is 4 inches and the height is 16 inches. **19.** The base is 14 meters and the height is 6 meters.

Practice Exercises

21. The integers are 7 and 8. **23.** The noon temperature was $-8°$. **25.** The integers are 2 and 3. **27.** She needs 34 in. of gold string. **29.** The sides are 6, 8, and 10 units. **31.** The sides are 16, 30, and 34 units. **33.** The dimensions are 8 mi by 14 mi. **35.** The dimensions are 15 cm by 13 cm. **37.** The base is 40 feet and the height is 10 feet. **39.** It is 6 units. **41.** They are 10 and 15, but we can't determine who is which.

REVIEW PROBLEMS CHAPTER 4

1. $(z + 4t)(z - 4t)$ **3.** prime **5.** $(x + t)(t + 2)$ **7.** $(2u - v)(4u^2 + 2uv + v^2)$ **9.** $a^2b^2c^3(bc - a + bc^2)$ **11.** $4(3d + 1)$ **13.** $(a + b)(z^2 - 5)$ **15.** $2(x - 3)(x + 3)$ **17.** $(x + 3)^2$ **19.** $(xy - 4)(x^2y^2 + 4xy + 16)$ **21.** Factored completely **23.** $(x + 7)(x + 2)$ **25.** $(k - 5j)(k^2 + 5jk + 25j^2)$ **27.** $(1 - 2n)(x^2 + 4)$ **29.** $b(2b + 1)(b - 3)$ **31.** $\{3, 7\}$ **33.** $\{\pm\sqrt{15}\}$ **35.** $\{0, 3\}$ **37.** $\{-2, 5\}$ **39.** $\{\pm 4\}$ **41.** The integers are -22 and -23. **43.** The dimensions are 11 by 16 units.

Let's Not Forget

45. x^2 **46.** 0 **47.** $6t + 1$ **48.** $y^2 - 2y + 1$ **49.** $x - 5$ **50.** -9 **51.** $3x - 2$ **52.** $x^2 + x - 2$ **53.** $x^2 + 2x - 1$ **54.** term **55.** factor **56.** term **57.** x **58.** $4x + 28$ **59.** $6y + 8$ **60.** expression; $x^2 + x - 6$ **61.** equation; $\{-3, 2\}$ **62.** inequality; $\{x \mid x < -4\}$ **63.** expression; $5x + 13$ **64.** $y(a + 1)$ **65.** $(x + 1)(x^2 - x + 1)$ **66.** factored **67.** factored

CHAPTER 4 TEST

1. B **2.** C **3.** B **4.** D **5.** D **6.** $(2a - 3)(4a^2 + 6a + 9)$ **7.** $(2x - 1)(x + 5)$ **8.** $(a - b)(x + 2)$ **9.** $(p - q)(x - y)$ **10.** $(a - 4)(a + 2)$ **11.** $(x + 7)^2$ **12.** $-3cd(2c + d)$ **13.** $(2y + 5)(2y - 5)$ **14.** $9a^2b^2(3b - 4a + 5a^2b^2)$ **15.** prime **16.** $\{\pm 9\}$ **17.** $\{-2, 6\}$ **18.** $\{-3, 0, 1\}$ **19.** The dimensions are 13 yd by 18 yd. **20.** The sides are 5 m, 12 m, and 13 m.

CHAPTER 5

PROBLEM SET 5.1

Warm-ups

1. -1 **3.** $\frac{1}{3}$ **5.** 0 **7.** 0 **9.** 3 **11.** -8 **13.** 2, 5 **15.** 4, 11 **17.** $-3, 3$ **19.** $\frac{5}{-x}, \frac{-5}{x}$ **21.** $\frac{-17}{x^2}, \frac{17}{-x^2}$ **23.** $\frac{-11x^3}{2x-3} \cdot \frac{11x^3}{2x-3}$ **25.** $-\frac{3x}{x+4}, \frac{3x}{-(x+4)}$ **27.** $\frac{-(x+6)}{x-3}, \frac{x+6}{3-x}$

Practice Exercises

29. 0 **31.** $-\frac{2}{3}$ **33.** $\frac{7}{4}$ **35.** no values of x **37.** 2 **39.** -5 **41.** -2 **43.** 2 **45.** 16

47. $-4, -2$ **49.** $-2, 2$ **51.** $-7, 7$ **53.** $-\frac{9}{2x}, \frac{-9}{2x}$ **55.** $-\frac{x-1}{x}, \frac{x-1}{-x}$ **57.** $\frac{3-x}{3+x}, \frac{x-3}{-(3+x)}$

59. $-\frac{9+8x}{x}, \frac{-(9+8x)}{x}$

Challenge Problems

61. $\frac{1}{2}$ **63.** $-\frac{1}{2}$ **65.** $\frac{4}{3}$ **67.** no value of x **69.** $-\sqrt{5}, \sqrt{5}$

PROBLEM SET 5.2

Warm-ups

1. $\frac{2x}{3y}$ **3.** $2x$ **5.** $\frac{x+2}{2y}$ **7.** $\frac{4x}{y}$ **9.** $\frac{1}{x+1}$ **11.** Cannot be reduced **13.** y **15.** $\frac{x+y}{x}$ **17.** $\frac{4}{3}$

19. $\frac{-x}{2(a+b)^3}$ **21.** $\frac{x}{2}$ **23.** $\frac{2}{x-4}$ **25.** $\frac{x^2(2z-1)}{4}$ **27.** $\frac{x+3}{x-3}$ **29.** $\frac{2x}{x+1}$ **31.** $\frac{3(3x-y)}{3x+y}$

33. $\frac{x-3}{x-1}$ **35.** -1 **37.** -2 **39.** $\frac{-1}{x-2}$ or $\frac{1}{2-x}$ **41.** $\frac{1}{9+3y+y^2}$ **43.** $\frac{(x-y)^2}{x^2+xy+y^2}$

45. $\frac{1+2x+4x^2}{1-2y}$ **47.** $\frac{21x^2}{6x^3}$ **49.** $\frac{10}{5(x-1)}$ **51.** $\frac{x^2-x-2}{x^2-4x+4}$

Practice Exercises

53. $\frac{4x}{3y}$ **55.** $\frac{2x^2}{t}$ **57.** $\frac{x-3}{3y}$ **59.** $\frac{-2}{x}$ **61.** $\frac{1}{(s+5)^2}$ **63.** Cannot be reduced **65.** $\frac{x+7}{x-7}$ **67.** $\frac{1}{2}$

69. $\frac{t^2}{3}$ **71.** $\frac{2}{s}$ **73.** $\frac{1-x}{3}$ **75.** $\frac{klm}{a+t}$ **77.** $\frac{x-1}{3}$ **79.** $\frac{x+2}{x+1}$ **81.** $\frac{x+1}{x-1}$ **83.** $\frac{x+3}{x+1}$

85. $\frac{-1}{t-1}$ or $\frac{1}{1-t}$ **87.** -1 **89.** $\frac{1}{s^2+s+1}$ **91.** $\frac{x-y}{7}$ **93.** $\frac{5x+y}{x+1}$ **95.** $\frac{10y}{5y^2}$ **97.** $\frac{14abx}{21a^2x^2}$

99. $\frac{-x^2-7x}{x^2-49}$ **101.** $\frac{x^2+x-2}{x^2+4x+4}$ **103.** $\frac{-10x+30}{6x^2-13x-15}$

Challenge Problems

105. $(x-2)(x^2+4)$ **107.** $-(x+1)(x^2+1)$ **109.** $\frac{2y^2-5}{y(2y^2-1)}$ **111.** $\frac{r-s}{r+s}$

PROBLEM SET 5.3

Warm-ups

1. $\frac{8x^2}{21y^2}$ **3.** $\frac{22m^4}{91n^3}$ **5.** $\frac{-3nx}{my}$ **7.** $-3x$ **9.** $\frac{x+5}{x+6}$ **11.** $\frac{5kx}{6}$ **13.** $\frac{2x-3}{2}$ **15.** $\frac{(2x+1)(x+1)}{(x-1)^2}$

17. $\frac{s+3}{(s+2)(s-3)}$ **19.** $\frac{(x+1)(x+4)}{(x-1)(x-4)}$ **21.** -2 **23.** $\frac{21a^3}{20b^2x}$ **25.** $\frac{14}{15}$ **27.** $\frac{-x}{40y^2}$ **29.** $\frac{2y^2}{7}$

31. $-(1+2x)$ **33.** $x-1$ **35.** $\frac{9x}{4}$ **37.** $\frac{(x-1)(2x+1)}{(x+1)^2}$ **39.** $\frac{y+2}{y+5}$ **41.** $-2(x+1)$

Practice Exercises

43. $\frac{15x^2}{28y^2}$ **45.** $\frac{27m^3}{55n^4}$ **47.** $\frac{4n^2y}{m^2x}$ **49.** $\frac{-9x^2}{4y^2}$ **51.** $\frac{3x}{5y^3}$ **53.** $\frac{-13x}{10y}$ **55.** $2y$ **57.** $\frac{x+5}{x-6}$

59. $\frac{-7kx^2}{12}$ **61.** $\frac{y^2}{4}$ **63.** $\frac{2x+1}{3x-2}$ **65.** $\frac{12w^2}{35}$ **67.** $\frac{-8xz^2}{5}$ **69.** $x-2$ **71.** $\frac{x}{5y(y-1)}$

73. $\frac{2-r}{2(r+2)}$ **75.** $x-3$ **77.** $\frac{(x-1)^2}{2x+1}$ **79.** $\frac{(x-1)(2x+1)}{(x+1)^2}$ **81.** $\frac{(x+1)(2x-1)}{(x-1)^2}$

83. $\frac{y+2}{(y+3)(y-2)}$ **85.** $\frac{3(x-2)}{4-x}$ **87.** $\frac{x+1}{x-1}$ **89.** $\frac{(2x-1)(x+2)}{(x+1)(x-1)}$

Challenge Problems

91. $-\frac{x+2}{x-2}$ or $\frac{x+2}{2-x}$ **93.** $\frac{(x-3)(x-7)}{(x+2)(4x+1)}$

PROBLEM SET 5.4

Warm-ups

1. $\frac{13}{x}$ **3.** 2 **5.** $\frac{-1}{x-1}$ **7.** $x+2$ **9.** $x-2$ **11.** 35 **13.** 27 **15.** 180 **17.** $x^2(x-2)$

19. $5x(x+5)$ **21.** $3(x+2)(x-3)$ **23.** $\frac{15y+2x}{3xy}$ **25.** $\frac{10-3x}{2x}$ **27.** $\frac{2x-7}{x(x+7)}$ **29.** $\frac{r+6}{(r+2)(r-2)}$

31. $\frac{2t^2}{(t+1)(t-1)}$ **33.** $\frac{2+x}{2x^2}$ **35.** $\frac{3-10a}{4a^2}$ **37.** $\frac{8y-9x}{30x^2y^2}$ **39.** $\frac{15x}{4(1-5x)}$ **41.** $\frac{k+9}{8(k-6)}$

43. $\frac{y-4}{(y+4)^2}$ **45.** $\frac{9d-10}{15(d+2)}$ **47.** $\frac{2x+1}{(x-1)^2}$ **49.** $\frac{2c^2+5c}{(c+4)(c-3)(c+1)}$ **51.** $\frac{x+7}{7}$ **53.** $\frac{y^2-2}{y}$

55. $\frac{x+5}{x+3}$ **57.** $\frac{-3x-2}{1+2x}$ **59.** $\frac{2x}{x-1}$ **61.** $\frac{x-1}{2x-1}$ **63.** $\frac{a-1}{b-1}$

Practice Exercises

65. 24 **67.** 8 **69.** 252 **71.** $x^2(x+3)$ **73.** $(2x+1)(x-3)(x+3)$ **75.** 1 **77.** $\frac{3y-1}{5-y}$

79. $\frac{8t}{3(2-3t)}$ **81.** $\frac{4y+3x^2}{6x^3y}$ **83.** $\frac{2x+4}{x(x-4)}$ **85.** $\frac{r^2+4r-4}{(r-2)(r+2)}$ **87.** $\frac{5a}{(a-3)(a+2)}$ **89.** $\frac{2t-9}{12t^3}$

91. $\frac{2y+5}{(y+5)^2}$ **93.** $\frac{1}{12(d+3)}$ **95.** $\frac{-3x-5}{(x+1)^2}$ **97.** $\frac{3c}{(c-4)(c+3)(c-1)}$ **99.** $\frac{x}{x-2}$ **101.** $\frac{3-4x}{x}$

103. $\frac{1}{s-11}$

Challenge Problems

105. $\frac{x+1}{x^2+5x+25}$

PROBLEM SET 5.5

Warm-ups

1. $\frac{1}{3}$ **3.** $\frac{3}{2}$ **5.** $\frac{1}{3y}$ **7.** $\frac{2}{9}$ **9.** $\frac{3}{4w}$ **11.** $\frac{x}{4}$ **13.** $\frac{5x}{14}$ **15.** $\frac{2(x+1)}{x-1}$ **17.** $\frac{1}{11}$ **19.** $\frac{1+x}{1-x}$

21. $\frac{2z+3}{4z-5}$ **23.** $\frac{2}{2s+1}$ **25.** $\frac{y+xy}{x+xy}$ **27.** $\frac{x+2}{5x-3}$

Practice Exercises

29. $\frac{5}{4}$ **31.** $\frac{5}{11}$ **33.** $\frac{-2}{17}$ **35.** $\frac{3y}{4}$ **37.** $\frac{2}{9}$ **39.** $\frac{2w}{5}$ **41.** $\frac{x^2}{3}$ **43.** $\frac{7x^2}{10}$ **45.** $\frac{x+2}{x-2}$
47. $\frac{3-4y}{5+6y}$ **49.** $6-z$ **51.** $-\frac{2k+5}{2}$ **53.** $\frac{1}{y-1}$

Challenge Problems

55. no

PROBLEM SET 5.6

Warm-ups

1. $\{-1\}$ **3.** $\{-1\}$ **5.** $\{10\}$ **7.** $\{3\}$ **9.** $\{7\}$ **11.** $\left\{\frac{14}{3}\right\}$ **13.** $\{2\}$ **15.** $\{1\}$ **17.** $\left\{\frac{1}{2}\right\}$ **19.** $\varnothing$
21. $\varnothing$ **23.** $\varnothing$ **25.** $\frac{x}{3}$ **27.** $\frac{1-x}{x+1}$ **29.** $\{-3, -1\}$ **31.** $\{-3, -2\}$

Practice Exercises

33. $\{10\}$ **35.** $\left\{\frac{1}{2}\right\}$ **37.** $\{2\}$ **39.** $\{-1\}$ **41.** $\{3\}$ **43.** $\{15\}$ **45.** $\{8\}$ **47.** $\{2\}$ **49.** $\frac{6-x}{6x}$
51. $\{0\}$ **53.** $\{4\}$ **55.** $\{-3, 1\}$ **57.** $\varnothing$ **59.** $\{0\}$ **61.** $\varnothing$

Challenge Problems

63. $\{2\}$ **65.** $\{Q-R\}$ **67.** $\{R\}$ **69.** $\{10\}$

PROBLEM SET 5.7

Warm-ups

1. (a) 16:12 or 4:3 (b) 12:16 or 3:4 (c) 8:7 (d) 16:8 or 2:1 (e) 12:28 or 3:7 **3.** 56 **5.** 25
7. There are 200 in. **9.** Five tomato plants can be planted. **11.** She used 9,500 gal. **13.** 3
15. The pieces are 12 m and 18 m. **17.** The numbers are 2 and 8. **19.** Frank ran 10 km/hr.

Practice Exercises

21. 8 ft. **23.** There are 15 lbs in the lighter and 25 lbs in the heavier.
25. The investment contained 130 shares. **27.** The number is -4. **29.** The speeds are 40 mph and 50 mph.

REVIEW PROBLEMS CHAPTER 5

1. 0 **3.** $-2, 3$ **5.** $\frac{84x}{4x^3}$ **7.** $\frac{x^2-2x+1}{x^2+x-2}$ **9.** $\frac{3t}{5}$ **11.** $\frac{x}{x+2}$ **13.** $\frac{x}{x+3}$ **15.** $\frac{-3x^2}{2y}$
17. $\frac{3x+1}{2(3x-2)}$ **19.** $\frac{1}{6(2x+3)}$ **21.** $\frac{-1}{t^2(4t+1)}$ **23.** $\frac{2x^2-3}{2x}$ **25.** $\frac{1}{w-2}$ **27.** -1 **29.** $\frac{1}{5}$
31. $\frac{y+1}{y-1}$ **33.** $\{3\}$ **35.** $\{8\}$ **37.** $\varnothing$ **39.** $\{-1, 3\}$ **41.** There are 40 wolves in the area.
43. The numbers are -1 and 1.

Let's Not Forget

45. -25 **46.** $x^2-10x+25$ **47.** 25 **48.** undefined **49.** x **50.** $\frac{7}{x-1}$ **51.** $-5x+14$
52. $3x^2-6x-7$ **53.** $3x^2-13x+14$ **54.** $-5x-7$ **55.** $\frac{2}{y}$ **56.** $\frac{2}{y(x-1)}$ **57.** term **58.** term
59. factor **60.** factor **61.** 1 **62.** $2x^2+14x$ **63.** $4x+6$ **64.** expression; $-(x+3)$

65. equation; $\{1\}$ **66.** inequality; $\left\{x \mid x < \frac{1}{5}\right\}$ **67.** expression; $\frac{11}{4x}$ **68.** factored **69.** $(x - 3)(x + 3)$
70. $(x - 3)^2$ **71.** $(x + 3)(x^2 - 3x + 9)$ **72.** factored **73.** $3(x - 3)(x^2 + 3x + 9)$

CHAPTER 5 TEST

1. C **2.** B **3.** C **4.** D **5.** A **6.** D **7.** B **8.** A **9.** $-\frac{5}{2}$ **10.** $\frac{1 + x^2}{x(1 - x)}$
11. $\frac{t - 4}{(t + 4)(t + 2)}$ **12.** $x + 3$ **13.** $\frac{x + 5}{x + 3}$ **14.** $\frac{s + 4}{s - 4}$ **15.** $\{1\}$ **16.** $\varnothing$ **17.** $\{-1, 7\}$
18. (a) 3:2 (b) 2:5 **19.** There are 187.5 miles between the cities. **20.** The numbers are 3 and 12.

CHAPTER 6

PROBLEM SET 6.1

Warm-ups

Answers to Problems 13–19 may vary.

1. (2, 3)

3.

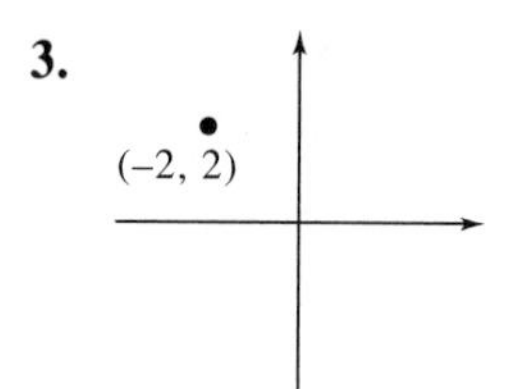

5. I **7.** IV **9.** (2, 0) **11.** (0, 0), (1, −1)
13. (0, 4), (4, 0), (2, 2) **15.** (0, 0), (2, 1), (4, 2)
17. (0, 4), (1, 4), (2, 4) **19.** $\left(\frac{7}{3}, 0\right), \left(\frac{7}{3}, 1\right), \left(\frac{7}{3}, 2\right)$

21.

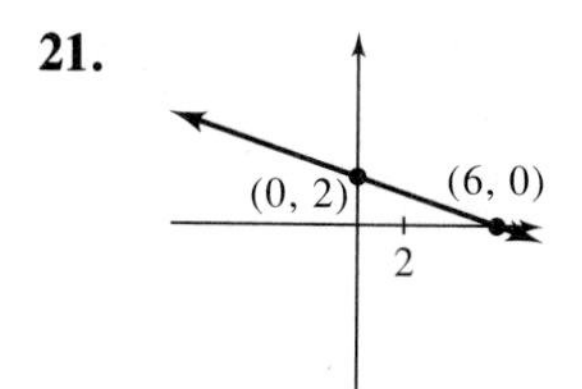

23.

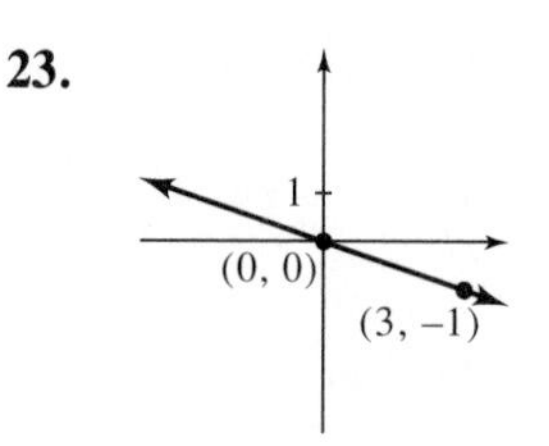

25.

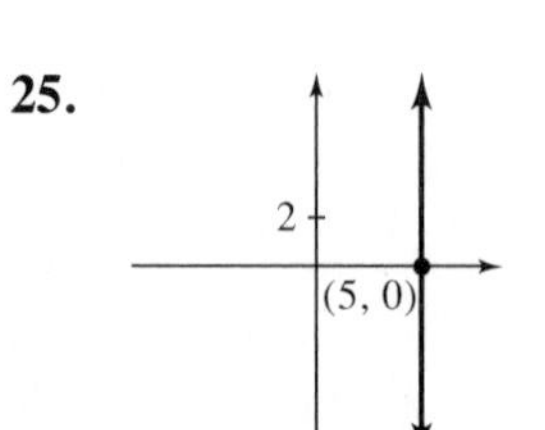

27.

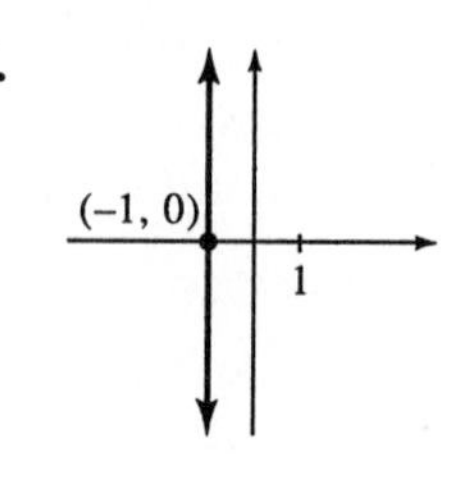

Practice Exercises

Answers to Problems 51–57 may vary.

29.

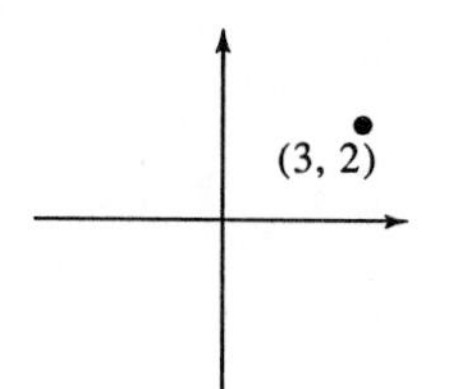

31.

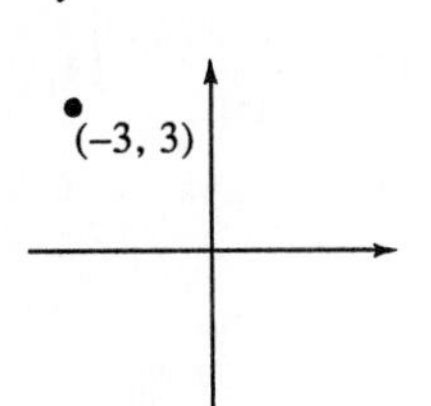

33.

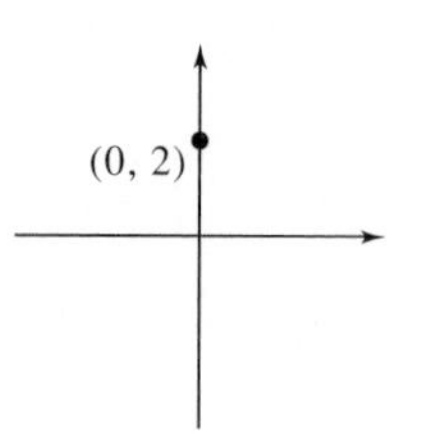

35.

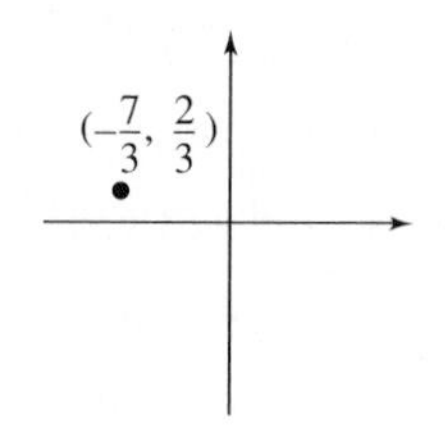

37.

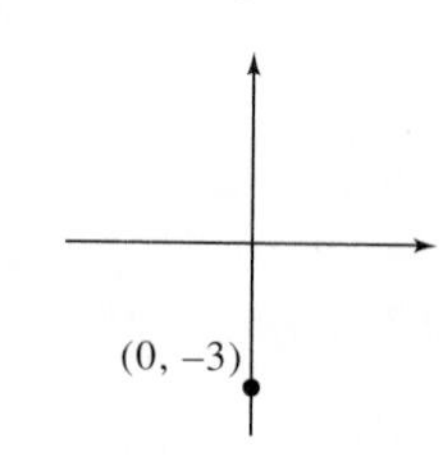

39. IV **41.** II **43.** (0, −4) **45.** (0, 0), (1, 1) **47.** (0, −3), (−1, −8)
49. (7, 7), (7, 3) **51.** (3, 0), (0, 3), (1, 2) **53.** (0, 0), (4, 1), (−4, −1)
55. (0, −3), (1, −3), (2, −3) **57.** (8, 0), (8, 1), (8, 2)

59.

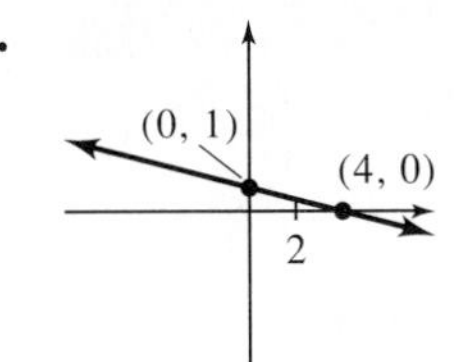

61.

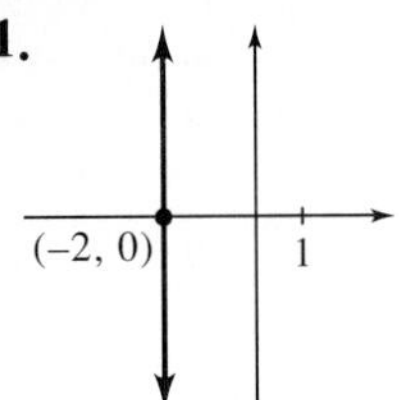

63.

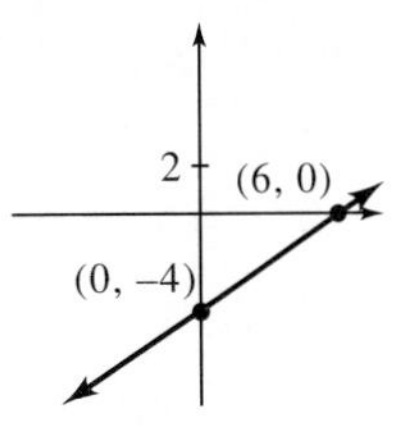

65.

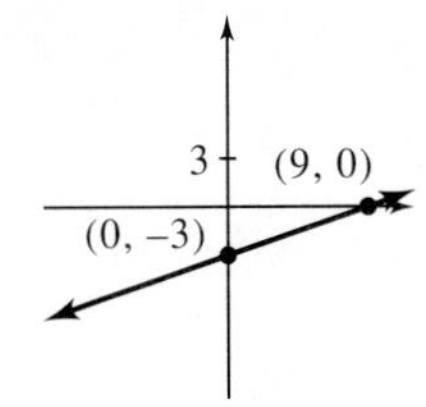

67.

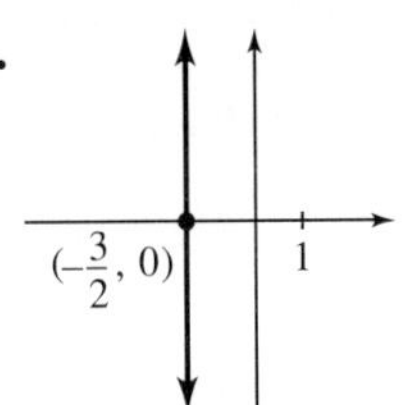

69.

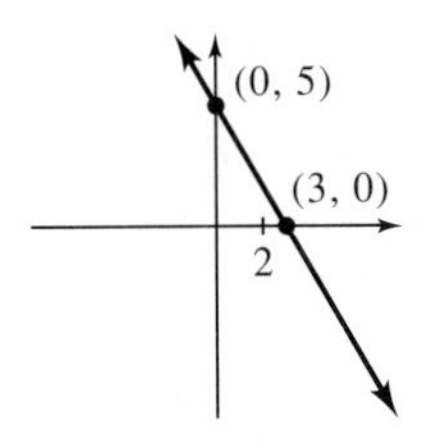

71.

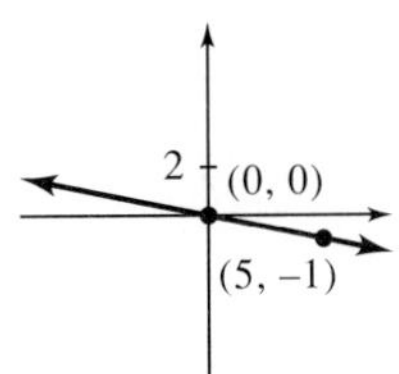

73.

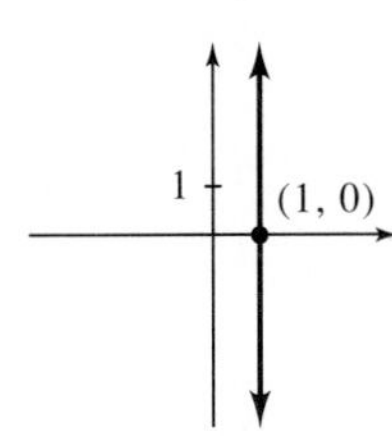

75.

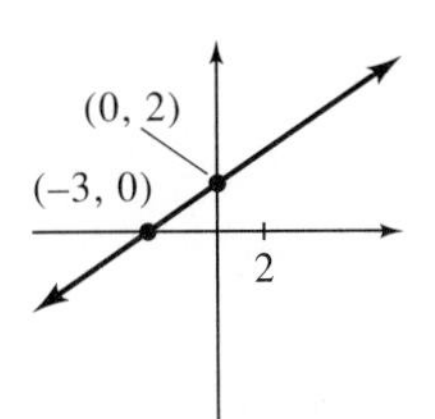

77.

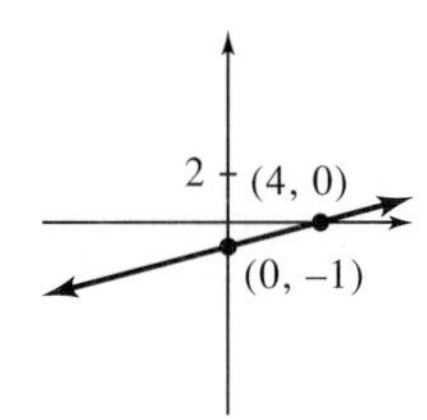

79.

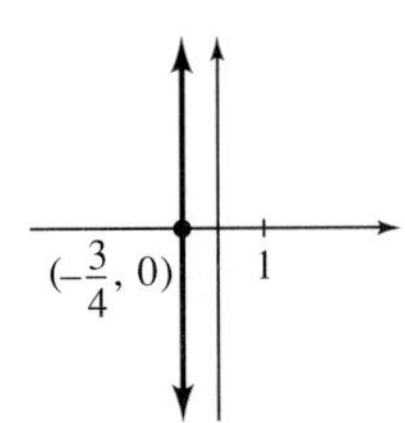

81.

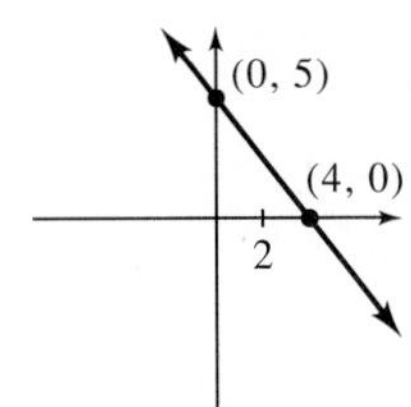

Challenge Problems

83. $x - y = 0$ **85.** $2x - y = 0$

PROBLEM SET 6.2

Warm-ups

1. 1 **3.** $\frac{1}{4}$ **5.** 1 **7.** -7 **9.** $-\frac{3}{5}$ **11.** -2 **13.** 0 **15.** no slope **17.** 3 **19.** $-\frac{2}{3}$

21. 0 **23.** $-4; \frac{1}{4}$ **25.** $-\frac{5}{3}; \frac{3}{5}$ **27.** $1; -1$

Practice Exercises

29. 1 **31.** $\frac{1}{5}$ **33.** $-\frac{3}{5}$ **35.** 1 **37.** -1 **39.** 0 **41.** no slope **43.** -5 **45.** 1

47. $-\frac{1}{2}$ **49.** $\frac{1}{3}$ **51.** $-\frac{1}{2}$ **53.** $\frac{1}{3}; -3$ **55.** $\frac{1}{2}; -2$ **57.** 0; no slope **59.** $-1; 1$ **61.** $2; -\frac{1}{2}$

63. $\frac{3}{5}; -\frac{5}{3}$ **65.** The slope should be $-\frac{1}{16}$. **67.** Its slope is $-\frac{11}{80}$. **69.** $-\frac{7}{18}$

Challenge Problems

71. no slope **73.** 0

PROBLEM SET 6.3

Warm-ups

1.

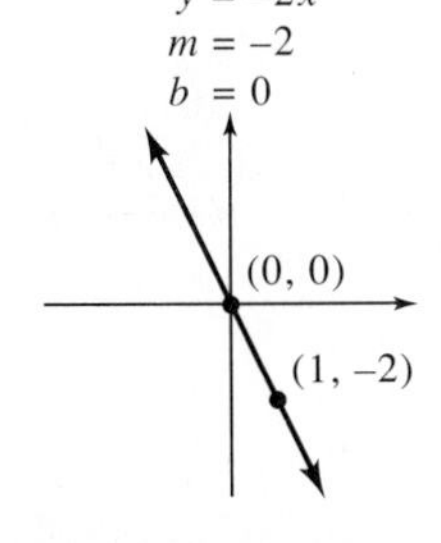

3.

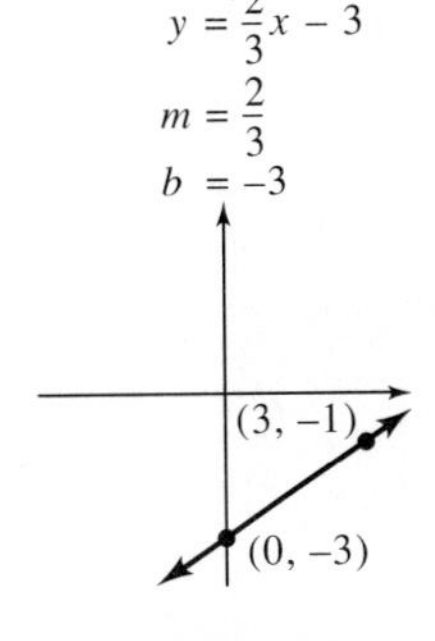

5.

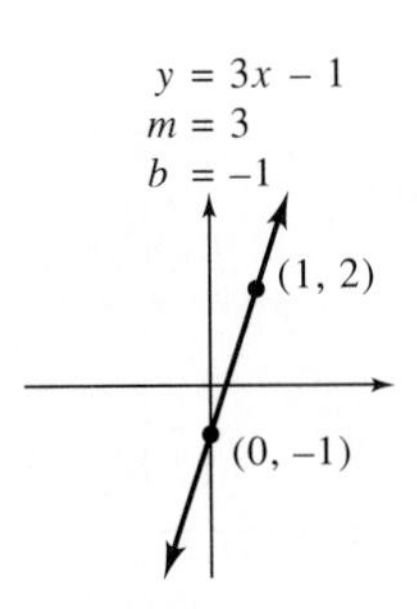

7. $x - 2y = 8$ **9.** $x - y = 0$ **11.** $3x - 4y = 29$ **13.** $3x - 2y = -6$ **15.** $y = \frac{2}{3}x - 1$ **17.** $x = 1$
19. $y = -2$

Practice Exercises

21. $y = -2x + 2$
$m = -2$
$b = 2$

23. $y = x - 3$
$m = 1$
$b = -3$

25. $y = \frac{2}{3}x + 2$
$m = \frac{2}{3}$
$b = 2$

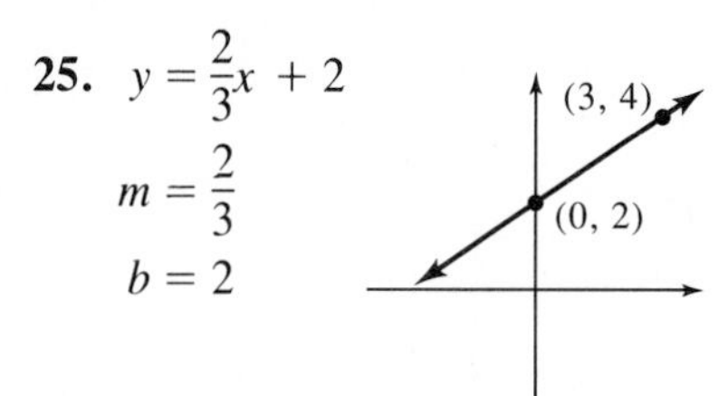

27. $y = -\frac{1}{3}x + 1$
$m = -\frac{1}{3}$
$b = 1$

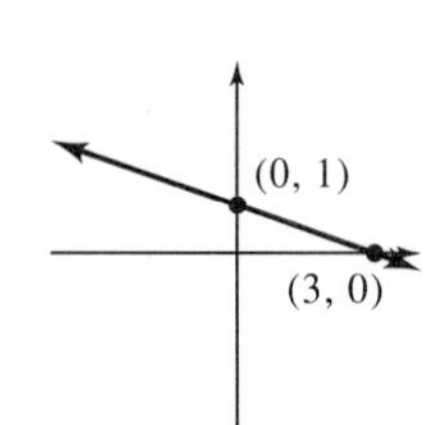

29. $y = \frac{3}{4}x - 1$
$m = \frac{3}{4}$
$b = -1$

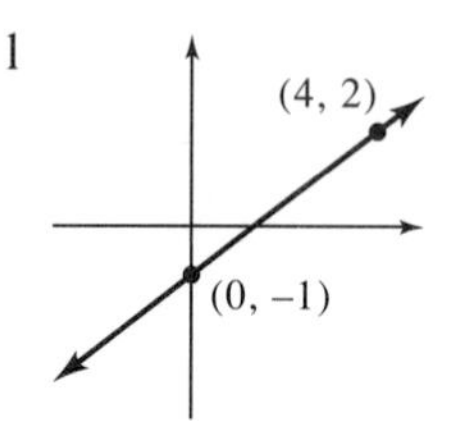

31. $x + y = -2$ **33.** $3x + 2y = 3$ **35.** $y = 2x - 2$ **37.** $x = -1$ **39.** $x = 2$ **41.** $x + 7y = 26$
43. $x - 3y = 22$ **45.** $x = 1$ **47.** $3x - 2y = 6$

Challenge Problems

49. $x = 0$ **51.** $y = 2x$ **53.** $x - 2y = -5$ **55.** $y = 2$

PROBLEM SET 6.4

1.

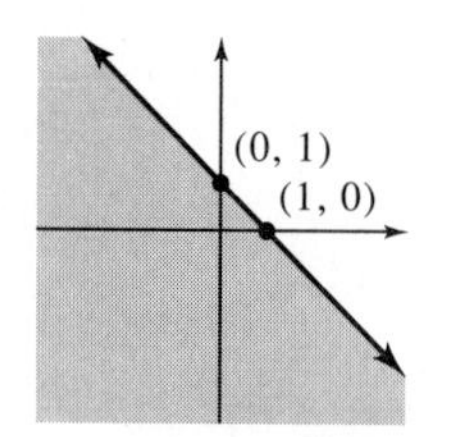

3.

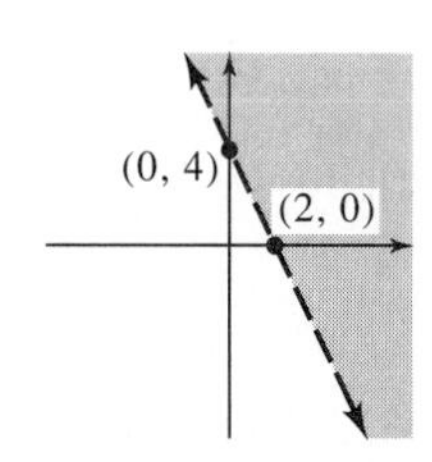

5.

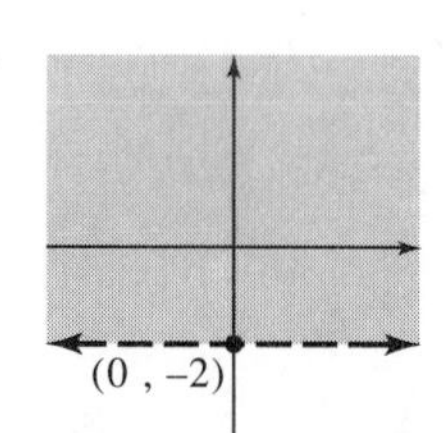

7.

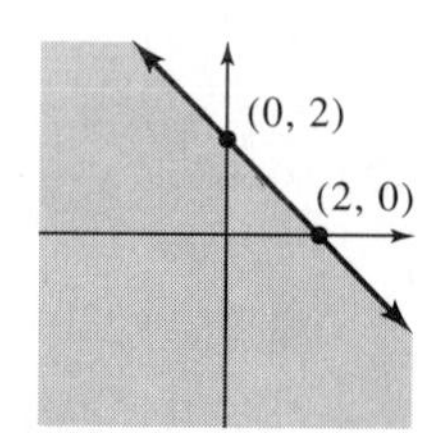

9.

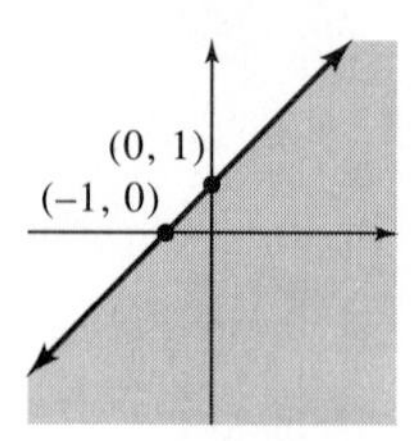

11.

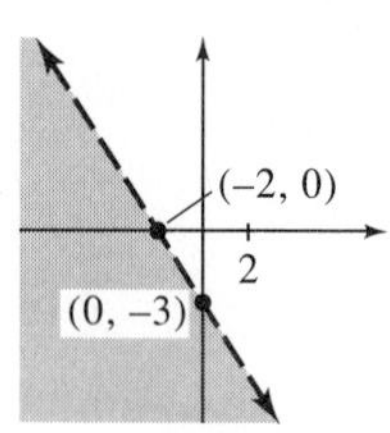

13.

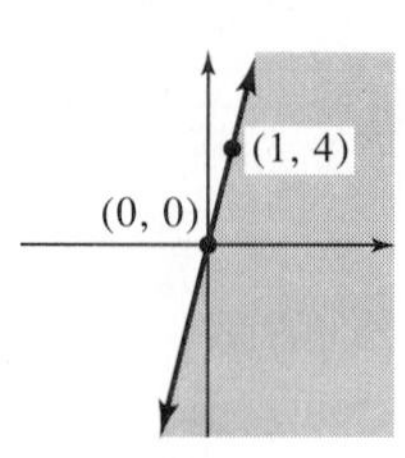

15.

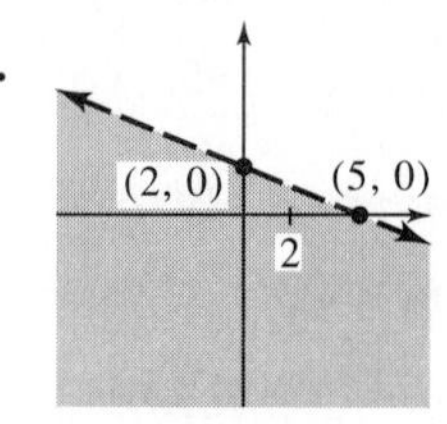

17.

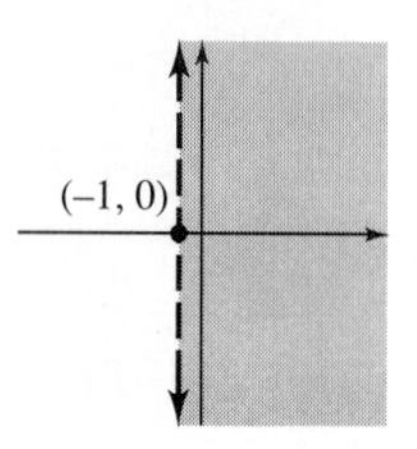

19.

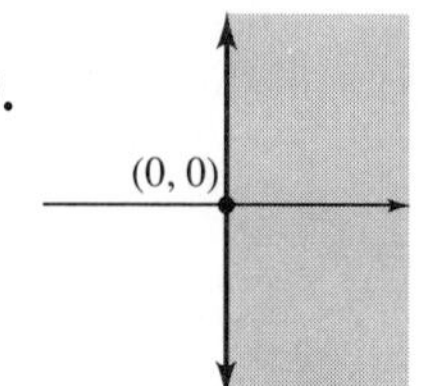

21.

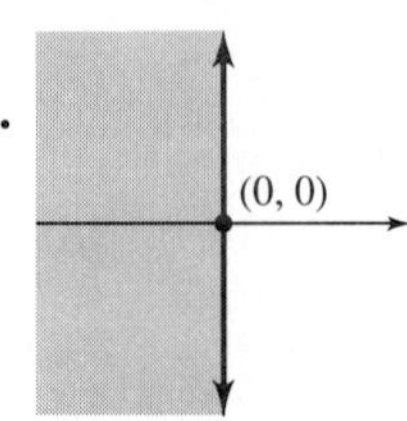

23.

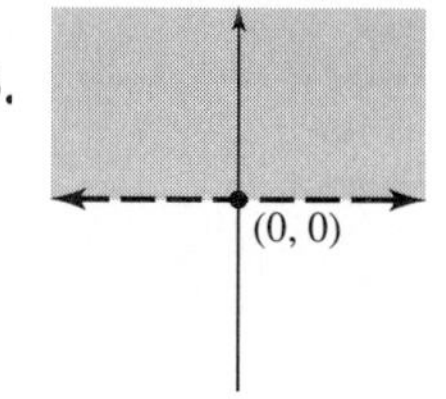

25.
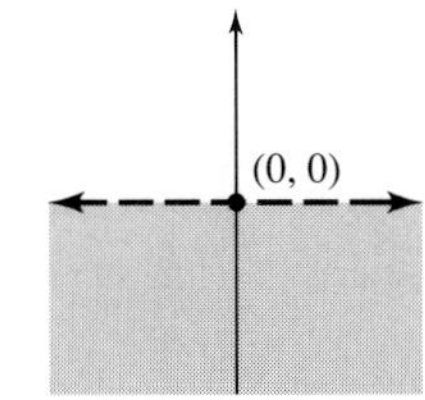

27.
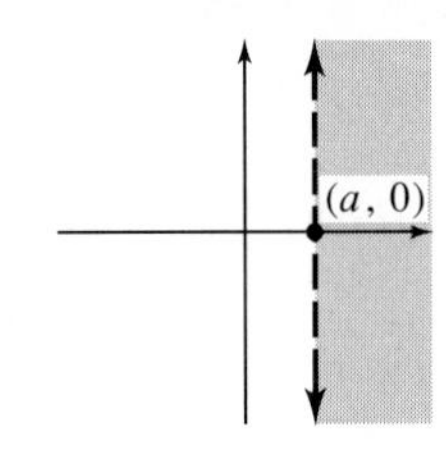

29.
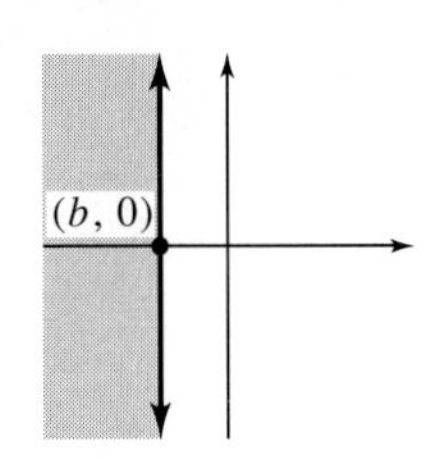

PROBLEM SET 6.5

Warm-ups

1. function **3.** not a function **5.** $f = \{(-1, 3), (0, 2), (1, 3)\}$; $\{2, 3\}$
7. $f(-1) = 2; f(0) = 0; f(1) = 0; f(7) = 42$ **9.** $h(0) = 1; h(3) = 1; h(-1) = 5; h(-3) = 19$

11. $\{x \mid x \neq 2; x \neq 3\}$ **13.** $\{x \mid x \geq 5\}$ **15.** $g(x) = -\frac{2}{3}x + 1$ **17.** not a function

19. domain: $\{x \mid -\pi < x < \pi\}$;
range: $\{y \mid -1 \leq y \leq 1\}$
21. domain: {all real numbers};
range: {all real numbers}

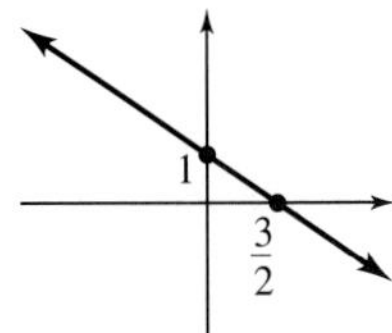

Practice Exercises

23. $g = \left\{(1, 12), (2, 6), (3, 4), (4, 3), \left(5, \frac{12}{5}\right)\right\}$; $\left\{\frac{12}{5}, 3, 4, 6, 12\right\}$ **25.** $G = \left\{\left(-1, \frac{1}{2}\right), (0, 0), \left(1, \frac{1}{2}\right)\right\}$; $\left\{0, \frac{1}{2}\right\}$

27. $f(-2) = 8; f(-1) = 3; f(0) = 0; f(1) = -1; f(2) = 0$ **29.** $h(-3) = 4; h(0) = 1; h(1) = 2$

31. {all real numbers} **33.** $\{t \mid t \neq -4\}$ **35.** $f(x) = -2x + 1$ **37.** $h(x) = \frac{3}{2}x + 2$

39. $G(x) = 4 - \frac{1}{3}x$

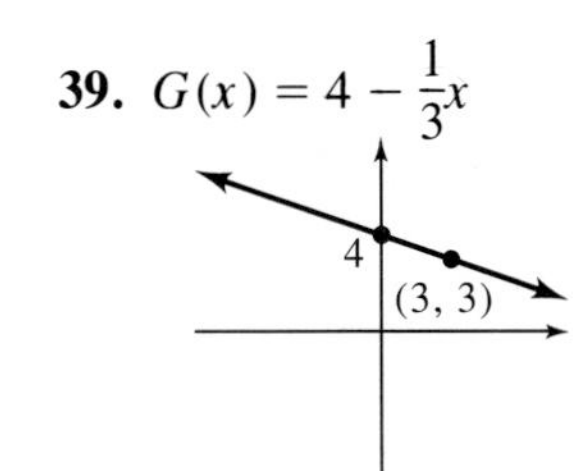

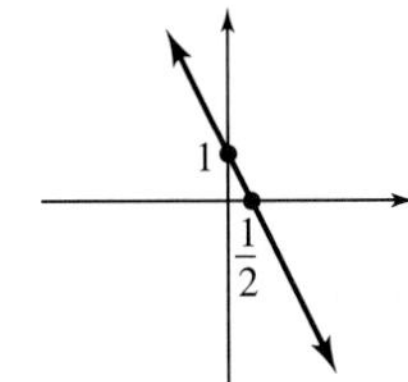

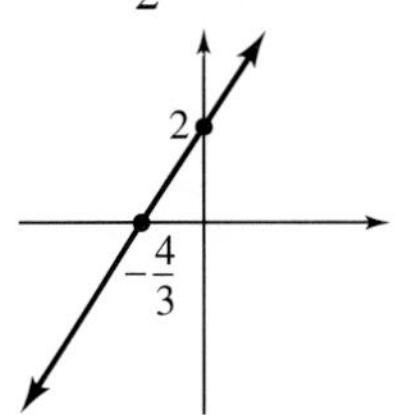

41. function **43.** not a function **45.** not a function **47.** domain: $\{x \mid -2 \leq x \leq 2\}$; range: $\{y \mid -1 \leq y \leq 1\}$
49. domain: $\{x \mid -4 \leq x \leq 4\}$; range: $\{y \mid -3 \leq y \leq 3\}$ **51.** domain: $\{x \mid x > 0\}$; range: $\{y \mid y > 0\}$
53. domain: {all real numbers}; range: {all real numbers}

Challenge Problems

55. (a) yes (b) domain: $\{t \mid 0 \leq t \leq 24\}$; range: $\{T \mid -10 \leq T \leq 20\}$ (c) 0°F; 20°F

REVIEW PROBLEMS CHAPTER 6

1. I **3.** III **5.** yes **7.** yes **9.** yes **11.** $(-2, 0), (0, -2)$ **13.** $(1, 0), (0, -3)$ **15.** $(0, -4)$

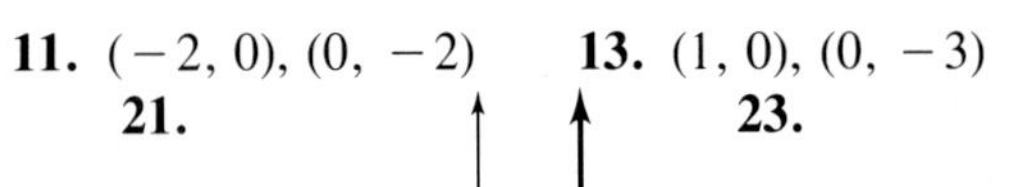
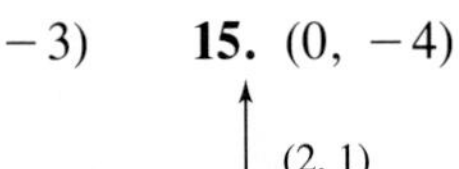

17.
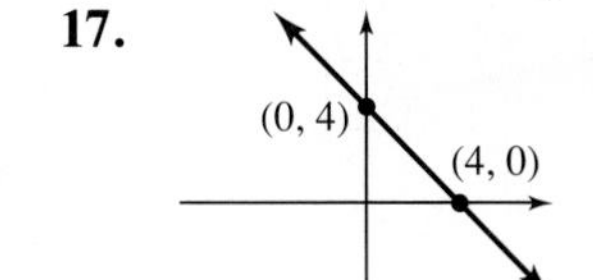

19.
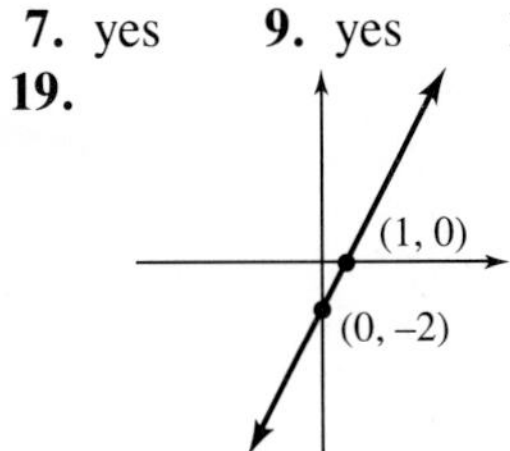

21.
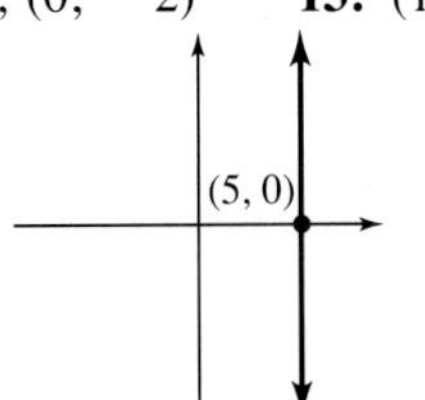

23.
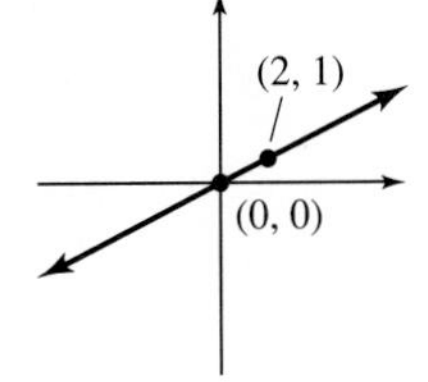

25. $-\frac{1}{3}$ **27.** no slope **29.** 0 **31.** $y = -2x + 4; m = -2; b = 4$ **33.** $y = -\frac{2}{3}x + 2; m = -\frac{2}{3}; b = 2$

35. $y = \frac{2}{3}x; m = \frac{2}{3}; b = 0$ **37.** $4x - y = 5$ **39.** $2x + y = 3$ **41.** $y = -3$ **43.** $x = -1$

45.

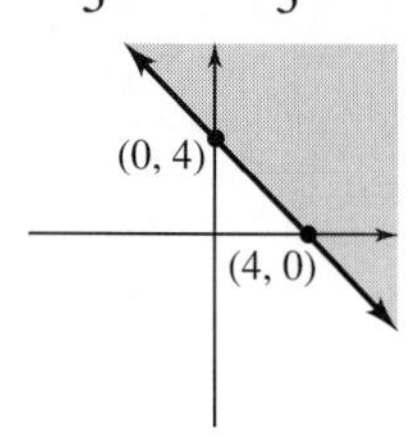

47.

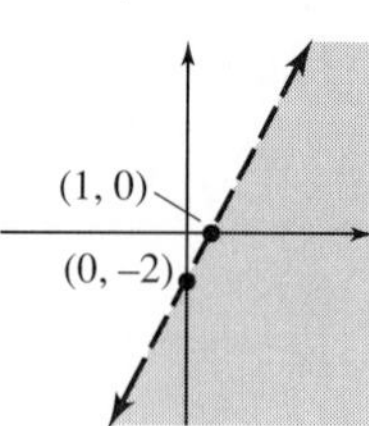

49.

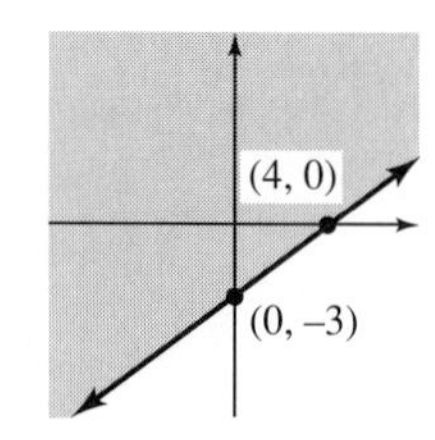

51. $f(2) = 2; f(-1) = -10$ **53.** $f(2) = 3; f(-1) = 0$

Let's Not Forget

54. -4 **55.** 9 **56.** $-4x - 3$ **57.** $-(a + b)$ **58.** 0 **59.** $\frac{x^2}{x + 2}$ **60.** $x^2 + 4xy + 4y^2$

61. $a^2 - 16$ **62.** $x^2 + 2x - 2$ **63.** $x^2 - 4$ **64.** $\frac{x}{x + 1}$ **65.** term **66.** term **67.** factor

68. factor **69.** 1 **70.** $\frac{2}{x + 1} + 5$ **71.** $x^2 - 1$ **72.** $y(4 + y)$ **73.** $(x + y)(x - y)$ **74.** factored

75. factored **76.** $(x + y)^2$ **77.** $(x - 2)(x^2 + 2x + 4)$ **78.** expression; $\frac{-4x - 15}{x + 4}$ **79.** equation; $\{-3\}$

80. inequality; $\{x \mid x \geq -15\}$ **81.** expression; $\frac{2 + x^2}{1 + 3x}$ **82.** expression; $3x - 7$

CHAPTER 6 TEST

1. C **2.** B **3.** A **4.** A **5.** D **6.** B **7.** A

8.

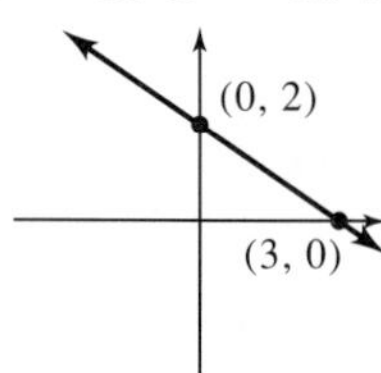

9.

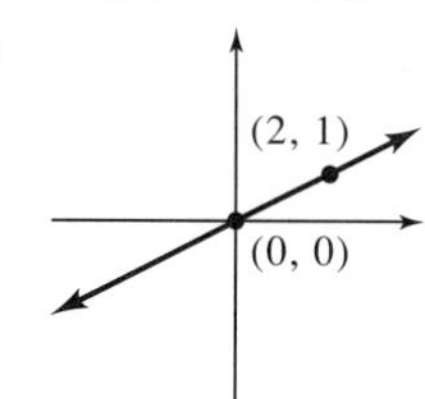

10.

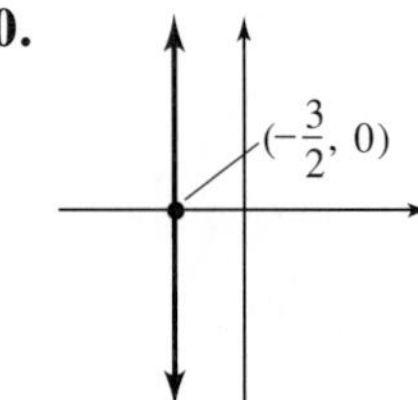

11.

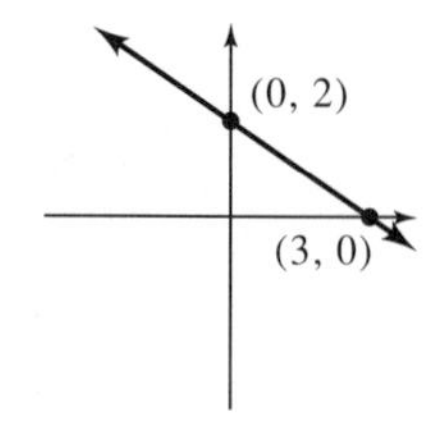

12.

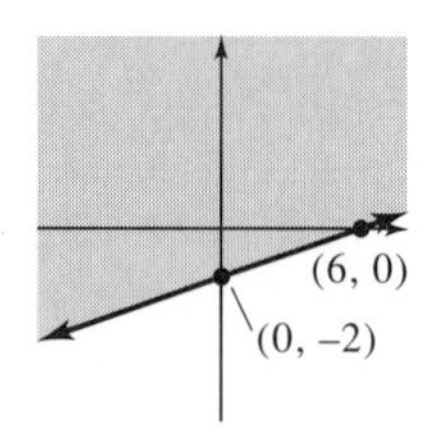

13.

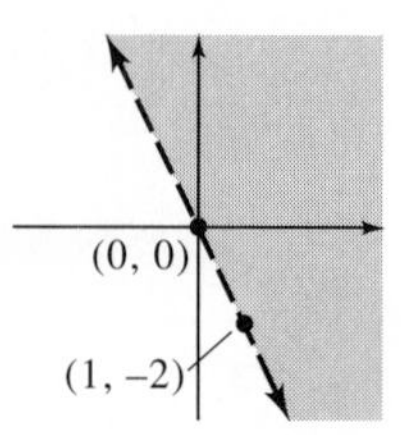

14. $3x - y = 5$ **15.** $y = -3$ **16.** 27 **17.** *a, b,* and *e*

PROBLEM SET 7.1

Warm-ups

1.

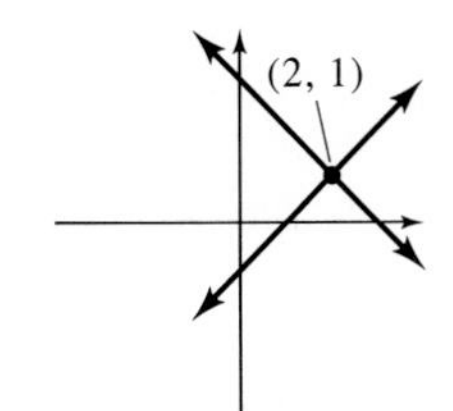

3.

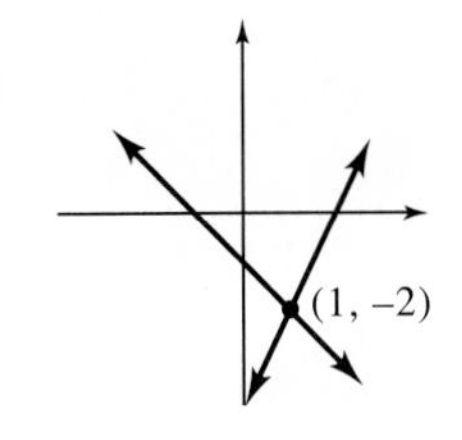

5.

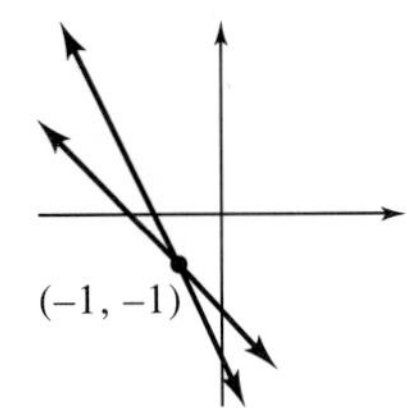

7.

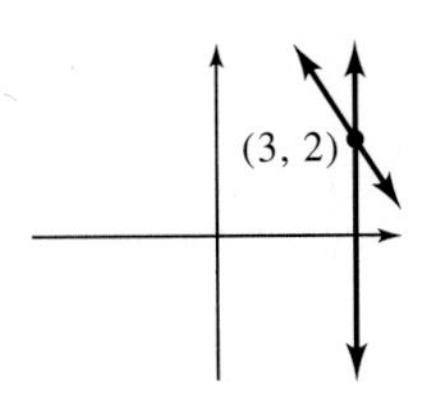

9.

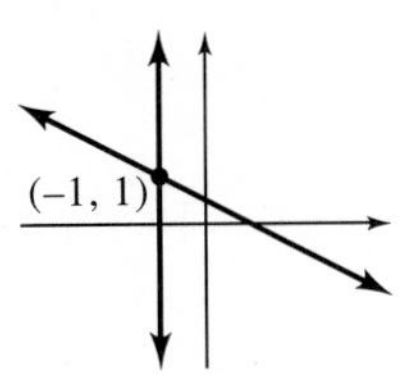

11. independent **13.** independent **15.** independent **17.** dependent

Practice Exercises

19.

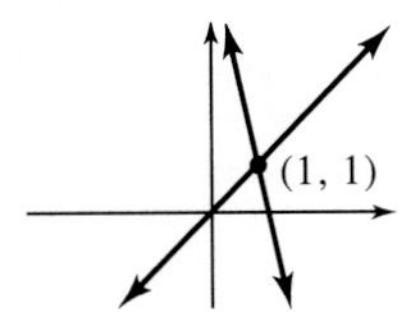

21.

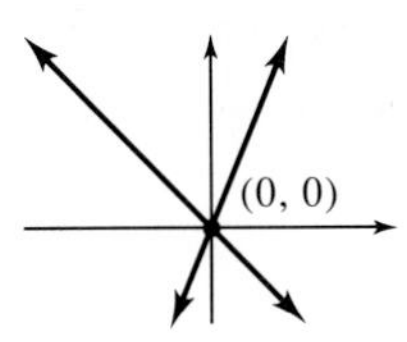

23.

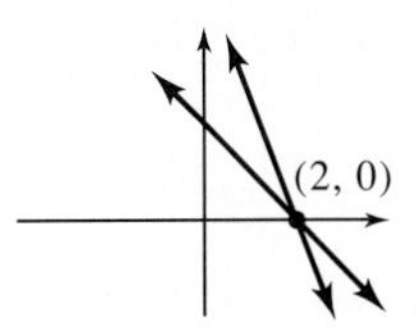

25.

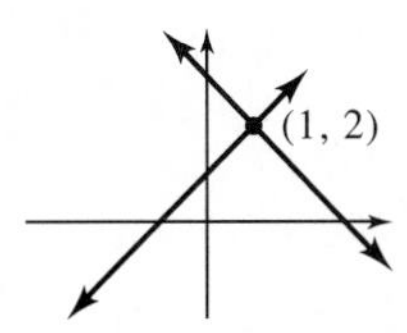

27.

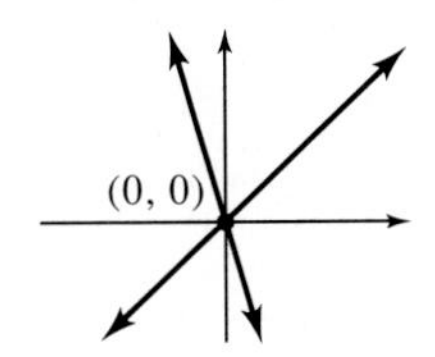

29.

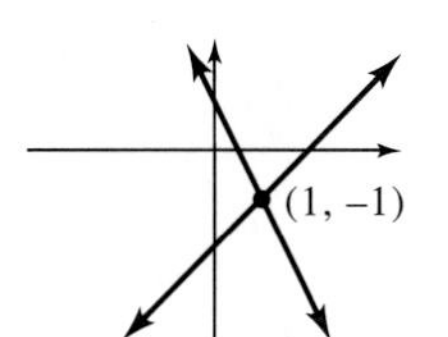

31.

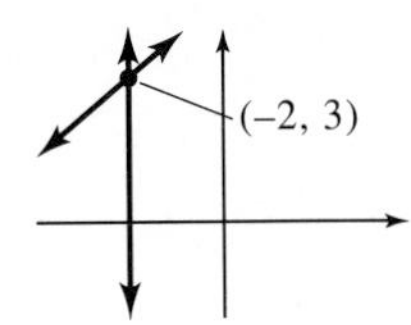

33.

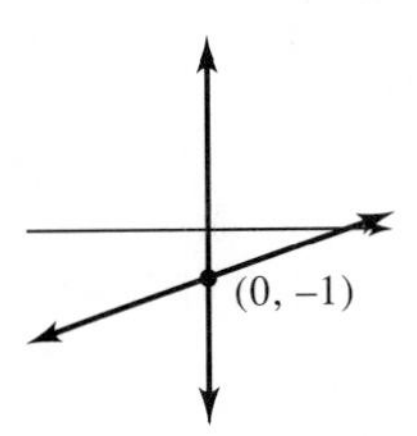

35. independent **37.** independent **39.** independent **41.** dependent

Challenge Problems

43. $\{(a, b)\}$ **45.** $\{(a, b)\}$

PROBLEM SET 7.2

Warm-ups

1. $\{(1, 1)\}$ **3.** $\{(0, 0)\}$ **5.** $\{(-1, 5)\}$ **7.** $\{(3, -1)\}$ **9.** $\{(1, -1)\}$ **11.** $\{2, -5)\}$ **13.** $\{(1, -2)\}$ **15.** $\{(-6, -1)\}$ **17.** dependent

Practice Exercises

19. $\{(1, 3)\}$ **21.** dependent **23.** $\{(-1, 7)\}$ **25.** $\{(5, 2)\}$ **27.** $\varnothing$ **29.** dependent **31.** $\{(11, 1)\}$ **33.** $\{(5, 3)\}$ **35.** $\{(-4, -7)\}$ **37.** $\{(2, -2)\}$ **39.** $\{(1, 0)\}$ **41.** $\{(2, 3)\}$

Challenge Problems

43. $\{(4-k, k)\}$ **45.** $\{(1, 3), (-1, -3)\}$

PROBLEM SET 7.3

Warm-ups

1. $\{(4, 0)\}$ **3.** $\{(2, 2)\}$ **5.** $\{(2, -1)\}$ **7.** $\{(3, -2)\}$ **9.** $\{(3, -1)\}$ **11.** dependent

Practice Exercises

13. $\{(3, 0)\}$ **15.** $\{(3, 3)\}$ **17.** $\{(3, -5)\}$ **19.** $\varnothing$ **21.** $\varnothing$ **23.** $\{(8, 2)\}$ **25.** $\{(-5, 4)\}$ **27.** $\{(3, -7)\}$
29. $\{(-1, -5)\}$ **31.** $\{(a, 0)\}$ **33.** $\left\{\left(\frac{2a+b}{3}, \frac{a-b}{3}\right)\right\}$

PROBLEM SET 7.4

Warm-ups

1. 16 and 32 **3.** 3,270 **5.** 15 m by 21 m **7.** 25 yds by 75 yds **9.** 1,250 attended Friday and 1,600 Saturday
11. $2.19 **13.** $9,000 at 10% and $6,000 at 14%

Practice Exercises

15. -6 and 12 **17.** They are $15 and $20. **19.** 1,200 **21.** $1.99 **23.** 99¢
25. 37 classic and 47 caffeine free **27.** 7 cm by 20 cm **29.** 40 white and 80 black **31.** 12 ft and 8 ft
33. 1 lb of peppermints and 3 lb of chocolates. **35.** Pencils are 29¢. **37.** Wool is $7.98 and cotton $2.99.
39. $200 at 16% and $600 at 12%

Challenge Problems

41. $4,770

PROBLEM SET 7.5

Warm-ups

1.

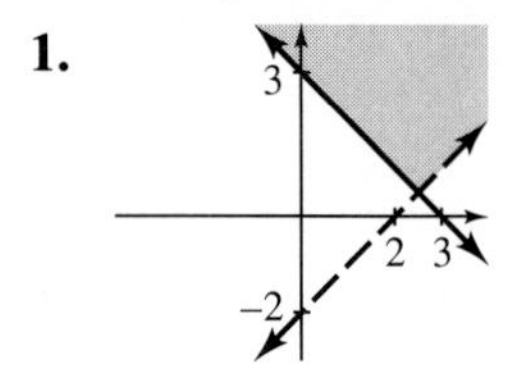

3.

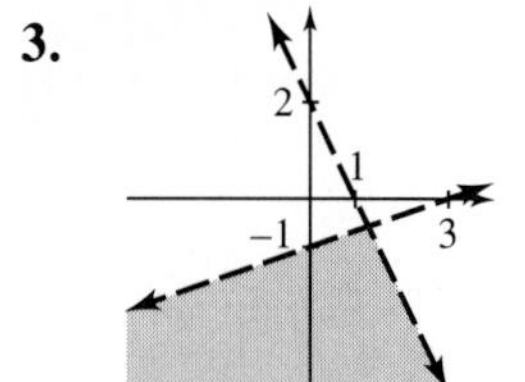

5.

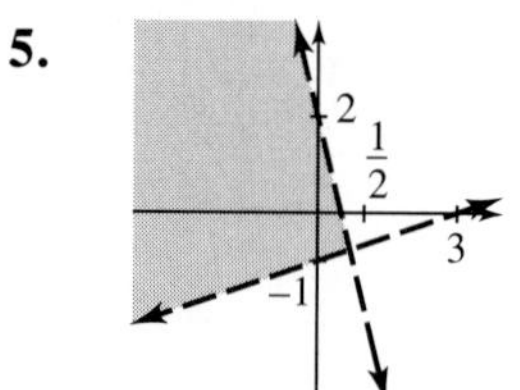

7.

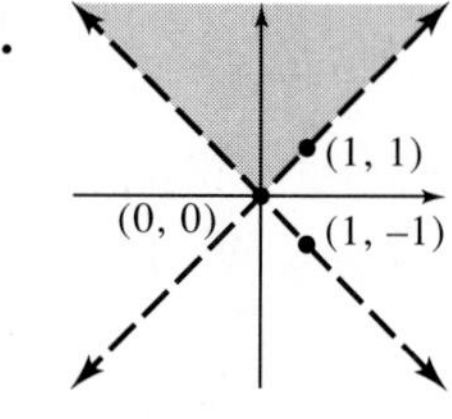

9.

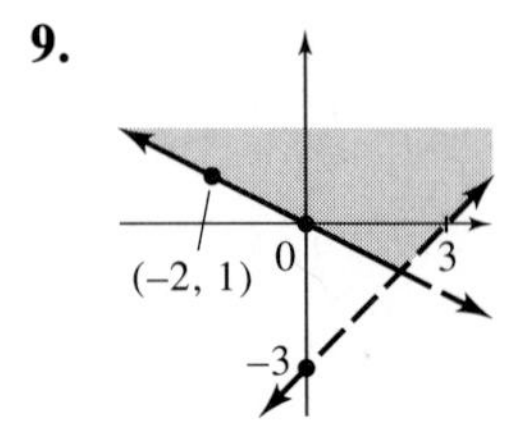

11.

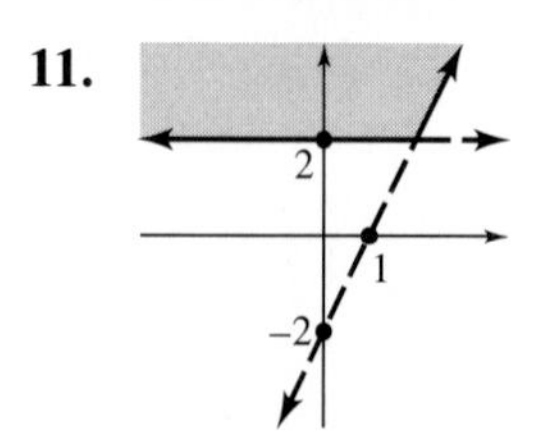

13.

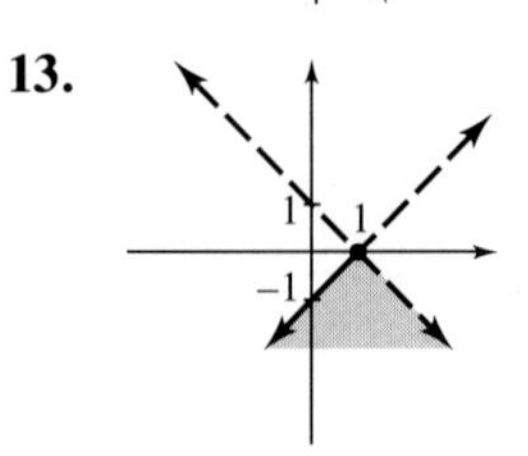

15.

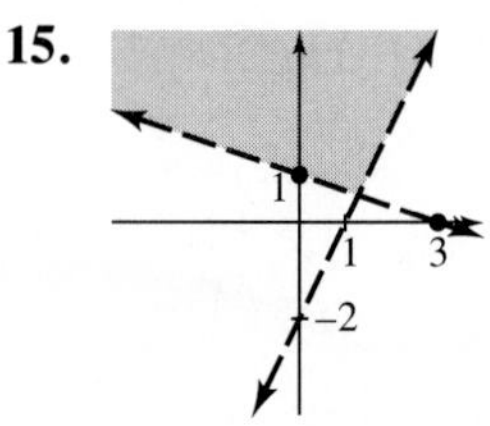

17.

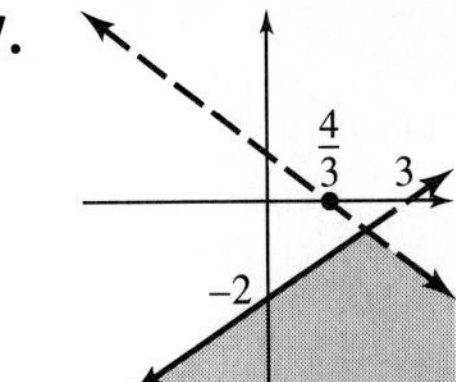

19.

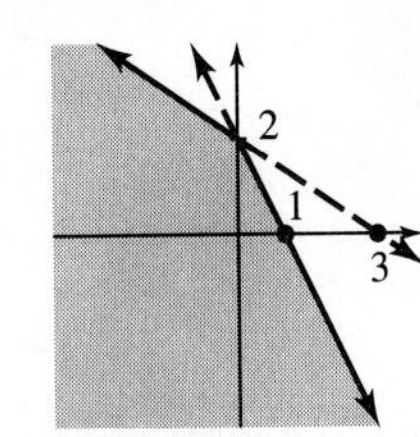

21.

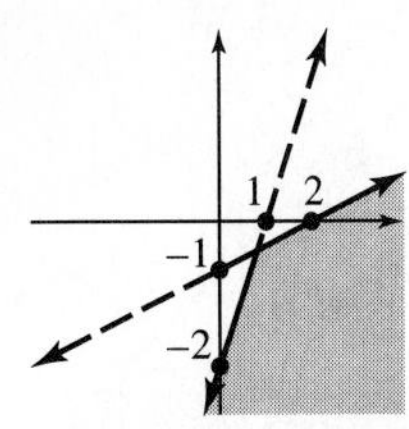

23.

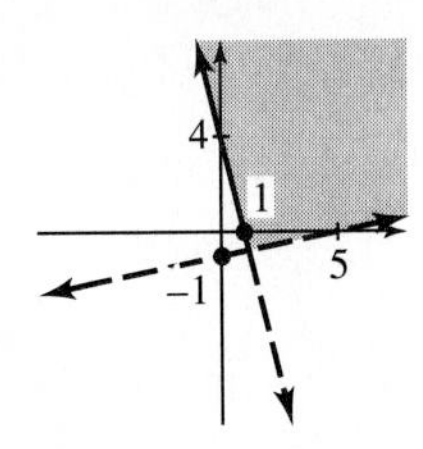

25.

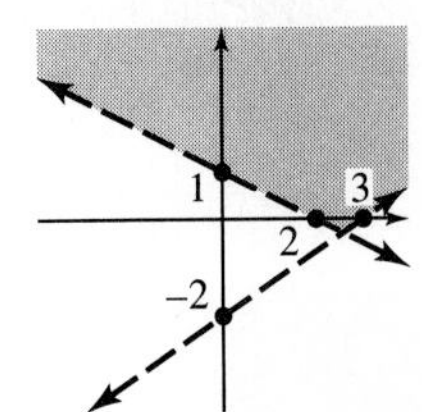

27.

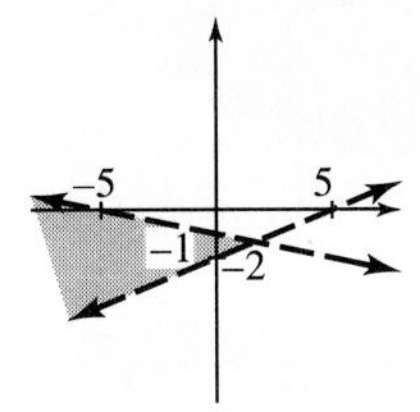

29.

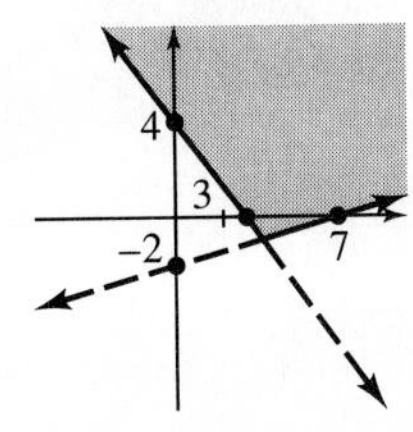

31.

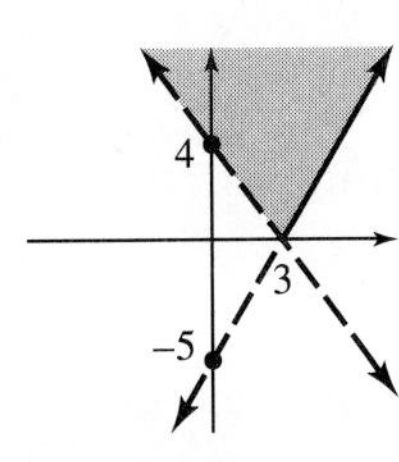

33.

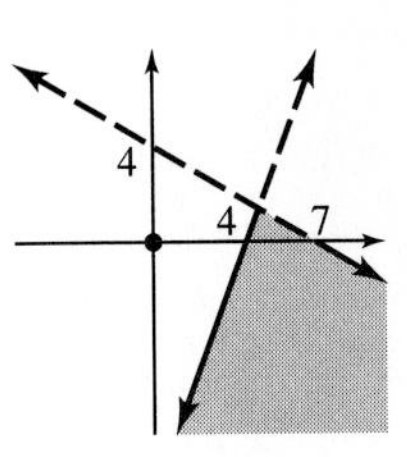

35.

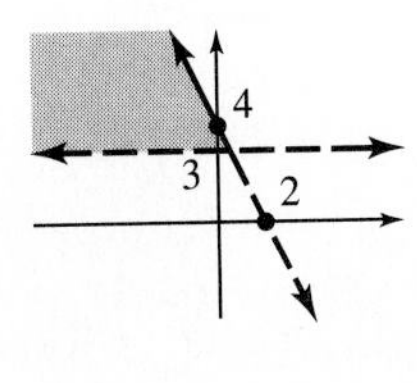

37.

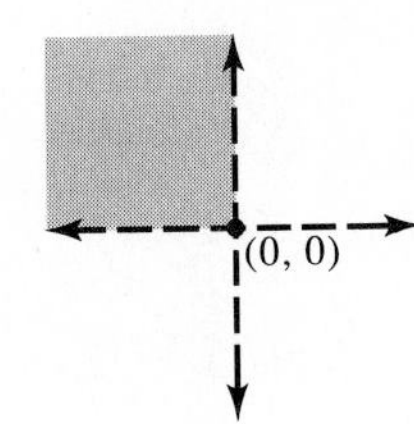

39.

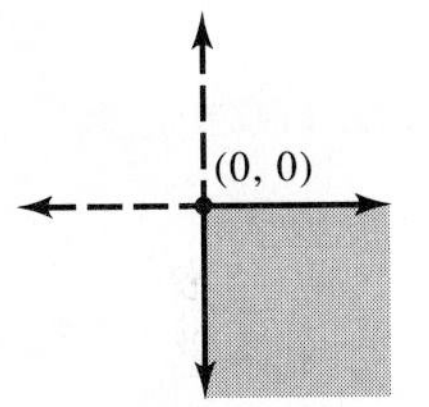

REVIEW PROBLEMS CHAPTER 7

1. independent **3.** dependent **5.** independent **7.** $\left\{\left(\frac{3}{2}, \frac{3}{2}\right)\right\}$ **9.** $\{(-1, -2)\}$ **11.** $\{(-1, 1)\}$

13. $\{(-2, 1)\}$ **15.** $\{(1, 1)\}$ **17.** $\{(0, 0)\}$

19.

21.

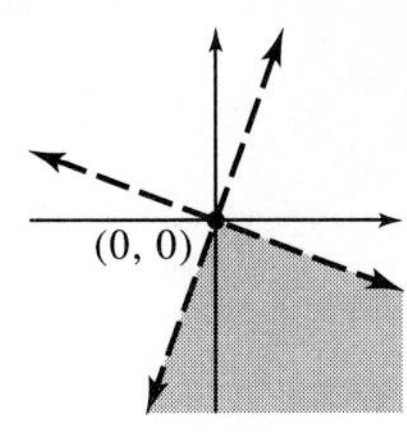

23.

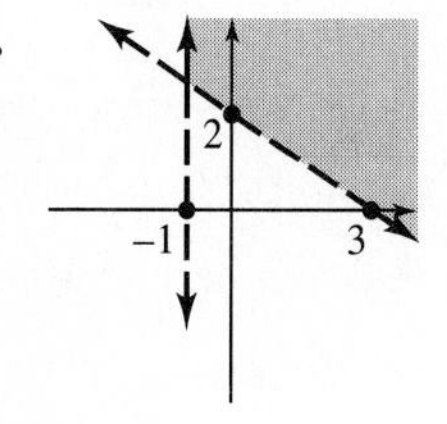

25. Joan has 27 and Sue has 9. **27.** 7 m by 12 m

Let's Not Forget

29. -16 **30.** 16 **31.** $5x - 3y$ **32.** $-a$ **33.** $9x^2 - 12x + 4$ **34.** $4x^2 - 25$ **35.** $6a^2 - ab - b^2$

36. $\frac{1}{x^3y}$ **37.** $2x + 2xy - y^2$ **38.** $2x^2 - xy - y^2$ **39.** $\frac{-2}{(x + 1)(x - 1)}$ **40.** term **41.** factor

42. term **43.** term **44.** 1 **45.** $\frac{2x}{x - 2}$ **46.** $1 + \frac{x - 2}{x + 2}$ **47.** factored **48.** $(x + 2z)(x - 2z)$

49. factored **50.** $b(a^2 - bc)$ **51.** prime **52.** $x\left(\frac{1}{2} + x\right)$ **53.** expression; $\frac{-3x - 1}{x + 1}$ **54.** equation; ∅

55. inequality; $\left\{x \mid x < -\frac{3}{2}\right\}$ **56.** expression; $\frac{2}{y}$ **57.** expression; $\frac{x + 2}{x}$

CHAPTER 7 TEST

1. C **2.** D **3.** C **4.** B **5.** A **6.** $\{(3, 1)\}$ **7.** $\varnothing$ **8.** $\{(2, -3)\}$ **9.** $\{(-1, 0)\}$

10.

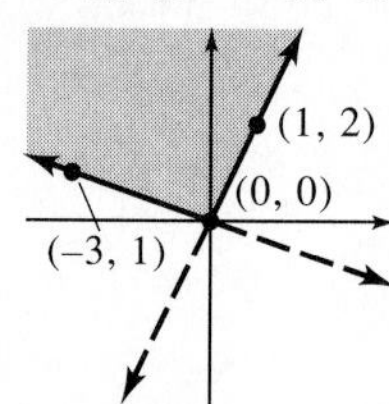

11.

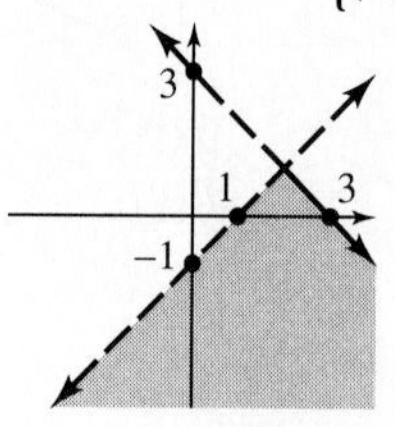

12. **13.**

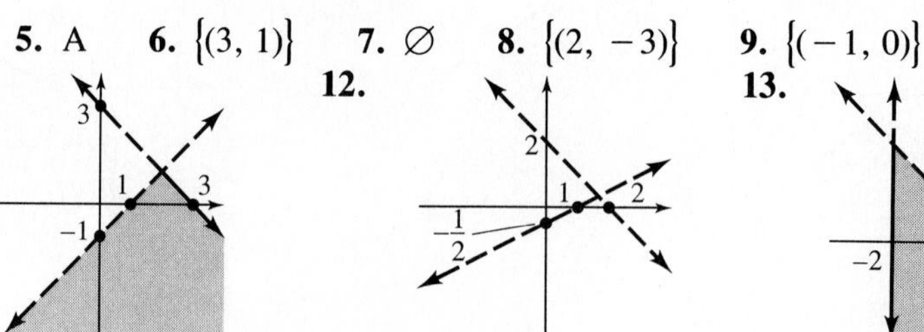

14. 8 ft by 15 ft **15.** 69¢

CHAPTER 8

PROBLEM SET 8.1

Warm-ups

1. 5 **3.** -6 **5.** $-\frac{2}{3}$ **7.** not a real number **9.** 2.83 **11.** 4 **13.** -4 **15.** $-\frac{1}{4}$

17. not a real number **19.** $\frac{1}{2}$ **21.** 2 **23.** -2 **25.** $-\frac{1}{2}$ **27.** y **29.** z^6 **31.** $2x^4$ **33.** $3t^3$

35. $2z$

Practice Exercises

37. not a real number **39.** -4 **41.** $-\frac{2}{9}$ **43.** not a real number **45.** not a real number **47.** 7

49. 1 **51.** $-\frac{1}{2}$ **53.** $\frac{8}{7}$ **55.** 10 **57.** -2 **59.** $-\frac{2}{3}$ **61.** -1 **63.** $-\frac{1}{5}$ **65.** 3 **67.** -5

69. $-\frac{1}{3}$ **71.** -3 **73.** $\frac{1}{3}$ **75.** 3 **77.** 0 **79.** not a real number **81.** y^5 **83.** z^{10} **85.** $4x^3$

87. $7t$ **89.** $5y^2z$ **91.** $11x^2yz^9$ **93.** $13abc$ **95.** $4x^2y$ **97.** $12s^2t$ **99.** $x^4y^2z^5$

101. not a real number **103.** 3.16

Challenge Problems

105. 4 **107.** 3 **109.** -2 **111.** -3

PROBLEM SET 8.2

Warm-ups

1. $2\sqrt{2}$ **3.** $5\sqrt{2}$ **5.** $2\sqrt{10}$ **7.** $3\sqrt[3]{2}$ **9.** $2\sqrt[3]{5}$ **11.** $2y\sqrt[3]{2}$ **13.** $-2x^2\sqrt[3]{3}$ **15.** $\frac{\sqrt{6}}{7}$ **17.** $\frac{3\sqrt{3}}{2}$

Practice Exercises

19. $5\sqrt{5}$ **21.** $10\sqrt{6}$ **23.** $6\sqrt{3}$ **25.** $2\sqrt{17}$ **27.** $3\sqrt{11}$ **29.** $4\sqrt{7}$ **31.** $10\sqrt{3}$ **33.** $4\sqrt{2}$

35. $3\sqrt{6}$ **37.** $4\sqrt{6}$ **39.** $5\sqrt[3]{2}$ **41.** $-2\sqrt[3]{2}$ **43.** $\frac{\sqrt[3]{5}}{2}$ **45.** $\frac{2\sqrt[3]{4}}{3}$ **47.** $3\sqrt[3]{4}$ **49.** $2\sqrt[3]{7}$

51. $5\sqrt[3]{4}$ **53.** $-2\sqrt[3]{18}$ **55.** $2\sqrt[3]{20}$ **57.** $-2\sqrt[3]{13}$ **59.** $3y^2\sqrt[3]{2}$ **61.** $-2x\sqrt[3]{4}$ **63.** $2x^2y\sqrt[3]{9}$

65. not a real number

Challenge Problems

67. The horizon is 15 mi away. **69.** $2\sqrt[4]{2}$ **71.** $2\sqrt[6]{2}$

PROBLEM SET 8.3

Warm-ups

1. $9\sqrt{5}$ **3.** $11\sqrt[3]{6}$ **5.** $-\sqrt{3}$ **7.** $3\sqrt{2}$ **9.** 0 **11.** $4\sqrt{5}$ **13.** $3\sqrt{2x}$ **15.** $-6\sqrt{x}$ **17.** $3x\sqrt[3]{2}$

Practice Exercises

19. $3\sqrt{3}$ **21.** $8\sqrt[3]{11}$ **23.** $5\sqrt{5}$ **25.** $4\sqrt{2}$ **27.** $\sqrt{3}$ **29.** $4\sqrt{5}$ **31.** $-3\sqrt{2}$ **33.** $23\sqrt{x}$
35. $-4\sqrt[3]{2x}$ **37.** $-17\sqrt[3]{2}$ **39.** $-\sqrt[3]{2}$ **41.** $5\sqrt{5x}$ **43.** $-7\sqrt{x}$ **45.** $29\sqrt{x}$ **47.** $-4\sqrt[3]{3x}$
49. $3x\sqrt[3]{3}$

PROBLEM SET 8.4

Warm-ups

1. $\sqrt{35}$ **3.** $\sqrt[3]{15}$ **5.** $8\sqrt{15}$ **7.** 60 **9.** $64x$ **11.** $\sqrt{42} - \sqrt{7}$ **13.** $t - \sqrt{ty}$ **15.** $7 + \sqrt{10}$
17. 2 **19.** $74 + 40\sqrt{3}$ **21.** $\sqrt{5}$ **23.** $\sqrt[3]{6}$ **25.** 3 **27.** -3 **29.** $\sqrt{6}$ **31.** $\frac{\sqrt{6}}{8}$
33. $\frac{x + \sqrt{xy}}{x}$ **35.** $-3(\sqrt{2} + \sqrt{3})$ **37.** $-3 - 2\sqrt{2}$

Practice Exercises

39. $\sqrt{15}$ **41.** $\sqrt[3]{28}$ **43.** $3\sqrt{2}$ **45.** 3 **47.** $8\sqrt{6}$ **49.** $6\sqrt[3]{18}$ **51.** $24\sqrt{6}$ **53.** $10y$ **55.** 7
57. 18 **59.** $36z$ **61.** $\sqrt{15} - \sqrt{5}$ **63.** $x - \sqrt{xy}$ **65.** $2\sqrt{10} - 13$ **67.** $x - 4\sqrt{xy} + 3y$ **69.** 1
71. 3 **73.** $11 + 6\sqrt{2}$ **75.** $14 + 8\sqrt{3}$ **77.** $58 + 24\sqrt{5}$ **79.** 2 **81.** $\sqrt[3]{3}$ **83.** 5 **85.** -2
87. $3\sqrt{2}$ **89.** $-3\sqrt[3]{2}$ **91.** $\sqrt{5}$ **93.** $\frac{5\sqrt{3}}{6}$ **95.** $\frac{\sqrt{15} - 3}{6}$ **97.** $\frac{\sqrt{xy} - y}{y}$ **99.** $-3(\sqrt{2} - \sqrt{3})$
101. $\frac{7 + 2\sqrt{10}}{-3}$ **103.** $2\sqrt{30} - 11$ **105.** $\frac{3s + 7\sqrt{st} - 2t}{9s - t}$ **107.** $5, \frac{5\sqrt{3}}{3}, \frac{10\sqrt{3}}{3}$ **109.** $\sqrt{2}, \sqrt{2}, 2$
111. $\frac{1}{2}, \frac{\sqrt{2}}{2}, \frac{\sqrt{2}}{2}$ **113.** $\sqrt{6}, \sqrt{3}, \sqrt{3}$

PROBLEM SET 8.5

Warm-ups

1. 4 **3.** 5 **5.** -2 **7.** 9 **9.** 27 **11.** -4

Practice Exercises

13. 7 **15.** 9 **17.** 6 **19.** 7 **21.** -5 **23.** -10 **25.** -7 **27.** -6 **29.** 125 **31.** 4
33. -27 **35.** -729

Challenge Problems

37. $\frac{1}{4}$ **39.** $-\frac{1}{16}$ **41.** $\frac{1}{5}$

REVIEW PROBLEMS CHAPTER 8

1. $8\sqrt{2}$ **3.** $2 + \sqrt{6}$ **5.** $16 + 6\sqrt{7}$ **7.** $8x$ **9.** 36 **11.** $6 + 3\sqrt{2} - 2\sqrt{3} - \sqrt{6}$ **13.** 10 **15.** 2
17. $26 - 8\sqrt{3}$ **19.** $\sqrt[3]{2}$ **21.** $2\sqrt{3}$ **23.** $2\sqrt[3]{3}$ **25.** $2\sqrt{6} - 8$ **27.** $\sqrt{6}$ **29.** $2\sqrt{6}$
31. $2\sqrt{2}$ **33.** $\frac{2\sqrt{x}}{x}$ **35.** $\frac{\sqrt{5}}{15}$ **37.** $4 + 2\sqrt{2}$ **39.** $\frac{16 + 11\sqrt{6}}{10}$ **41.** -4 **43.** $-\frac{1}{8}$ **45.** 125

Let's Not Forget

46. -9 **47.** 9 **48.** -2 **49.** -1 **50.** $8 - 2\sqrt{15}$ **51.** 2 **52.** $\frac{a-1}{a^3}$ **53.** $10 + 2\sqrt{3}$
54. $\frac{-2}{x+1}$ **55.** $\frac{1}{x+1}$ **56.** $\frac{x^2 - 3x}{(x+1)(x-1)}$ **57.** term **58.** factor **59.** term **60.** term **61.** 12
62. $4\sqrt{3} + 6$ **63.** $2 + \sqrt{3}$ **64.** $\sqrt{3}(a+2)(a-2)$ **65.** factored **66.** factored **67.** factored
68. $(x + y)(x^2 - xy + y^2)$ **69.** expression; $\frac{x+7}{(x-1)(x+3)}$ **70.** equation; $\{7\}$ **71.** inequality; $\{x \mid x \geq -9\}$
72. expression; $\frac{2(x+3)}{x-1}$

CHAPTER 8 TEST

1. B **2.** D **3.** C **4.** D **5.** B **6.** A **7.** $12\sqrt{35}$ **8.** $4\sqrt{3} - 3\sqrt{2}$ **9.** -25
10. $31 + 4\sqrt{21}$ **11.** $\sqrt{15} - 5 + 6\sqrt{2} - 2\sqrt{30}$ **12.** -3 **13.** 0 **14.** $\frac{\sqrt{3}}{2}$
15. $-\sqrt{6} + 3 + 2\sqrt{2} - 2\sqrt{3}$ **16.** -7 **17.** 25

CHAPTER 9

PROBLEM SET 9.1

Warm-ups

1. $\{\pm 5\}$ **3.** $\{\pm 10\}$ **5.** $\{\pm 11\}$ **7.** $\{\pm 15\}$ **9.** $\{\pm\sqrt{3}\}$ **11.** $\{\pm 2\sqrt{6}\}$ **13.** $\{\pm 4\sqrt{2}\}$
15. $\{\pm 3\sqrt{6}\}$ **17.** $\{\pm 1\}$ **19.** $\left\{\pm\frac{3}{2}\right\}$ **21.** $\{\pm 7\}$ **23.** $\{-5, -1\}$ **25.** $\{-7, 1\}$ **27.** $\{-2 \pm \sqrt{5}\}$
29. $\{-5 \pm 2\sqrt{3}\}$ **31.** $\left\{\frac{1}{6}, \frac{5}{6}\right\}$ **33.** $\left\{\frac{-20}{21}, \frac{8}{21}\right\}$ **35.** $\left\{\frac{1}{3} \pm \frac{\sqrt{3}}{2}\right\}$ **37.** The hypotenuse is $\sqrt{2}$ ft.
39. The ladder is $8\sqrt{6}$ ft above the ground. **41.** The side is 7 in.

Practice Exercises

43. $\{\pm 4\}$ **45.** $\{\pm 17\}$ **47.** $\{\pm 8\}$ **49.** $\{\pm 18\}$ **51.** $\{\pm 22\}$ **53.** $\left\{\pm\frac{9}{2}\right\}$ **55.** $\{\pm\sqrt{23}\}$
57. $\{\pm 3\sqrt{7}\}$ **59.** $\{4, 12\}$ **61.** $\{-10, 6\}$ **63.** $\{5 \pm \sqrt{3}\}$ **65.** $\{-7 \pm \sqrt{11}\}$ **67.** $\{5 \pm 2\sqrt{6}\}$
69. $\{3 \pm 5\sqrt{5}\}$ **71.** $\left\{\frac{1}{15}, \frac{11}{15}\right\}$ **73.** $\left\{-\frac{51}{20}, \frac{11}{20}\right\}$ **75.** $\left\{-\frac{2}{3} \pm \frac{\sqrt{5}}{2}\right\}$ **77.** The legs are $3\sqrt{2}$ m.
79. There are two such numbers, -7 and 3. **81.** The radius is 90 in. **83.** Warren should buy $9\sqrt{29}$ m.
85. The dimensions of the place mat are $6\sqrt{5}$ in. by 12 in. **87.** The radius is $8\sqrt{2}$ ft. **89.** The radius is $\frac{3}{2}$ in.

Challenge Problems

91. $\{\pm\sqrt{c}\}$ **93.** $\{a \pm 2\}$

PROBLEM SET 9.2

Warm-ups

1. 4 **3.** 9 **5.** 36 **7.** $\frac{25}{4}$ **9.** $\frac{1}{4}$ **11.** $\frac{169}{4}$ **13.** $\frac{1}{16}$ **15.** $\frac{1}{25}$ **17.** $\frac{4}{9}$ **19.** $\{-5, 1\}$
21. $\{-7, 1\}$ **23.** $\{1\}$ **25.** $\{-2\}$ **27.** $\{-2, 7\}$ **29.** $\{-3 \pm \sqrt{11}\}$ **31.** $\{4 \pm \sqrt{5}\}$ **33.** $\{2 \pm \sqrt{2}\}$
35. $\{-2, 1\}$ **37.** $\{-4, 2\}$ **39.** $\{-\frac{1}{2}, 3\}$ **41.** $\left\{\frac{5}{2} \pm \frac{\sqrt{23}}{2}\right\}$

Practice Exercises

43. $\{-6, 2\}$ **45.** $\{-8, 2\}$ **47.** $\{2\}$ **49.** $\{2, 6\}$ **51.** $\{-4, 3\}$ **53.** $\{2, 3\}$ **55.** $\{-5\}$ **57.** $\{-1\}$
59. $\left\{-\frac{1}{2}, 2\right\}$ **61.** $\{-7, 2\}$ **63.** $\{-3, 1\}$ **65.** $\left\{\frac{-3}{4} \pm \frac{\sqrt{17}}{4}\right\}$ **67.** $\left\{\frac{5}{4} \pm \frac{\sqrt{33}}{4}\right\}$ **69.** $\left\{\frac{1}{2}, 1\right\}$ **71.** $\left\{\frac{-1}{3}, 1\right\}$

Challenge Problems

73. $\{-1 \pm \sqrt{c+1}\}$ **75.** $\left\{-\frac{b}{2} \pm \frac{\sqrt{b^2 - 4c}}{2}\right\}$

PROBLEM SET 9.3

Warm-ups

1. $a = 4; b = -5; c = 7$ **3.** $x^2 - 8x - 7 = 0; a = 1; b = -8; c = -7$ **5.** $a = 1; b = 0; c = 0$
7. $3y^2 - y + 5 = 0; a = 3; b = -1; c = 5$ **9.** $2x^2 + 4x - 7 = 0; a = 2; b = 4; c = -7$
11. $x^2 + x - 7 = 0; a = 1; b = 1; c = -7$ **13.** $y^2 - y - 20 = 0; a = 1; b = -1; c = -20$ **15.** $\{-2, -1\}$
17. $\{-1, 4\}$ **19.** $\left\{-3, -\frac{1}{2}\right\}$ **21.** $\left\{-\frac{2}{3}, \frac{1}{2}\right\}$ **23.** $\left\{-\frac{3}{2}, \frac{1}{3}\right\}$ **25.** $\{1\}$ **27.** $\left\{\frac{3 \pm \sqrt{13}}{2}\right\}$ **29.** $\left\{\frac{7 \pm \sqrt{17}}{4}\right\}$
31. $\left\{\frac{-5 \pm \sqrt{17}}{2}\right\}$ **33.** $\left\{\frac{3 \pm \sqrt{21}}{2}\right\}$ **35.** $2 \pm \sqrt{2}$ **37.** $-2 \pm \sqrt{2}$ **39.** $\{-1 \pm \sqrt{5}\}$ **41.** $\left\{1 \pm \frac{\sqrt{2}}{2}\right\}$
43. $\left\{-\frac{3}{2} \pm \frac{\sqrt{3}}{2}\right\}$ **45.** $\left\{-1 \pm \frac{\sqrt{6}}{3}\right\}$ **47.** $\left\{\frac{1}{2} \pm \frac{\sqrt{7}}{2}\right\}$

Practice Exercises

49. $\{-4, -1\}$ **51.** $\{-2, 4\}$ **53.** $\left\{-\frac{5}{3}, 1\right\}$ **55.** $\left\{-\frac{4}{3}, -\frac{1}{2}\right\}$ **57.** $\left\{\frac{5 \pm \sqrt{29}}{2}\right\}$ **59.** $\left\{\frac{1}{2}, 4\right\}$
61. $\left\{\frac{-5 \pm \sqrt{13}}{2}\right\}$ **63.** $\{2\}$ **65.** $\{-1 \pm \sqrt{2}\}$ **67.** $\left\{\frac{3}{2} \pm \frac{\sqrt{7}}{2}\right\}$ **69.** $\left\{-2 \pm \frac{\sqrt{6}}{2}\right\}$ **71.** $\left\{-1, \frac{1}{3}\right\}$
73. $\left\{\frac{1}{2} \pm \frac{\sqrt{3}}{2}\right\}$

Challenge Problems

75. $a = 1; b = 1; c = k$ **77.** $\pi x^2 - x - k = 0; a = \pi; b = -1; c = -k$ **79.** $\left\{\frac{-k \pm \sqrt{k^2 - 4}}{2}\right\}$
81. $\left\{\frac{-p \pm \sqrt{p^2 - 4q}}{2}\right\}$

PROBLEM SET 9.4

Warm-ups

1. $3i$ **3.** $\sqrt{5}i$ **5.** $2\sqrt{2}i$ **7.** $3 + 4i$ **9.** $8 - 7i$ **11.** $-4 - 2i$ **13.** $-1 + 5i$ **15.** $7 + 4i$
17. $7 - 24i$ **19.** 5 **21.** $\frac{1}{2} - \frac{1}{2}i$ **23.** $\frac{1}{5} + \frac{3}{5}i$ **25.** $\frac{4}{13} - \frac{7}{13}i$ **27.** $\{\pm i\}$ **29.** $\{\pm\sqrt{7}i\}$
31. $\{2 \pm 2\sqrt{3}i\}$ **33.** $\left\{-\frac{1}{2} \pm \frac{\sqrt{3}}{2}i\right\}$ **35.** $\left\{-\frac{2}{5} \pm \frac{1}{5}i\right\}$ **37.** $\{-1 \pm 2i\}$ **39.** $\left\{-\frac{1}{2} \pm \frac{\sqrt{7}}{2}i\right\}$

Practice Exercises

41. $-1 - 4i$ **43.** $8 + 7i$ **45.** $-2 - 2i$ **47.** $5 - i$ **49.** $-7 - 9i$ **51.** $-2i$ **53.** 10
55. $\frac{1}{2} + \frac{1}{2}i$ **57.** $\frac{3}{5} + \frac{6}{5}i$ **59.** $\frac{3}{13} - \frac{11}{13}i$ **61.** $\{\pm 6i\}$ **63.** $\{\pm\sqrt{5}i\}$ **65.** $\{3 \pm 2\sqrt{2}i\}$ **67.** $\left\{\frac{1}{2} \pm \frac{\sqrt{11}}{2}i\right\}$
69. $\left\{\frac{2}{5} \pm \frac{1}{5}i\right\}$ **71.** $\{1 \pm 2i\}$ **73.** $\left\{-1 \pm \frac{\sqrt{6}}{2}i\right\}$

Challenge Problems

75. $(x - 3i)(x + 3i)$

PROBLEM SET 9.5

Warm-ups

1. $\{\pm\sqrt{5}\}$ **3.** $\{7 \pm \sqrt{5}i\}$ **5.** $\{8 \pm 2\sqrt{3}i\}$ **7.** $\{\pm 2\sqrt{2}\}$ **9.** $\{-5, 0\}$ **11.** $\{-2, 5\}$ **13.** $\left\{-4, -\frac{1}{2}\right\}$
15. $\left\{\frac{1}{2} \pm \frac{\sqrt{3}}{2}i\right\}$ **17.** $\{1 \pm \sqrt{11}i\}$ **19.** $\left\{\frac{1}{3} \pm \frac{\sqrt{2}}{3}i\right\}$

Practice Exercises

21. $\{\pm\sqrt{5}\}$ **23.** $\{-7 \pm \sqrt{3}i\}$ **25.** $\{-6, 0\}$ **27.** $\{-7, 4\}$ **29.** $\left\{\frac{1}{2} \pm \frac{\sqrt{5}}{2}\right\}$ **31.** $\left\{\pm\frac{7}{3}\right\}$ **33.** $\left\{\frac{1}{2}, 4\right\}$
35. $\{-1 \pm \sqrt{6}\}$ **37.** $\{1 \pm 2\sqrt{6}i\}$ **39.** $\left\{\frac{1}{4} \pm \frac{\sqrt{11}}{4}i\right\}$ **41.** The walk is 3 m wide.
43. The walk is 2 yd wide.

Challenge Problems

45. $t = \pm\sqrt{\frac{2s}{g}}$

PROBLEM SET 9.6

Warm-ups

1. $\{14\}$ **3.** $\{3\}$ **5.** $\{4\}$ **7.** $\varnothing$ **9.** $\varnothing$ **11.** $\{1\}$ **13.** $\{-4\}$ **15.** $\{4\}$ **17.** $\{1\}$ **19.** $\{-2\}$
21. $\{4\}$ **23.** $\{5\}$ **25.** $(-3, -1)$ **27.** $\{-1, 7\}$

Practice Exercises

29. $\{2\}$ **31.** $\varnothing$ **33.** $\{5\}$ **35.** $\{-4\}$ **37.** $\varnothing$ **39.** $\{7\}$ **41.** $\{-1\}$ **43.** $\{-1\}$ **45.** $\{0\}$ **47.** $\{2\}$
49. $\{-1\}$ **51.** $\varnothing$ **53.** $\{5\}$ **55.** $\{-1\}$ **57.** $\{7\}$ **59.** $\{3\}$ **61.** $\{\pm 1\}$ **63.** $\{-1, 3\}$

Challenge Problems

65. $\left\{\frac{b^2}{a}\right\}$

REVIEW PROBLEMS CHAPTER 9

1. $\{3, 7\}$ **3.** $\{-6, 0\}$ **5.** $\left\{-7, -\frac{4}{3}\right\}$ **7.** $\left\{0, \frac{9}{5}\right\}$ **9.** $\{\pm\sqrt{5}\}$ **11.** $\{-5, -2\}$ **13.** $\{-6, 4\}$ **15.** $\{0, 2\}$
17. $\left\{-1, \frac{1}{3}\right\}$ **19.** $\{\pm 13\}$ **21.** $\{\pm 7i\}$ **23.** $\{2 \pm \sqrt{7}\}$ **25.** $\{-4, -2\}$ **27.** $\{-4, 2\}$
29. $\left\{-\frac{3}{2} \pm \frac{\sqrt{21}}{2}\right\}$ **31.** $\left\{-\frac{1}{2} \pm \frac{\sqrt{13}}{2}\right\}$ **33.** $\{-4, -1\}$ **35.** $\left\{\frac{1}{4} \pm \frac{\sqrt{7}}{4}i\right\}$ **37.** $\left\{\frac{1}{2}, 1\right\}$ **39.** $\left\{\frac{7}{10} \pm \frac{\sqrt{29}}{10}\right\}$
41. $\{-4, 3\}$ **43.** $\{3, 7\}$ **45.** $\left\{\frac{3}{2} \pm \frac{\sqrt{17}}{2}\right\}$ **47.** $\{-8 \pm 2\sqrt{2}i\}$ **49.** $\left\{-\frac{1}{2} \pm \frac{\sqrt{13}}{2}\right\}$ **51.** $\{0, 3\}$
53. $\left\{-\frac{2}{3} \pm \frac{\sqrt{10}}{3}\right\}$ **55.** $\left\{\frac{2}{3}\right\}$ **57.** $\left\{-1 \pm \frac{\sqrt{6}}{3}\right\}$ **59.** $\left\{-\frac{7}{3}, \frac{4}{5}\right\}$ **61.** $\left\{-\frac{1}{2}, \frac{7}{3}\right\}$ **63.** $\left\{\frac{1}{7}, 1\right\}$ **65.** $\left\{0, \frac{5}{3}\right\}$
67. $\left\{-\frac{7}{2}, 5\right\}$ **69.** $\left\{-\frac{7}{6} \pm \frac{\sqrt{13}}{6}\right\}$ **71.** $\{\pm 11\}$ **73.** $\{-1 \pm 2\sqrt{2}\}$ **75.** $\{\pm 12\}$ **77.** $\{21\}$ **79.** $\{3\}$
81. $\varnothing$ **83.** $\{-4, -2\}$ **85.** $\varnothing$ **87.** The integers are -18 and -17.
89. The dimensions are 12 by 17 units.

Let's Not Forget

91. -16 **92.** $5 + 2\sqrt{6}$ **93.** $\frac{2y^2}{x}$ **94.** -4 **95.** 0 **96.** -3 **97.** $\sqrt{2} + \sqrt{10} - 5$ **98.** $\frac{-2}{x+3}$
99. Term **100.** Factor **101.** Term **102.** Term **103.** $-\frac{2}{x}$ **104.** $3x - 21$ **105.** $\frac{x^3}{7} + x^2$
106. Expression; $\frac{x+5}{(x+1)(x+3)}$ **107.** Equation; $\{5\}$ **108.** Inequality; $\{x \mid -3 \le x \le -1\}$
109. Expression; $\frac{2x+6}{x+1}$ **110.** Equation; $\{2\}$ **111.** $\sqrt{3}\,(y + x)$ **112.** $(x + y)(x^2 - xy + y^2)$
113. Factored **114.** Factored

CHAPTER 9 TEST

1. B **2.** C **3.** A **4.** C **5.** B **6.** $\{-3, 7\}$ **7.** $\{5 \pm \sqrt{7}i\}$ **8.** $\{-1, 9\}$ **9.** $\left\{\frac{3}{4} \pm \frac{\sqrt{7}}{4}i\right\}$
10. $\{\pm 10\}$ **11.** $\left\{\frac{1}{2} \pm \frac{\sqrt{3}}{2}\right\}$ **12.** $\{0, 6\}$ **13.** $\{1 \pm \sqrt{2}\}$ **14.** $3 - 10i$ **15.** $\frac{1}{2} - \frac{1}{2}i$
16. The legs are 5 m and 12 m long.

Index